Quantifying Morphology and Physiology of the Human Body Using MRI

Series in Medical Physics and Biomedical Engineering

Series Editors: John G Webster, Alisa Walz-Flannigan, Slavik Tabakov, and Kwan-Hoong Ng

Series in Medical Physics and Biomedical Engineering

Quantifying Morphology and Physiology of the Human Body Using MRI

Edited by

L. Tugan Muftuler

Department of Neurosurgery
Medical College of Wisconsin, Milwaukee, USA

CRC Press
Taylor & Francis Group
Boca Raton London New York

CRC Press is an imprint of the
Taylor & Francis Group, an **informa** business
A TAYLOR & FRANCIS BOOK

CRC Press
Taylor & Francis Group
6000 Broken Sound Parkway NW, Suite 300
Boca Raton, FL 33487-2742

First issued in paperback 2019

© 2013 by Taylor & Francis Group, LLC
CRC Press is an imprint of Taylor & Francis Group, an Informa business

ISBN-13: 978-1-4398-5265-1 (hbk)
ISBN-13: 978-0-367-38009-0 (pbk)

Library of Congress Cataloging-in-Publication Data

Quantifying morphology and physiology of the human body using MRI / [edited by] L. Tugan Muftuler.
    p. ; cm. -- (Series in medical physics and biomedical engineering)
    Includes bibliographical references and index.
    ISBN 978-1-4398-5265-1 (hardcover : alk. paper)
    I. Muftuler, L. Tugan. II. Series: Series in medical physics and biomedical engineering.
    [DNLM: 1. Magnetic Resonance Imaging--methods. 2. Anatomy. 3. Data Interpretation, Statistical. 4. Image Interpretation, Computer-Assisted--methods. 5. Neuroimaging. WN 185]

616.07'548--dc23   2012042675

Visit the Taylor & Francis Web site at
http://www.taylorandfrancis.com

and the CRC Press Web site at
http://www.crcpress.com

Contents

Part I Brain

Part II Body

About the Series

The *Series in Medical Physics and Biomedical Engineering* describes the applications of physical sciences, engineering, and mathematics in medicine and clinical research.

The series seeks (but is not restricted to) publications in the following topics:

- Artificial organs
- Assistive technology
- Bioinformatics
- Bioinstrumentation
- Biomaterials
- Biomechanics
- Biomedical engineering
- Clinical engineering
- Imaging
- Implants
- Medical computing and mathematics
- Medical/surgical devices
- Patient monitoring
- Physiological measurement
- Prosthetics
- Radiation protection, health physics, and dosimetry
- Regulatory issues
- Rehabilitation engineering
- Sports medicine
- Systems physiology
- Telemedicine
- Tissue engineering
- Treatment

The *Series in Medical Physics and Biomedical Engineering* is an international series that meets the need for up-to-date texts in this rapidly developing field. Books in the series range in level from introductory graduate textbooks and practical handbooks to more advanced expositions of current research.

The *Series in Medical Physics and Biomedical Engineering* is the official book series of the International Organization for Medical Physics (IOMP).

The International Organization for Medical Physics

The IOMP, founded in 1963, is a scientific, educational, and professional organization of 76 national adhering organizations, more than 16500 individual members, several corporate members, and four international regional organizations.

IOMP is administered by a council, which includes delegates from each of the adhering national organizations. Regular meetings of the council are held electronically as well as every three years at the World Congress on Medical Physics and Biomedical Engineering. The president and other officers form the executive committee, and there are also committees covering the main areas of activity, including education and training, scientific, professional relations, and publications.

Objectives

- To contribute to the advancement of medical physics in all its aspects
- To organize international cooperation in medical physics, especially in developing countries
- To encourage and advise on the formation of national organizations of medical physics in those countries which lack such organizations

Activities

Official journals of the IOMP are *Physics in Medicine and Biology* and *Medical Physics and Physiological Measurement*. The IOMP publishes a bulletin *Medical Physics World* twice a year, which is distributed to all members.

A World Congress on Medical Physics and Biomedical Engineering is held every three years in cooperation with IFMBE through the International Union for Physics and Engineering Sciences in Medicine (IUPESM). A regionally based international conference on medical physics is held between world congresses. IOMP also sponsors international conferences, workshops, and courses. IOMP representatives contribute to various international committees and working groups.

The IOMP has several programs to assist medical physicists in developing countries. The joint IOMP Library Programme supports 69 active libraries in 42 developing countries, and the Used Equipment Programme coordinates equipment donations. The Travel Assistance Programme provides a limited number of grants to enable physicists to attend the world congresses. The IOMP Web site is being developed to include a scientific database of international standards in medical physics and a virtual education and resource center.

Information on the activities of the IOMP can be found on its Web site at www.iomp.org.

Preface

Magnetic resonance imaging (MRI) remains a truly unique medical imaging system because of the breadth of information that one can acquire from it. The macromolecular makeup of tissues can be studied indirectly by way of relaxation mechanisms (T_1, $T_{1\rho}$, T_2, T_2^*) or magnetization-transfer phenomenon. Diffusion imaging detects microstructural changes. Flow and perfusion imaging, which are powerful modalities, aid in the evaluation of cardiovascular diseases. However, for the most part, the information is analyzed qualitatively, using a particular mechanism to modulate the image voxel intensities. For instance, the relaxation mechanisms serve as weighting factors to generate contrast in images; they are rarely quantified. Similarly, the amount of molecular diffusion is utilized as a contrast mechanism in *diffusion-weighted imaging*, and the quantitative measurements were not used in most cases. While visual assessments by physicians have been effective, quantitative measurements from tissues could complement the existing practices in achieving more accurate diagnoses or determining disease stages. It is interesting that a metric similar to the Hounsfield unit has not been established for clinical MRI technology. It could prove highly useful sometimes in making a challenging decision.

Researchers in the field recognize the growing need for quantitative MRI applications. Consequently, recent international meetings have placed greater emphasis on this topic. For instance, several International Society of Magnetic Resonance in Medicine (ISMRM) sessions and symposia focused on this. Also, the 2012 Radiological Society of North America (RSNA) meeting features about 17 sessions with "Quantitative Imaging" in their titles. MRI system manufacturers also have noticed the growing importance of quantitative MRI. They have started to include new data acquisition techniques with easy-to-use software interfaces for processing and analysis. This considerable growing interest in the field has driven the need for a comprehensive body of knowledge for the scientific community. In addition, clinicians and researchers who would like to use these novel techniques in their practices need a good reference manual. Therefore, the idea for this book began to flourish shortly after I gave a talk on *Quantitative Brain Imaging* at the 51st annual meeting of the American Association of Physicists in Medicine (AAPM) in 2009.

In order to deliver the latest science on a broad range of topics, internationally renowned experts in respective fields participated in this book. Each chapter provides comprehensive information about data acquisition, processing, and analysis techniques. Then, various clinical applications are presented. Some of the techniques described in the book are uniquely tailored for the brain or other organs; therefore, the book is divided into two main parts: (1) *brain* and (2) *body*. This classification makes it easier for the readers to focus on techniques that are more relevant to their interests. However, some of the techniques, such as *diffusion-weighted imaging* and *diffusion tensor imaging*, have widespread applications. Thus, the readers can see their applications in brain imaging, cancer imaging, or musculoskeletal imaging. The last three chapters of this book are dedicated to newly emerging quantitative techniques that explore tissue properties other than the presence of protons (or other MRI-observable nuclei) and their interactions with the environment. These new techniques provide unique information about electromagnetic and mechanical properties of tissues. They introduce new frontiers of study into various disease mechanisms.

Because of the growing importance of quantitative MRI, we believe that this book is an indispensable reference for basic scientists and clinicians who would like to explore *in vivo* MRI techniques to quantify changes in the morphology and physiology of tissues caused by various disease mechanisms.

First of all, I express my deep gratitude to all the authors for their outstanding contributions. After I drafted the outline of each chapter, I contacted the world experts in their respective fields. They kindly accepted my invitation and put in an enormous amount of work to explain the details of their cutting-edge work for the readers. Finally, I thank John Navas, PhD, for giving me the encouragement and guidance to initiate this book project; Amber Donley and Rachel Holt for their help and guidance in editing the manuscript. I am also very grateful to Kim Pechman, PhD, for her help in proofreading the chapters. Her dedicated work on the manuscripts ensured that the text is free of major errors.

Color Figures Available for Download

Many figures were produced in color, but there are too many to fit into the book. There are 32 color figures that appear as grayscale in the text, but are clearly indicated as available for download. The reader may obtain these images from the CRC Press Web site at http://www.crcpress.com/product/isbn/9781439852651. Readers are urged to check this Web site periodically, as these figures will be updated as science develops.

L. Tugan Muftuler
Medical College of Wisconsin

List of Contributors

Murat Aksoy
Department of Radiology
Stanford University
Stanford, California

Roland Bammer
Center for Quantitative and
    Personalized Neuroimaging
    and Translational Neuroimaging
    Research
Department of Radiology
Stanford University
Stanford, California

John Biglands
Multidisciplinary Cardiovascular
    Research Centre & Leeds
    Institute of Genetics,
    Health and Therapeutics
University of Leeds
Leeds, United Kingdom

Nathan Bryant
Department of Radiology
    and Radiological Sciences
Vanderbilt University
Nashville, Tennessee

Amanda K.W. Buck
Department of Radiology
    and Radiological Sciences
Vanderbilt University
Nashville, Tennessee

Richard A.D. Carano
Biomedical Imaging Department
Genentech, Inc.
South San Francisco, California

Gang Chen
Department of Biophysics
Medical College of Wisconsin
Milwaukee, Wisconsin

Jeon-Hor Chen
Department of Radiological
    Sciences
University of California, Irvine
Irvine, California

and

Department of Radiology
China Medical University
    Hospital
and
Department of Medicine
China Medical University
Taichung, Taiwan

Daniel Coman
Magnetic Resonance Research
    Center (MRRC)
Yale University
New Haven, Connecticut

Bruce M. Damon
Department of Radiology
    and Radiological Sciences
Department of Biomedical
    Engineering
and
Department of Molecular
    Physiology and Biophysics
Vanderbilt University
Nashville, Tennessee

Zhaohua Ding
Department of Radiology
    and Radiological Sciences
Department of Biomedical
    Engineering
and
Department of Electrical
    Engineering and
    Computer Science
Vanderbilt University
Nashville, Tennessee

Richard L. Ehman
Department of Radiology
Mayo Clinic
Rochester, Minnesota

Peter Herman
Department of Diagnostic
    Radiology
Yale University
New Haven, Connecticut

Samantha Holdsworth
Department of Radiology
Stanford University
Stanford, California

Fahmeed Hyder
Department of Biomedical
    Engineering
and
Department of Diagnostic
    Radiology
Yale University
New Haven, Connecticut

Ulrich Katscher
Innovative Technologies,
    Research Laboratories
Philips Technologie GmbH
Hamburg, Germany

Ananth Kidambi
Multidisciplinary Cardiovascular
    Research Centre &
    Leeds Institute of Genetics,
    Health and Therapeutics
University of Leeds
Leeds, United Kingdom

Frithjof Kruggel
Department of Biomedical
    Engineering
University of California, Irvine
Irvine, California

Ke Li
Department of Radiology
    and Radiological Sciences
Vanderbilt University
Nashville, Tennessee

Shi-Jiang Li
Department of Biophysics
Medical College of Wisconsin
Milwaukee, Wisconsin

Xiaojuan Li
Department of Radiology
    and Biomedical Imaging
University of California,
    San Francisco
San Francisco, California

Tian Liu
Department of Radiology
Weill Cornell Medical College
New York, New York

Sharmila Majumdar
Department of Radiology
    and Biomedical Imaging
University of California,
    San Francisco
San Francisco, California

Yogesh K. Mariappan
Department of Radiology
Mayo Clinic
Rochester, Minnesota

Kiaran P. McGee
Department of Radiology
Mayo Clinic
Rochester, Minnesota

Manish Motwani
Multidisciplinary Cardiovascular
    Research Centre & Leeds
    Institute of Genetics,
    Health and Therapeutics
University of Leeds
Leeds, United Kingdom

Rajakumar Nagarajan
Department of Radiological Sciences
University of California, Los Angeles
Los Angeles, California

James P.B. O'Connor
Imaging Sciences Research Group
University of Manchester
Manchester, United Kingdom

Jane H. Park
Department of Radiology
    and Radiological Sciences
and
Department of Molecular
    Physiology and Biophysics
Vanderbilt University
Nashville, Tennessee

Sven Plein
Multidisciplinary Cardiovascular
    Research Centre & Leeds
    Institute of Genetics,
    Health and Therapeutics
University of Leeds
Leeds, United Kingdom

Douglas L. Rothman
Department of Biomedical
    Engineering
Yale University
New Haven, Connecticut

Basavaraju G. Sanganahalli
Department of Diagnostic
    Radiology
Yale University
New Haven, Connecticut

Manoj K. Sarma
Department of Radiological
    Sciences
University of California, Los Angeles
Los Angeles, California

Christina Y. Shu
Department of Biomedical
    Engineering
Yale University
New Haven, Connecticut

Min-Ying Su
Department of Radiological
    Sciences
University of California, Irvine
Irvine, California

Peter Swoboda
Multidisciplinary Cardiovascular
    Research Centre & Leeds
    Institute of Genetics,
    Health and Therapeutics
University of Leeds
Leeds, United Kingdom

M. Albert Thomas
Department of Radiological
    Sciences
University of California,
    Los Angeles
Los Angeles, California

Theodore Towse
Department of Radiology
   and Radiological Sciences
Vanderbilt University
Nashville, Tennessee

Anh Tu Van
Department of Radiology
Stanford University
Stanford, California

Tobias Voigt
Innovative Technologies,
   Research Laboratories
Philips Technologie GmbH
Hamburg, Germany

Yi Wang
Department of Biomedical
   Engineering
Cornell University
Ithaca, New York
and
Department of Radiology
Weill Cornell Medical College
New York, New York

Part I

Brain

1

Quantifying Brain Morphology Using Structural Imaging

Frithjof Kruggel

University of California, Irvine

CONTENTS

1.1 Introduction

Quantitative MRI techniques have become an important tool in neurobiology and clinical neuroscience. Today, broad availability of modern MRI scanners and the maturity of MRI protocols offer easy access to high-resolution structural scans of the human head and brain. Acquired datasets are a valuable resource to address important questions in basic and clinical research. Anatomical imaging provides a "snapshot" of the macroscopic brain morphology. Example applications are as follows:

1. The brain compartments and substructures can be quantified and characterized in terms of their *volume, shape,* and *textural properties.* Resulting numbers may be included in statistical models and related to demographic, genetic, metabolic, or clinical data of single subjects or groups.

2. Different MRI protocols are sensitive to specific properties of tissues under study, as detailed in Section 1.2. The *quantitative analysis of multimodal images* helps discriminate healthy tissues from those

affected by pathological processes. A quantification of the amount and severity of changes can be related to the clinical status of a patient, and can provide insight into the functional correlates of structural changes.

3. The *quantitative description of changes in time-series images* (e.g., from longitudinal studies) reveals important information about the effect of aging on brain structures, and allows the discrimination of healthy from pathological aging. Quantitative measures of changes in the amount and severity of pathological processes can serve as biomarkers to rate the progression of brain diseases and the efficiency of therapeutic interventions.

4. Anatomical images are often used as a *structural reference* for quantities obtained from other neuroimaging modalities such as functional MRI (fMRI), positron emission tomography (PET), diffusion-weighted imaging (DWI), electroencephalography (EEG), and magnetoencephalography (MEG). The metabolic information obtained from fMRI or PET imaging is registered with the anatomical image in order to address focal metabolic activity to specific brain structures. The complex data analysis procedures applied to diffusion-weighted images finally lead to the definition of fiber tracts of the brain's white matter. Tracts are identified by relating them to an anatomical image of the same subject. Sources of electromagnetic brain activity as revealed by EEG and MEG data analysis are found as positions in space, and are addressed to specific brain structures by mapping the source localization into an anatomical image.

5. In neurosurgery, *interactive models of brain structures* based on structural and functional images of a patient have become an indispensable tool for planning complex interventions and helped optimize the functional outcome, for instance, after tumor removal.

Quantitative MRI techniques make the content of neuroimages available for further statistical evaluation. The intent is to provide useful biomarkers that describe the structural status of a brain and its changes due to various conditions, thus, supporting other evidence leading to diagnostic or therapeutic decisions. At this time, little research is performed to feed this information into computer-aided diagnostic systems—thus, such attempts are not discussed here.

An experienced human operator still outperforms computer-based techniques in difficult discrimination and segmentation tasks, such as outlining the border of a highly malignant tumor or a complex-shaped ischemic brain lesion. Automated approaches provide complementary tasks, such as (non) linear registration, that are difficult (or impossible) for humans to perform. Often, even a known bias of an automated method is acceptable because such techniques yield highly reproducible and consistent results, and easily cope with the vast amount of data such as those acquired in multicenter studies. Thus, we deliberately focus on the discussion of automated techniques for quantitative MRI. Note that any automated procedure may fail to produce an expected output without indication. Therefore, we discuss if and when checks by a human observer are advised.

Sections 1.2 through 1.8 provide an overview of current methods for quantifying morphology as revealed by structural imaging. We start by discussing issues of the *data acquisition* (Section 1.2), including the composition of the study protocol, data conversion and storage, measures of image quality, and possible artifacts. Software packages for neuroimaging data analysis are introduced in Section 1.3. Next, we discuss *data preprocessing* (Section 1.4) that is concerned with the standardization of images in terms of spatial resolution, intensity, and contrast, and the introduction of a common coordinate system, within and across different imaging modalities. Sections 1.5 through 1.7 describe techniques to quantify compartment *volumes* and their substructures, measures of *shape*, and *textural properties* of brain tissues. Finally, in Section 1.8 we focus on methods to analyze and quantify *disease processes* in the brain that lead to changes in tissue properties detectable at a macroscopic level.

1.2 Data Acquisition

Protocols for MR data acquisition are the result of an optimization process that includes (1) the specification of the scanner hardware, (2) the physical property under study, (3) the extent and resolution of the volume to be examined, (4) a sufficient signal-to-noise ratio (SNR), and (5) the amount of scanning time available. Obviously, scanning time is limited, and given (1) and (2), dictate the possible spatial resolution and expected SNR. The term "protocol" refers to the set of all parameters that are required to reproduce an MR experiment. The obtained tissue contrast predominately depends on (2), and differentiates protocols into "weightings." A discussion of the detail of the physical effects underlying MRI and the design of MRI sequences is considered as beyond the scope of this book—see Reiser et al. (2005) for an excellent reference.

Structural imaging was one of the first achievements of MR physics. Thus, clinically available imaging protocols are well understood and mature. However, continued advance in scanner technology, most notably, the development of scanners operating at high fields (3 T and above) and the introduction of parallel receiver coils yield an improved SNR and an increased spatial resolution. We now discuss the most commonly used protocols for structural imaging.

T_1-weighted scans are almost synonymous with structural imaging. Typically, a gradient echo (GRE) sequence with a short echo time T_E and a short repetition time T_R is used. Acquired images show water with a low intensity, soft tissues relatively bright, and fat as very bright. Because the repetition time is short, images can be acquired in 3D at a high resolution in a reasonable time. Today, a full brain scan with a field of view (FOV) of 240 mm × 240 mm × 192 mm at a resolution of about 1 mm can be acquired in 6–9 min. In order to improve the discrimination of gray matter (GM) and white matter (WM) in the human brain, an inversion recovery (IR) sequence can be used. IR protocols are especially useful if the segmentation of small GM structures, for example, the amygdala or hippocampus, is sought for. Compared with GRE, IR protocols need a longer acquisition time, and typically have a lower SNR. Using an inversion pulse in a GRE sequence leads to the magnetization-prepared rapid gradient-echo (MP-RAGE) imaging sequence, the most widely used T_1-weighted protocol today (Mugler and Brookeman 1990). Especially at higher field strengths (3 T and above), the modified driven equilibrium Fourier transform (MDEFT) protocol is less susceptible to intensity inhomogeneities and offers a better SNR and contrast (Deichmann et al. 2004) and offers advantages over the MP-RAGE protocol for the purpose of brain segmentation in quantitative MRI (Tardif et al. 2009).

Several other protocols offer additional information over T_1-weighted scans, and are especially useful if pathologic structures must be detected and quantified (Parizel et al. 2010):

- T_2-weighted scans require a long echo time T_E and a long repetition time T_R. Acquired images show water very bright, soft tissues with a relatively low intensity, and fat as dark. Therefore, the acquisition of T_2-weighted images is most useful to detect an increase of water in brain tissue, for example, in diffuse WM disease, edema due to tumors, infections, or cerebral infarction. Due to the long T_R, rather a stack of slices at an anisotropic resolution (e.g., 48 axial slices with a 3 mm slice thickness) are acquired that typically do not cover the full brain.

- Proton density (PD)-weighted scans show an intensity that is roughly proportional to the local proton concentration. Thus, water, soft tissue, and fat have a relatively uniform bright intensity while bone and air-filled cavities are dark. PD-weighted scans require a short echo time T_E and a long repetition time T_R, and are often acquired by *dual-echo protocols* while recording T_2-weighted scans. For quantitative MRI, PD-weighted images are most useful for a precise segmentation of the bright intracranial compartment from the surrounding dark skull.

- The fluid-attenuated inversion recovery (FLAIR) sequence (De Coene et al. 1992) is standard for WM lesion detection, for example, due to multiple sclerosis or diffuse WM disease. FLAIR scans offer an advantage over T_2-weighted images due to a better discrimination of lesions close to ventricles (Piguet et al. 2005).

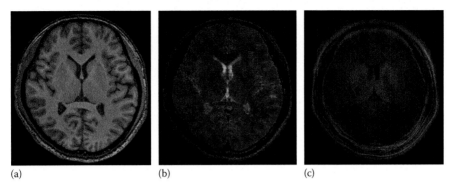

(a)   (b)   (c)

FIGURE 1.1
Axial slices of a T_1-weighted (a), T_2-weighted (b), and ultra-short TE (UTE) (c) protocols used for the segmentation of head compartments scalp, skull, cerebrospinal fluid, white, and gray matter.

- Ultra-short T_E (UTE) sequences (Robson et al. 2003) use a very short T_E (typically 50–100 µs) to capture the MR signal from bones, allowing imaging of the skull. The acquisition of UTE images is most useful if a precise segmentation of the skull compartment is sought for (Johansson 2010), for example, in forward models for bionumerical simulations of the biomechanical and electromagnetic properties of the human head (see Figure 1.1).

In summary, for the purpose of quantifying healthy brain structures [items (1) and (4) in Section 1.1], the acquisition of T_1-weighted images may be sufficient. If a precise segmentation of all major tissue compartments in the head is needed, the additional information contained in PD and UTE datasets is useful. If segmentation and quantification of brain lesions is sought for— for example, items (2), (3), and (5)—FLAIR and T_2-weighted scans should be acquired in addition to T_1-weighted images. Constraints on the total acquisition time of multiprotocol imaging sessions will typically result in a lower spatial resolution and/or limited coverage of sequences with a long repetition time, that is, T_2, PD, and FLAIR. However, the additional information provided by these protocols is highly useful for lesion segmentation and quantification (Figure 1.2). Tools for the integration and segmentation of multimodal images will be discussed in Sections 1.4 and 1.5.

Assessment of image quality at early stages and throughout the study is important to ensure that acquired images may be used for their intended purpose—the quantification of image content. Several effects and artifacts must be considered:

- The *SNR* is often computed as the ratio of the mean intensity in the foreground (tissue compartment) and the standard deviation of the intensity distribution in the background (air). [Since the noise in the background is rectified due to magnitude operation, it is usually

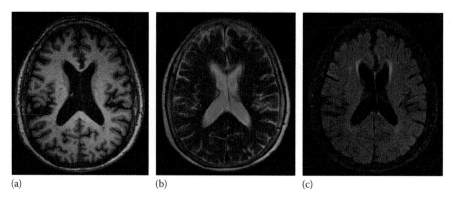

(a)   (b)   (c)

FIGURE 1.2
Axial slices of a T_1-weighted (a), T_2-weighted (b), and FLAIR (c) protocols used for the segmentation of brain lesions in the white matter.

recommended to multiply the standard deviation in the background by 1.53 to estimate the noise in the brain compartments (Gudbjartsson and Patz 1995).] Since noise broadens the intensity distribution of tissue compartments, a low SNR makes image segmentation difficult.

- The *contrast-to-noise ratio* (CNR) in brain images is defined as the intensity difference of WM and GM, divided by the standard deviation of the intensity distribution in the background (air). An acceptable CNR, such as the one provided by IR and MDEFT sequences, is required to reliably segment GM and WM structures. This is especially important in brain areas with a complex spatial arrangement of GM/WM structures, for example, in the mesial temporal lobe, or where GM structures contain an increased proportion of fibers, for example, in the pallidum.

- The spatial resolution of current imaging protocols is insufficient to reliably resolve tissues at boundaries. Thus, volume elements (voxels) at the GM/WM interface contain a mixture of both compartments, called the *partial volume effect* (PVE). This effect also broadens the intensity distribution of tissue compartments, especially for the GM compartment.

- The point spread function (PSF) (Robson et al. 1997) describes the imaged equivalent of a (infinitesimal) point in space, and thus relates the desired image resolution as defined in the imaging protocol to the true resolution. A wider PSF is perceived as an increased image blur.

- Imperfections of the scanner hardware and the presence of material boundaries with different magnetization properties lead to *geometrical distortions* in the image. While the former might be corrected using phantom measurements (Wang et al. 2004), the latter is most noticeable in the vicinity of metallic objects (e.g., clips and dental

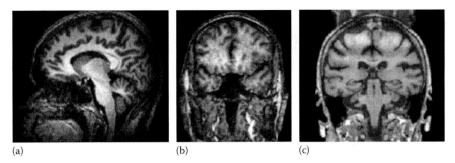

(a) (b) (c)

FIGURE 1.3
Examples of artifacts in structural MR images: intensity inhomogeneities (a), blur due to subject motion (b), and folding artifact (c).

implants) and can be difficult to be corrected for. If possible, such regions should be marked as unreliable and excluded from further evaluation.

- Imperfections of the scanner hardware and tissue dielectric effects, especially at higher field strengths, lead to *intensity inhomogeneities* in the imaged volume (see Figure 1.3a). At 1.5 T field strength, inhomogeneities are on the order of 5% of the mean intensity and may be considered as minor. At 3.0 T field strength, they often approach 20%, and must be corrected to avoid segmentation errors. At higher field strengths, intensity inhomogeneities may become so dominant that current correction approaches fail. Techniques for the correction of this artifact are discussed in Section 1.4.

- *Subject motion* leads to blur in the image. Head motion due to breathing, swallowing, and the cardio-ballistic effect is typically sufficiently suppressed by restraining a subject's head. However, noncooperative and anxious patients may show considerable motion blur in the acquired images and render them as insufficient for further analysis (see Figure 1.3b). For high-resolution anatomical imaging (<0.5 mm), the use of prospective motion correction schemes should be considered (Qin et al. 2009).

- *Aliasing* occurs when the FOV is smaller than the body part under study. The region outside the FOV is projected on the other side of the image. This artifact can be avoided by oversampling the data in the direction of the aliasing. This "ghost" artifact leads to segmentation errors that typically cannot be corrected (see Figure 1.3c).

Acquired images should be assessed for quality by a neuroradiologist or experienced project scientist because any unexpected image content will most likely cause subsequent processing steps to fail, often without generating a user-level notification. Even if a standardization of the imaging protocol was implemented, quantitative MRI results from multicenter studies may show considerable differences due to differences in the scanner hardware and specifics of the implementation of the MRI protocol (Kruggel et al. 2010). Most available image processing software require conversion from the native scanner data format into some proprietary format. Typically, this involves anonymization of patient data by addressing a study-internal code. Thus, it is advisable to archive and store original data separate from later processing stages, as mandated by law concerning privacy issues of personal health-care data. It is useful if the data format retains a study-related subject identifier inside the file, for example, as an image tag, so that the originating subject can be identified after some possibly erroneous file copy and renaming operations.

Another common error must also be avoided, best by careful provisions in the study protocol: Because the head is approximately mirror-symmetric with respect to the midsagittal plane, it is easy to confuse the left and right body side in an image. While the original scanner format typically adheres to the radiologic convention (the left image side corresponds to the right body side), most image processing software employ the natural convention (vice versa). Care should be taken especially if multimodal information from different devices is integrated. A small MR-observable bead attached to a subject's (left) forehead during imaging can help avoid possible confusion.

1.3 Software for Analyzing Structural Images

The availability of analysis software is crucial for quantitative MRI. Tremendous effort was spent in recent years to develop reliable and robust methods. Four major, publicly available software packages (for noncommercial use) are briefly introduced below, and underline the relevance of medical image analysis for advanced neuroimaging research. The software and documentation can be accessed on the servers provided in the references.

1. *Statistical parametric mapping* (SPM 2012) was originally developed for the analysis of brain imaging data sequences (i.e., from PET and fMRI). Later, add-on packages (called toolboxes) for processing EEG, MEG data, and structural MRI were included. Most notably, the toolbox for "voxel-based morphometry" or VBM is widely used.

2. *FreeSurfer* (2012) started as a software package for the reconstruction of brain surfaces from structural MRI data, in order to map results from functional experiments (first EEG, later fMRI) onto the surface. Later, modules were added for structural analysis (e.g., segmentation of basal ganglia, neocortical parcellation).

3. *BrainVisa* (2012) has its roots in the structural analysis of MRI data, with special attention to the problem of analyzing the individual neocortical structure and brain development. Toolboxes for analyzing functional data are available as external packages.

4. The *FMRIB Software Library* (FSL 2012) contains state-of-the-art software modules for processing anatomical and functional MR images. Some of these modules are integrated with FreeSurfer and BrainVisa.

All of these software packages offer tools for visualization. The most comprehensive tool for annotating and editing data, called "Anatomist," is included in BrainVisa. Besides these major packages, an increasing number of smaller or specialized packages are publicly available. Refer to the Neuroimaging Informatics Tools and Resources Clearinghouse (NITRC 2012) for a catalog.

All current MR scanners provide information in the ISO-standardized Digital Imaging and Communication in Medicine (DICOM) data format (2012). This format defines a set of tags that describe image content (i.e., patient identifiers, examination date, modality, protocol parameters). DICOM also standardizes communication, and is widely used in clinics to exchange and store medical images. Unfortunately, only a 2D storage format is defined, so volumetric data are stored as a set of slices in separate files. This renders image processing based on DICOM cumbersome due to the large number of files involved. Therefore, most packages discussed above developed proprietary 3D image formats. To improve the information exchange between different packages, the Neuroimaging Informatics Technology Initiative (NIfTI)-1 data format (2012) was developed and became a quasi-standard for storing 1D–4D (medical) images. Note that the NIfTI format does not allow for user-defined tags, so subject-, study-, and processing-related information has to be maintained separately. Similar standards do not exist for surface or volumetric meshes.

1.4 Data Preprocessing

The purpose of data preprocessing is to standardize structural images acquired in groups of subjects (using different protocols, on different scanners), and to minimize the influence of some artifacts discussed above.

Often, the first step will introduce a *standard coordinate system* in the high-resolution T_1-weighted datasets, removing pose differences between scans. For the brain, the widely used stereotaxic coordinate system adopted from minimally invasive neurosurgery (Gildenberg 1988) is defined as follows: (1) Find the anterior and posterior commissure (AC, PC) crossing the midsagittal plane (Cho 2010), and draw a line through the center of both fiber bundles in the midsagittal plane (see Figure 1.4). The center between both

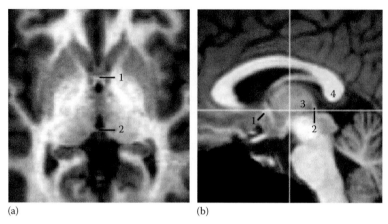

(a)   (b)

FIGURE 1.4
Reference structures of the stereotaxic coordinate system. (a) Axial plane showing the anterior (1) and posterior (2) commissure crossing the midsagittal plane. (b) Midsagittal plane with anterior commissure (1), posterior commissure (2), adhesio interthalamica (3), and splenium (4). (From Kruggel, F. and von Cramon, D.Y., *Med Image Anal* 3, 175–185, 1999. With permission.)

commissures defines the origin of the coordinate system. The line corresponds to the y-axis, and the positive half-axis points toward the front of the head. (2) The z-axis is perpendicular to the y-axis in the midsagittal plane, and the positive half-axis points toward the top of the head. (3) Finally, the x-axis is perpendicular to the y- and z-axes, and the positive half-axis points toward the right body side.

Note that this coordinate system retains the acquisition space: Distances, areas, and volumes correspond to the "real" values of this subject. For the "manual" method of introducing this coordinate system, the position of AC and PC and the three rotation angles are determined using an image editor, and software computes the transformation matrix and applies it to the dataset. Typically, data are interpolated to an isotropic resolution of 1 mm in each spatial direction. A simple approach to introduce this coordinate system automatically uses an already aligned reference. An object dataset is aligned with the reference using linear registration. If object and reference are sufficient similar (i.e., acquired in healthy subjects using the same protocol), the expected error on the origin is in the order of 3 mm, and 5° around the x-axis, less around the y- and z-axes. A higher precision is obtained if the midsagittal plane is detected, and the position of AC and PC is detected automatically (Kruggel and von Cramon 1999, Verard et al. 1997). However, automated procedures are not robust if considerable deviations from a "healthy" brain exist, so a visual check of the output is advised.

The other widely used system is Talairach coordinates (Talairach and Tournoux 1993), which uses a linear mapping of the stereotaxic coordinates into a standard frame: (1) The origin is shifted to the crossing of AC in the midsagittal plane. (2) The x-axis runs from the leftmost point of the brain (-64 mm) to the rightmost point ($+64$ mm). (3) The y-axis extends from the back (-104 mm) to the front ($+68$ mm). (4) Finally, the z-direction defines the topmost point as $+72$ mm. A Web-based application is available to convert coordinates into Talairach space (Research Imaging Center 2012). Due to the scaling, measures do not directly correspond to the real values of a subject. This coordinate system was adopted by the "International Consortium for Brain Mapping" (ICBM 1993) as a standard to communicate brain locations. This consortium also has provided the *ICBM 452 T1 atlas*, an average of T_1-weighted brain images acquired in 452 healthy subjects, mapped into Talairach space (Mazziotta et al. 1995). This atlas serves as a reference for several procedures in quantitative MRI, as detailed in Section 1.5. However, it is now commonly realized that a common reference is better generated from the study samples itself (Mazziotta et al. 2009). For a discussion of the relation between different coordinate systems in use at this time, refer to Brett (2002).

The basic procedure, *registration*, underlying these coordinate transformations deserves a deeper explanation. The registration problem is written (Maintz and Viergever 1998) as follows:

$$\hat{T} = \arg\max_{T}\left(S\left(T(O),R\right)\right), \tag{1.1}$$

where O corresponds to the object image, $T(\cdot)$ to the transformation, R to the reference image, and $S(\cdot)$ to a similarity metric between the reference and transformed object. For the purpose of an affine mapping to Talairach space, T corresponds to a 4×3 linear transformation matrix. For a mapping into stereotaxic space, T is restricted to rotation, translation, and scaling only. If object and reference images have a similar intensity distribution (e.g., both are T_1-weighted), image cross-correlation is used as a similarity metric (Maintz and Viergever 1998). If intensity distributions differ (e.g., different MRI weightings), an information-based criterion, normalized mutual information (NMI), is preferred as a similarity criterion (Maes et al. 1997). Optimal transformation parameters $\hat{T}$ are found at the maximal similarity between the transformed object and reference image. The final transformation uses a fourth-order b-spline interpolation (Thevenaz et al. 2000) in order to minimize the low-pass filtering effect due to interpolation. Finding optimal transformation parameters can be difficult if the object image consists only of a few slices. Robustness can be improved by using a multiresolution approach. Approximate parameters are determined at a coarse resolution level and refined at higher resolution levels. All commonly available software packages for quantitative MRI provide elaborate and mature modules for linear registration.

Once the T_1-weighted dataset of a subject is aligned with a coordinate system, other acquired modalities can be transformed into the same space, using linear regression and NMI as a similarity metric. As a result of this procedure, all datasets of a study reside in the same (individual or normalized) coordinate space. The next preprocessing step aims at correcting intensity inhomogeneities in the T_1-weighted images (Belaroussia et al. 2006). As discussed above, inhomogeneities are consequences of imperfections in the scanner's RF transmit (B_1^+) field and the nonuniform sensitivity of receiver coils (B_1^-). Typically, inhomogeneities are modeled as a multiplicative factor on the intensity that is independent of the local material and varies smoothly in space. In most cases, these inhomogeneities can be simply corrected if they were measured in the study protocol through so-called B_1 mapping (Mihara et al. 1998). Although B_1 maps are acquired in less than a minute on modern scanners, they are often not included in study protocols. Thus, a *post hoc* correction is often necessary. The most widely used approach is called nonparametric nonuniform intensity normalization (Sled et al. 1998: referred to as *n3* method). The inhomogeneity field acts as a low-pass filter on the intensity histogram. This iterative correction scheme finds iteration-wise better approximations of the true intensity distribution by inverse (high-pass) filtering of the intensity histogram. Then, the inhomogeneity field is estimated from the approximated intensity distribution and the measured image. An alternate approach estimates the local bias by comparing the local intensity histogram with the global histogram [Shattuck et al. 2001: referred to as bias field corrector (BFC) method]. Both methods performed best in a comparative evaluation of six approaches (Arnold et al. 2001). More recent techniques integrate image segmentation with the estimation of the bias field, and are discussed in Section 1.5.

Another (optional) preprocessing step aims at intensity correction of scans with the same weighting across subjects of a study. Such a correction is necessary if absolute intensities are compared across scans, for example, in longitudinal studies of the same subject. A simple linear scaling of the intensity histogram can be used if data were acquired on the same scanner using the same protocol. Images acquired in the same subject using the same protocol but on different scanners may show considerable differences in their intensity distribution (Kruggel et al. 2010), and thus are not easily corrected.

1.5 Volumetric Analysis

A reliable and robust segmentation of brain compartments (e.g., WM and GM) and substructures (e.g., hippocampus, thalamus) is required for a volumetric analysis. First, the outer hulls of the brain are removed in MR images of the head. This processing step is commonly called *brain extraction* (see Figure 1.5). Because further processing steps require an extracted brain as input, a considerable number of approaches were developed (Pham et al. 2000). Two basic strategies can be distinguished:

1. In T_1-weighted images, bone and cerebrospinal fluid (CSF) show a low signal compared with brain tissue. Few structures connect the intracranial compartment with extracranial structures (e.g., optic nerves, cerebral arteries). Techniques for brain extraction use this fact to either (1) develop a voxel-based stencil using a distance transform and morphological operators [Kruggel and von Cramon 1999 (see Figure 1.5); Mangin et al. 1998: implemented in BrainVisa] or (2) "Shrink-wrap" a smooth surface around the brain and extract the enclosed object (Smith 2002: implemented in FSL; Dale et al. 1999: implemented in FreeSurfer). Note that these techniques extract the brain only.

2. A segmentation of the intracranial compartment (brain + CSF) can be obtained from a segmentation of PD-weighted images. As discussed before, the proton-rich intracranial compartments are easily distinguished from the bone. The resulting binary mask is used to extract all intracranial compartments. If PD-weighted images are not available, a binary mask can be generated from a model: A nonlinear registration is computed from a T_1-weighted reference to an object image. The resulting deformation field is used to transform an intracranial mask of the reference into object space, and the transformed mask is used to extract the intracranial compartment in the object image (Hentschel and Kruggel 2004). If datasets were already transferred into Talairach space, an intracranial mask in this space can easily be applied (Ashburner and Friston 1997: implemented in SPM).

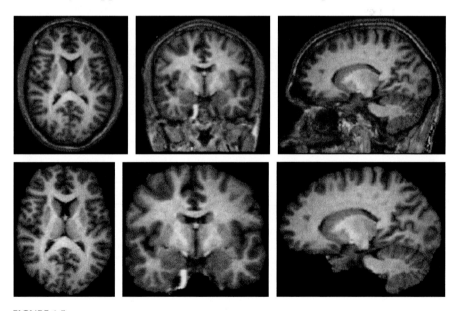

FIGURE 1.5
Top row: T_1-weighted image of the human head, aligned with the stereotaxic coordinate system. *Bottom row*: extracted brain.

The extracted brain is used for visualization purposes or forms the basis for further *brain segmentation*. Typically, the extracted brain is assumed to consist of three ($K = 3$) compartments: WM, GM, and CSF. Note that other tissues (e.g., meninges, vessels, plexus choroideus) are neglected here. Most segmentation approaches use the voxel-wise intensity v_i as the classification criterion, and model the intensities of each compartment k by a Gaussian distribution $\mathcal{N}_k(\mu_k, \sigma_k^2)$. Depending on the tissue contrast, distributions are not well separated and mix due to the PVE. Therefore, current approaches provide a probabilistic segmentation of voxel-wise intensities v_i into voxel-wise vectors of class-wise probabilities $u_{i,k}$:

$$\hat{\mu}, \hat{\sigma} = \arg\min_{\mu,\sigma} \left(\sum_i \sum_{k=1}^{K} u_{i,k} \exp\left[\frac{(v_i - \mu_k)^2}{\sigma_k^2} \right] \right). \tag{1.2}$$

Class-wise distribution parameters μ_k and σ_k of this "Gaussian mixture model" (GMM) are typically estimated using the iterative "expectation maximization" (EM) algorithm (McLachlan and Basford 1988). Robustness against noise is achieved by incorporating neighborhood information through the formalism of a "Markov random field" (MRF) (Li 2009). More specifically, two basic strategies can be distinguished:

1. One of the most widely used approaches (implemented in FSL: Zhang et al. 2001) uses a sequence of (1) parameter estimation of the GMM, (2) refining classification decision by an MRF, and (3) estimation of the bias field using the n3 algorithm in each iteration of the EM algorithm. An alternative approach combines the mixture model, neighborhood information, and bias field distortion in a common estimation model (Pham 2001; also see Figure 1.6). An even better analysis of voxels at tissue boundaries can be achieved if gradient information is included in the classification process (Chiverton 2006). Note that these approaches are typically applied in the individual space of a subject. Volumes of the three compartments are determined by simply summing up probability information within a class. The sum of all compartments corresponds to the volume of the intracranial compartment.

2. The second strategy uses a presegmented atlas (in Talairach space) to initialize the classification. Here, a (non)linear registration with the atlas must be included (van Leemput et al. 1999, implemented in SPM: Ashburner and Friston 1997, 2005). The use of an atlas provides strong prior information, especially for the estimation of the bias field. However, the more an object brain differs from the atlas, the higher the possibility that the prior is wrong. Results are typically provided in normalized (Talairach) space.

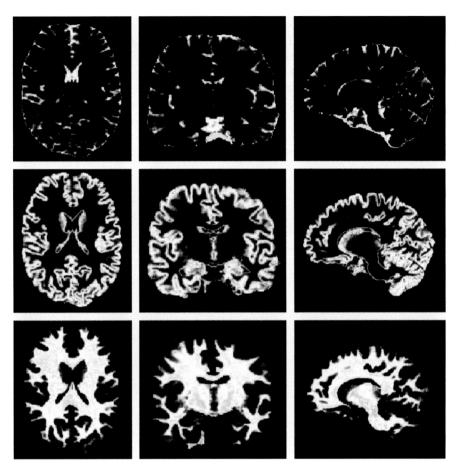

FIGURE 1.6
T_1-weighted image of the human brain, segmented into the CSF (top row), GM (middle row), and WM (bottom row) compartments.

An extension of the univariate GMM to the multivariate case is possible (He et al. 2008, van Leemput et al. 1999), although the relatively low GM/WM contrast in the other modalities provides little additional information. These approaches are most useful in the discrimination of pathological structures (see below).

Next, a segmentation of *brain substructures* can be performed. Until recently, the only reliable method was obtained by manually outlining contours of a structure using an image editor. Although tedious and time-consuming, this method is still considered as superior, especially in small structures that are not well separated from neighboring structures (e.g., amygdala and hippocampus). Note that the intra- and interrater variance can be high: consider a structure with a size of 10^3 versus 11^3 voxels: although the boundary differs by just 1 voxel, the volume differs by 33%. Due to this variance, the statistical power of such measures may be low. To improve the precision, the inclusion of an imaging protocol with good tissue contrast (e.g., IR, MDEFT) at a higher resolution (0.5–0.7 mm in plane) might be considered. Protocols are available that guide the outlining process (e.g., Entis et al. 2012, Pruessner et al. 2000). Recently, techniques for the automated segmentation of basal ganglia were developed that approach the precision of manual procedures. Typically, an image-based (iconic) atlas is used and adapted to an object image using nonlinear registration (Khan et al. 2008: implemented in FreeSurfer). An even higher precision is achieved if statistical information about the position and shape variation is included (Patenaude et al. 2011: implemented in FSL). Reported overlap measures with manual segmentations reach 70% up to 90%. This relatively lower precision may be accepted in a morphometric study, because automated algorithms typically have a high reproducibility and consistency.

Similarly, the reference method for the *segmentation of neocortical structures* is still manual outlining. For a list of protocols, refer to Shattuck et al. 2008; (see also supplement 1: document NIHMS39232-supplement-01.pdf). Automated segmentation and labeling approaches have been the focus of much research. Three methods are available at this time:

1. An example brain in Talairach space was manually segmented into 116 regions of interest (ROI). An extracted brain is mapped into Talairach space using nonlinear registration. The resulting deformation field is applied to map the probabilistic GM segmentation into normalized space. Then, the automated anatomical labeling (AAL) template is used to sample voxel-wise GM probabilities within ROIs. The resulting ROI-wise information can be interpreted as GM "concentration" (implemented in SPM: Tzourio-Mazoyer et al. 2002). This procedure is conceptually simple and robust (see Figure 1.7). However, the nonlinear registration only has a limited precision, and neocortical structures are not identified *per se*.

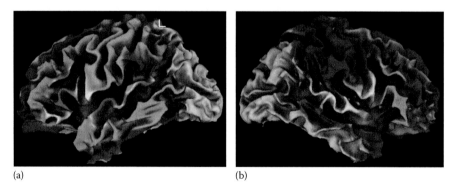

(a)   (b)

FIGURE 1.7
(See color insert.) Results from a study in healthy brain aging using AAL: region-specific significant loss of GM with age, in the left (a) and right (b) brain hemispheres; z-scores are color-coded from −3 (magenta) to −10 (red).

2. A triangulated surface of the interface between WM and GM is computed from the probabilistic brain segmentation (see above) of the object image. Local surface curvature and geodesic depth (to a smooth outer brain surface) are determined. These properties are nonlinearly registered with a surface-based atlas, and predefined atlas labels mapped onto the object surface (implemented in FreeSurfer: Desikan et al. 2006). This atlas contains 34 ROIs per hemisphere. Due to the surface-based matching process, a higher precision with respect to the AAL method is expected, at the expense of several added processing steps. However, the considerable variability of neocortical structures renders the use of a single atlas only an approximation.

3. Sulcal medial surfaces were computed and 60 consistent structures per hemisphere were labeled in a reasonably sized database. A statistical shape model is used to capture the intersubject variability. A mean recognition rate of 86% was obtained through a combined global and local joint labeling (implemented in BrainVisa: Perrot et al. 2011).

Note that the latter two methods achieve neocortical segmentation into patches and other region-wise measures (e.g., neocortical thickness, volume, curvature) can be sampled within these regions. All three methods were developed using datasets of healthy subjects and represent the "normal" brain morphology. A careful check of results is advised if these methods are applied to patient data that considerably differ from the normal morphology.

A core operation to some of the procedures outlined above is nonlinear registration. The development of approaches for nonlinear registration has been a "hot topic" in medical image analysis during the last two decades. Compared with Equation 1.1, this problem is written as follows:

$$\hat{T} = \arg\max_{T}\left(S\big(T(O),R\big) + \kappa E\big(T\big)\right), \tag{1.3}$$

where the transformation T now corresponds to a 3D field of shift vectors that deform the object O to match the reference R. Although mathematically, a very large number of solutions (deformation fields $\hat{T}$) are valid, we aim at a biophysically valid solution, in which neighborhood relationships are retained, a so-called diffeomorphic transformation (Miller et al. 2002). A key criterion here is the value of Jacobian (the determinant of the first partial derivatives of the deformation field): this value must be greater than 0 everywhere in the deformation field to maintain topology. Therefore, an additional "energy" term $E(T)$ is included that regularizes the transformation T weighted by κ. The iterative estimation of the high-dimensional deformation is computationally costly, so efficiency is a key issue. Starting from the efficient "demons" algorithm (Thirion et al. 1998), a better understanding of its theoretical underpinnings led to the development of a diffeomorphic variant (Vercauteren et al. 2008). An increased robustness is provided if both deformation fields $O \rightarrow R$ and $R \rightarrow O$ are estimated together (symmetric demons). Using cross-correlation as a similarity metric (instead of the intensity difference) is advised if O and R were acquired by the same imaging protocol but on different scanners, or if R corresponds to a group average (Avants et al. 2008). For a quantitative comparison of recent approaches, refer to Klein et al. (2009). Efficient implementations are now provided in most available software packages (e.g., FSL, FreeSurfer, SPM). A variant of the "symmetric demons" algorithm can be used for nonlinear registration of surfaces (Yeo et al. 2011), as used in the atlas-adoption procedure discussed above. Note that these algorithms can only be applied to intramodal (e.g., T_1- vs. T_2-weighted) registration problems. The best intermodal similarity metric, NMI, is global (i.e., computed over the whole image domain), and does not easily drive the iterative estimation of the deformation field. This problem is still active area of research.

The region-wise approaches for brain morphometry discussed above are applicable in single subjects, longitudinal studies of subjects, and subject groups. As an alternative, two voxel-based schemes found widespread application:

1. *VBM* (Ashburner and Friston 2000) is a technique to evaluate the voxel-wise GM (or WM) concentration in a multivariate regression context. Results of the probabilistic segmentation in object space are mapped into a common (Talairach) space using nonlinear registration as discussed above. If this transformation were perfect, a specific voxel would contain information from the homologue location of all brains included. The vector of voxel-wise measurements is then evaluated against a set of independent variables (e.g., group designation, clinical status, age). As a result, z-scores corresponding to the importance of a specific regressor are obtained and compiled in a statistical map. Because this approach is conceptually simple, thousands of clinical studies used this method. However, four technical details require careful attention: (1) Because the registration is imperfect, authors suggest blurring transformed images to increase the amount of overlap. Using the modern registration techniques discussed above, Gaussian filtering with $\sigma = 2$ is suggested here. (2) Mapping the object image into Talairach space also leads to a scaling. Authors suggest to take scaling into account by weighting the voxel-wise concentrations with the Jacobian (see previous paragraph). This value is less than 1 for a local contraction, greater than 1 for a local expansion, and equal to 1 for a local shift (or no change).

However, if large deformations occur in the transformation, using this "modulation" is conceptually questionable. As an (statistically weaker) alternative, the total brain volume may be included in the regressor matrix. (3) Voxel-wise concentrations are not Gaussian-distributed, especially if the WM is analyzed. Instead logit or power-transformed values should be used (Viviani et al. 2007). (4) The resulting statistical maps must be corrected for multiple testing. The choice of the correction approach depends on adopted error criterion for false positive detections. Authors propose to use the same criterion widely applied in fMRI analysis here: Clusters above a specified z-score threshold are evaluated for their significance based on the theory of excursion sets (Cao 1999). A processing chain for VBM is implemented in SPM. For a review of results achieved by this method, refer to Mechelli et al. (2005).

2. In contrast, *tensor-based morphometry* (TBM, Chung et al. 2001) analyzes the deformation field itself. As discussed above, the Jacobian encodes information about the local tissue expansion or contraction of the object versus reference image. As in VBM, this voxel-wise measure can be evaluated in a regression context to infer about underlying structural changes. The same caveats (1)–(4) listed above apply to this technique as well. Authors suggest using log-transformed data (Leow et al. 2006) and typically use a correction for multiple testing based on a control of the false discovery rate, for example, by the Benjamini–Hochberg procedure (1995). For a recent example study using this method, refer to Hua et al. (2008).

A critical issue with both VBM and TBM is the choice (or generation) of a suitable reference image (Senjem et al. 2005). While most early studies used the ICBM template as a reference (e.g., as included in SPM), a study-internal, "customized" reference yields better results for two reasons: (1) Subjects included in the ICBM template were mostly adults. In studies of children or elderly subjects, the properties of this reference may be too different from within-study average. (2) The intensity distribution of MRI protocols differs considerably. Therefore, the development of a reference based on the actual imaging protocol is preferable. The following procedure for generating a customized template is generally accepted: (1) Select a set of T_1-weighted images (best aligned with the stereotaxic coordinate system) acquired in the study. (2) Map images into Talairach space using nonlinear registration with the ICBM reference. (3) Average mapped images to yield an intermediate template. (4) Map images of set (1) into Talairach space again, using the intermediate template as a reference. (5) Average mapped images to yield the final customized template. The discussion has not yet settled which and how many images should be included in the template generation (Avants et al. 2010).

1.6 Surface and Shape Analysis

Besides volumetric measures, brain compartments and their substructures may be characterized by their surface properties. Of course, the first and necessary step is to compute a surface from a segmented brain structure. Due to noise and the PVEs, this segmented structure may have topological errors (e.g., holes), such that the expected genus does not match with the segmented object. For a thorough discussion of topological properties in binary images, refer to Toriwaki and Yonekura (2002). Typically, a simplifying assumption is made that the expected genus is zero, for example, that no holes are found in the object. The most important surfaces in the brain are the GM/WM and the GM/CSF interfaces that are computed using the WM (or union of WM and GM) segmentation. Three major approaches were developed:

1. *Deformable models:* A surface of known topology (e.g., a sphere) is shrink-wrapped or inflated to match an object of interest (Davatzikos and Prince 1995). An image-based force drives the surface toward the object boundary, while a smoothness constraint on the surface achieves stability against local segmentation errors. This approach works best for simple-shaped, compact objects such as the basal ganglia (Collins et al. 1995). For complex-shaped structures (e.g., the GM/CSF interface), this method is less successful, because it is difficult to evolve the surface into detailed structures (e.g., neocortical folds).

2. *Surface correction:* After a surface was generated from an object with segmentation errors, topology violations can be detected as local areas of high curvature, and mended by remeshing (implemented in FreeSurfer, Fischl et al. 2001). However, this approach is computationally demanding.

3. *Region growing:* A structure is initialized from a seed of matching topology. In the case of genus zero, this seed can be a single voxel. Then, voxels on the boundary of the structure are added so that the overall topology is retained (Kriegeskorte and Goebel 2001). A local criterion exists for these so-called simple points (Bertrand 1995). More conveniently, the region growing process is steered by the evolution of a level-set curve that allows including a smoothness constraint on the object's boundary. Finally, the surface is directly generated from the level-set curve (Han et al. 2005).

Once a topologically correct segmentation of the GM/WM interface is obtained, the GM/CSF interface can be generated by inflating the surface (Dale et al. 1999). Because surface self-intersections must be avoided in this process, time-consuming checks must be performed. It is easier to continue the region growing process after adding the GM compartment. However,

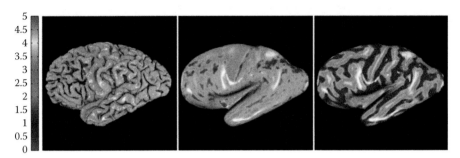

FIGURE 1.8
(See color insert.) Lateral view of pial surfaces from a high-resolution dataset: neocortical thickness (left), inflated thickness maps (middle), and inflated convexity maps (right). (From Osechinskiy, S. and Kruggel, F., *Int J Biomed Imaging*, doi:10.1155/2012/870196, 2012.)

the PVE at the GM/CSF boundary requires the definition of a delicate stopping criterion in the CSF compartment (Han et al 2004, Osechinskiy and Kruggel 2012; see Figure 1.8). Note that for the surface correction and the region growing approach, inner cavities of the WM segmentation (e.g., the basal ganglia and inner ventricles) must be filled first.

All the above approaches require a method to generate a surface from a 3D image, given an intensity threshold. Most commonly, the so-called marching cubes algorithm is used here (Lorensen and Cline 1987), which generates a mesh of triangles on output. Although the original formulation was straightforward to implement, it took 16 years of research to resolve topological ambiguities in the surface (Nielson 2003). A brain surface generated by this approach consists of about 500000 triangles at a submillimeter resolution. For most applications, a lower spatial resolution is sufficient. Often, hemispheric meshes of 50000 triangles at ~2 mm spatial resolution are used. Mesh decimation algorithms were developed that adapt to the local curvature and preserve topology (Luebke 2001).

Then, surfaces representing the GM/WM and the GM/CSF interfaces can be used in further processing or to determine the following properties:

- *Neocortical thickness:* This measure can be determined in image space directly from the segmented GM compartment. However, due to noise and the PVE in image data, using the (shortest) local distance between the GM/WM and the GM/CSF interfaces yields more reliable estimates (MacDonald et al. 2000). Regional measures can be obtained by applying the AAL template or the surface-based atlas incorporated in FreeSurfer (Fischl and Dale 2000). Gender-related (Im et al. 2006a, Sowell et al. 2007) and interhemispheric differences (Luders et al. 2006) were studied, as well as developmental aspects

(O'Donnell et al. 2005). Changes in regional neocortical thickness were documented in several atrophying diseases (e.g., focal epilepsy: Lin et al. 2007, Alzheimer's disease: Lerch et al. 2005).

- *Fractal surface measures:* The neocortical surface can be viewed as a 2D sheet folded in 3D space. Thus, the fractal dimension is a local measure of the "convolutedness" of the neocortex, ranging between 2 and 3 (Lopes and Betrouni 2009). This measure can be used to describe morphometric properties of brain regions in relation to cognitive abilities (Im et al. 2006b, Zhang et al. 2008).

- The determination of *surface curvature* is of importance in different applications such as smoothing, edge enhancement, mesh decimation, and quality checking [for a comprehensive treatment, refer to Meyer et al. (2003)]. The adaption of a local parabolic patch (Stokely and Wu 1992) allows a robust estimation of the local Gaussian and mean curvature. Because sulcal areas are concave (and gyral areas are convex), the local mean curvature can be used to segment the neocortical surface into patches. Normalization of the curvature by the local bending radius achieves a better stability against small surface features (Yang and Kruggel 2008). Geodesic depth (the distance from the surface to the convex hull enclosing the brain) can be used to distinguish primary from secondary sulci: The former are deeper and longer.

- Brain surfaces generated by the procedures discussed previously have a topological genus of zero, that is, they are homologue to a sphere. The process of mapping a convoluted surface mesh on a sphere is called *surface flattening* (Floater and Hormann 2005). Such a mapping can be either angle- or area-preserving but not isometric (both)—which is most desirable because all properties of the surface are represented in the parameter domain. Finding an angle-preserving (conformal) mapping yields an almost unique solution: All mappings form a so-called Mobius group (Gu et al. 2004). Equiareal mappings (Sheffer et al. 2004) are substantially different because from the point of view of uniqueness, there are many more of them. A well-behaved mapping aims for area preservation with a minimization of angular distortion [e.g., Kruggel 2008 (see Figure 1.9), Yotter et al. 2011].

- *Spatial correspondence:* Spherical maps as obtained above have the advantage that a coordinate system can readily be introduced on the surface. Thus, via their intermediate representation of spherical meshes, a spatial correspondence between brain surfaces can be established (Davatzikos 1995).

- Spherical maps were used to develop a *surface-based atlas* of the human neocortex (Fischl et al. 1999). Local surface features (i.e., curvature and geodesic depth) were incorporated in the spherical map.

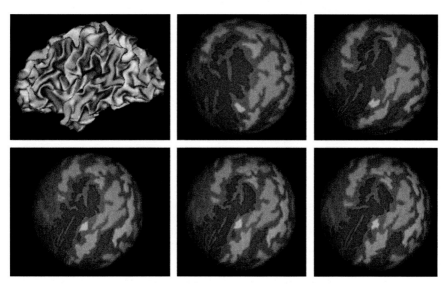

FIGURE 1.9
(Color version available from crcpress.com; see Preface.) Mapping of surface structures onto a spherical mesh: surface mesh representing the left brain hemisphere with color-coded segmentation of brain sulci (top left). Corresponding spherical maps with angle-preserving properties (top middle) to area-preserving properties (bottom right). (From Kruggel, F., *Med Image Anal* 12, 291–299, 2008. With permission.)

A method for the nonlinear registration of these feature maps on the sphere (Yeo et al. 2011) led to the development of a precise cortical parcellation scheme (Desikan et al. 2006) that is available with FreeSurfer. However, it should be noted that this atlas contains a fixed number of structures, and thus does not accurately represent the structural variability of the neocortex. This approach is also not stable in the presence of cortical brain lesions.

- *Surface representation:* Spherical surfaces, such as brain surface meshes with genus zero, can be represented by the coefficients of a spherical harmonics transformation (SHT, Shen et al. 2009). Loosely called a "Fourier transform on the sphere," coefficients with increasing degree provide a coarse-to-fine description of brain surface. This approach has several attractive properties: (1) The complex and highly individual surface of a hemisphere is described by a sparse set of coefficients. (2) Spherical harmonics form a set of orthonormal functions. Thus, they span a shape space with a metric for comparing and classifying shapes. (3) Surfaces can be easily reconstructed from parameters and point-to-point correspondences between different brain surfaces defined (Brechbuehler et al. 1995). Note that the SHT provides a global shape representation, and it is difficult to detect and describe regional differences. Thus, SHT representations were integrated with "active shape models" (Cootes et al. 1995) to provide a better local adaptivity (Styner et al. 2006).

Surfaces also provide a "natural" choice for visualizing quantitative data (e.g., local measurements, statistical results, group differences), and allow representing complex measurements in an intuitive, easily approachable way.

1.7 Texture Analysis

Biomedical tissue contains rich texture, due to the composite nature of tissue compartments in the brain and its outer hulls. Much different tissue patterns are found at different scales. Consider the neocortex: at the microlevel, a complex pattern of cells, dendrites, and axons is found; at the mesolevel, the typical six-layered organization of the neocortex is revealed; and at the macrolevel, the cortex is found as a highly complex, thin sheet on the outer surface of the brain. With the increasing fidelity of MR technology, more texture levels can be studied with more detail, so texture analysis is just becoming an interesting new option for the characterization of tissue properties.

Texture analysis is concerned with the description of repeated local intensity patterns, either in the spatial or in the frequency domain [for an introduction, refer to Mirmehdi et al. (2008)]. Texture descriptions are often multilevel: the lower or syntactic level characterizes a single entity (e.g., a cell), while a higher or semantic level describes the organization of cells in a tissue. At the syntactic level, statistical approaches (e.g., using the autocorrelation function, co-occurrence matrices, texture transformations) are better suited, while for the semantic level, structural approaches (e.g., using periodicity, adjacency, homogeneity) provide better measures (Haralick 1979).

The use of texture analysis in MRI and its prerequisites for imaging protocols are well studied (Lerski et al. 2004, Schad 2002). The most straightforward and flexible approach for the statistical analysis of texture properties is provided through the use of co-occurrence matrices. These matrices capture the relative frequency of neighboring intensity patterns in a specific direction (Haralick et al. 1973). This concept was later extended to three dimensions, and enhanced by the inclusion of gradient patterns and gradient angles in so-called multisort co-occurrence matrices (Kovalev and Petrou 1996). Texture measures derived from these matrices were demonstrated to be highly useful in the detection of disease patterns (Kovalev et al. 2001), in the description of gender-related differences and age-related changes (Kovalev et al. 2003; see Figure 1.10), and in the age-related decline of WM fiber structure (Kovalev and Kruggel 2007).

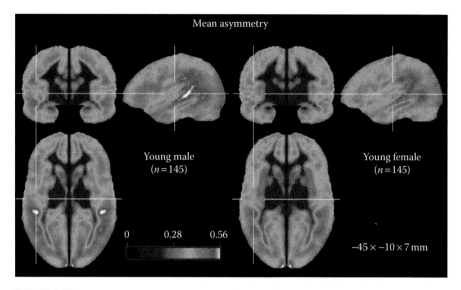

FIGURE 1.10
(See color insert.) Orthogonal sections of mean asymmetry maps for young males (left panel, $n = 145$, 18–32 years) and females (right panel, $n = 145$, 18–32 years). Note the prominent asymmetry in males apparent in the vicinity of Heschl's gyrus, the superior temporal gyrus, post-central gyrus, and the optic radiation. In females (as in males), the only prominent asymmetry is found at the occipital pole, and much less at the postcentral gyrus. (From Kovalev, V.A., et al., *Neuroimage* 19, 895–905, 2003. With permission.)

The rich local information contained in texture feature vectors can be exploited in difficult segmentation problems. Thus, texture-based segmentation was used to detect diffuse WM lesions (Kruggel et al. 2007), brain tumors (Herlidou-Meme et al. 2003, Mahmoud-Ghoneim et al. 2003, Zacharaki et al. 2009), and focal dysplasia (Antel et al. 2003). With increasing spatial resolution (and computational power), texture analysis appears as a promising approach awaiting further consideration. Little attention was spent for the texture analysis from multimodal images.

1.8 Analysis of Brain Diseases

MRI is a standard method in clinical practice to reveal focal and diffuse pathological processes in the brain, such as caused by degenerative diseases, trauma, intracerebral hemorrhage, cerebral infarctions, inflammatory diseases, and neoplasms. Most typically, T_1-weighted images are acquired along with other modalities that are sensitive to correlates of the disease process (e.g., T_2-, FLAIR-weighted images). An increased permeability of the blood–brain barrier may be detected in contrast-enhanced images using gadolinium-DTPA as a tracer (Tofts and Kermode 1991). While a *semiquantitative* analysis of MR tomograms based on visual inspection (e.g., using diagnostic scales) is common today in clinical protocols, tools for a *quantitative* analysis are still rarely available. One of the reasons for this deficiency is that the validity of their results is often difficult to assess, and even arguable in their meaning, as we will discuss below. The variety in the appearance of pathological processes conveys the notion that there is no single best method to detect disease processes in the brain. Rather, specific methods were developed that capture specific properties of a known disease process. Another added difficulty is that most disease processes are individual in terms of the position, extent, and severity of affected tissues. Thus, methods to analyze single cases, perhaps in a longitudinal series, are of interest here. For describing group-related changes due to diffuse diseases (e.g., Alzheimer's disease, schizophrenia), tools discussed in the previous sections may be sufficient.

The quantitative description of brain lesions may be used to (1) relate the position, extent, and severity of a lesion to the clinical status of a patient (e.g., Kincses et al. 2011, Wen et al. 2004, Wilde et al. 2005); (2) detect and quantify the amount of progression (and remission) in chronic inflammatory (e.g., Meier and Guttmann 2003) or degenerative brain diseases (e.g., Misra et al. 2008); (3) rate and understand the efficiency of therapeutic interventions (e.g., Godin et al. 2011); and (4) understand the structural and functional reorganization processes after focal brain damage (e.g., Levin 2003).

First, let us consider the case of *compact, uniform lesions* that are well separated from the surrounding tissue by their intensity characteristics. Examples of such lesions are found with multiple sclerosis, low-grade glioma and lacunar infarctions. Such lesions may be detected as compact regions of outlier voxels in terms of the intensity distribution of normal tissues. van Leemput et al. (2001) proposed a method to segment multimodal brain datasets (here, T_1-, T_2-, and PD-weighted images) into four classes denoted as GM, WM, CSF, and "other" based on a multivariate GMM and an atlas prior. Voxels of the latter class are then examined for their intensity and context. "Bright" voxels (in T_2) in overprojection with the WM are denoted as lesions. A variant of this approach (Yang et al. 2010) provided a more rigorous statistical treatment of outlier voxels that uses the theory of excursion sets (see above) to increase the specificity of lesion detection. Depending on the spatial resolution and image contrast, lesions over 50–100 mm^3 may be considered as a significant detection. Diffuse WM lesions pose a more difficult segmentation problem, due to their inhomogeneous pattern and faint boundaries. Local texture descriptors (e.g., co-occurrence feature vectors as defined above) offer an exhaustive assessment of tissue properties. Feature vectors are discriminated into "healthy" and "affected" classes based on a sensible choice of a distance measure in high-dimensional space (Kruggel et al. 2007; see Figure 1.11). Longitudinal studies, for example, in patients with multiple

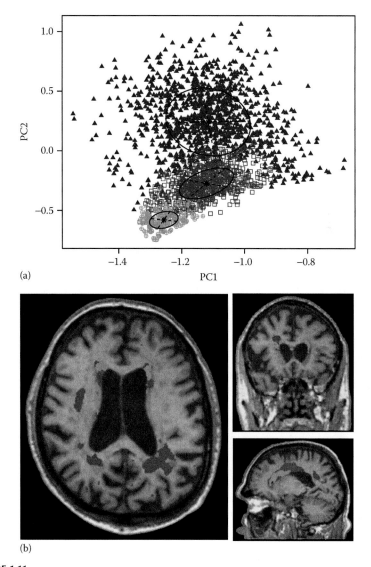

FIGURE 1.11
(See color insert.) (a) Discrimination of feature vectors corresponding to WM (blue), diffuse white matter hyperintensities (DWMH, red), and periventricular lesions (PVL, green). (b) Overlay of classification result onto the T_1-weighted image with DWMH (red) and PVL (green). (From Kruggel, F., et al., *Neuroimage* 39, 987–996, 2007. With permission.)

sclerosis, pose an additional problem: Because scanners have a limited long-term stability, images acquired on the same scanner using the same protocol but at a certain interval (e.g., 6–12 months) may show considerable differences in the intensity range and distribution. Thus, across-scan intensity normalization is required here (Meier and Guttmann 2003).

The problem of detecting *focal, nonuniform lesions* is considered as ill-posed for the following reasons: (1) Often, predisease information is not available, so how can the extent of a (partly) missing brain region be determined? How can the boundary between a cortical lesion and the CSF compartment be determined? (2) MR intensities of a lesion range between values of undamaged tissue and values similar to CSF, indicating a completely damaged area. How can a reliable and consistent decision be made about the lesion border drawn based on imaging data? (3) Properties of a lesion are often inhomogeneous in space, often with completely damaged core parts and minor damage in peripheral portions. How can a meaningful assessment of an inhomogeneous lesion be performed? (4) It is well known that the brain undergoes restoration processes following a focal damage. Thus, the time lapsed since the event is an important parameter to be included in the analysis. Even more, a complex geometric reorganization of the brain's structural "skeleton" is often found with large lesions: while the core lesion area decreases in extent, adjacent inner and outer ventricles enlarge, and even remote sulci widen. (5) When multiple lesions are found in a brain, it is often unclear, whether they should be addressed to a single or multiple events. (6) In addition to the disease mechanisms listed above, other chronic or acute conditions may add to the amount of structural brain damage [e.g., neurodegenerative diseases (mostly) affecting the GM compartment and small vessel diseases (as a consequence of diabetes or high blood pressure), (mostly) affecting the WM]. In short, if no prior information is available, the results of lesion segmentation have to be evaluated with respect to the properties of the segmentation method and the known clinical history of a specific patient. The following general approaches may be distinguished:

- *Determination of the total brain atrophy:* It is well known that brain growth drives skull growth in development. The maximum occupancy of brain in the intracranial space is found in early adulthood (i.e., 25 years) and amounts to 87.5% (Kruggel 2006). Given a normal brain atrophy rate of −0.12% per year of age, the premorbid brain volume can be estimated and compared with the actual volume: the difference corresponds to the estimated total brain loss.

- *Lesion segmentation using boundary features:* Ischemic lesions that differ in intensity from the surrounding tissue can be segmented by a region growing process. A seed inside the lesion is selected, and surrounding voxels with similar intensities are absorbed into a lesion region. The contrast at the region boundary is monitored. The process stops when the peripheral contrast is maximal (Hojjatoleslami and Kruggel 2000). This method detects arbitrarily shaped but compact regions as lesions. However, if the boundary between the lesion and the CSF compartment is not well delineated (e.g., in the case of a cortical infarction), this algorithm may fail.

- *Lesion segmentation using symmetry:* As a first approximation, the brain is a mirror-symmetric organ. Assuming that a lesion is confined to a single hemisphere with a healthy region on the contralateral side, it may be detected as a compact region with a significantly different intensity statistic. Small windows at mirror-symmetric locations in both hemispheres are defined, and their intensities compared using a Hotelling T^2-test. Voxel-wise T^2 measures are converted into z-scores and collected in a lesion probability map. Local regions of high z-scores are candidates for lesion regions. "Natural" nonsymmetric regions are rather small (<0.8 ml) and are removed from this map (Kruggel 2004; see Figure 1.12). Note that this method also detects any consecutive enlargements of other compartments (e.g., inner ventricles, ipsilateral cortical folds) as "lesions." A similar approach uses symmetry violations to detect a tumor, and a deformable model to segment the boundary (Khotanlou et al. 2009; Figure 1.12).

- *Model-based lesion segmentation:* If a strong prior assumption about the shape of a lesion is possible, using a structural model for lesion segmentation is a viable approach. A model can be constructed from training data and evolved in the actual image (Tsai et al. 2003).

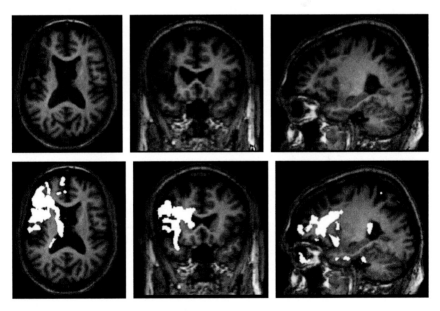

FIGURE 1.12
Top row: orthogonal sections of a T_1-weighted image of a patient with an ischemic lesion in the anterior and middle territory of the left cerebral artery. *Bottom row*: symmetry-based segmentation of the lesion area. Note that the lesion-related consecutive enlargement of the inner and outer ventricles is detected and counted as "lesion."

A shape model may be used to build a matched filter. From candidate objects found by this filter, subsets are selected that match a predefined configuration. This approach was used to segment enlarged perivascular spaces in the brain (Descombes and Kruggel 2004).

- *Atlas-guided lesion segmentation:* The concept of outlier detection was put forward for the segmentation of inhomogeneous lesions. An atlas is included to detect pathologic regions and elaborate postprocessing techniques are required to segment the lesion (Prastawa et al. 2004, Corso et al. 2008).

Unfortunately, tools for lesion segmentation are currently not available with any of the major software packages. In the end, manual lesion segmentation is still considered as the "gold standard." While this method obviously produces the most reliable results, it is time consuming and tedious. Studies comparing results of manually and automatically segmented lesions rarely reach 90% correspondence (Clark et al. 1998, Joe et al. 1999).

References

Antel, S.B., Collins, D.L., Bernasconi, N., et al. 2003. Automated detection of focal cortical dysplasia lesions using computational models of their MRI characteristics and texture analysis. *Neuroimage* 19, 1748–1759.

Ashburner, J. and Friston, K.J. 1997. Multimodal image coregistration and partitioning—A unified framework. *Neuroimage* 6, 209–217.

Ashburner, J. and Friston, K.J. 2000. Voxel-based morphometry—The methods. *Neuroimage* 11, 805–821.

Ashburner, J. and Friston, K.J. 2005. Unified segmentation. *Neuroimage* 26, 839–851.

Arnold, J.B., Liow, J.S., Schaper, K.A., et al. 2001. Qualitative and quantitative evaluation of six algorithms for correcting intensity nonuniformity effects. *Neuroimage* 13, 931–943.

Avants, B.B., Epstein, C.L., Grossman, M., and Gee, J.C. 2008. Symmetric diffeomorphic image registration with cross-correlation: Evaluating automated labeling of elderly and neurodegenerative brain. *Med Image Anal* 12, 26–41.

Avants, B.B., Yushkevich, P., Pluta, J., et al. 2010. The optimal template effect in hippocampus studies of diseased populations. *Neuroimage* 49, 2457–2466.

Belaroussia, B., Milles, J., Carmec, S., et al. 2006. Intensity non-uniformity correction in MRI: Existing methods and their validation. *Med Image Anal* 10, 234–246.

Benjamini, Y. and Hochberg, Y. 1995. Controlling the false discovery rate: A practical approach to multiple testing. *J R Statist Soc* 57, 289–300.

Bertrand, G. 1995. A Boolean characterization of three-dimensional simple points. *Pattern Recognit Lett* 17, 115–124.

BrainVisa. 2012. Institut Federatif de Recherche 49. http://brainvisa.info/ (accessed Feb 27, 2012).

Brechbuehler, G., Gerig, G., and Kuebler, O. 1995. Parametrization of closed surfaces for 3-D shape description. *Comput Vis Graph Image Process* 61, 154–170.

Brett, M. 2002. The MNI brain and the Talairach atlas. http://imaging.mrc-cbu.cam. ac.uk/imaging/MniTalairach (accessed Feb 27, 2012).

Cao, J. 1999. The size of the connected components of excursion sets of χ^2, t and F fields. *Adv Appl Probab* 31, 579–595.

Chiverton, J.P. 2006. Probabilistic partial volume modelling of biomedical tomographic image data. Ph.D. Thesis, University of Surrey, Surrey.

Cho, Z.H. 2010. *7.0 Tesla MRI Brain Atlas*. Springer, Berlin.

Chung, M.K., Worsley, K.J., Paus, T., et al. 2001. A unified statistical approach to deformation-based morphometry. *Neuroimage* 14, 595–606.

Clark, L.P., Velthuizen, R.P., Clark, M., et al. 1998. MRI measurement of brain tumor response: Comparison of visual metric and automatic segmentation. *Magn Reson Imaging* 16, 271–279.

Collins, D.L., Holmes, C.J., Peters, T.M., et al. 1995. Automatic 3-D model-based neuroanatomical segmentation. *Hum Brain Mapp* 3, 190–208.

Cootes, T.F., Taylor, C.J., Cooper, D.H., and Graham, J. 1995. Active shape models—Their training and application. *Comput Vis Image Underst* 61, 38–59.

Corso, J.J., Sharon, E., Dube, S., et al. 2008. Efficient multilevel brain tumor segmentation with integrated Bayesian model classification. *IEEE Trans Med Imaging* 27, 629–640.

Dale, A.M., Fischl, B., and Sereno, M.I. 1999. Cortical surface-based analysis I: Segmentation and surface reconstruction. *Neuroimage* 9, 179–194.

Davatzikos, C.A. 1995. Spatial normalization of 3D brain images using deformable models. *J Comput Tomogr* 20, 656–665.

Davatzikos, C.A. and Prince, J.L. 1995. An active contour model for mapping the cortex. *IEEE Trans Med Imaging* 14, 65–80.

De Coene, B., Hajnal, J.V., Gatehouse, P., et al. 1992. MR of the brain using fluid-attenuated inversion recovery (FLAIR) pulse sequences. *Am J Neuroradiol* 13, 1555–1584.

Deichmann, R., Schwarzbauer, C., and Turner, R. 2004. Optimisation of the 3D MDEFT sequence for anatomical brain imaging: Technical implications at 1.5 and 3 T. *Neuroimage* 21, 757–767.

Descombes, X. and Kruggel, F. 2004. An object-based approach for detecting small brain lesions: Application to Virchow-Robinson spaces. *IEEE Trans Med Imaging* 23, 246–255.

Desikan, R.S., Segonne, F., Fischl, B., et al. 2006. An automated labeling system for subdividing the human cerebral cortex on MRI scans into gyral-based regions of interest. *Neuroimage* 31, 968–980.

Digital Imaging and Communication in Medicine (DICOM). 2012. http://medical. nema.org/ (accessed Feb 27, 2012).

Entis, J.J., Doerga, P., Barrett, L.F., et al. 2012. A reliable protocol for the manual segmentation of the human amygdala and its subregions using ultra-high resolution MRI. *Neuroimage* 60, 1226–1235.

Fischl, B. and Dale, A.M. 2000. Measuring the thickness of the human cerebral cortex from magnetic resonance images. *Proc Natl Acad Sci* 97, 11050–11055.

Fischl, B., Liu, A., and Dale, A.M. 2001. Automated manifold surgery: Constructing geometrically accurate and topologically correct models of the human cerebral cortex. *IEEE Trans Med Imaging* 20, 70–80.

Fischl, B., Sereno, M.I., and Dale, A.M. 1999. Cortical surface-based analysis II: Inflation, flattening, and a surface-based coordinate system. *Neuroimage* 9, 195–207.

Floater, M.S. and Hormann, K. 2005. Surface parameterization: A tutorial and survey. In: Dodgson, N.A., Floater, M.S., and Sabin, M.A. (eds.), *Advances in Multiresolution for Geometric Modelling*. Springer, Berlin, pp. 157–186.

FMRIB Software Library (FSL). 2012. FMRIB Analysis Group. http://www.fmrib.ox.ac.uk/fsl/ (accessed Feb 27, 2012).

FreeSurfer. 2012. Martinos Center for Biomedical Imaging. http://surfer.nmr.mgh.harvard.edu/ (accessed Feb 27, 2012).

Gildenberg, P.L. 1988. Stereotactic surgery: Present and past. In: Heilbrun, M.P. (ed.), *Stereotactic Neurosurgery*. Williams and Wilkins, Baltimore, pp. 1–15.

Godin, O., Tzourio, C., Maillard, P., et al. 2011. Antihypertensive treatment and change in blood pressure are associated with the progression of white matter lesion volumes: The three-city (3C)-Dijon magnetic resonance imaging study. *Circulation* 25, 266–273.

Gu, X., Wang, Y., Chan, T.F., et al. 2004. Genus zero surface conformal mapping and its application to brain surface mapping. *IEEE Trans Med Imaging* 23, 949–958.

Gudbjartsson, H. and Patz, S. 1995. The Rician distribution of noisy MRI data. *Magn Reson Med* 34, 910–914.

Han, X., Pham, D.L., Tosun, D., et al. 2004. CRUISE: Cortical reconstruction using implicit surface evolution. *Neuroimage* 23, 997–1012.

Han, X., Xu, P., and Prince, J.L. 2005. A topology preserving level set method for geometric deformable models. *IEEE Trans Pattern Anal Mach Intell* 25, 755–768.

Haralick, R.M. 1979. Statistical and structural approaches to texture. *Proc IEEE* 67, 786–804.

Haralick, R.M., Shanmugam, K., and Dinstein, J. 1973. Textural features for image classification. *IEEE Trans Syst Man Cybern* 6, 610–662.

He, R., Datta, S., Sajja, B.R., et al. 2008. Generalized fuzzy clustering for segmentation of multi-spectral magnetic resonance images. *Comput Med Imaging Graph* 32, 353–366.

Hentschel, S. and Kruggel, F. 2004. Determination of the intracranial compartment: A registration approach. In: Jiang, T. (ed.), *Lecture Notes in Computer Science*, Vol. 3150. Springer, Berlin, pp. 253–260.

Herlidou-Meme, S., Constans, J.M., Carsin, B., et al. 2003. MRI texture analysis on texture test objects, normal brain and intracranial tumors. *Magn Reson Imaging* 21, 989–993.

Hojjatoleslami, S.A. and Kruggel, F. 2000. Segmentation of large brain lesions. *IEEE Trans Med Imaging* 20, 666–669.

Hua, X., Leow, A.D., Parikshak, N., et al. 2008. Tensor-based morphometry as a neuroimaging biomarker for Alzheimer's disease: An MRI study of 676 AD, MCI, and normal subjects. *Neuroimage* 43, 458–469.

Im, K., Lee, J.M., Lee, J., et al. 2006a. Gender difference analysis of cortical thickness in healthy young adults with surface-based methods. *Neuroimage* 31, 31–38.

Im, K., Lee, J.M., Yoon, U., et al. 2006b. Fractal dimension in human cortical surface: Multiple regression analysis with cortical thickness, sulcal depth, and folding area. *Hum Brain Mapp* 27, 994–1003.

The International Consortium for Brain Mapping (ICBM). 2012. http://www.loni.ucla.edu/ICBM/ (accessed Feb 27, 2012).

Joe, B.N., Fukui, M.B., Meltzer, C.C., et al. 1999. Brain tumor volume measurement: Comparison of manual and semiautomated methods. *Radiology* 212, 811–816.

Johansson, A. 2010. Evaluation of bone contrast enhanced MRI sequences and voxel-based segmentation. M.S. thesis, Umea University, Sweden.

Khan, A.R., Wang, L., and Beg, M.F. 2008. FreeSurfer-initiated fully automated subcortical brain segmentation in MRI using large deformation diffeomorphic metric mapping. *Neuroimage* 41, 735–746.

Khotanlou, H., Colliot, O., Atif, J., et al. 2009. 3D brain tumor segmentation in MRI using fuzzy classification, symmetry analysis and spatially constrained deformable models. *Fuzzy Sets Syst* 160, 1457–1473.

Kincses, Z.T., Ropele, S., Jenkinson, M., et al. 2011. Lesion probability mapping to explain clinical deficits and cognitive performance in multiple sclerosis. *Mult Scler* 17, 681–689.

Klein, A., Anderson, J., Ardekani, B.A., et al. 2009. Evaluation of 14 nonlinear deformation algorithms applied to human brain MRI registration. *Neuroimage* 46, 768–802.

Kovalev, V. and Kruggel, F. 2007. Texture anisotropy of the brain's white matter as revealed by anatomical MRI. *IEEE Trans Med Imaging* 26, 678–685.

Kovalev, V.A., Kruggel F., and von Cramon, D.Y. 2003. Gender and age effects in structural brain asymmetry as measured by MRI texture analysis. *Neuroimage* 19, 895–905.

Kovalev, V.A., Kruggel, F., von Cramon, D.Y., et al. 2001. 3D texture analysis of MR brain datasets. *IEEE Trans Med Imaging* 20, 424–433.

Kovalev, V.A. and Petrou, M. 1996. Multidimensional co-occurrence matrices for object recognition and matching. *Graph Models Image Process* 58, 187–197.

Kriegeskorte, N. and Goebel, R. 2001. An efficient algorithm for topologically correct segmentation of the cortical sheet in anatomical MR volumes. *Neuroimage* 14, 329–346.

Kruggel, F. 2004. Segmentation of focal brain lesions. In: Jiang, T. (ed.), *Lecture Notes in Computer Science*, Vol. 3150. Springer, Berlin, pp. 10–18.

Kruggel, F. 2006. MRI-based volumetry of head compartments: Normative values of healthy adults. *Neuroimage* 30, 1–11.

Kruggel, F. 2008. Robust parametrization of brain surface meshes. *Med Image Anal* 12, 291–299.

Kruggel, F., Paul, J.S., and Gertz, H.J. 2007. Texture-based segmentation of diffuse lesions of the brain's white matter. *Neuroimage* 39, 987–996.

Kruggel, F., Turner, J., and Muftuler, L.T. 2010. Impact of scanner hardware and imaging protocol on image quality and compartment volume precision in the ADNI cohort. *Neuroimage* 49, 2123–2133.

Kruggel, F. and von Cramon, D.Y. 1999. Alignment of MR brain datasets with the stereotactical coordinate system. *Med Image Anal* 3, 175–185.

Leow, A.D., Chiang, M.C., Huang, S.C., et al. 2006. Realizing unbiased deformation: A theoretical consideration. Workshop on Mathematical Foundations of Computational Anatomy: Geometrical, Statistical and Registration Methods for Modeling Biological Shape Variability, Oct 1, pp. 174–181. http://www-sop.inria.fr/asclepios/events/MFCA06/ (accessed Oct 1, 2006).

Lerch, J.P., Pruessner, J.C., Zijdenbos, A., et al. 2005. Focal decline of cortical thickness in Alzheimer's disease identified by computational neuroanatomy. *Cereb Cortex* 15, 995–1001.

Lerski, R.A., Straughan, K., Schad, L.R., et al. 2004. MR image texture analysis—An approach to tissue characterization. *Magn Reson Imaging* 11, 873–887.

Levin, H.S. 2003. Neuroplasticity following non-penetrating traumatic brain injury. *Brain Inj* 17, 665–674.

Lin, J.J., Salamon, N., Lee, A.D., et al. 2007. Reduced neocortical thickness and complexity mapped in mesial temporal lobe epilepsy with hippocampal sclerosis. *Cereb Cortex* 17, 2007–2018.

Lopes, R. and Betrouni, N. 2009. Fractal and multifractal analysis: A review. *Med Image Anal* 13, 634–649.

Lorensen, W.E. and Cline, H.E. 1987. Marching cubes: A high resolution 3D surface construction algorithm. *Comput Graph* 21, 163–169.

Li, S.Z. 2009. *Markov Random Field Modeling in Image Analysis*. Springer, Berlin.

Luders, E., Narr, K.L., Thompson, P.M., et al. 2006. Hemispheric asymmetries in cortical thickness. *Cereb Cortex* 16, 1232–1238.

Luebke, D.P. 2001. A developer's survey of polygonal simplification algorithms. *IEEE Comput Graph Appl* 21, 24–35.

MacDonald, D., Kabani, N., Avis, D., et al. 2000. Automated 3-D extraction of inner and outer surfaces of cerebral cortex from MRI. *Neuroimage* 12, 340–356.

Maes, F., Collignon, A., Vandermeulen, D., et al. 1997. Multimodality image registration by maximization of mutual information. *IEEE Trans Med Imaging* 16, 187–198.

Mahmoud-Ghoneim, D., Toussaint, G., Constans, J.M., et al. 2003. Three-dimensional texture analysis in MRI: A preliminary evaluation in gliomas. *Magn Reson Imaging* 21, 983–987.

Maintz, J.B.A. and Viergever, M.A. 1998. A survey of medical image registration. *Med Image Anal* 2, 1–36.

Mangin, J.F., Coulon, O., and Frouin, V. 1998. Robust brain segmentation using histogram scale-space analysis and mathematical morphology. In: Wells, W.M. et al. (eds.), *Lecture Notes in Computer Science*, Vol. 1496. Springer, Berlin, pp. 1230–1241.

Mazziotta, J.C., Toga, A.W., Evans, A., et al. 1995. A probabilistic atlas of the human brain: Theory and rationale for its development: The International Consortium for Brain Mapping (ICBM). *Neuroimage* 2, 89–101.

Mazziotta, J.C., Woods, R., Iacobini, M., et al. 2009. The myth of the normal, average human brain—The ICBM experience: (1) Subject screening and eligibility. *Neuroimage* 44, 914–922.

McLachlan, G.J. and Basford, K.E. 1988. *Mixture Models: Inference and Applications to Clustering*. Marcel Dekker, New York.

Mechelli, A., Price, C.J., Friston, K.J., et al. 2005. Voxel-based morphometry of the human brain: Methods and applications. *Curr Med Imaging Rev* 1, 105–113.

Meier, D.S. and Guttmann, C.R.G. 2003. Time-series analysis of MRI intensity patterns in multiple sclerosis. *Neuroimage* 20, 1193–1209.

Meyer, M., Desbrun, M., Schroeder, P., et al. 2003. Discrete differential-geometry operators for triangulated 2-manifolds. *Vis Math* III, 35–57.

Mihara, H., Iriguchi, N., and Ueno, S. 1998. A method of RF inhomogeneity correction in MR imaging. *Magn Reson Mater Phys Biol Med* 7, 115–120.

Miller, M.I., Trouve, A., and Younes, L. 2002. On the metrics and Euler–Lagrange equations of computational anatomy. *Ann Rev Biomed Eng* 4, 375–405.

Mirmehdi, M., Xie, X., and Sur, J. (eds.). 2008. *Handbook of Texture Analysis*. World Scientific Books, Singapore.

Misra, C., Fan, Y., and Davatzikos, C. 2008. Baseline and longitudinal patterns of brain atrophy in MCI patients, and their use in prediction of short-term conversion to AD: Results from ADNI. *Neuroimage* 44, 1415–1422.

Mugler III, J.P. and Brookeman, J.R. 1990. Three-dimensional magnetization-prepared rapid gradient-echo imaging (3D MP RAGE). *Magn Reson Med* 15, 152–157.

Neuroimaging Informatics Technology Initiative (NIfTI). 2012. http://nifti.nimh.nih.gov/ (accessed Feb 27, 2012).

Neuroimaging Informatics Tools and Resources Clearinghouse (NITRC). 2012. http://www.nitrc.org/ (accessed Feb 27, 2012).

Nielson, G.M. 2003. On marching cubes. *IEEE Trans Vis Comput Graph* 9, 283–297.

O'Donnell, S., Noseworthy, M.D., Levine, B., et al. 2005. Cortical thickness of the frontopolar area in typically developing children and adolescents. *Neuroimage* 24, 948–954.

Osechinskiy, S. and Kruggel, F. 2012. Cortical surface reconstruction from high-resolution MR brain images. *Int J Biomed Imaging*, doi:10.1155/2012/870196.

Parizel, P.M., van den Hauwe, L., De Belder, F., et al. 2010. Magnetic resonance imaging of the brain. In: Reimer, P. et al. (eds.), *Clinical MR Imaging*. Springer, Berlin, pp. 107–196.

Patenaude, B., Smith, S.M., Kennedy, D.N., et al. 2011. A Bayesian model of shape and appearance for subcortical brain segmentation. *Neuroimage* 56, 907–922.

Perrot, M., Riviere, D., and Mangin, J.-F. 2011. Cortical sulci recognition and spatial normalization. *Med Image Anal* 15, 529–550.

Pham, D.L. 2001. Robust fuzzy segmentation of magnetic resonance images. *IEEE Symposium on Computer-Based Medical Systems 2001*. July 26–27, Bethesda, MD. IEEE Computer Society, Washington, DC, pp. 127–131.

Pham, D.L., Xu, C., and Prince, J.L. 2000. Current methods in medical image segmentation. *Ann Rev Biomed Eng* 2, 315–337.

Piguet, O., Ridley, L.J., Grayson, D.A., et al. 2005. Comparing white matter lesions on T2 and FLAIR MRI in the Sydney Older Persons Study. *Eur J Neurol* 12, 399–402.

Prastawa, M., Bullitt, E., Ho, S., et al. 2004. A brain tumor segmentation framework based on outlier detection. *Med Image Anal* 8, 275–283.

Pruessner, J.C., Li, L.M., Serles, W., et al. 2000. Volumetry of hippocampus and amygdala with high-resolution MRI and three-dimensional analysis software: Minimizing the discrepancies between laboratories. *Cereb Cortex* 120, 433–442.

Qin, L., van Gelderen, P., Derbyshire, J.A., et al. 2009. Prospective head-movement correction for high-resolution MRI using an in-bore optical tracking system. *Magn Reson Med* 62, 924–934.

Reiser, M.F., Semmler, W., and Hricak, H. 2005. *Magnetic Resonance Tomography*. Springer, Berlin.

Research Imaging Center. 2012. Talairach coordinate space. http://www.talairach.org/ (accessed Feb 27, 2012).

Robson, M.D., Gatehouse, P.D., Bydder, M., et al. 2003. Magnetic resonance: An introduction to ultrashort TE (UTE) imaging. *J Comput Assist Tomogr* 27, 825–846.

Robson, M.D., Gore, J.C., and Constable, R.T. 1997. Measurement of the point spread function in MRI using constant time imaging. *Magn Reson Med* 38, 733–740.

Schad, L.R. 2002. Problems in texture analysis with magnetic resonance imaging. *Dialogues Clin Neurosci* 4, 235–242.

Senjem, M.L., Gunter, J.L., Shiung, M.M., et al. 2005. Comparison of different methodological implementations of voxel-based morphometry in neurodegenerative disease. *Neuroimage* 26, 600–608.

Shattuck, D.W., Mirza, M., Adisetiyo, V., et al. 2008. Construction of a 3D probabilistic atlas of human cortical structures. *Neuroimage* 39, 1064–1080.

Shattuck, D.W., Sandor-Leahy, S.R., Schaper, K.A., et al. 2001. Magnetic resonance image tissue classification using a partial volume model. *Neuroimage* 13, 856–876.

Sheffer, A., Gotsman, C., and Dyn, N. 2004. Robust spherical parametrization of triangular meshes. *Computing* 72, 185–193.

Shen, L., Farid, H., and McPeek, M.A. 2009. Modeling three-dimensional morphological structures using spherical harmonics. *Evolution* 63, 1003–1016.

Sled, J.G., Zijdenbos, A.P., and Evans, A.C. 1998. A nonparametric method for automatic correction of intensity nonuniformity in MRI data. *IEEE Trans Med Imaging* 17, 87–97.

Smith, S.M. 2002. Fast robust automated brain extraction. *Hum Brain Mapp* 17, 143–155.

Sowell, E.R., Peterson, B.S., Kan, E., et al. 2007. Sex differences in cortical thickness mapped in 176 healthy individuals between 7 and 87 years of age. *Cereb Cortex* 17, 1550–1560.

Statistical Parametric Mapping. 2012. Wellcome Trust Centre for Neuroimaging. 2012. http://www.fil.ion.ucl.ac.uk/spm/ (accessed Feb 27, 2012).

Stokely, E.M. and Wu, S.Y. 1992. Surface parametrization and curvature measurement of arbitrary 3D objects: Five practical methods. *IEEE Trans Pattern Anal Mach Intell* 14, 833–840.

Styner, M., Oguz, I., Xu, S., et al. 2006. Statistical shape analysis of brain structures using SPHARM-PDM. *Insight J*, 1–6.

Talairach, J. and Tournoux, P. 1993. *Referentially Oriented Cerebral MRI Anatomy: An Atlas of Stereotaxic Anatomical Correlations for Gray and White Matter.* Thieme Medical Publishers, New York.

Tardif, C.L., Collins, D.L., and Pike, G.B. 2009. Sensitivity of voxel-based morphometry analysis to choice of imaging protocol at 3 T. *Neuroimage* 44, 827–838.

Thevenaz, P., Blu, T., and Unser, M. 2000. Interpolation revisited. *IEEE Trans Med Imaging* 19, 739–758.

Thirion, J.P. 1998. Image matching as a diffusion process: An analogy with Maxwell's demons. *Med Image Anal* 2, 243–260.

Tofts, P.S. and Kermode, A.G. 1991. Measurement of the blood-brain barrier permeability and leakage space using dynamic MR imaging. 1. Fundamental concepts. *Magn Reson Med* 17, 357–367.

Toriwaki, J. and Yonekura, T. 2002. Euler number and connectivity indexes of a three dimensional digital picture. *Forma* 17, 183–209.

Tsai, A., Yezzi, A., Wells, W., et al. 2003. A shape-based approach to the segmentation of medical imagery using level sets. *IEEE Trans Med Imaging* 22, 137–154.

Tzourio-Mazoyer, N., Landeau, B., Papathanassiou, D., et al. 2002. Automated anatomical labeling of activations in SPM using a macroscopic anatomical parcellation of the MNI MRI single-subject brain. *Neuroimage* 15, 273–289.

van Leemput, K., Maes, F., Vandermeulen, D., et al. 1999. Automated model-based tissue classification of MR images of the brain. *IEEE Trans Med Imaging* 18, 897–908.

van Leemput, K., Maes, F., Vandermeulen, D., et al. 2001. Automated segmentation of multiple sclerosis lesions by model outlier detection. *IEEE Trans Med Imaging* 20, 677–688.

Verard, L., Allain, P., Travere, J.M., et al. 1997. Fully automatic indentification of AC and PC landmarks on brain MRI using scene analysis. *IEEE Trans Med Imaging* 16, 610–616.

Vercauteren, T., Pennec, X., Perchant, A., et al. 2008. Diffeomorphic demons: Efficient non-parametric image registration. *Neuroimage* 45, S61–S72.

Viviani, R., Beschoner, P., Erhard, K., et al. 2007. Non-normality and transformations of random fields, with an application to voxel-based morphometry. *Neuroimage* 35, 121–130.

Wang, D., Strugnell, W., Cowin, G., et al. 2004. Geometric distortion in clinical MRI systems: Part II: Correction using a 3D phantom. *Magn Reson Med* 22, 1223–1232.

Wen, H.M., Mok, V.C.T., Fan, Y.H., et al. 2004. Effect of white matter changes on cognitive impairment in patients with lacunar infarcts. *Stroke* 35, 1826–1830.

Wilde, E.A., Hunter J.V., Newsome, M.R., et al. 2005. Frontal and temporal morphometric findings on MRI in children after moderate to severe traumatic brain injury. *J Neurotrama* 22, 333–344.

Yang, F. and Kruggel, F. 2008. Automatic segmentation of human brain sulci. *Med Image Anal* 12, 442–451.

Yang, F., Shan, Z.Y., and Kruggel, F. 2010. White matter lesion segmentation based on feature joint occurrence probability and χ^2 random field theory from magnetic resonance (MR) images. *Pattern Recognit Lett* 31, 781–790.

Yeo, B.T.T., Sabuncu, M.R., Vercauteren, T. et al. 2011. Spherical demons: Fast diffeomorphic landmark-free surface registration. *IEEE Trans Med Imaging* 29, 650–668.

Yotter, R.A., Thompson, P.M., and Gaser, C. 2011. Algorithms to improve the reparameterization of spherical mappings of brain surface meshes. *J Neuroimaging* 21, e134–e147.

Zacharaki, E.I., Wang, S., Chawlaet S., et al. 2009. Classification of brain tumor type and grade using MRI texture and shape in a machine learning scheme. *Magn Reson Med* 62, 1609–1618.

Zhang, Y., Brady, M., and Smith, S. 2001. Segmentation of brain MR images through a hidden Markov random field model and the expectation-maximization algorithm. *IEEE Trans Med Imaging* 20, 45–57.

Zhang, Y., Jiang, J., Lin, L., et al. 2008. A surface-based fractal information dimension method for cortical complexity analysis. In: Jiang, T. (ed.), *Lecture Notes in Computer Science*, Vol. 5128. Springer, Berlin, pp. 133–141.

2

Quantifying Brain Morphology Using Diffusion Imaging

Murat Aksoy, Samantha Holdsworth, Anh Tu Van, and Roland Bammer

Stanford University

CONTENTS

2.1 Introduction

Diffusion-weighted imaging (DWI) is an MRI modality that is sensitive to the random Brownian motion of water molecules. By using this methodology, it is possible to probe tissue microstructure in the brain at a level that has not been possible using conventional MRI. Diffusion MRI has been used to assess various brain lesions, including ischemia, abscess, toxic–metabolic diseases, tumors, and others. DWI may also have a role in therapeutic monitoring and establishing prognosis in some of these diseases. DWI is particularly sensitive for the detection of hyperacute and acute ischemic stroke, and provides adjunctive information from other neuropathological processes (González et al., 1999; Lovblad et al., 1998; Marks et al., 1996). Given its important clinical role, DWI is now considered an essential part of diagnostic brain imaging in many hospitals.

Diffusion tensor imaging (DTI) is an extension of DWI whereby multiple diffusion-weighted acquisitions are performed to infer the directional diffusivity of water molecules in 3D (Basser et al., 1994). DTI uses the mathematics and physical interpretations of geometric quantities of diffusion known as tensors, and this information can be used as a surrogate to estimate the white matter connectivity in the brain (Basser et al., 2000). DTI data can be used to perform tractography whereby the direction information is exploited to follow neural tracts through the brain—a method that is especially useful in presurgical planning (Witwer et al., 2002).

In this chapter, we discuss the physical basics underlying diffusion acquisition in MRI and introduce diffusion-based quantitative metrics used in DTI, such as fractional anisotropy (FA) and mean diffusivity (MD). In addition, mathematical alternatives to the tensor model, such as q-space, high angular resolution diffusion imaging (HARDI), and q-ball will be explained. Image artifacts and challenges of DTI will also be described and potential remedies for them will be discussed.

2.2 Basic Principles of Diffusion-Weighted MRI

2.2.1 Physical Principles of Diffusion

Molecular diffusion is described as the random translational motion of molecules as a result of their thermal energy. This motion is also called "Brownian motion" (Einstein, 1905). In the case of isotropic (i.e., direction-independent) diffusion, the Brownian motion is described by a Gaussian probability density function (pdf) with mean-squared displacement of

$$\left\langle \mathbf{r}^2 \right\rangle = 6Dt, \tag{2.1}$$

where t is the diffusion time, $\mathbf{r}$ is the displacement, and $\langle \ \rangle$ represents the ensemble average over all molecules diffusing. In the brain, the average diffusion coefficient D is on the order of 1000×10^{-6} mm^2/s, and given a diffusion time of 50 ms, typical in MR diffusion experiments, the mean-squared displacement of molecules will be on the order of 20 μm. However, barriers such as axonal membranes, myelin sheets, and subcellular structures modulate this mean-squared displacement. Thus, measurement of water diffusion provides an indirect measure of the microarchitecture of the biological tissue. The strength of diffusion imaging lies in its ability to probe these small displacements of water molecules and, thus, to having a contrast that is altered by the microarchitecture at the level of cells.

2.2.2 Measurement of Diffusion Using MRI

Measurement of diffusion in the MRI environment can be performed using the Stejskal–Tanner diffusion preparation sequence depicted in Figure 2.1a. In the absence of any molecular diffusion, the spins refocus at time T_E, giving the full spin-echo signal $(= S_0)$. However, the random motion of molecules attenuates the MR signal. Thus, in the presence of Gaussian diffusion, the signal at time T_E is given by

$$S = S_0 e^{-bD}. \qquad (2.2)$$

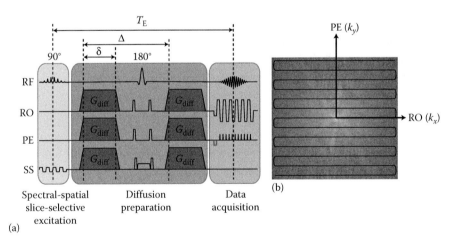

(a)

(b)

FIGURE 2.1
(Color version available from crcpress.com; see Preface.) Stejskal–Tanner sequence for measuring diffusion. The sequence consists of three parts: (1) spins are excited using an RF pulse. The excited signal must be free of fat signal to avoid artifacts due to fat chemical shift. In this figure, a "spectral-spatial" RF pulse that excites only water is shown. Another option is to excite the fat signal only using a selective RF pulse and dephase the fat signal using a dephasing gradient, followed by a standard 90° RF excitation. (2) The diffusion preparation part consists of two diffusion-weighting gradients (G_{diff}) separated by a 180° refocusing RF pulse. Diffusing spins will not be completely refocused and result in a signal loss proportional to the diffusion-induced displacement. (3) The data acquisition is accomplished using an EPI readout, which is the most common readout technique for diffusion acquisition. For clarity, the readout part in (a) is simplified and does not correspond to the k-space trajectory in (b). PE and RO represent phase-encode and readout axes, respectively.

Here, D is the diffusion coefficient as given in Equation 2.1 and b is called the b-value. Assuming rectangular gradient waveforms, b-value depends on the strength, duration, and separation of the diffusion-weighted gradients as

$$b = \gamma^2 G^2 \delta^2 \left(\Delta - \frac{\delta}{3} \right). \tag{2.3}$$

Here, G is the amplitude of the diffusion-weighting gradient, γ is the gyromagnetic ratio (= 42.58 MHz/T), Δ and δ are diffusion time and diffusion weighting gradient length, respectively, as shown in Figure 2.1a. Note that Equation 2.3 does not take into account the rising and falling edges of the diffusion gradients in Figure 2.1a and assumes perfect rectangles.

The diffusion coefficient D in Equation 2.3 is also called the "apparent diffusion coefficient," or ADC (Le Bihan et al., 1986). The reason D is called "apparent" is that biological tissues have a complicated structure, and the observed (i.e., the apparent) diffusion process is the sum of multiple incoherent motion processes. Thus, the diffusion in a biological tissue is different from that in, for example, a crystal. ADC maps are important because they provide increased specificity in many clinical settings. Figure 2.2 illustrates where ADC maps can aid in the differential diagnosis between an acute and a subacute infarct. Note that sometimes the "exponential" ADC (or eADC) is used, which is given by

$$\mathrm{eADC} = \mathrm{e}^{-b \cdot \mathrm{ADC}}. \tag{2.4}$$

Despite the ongoing research and the vast body of literature on alternative image acquisition methods for DWI, the most common and clinically used technique for reading out diffusion signals is echo-planar imaging (EPI) (Turner et al., 1990) (Figure 2.1b). The EPI technique traverses the k-space in a rectilinear fashion with alternating readouts that switch the direction for each line of k-space. The main motive behind using EPI is the fact that in diffusion imaging, due to the high motion sensitivity of the diffusion-sensitizing gradients, even small patient motion (such as bulk motion or cardiac motion) results in a phase in the image. Since this phase is different and random for each diffusion preparation period (or "excitation"), a segmented (aka "multishot") k-space acquisition (such as in conventional Fourier encoding where one line is acquired per excitation) will result in a mismatch between each k-space line. Since EPI is a single-shot method that acquires the whole k-space after one excitation (Figure 2.1b), EPI is much more robust to motion. Another important reason for using EPI is its high acquisition speed, which can keep DWI/DTI scan times at clinically acceptable levels.

Despite its robustness to motion and high signal-to-noise ratio (SNR) efficiency, there are significant challenges with single-shot EPI. Signal (i.e., T_2^*) decay and phase discrepancy induced by B_0 inhomogeneities that accumulate with the time spent traversing the EPI trajectory make single-shot EPI prone

isoDWI   isoADC   FLAIR

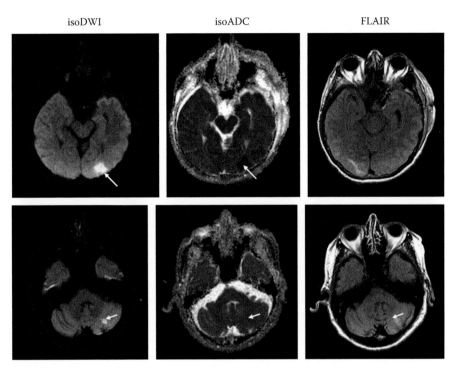

FIGURE 2.2
(Top row) Lesion that is bright on isoDWI and dark on isoADC compared to background tissue, indicative of an acute infarct. Note that the lesion is not seen on fluid attenuated inversion recovery (FLAIR) images. (Bottom row) A bright lesion on isoDWI and a more normalized ADC map (and bright on FLAIR), indicative of a subacute or evolving infarct. Note that the prefix "iso" stands for "isotropic," because maps for all three acquired gradient directions are combined.

to susceptibility artifacts—which manifest as geometric distortion, blurring, and signal dropouts. As such, single-shot EPI is generally ill-suited to high-resolution diffusion imaging, and alternative methods such as multishot EPI with phase navigation are typically used for high-resolution diffusion imaging to reduce the time spent traversing the EPI trajectory per excitation in k-space. Both single-shot EPI and some multishot EPI approaches will be described in more detail later in this chapter.

2.3 Diffusion-Weighted to Diffusion Tensor Imaging

In the previous section, the effect of diffusion on the MR signal was described in the context of isotropic diffusion as a mono-exponential signal attenuation that is proportional to the diffusion coefficient, D, and the magnitude of diffusion encoding, b (Equation 2.2). In the brain, however, the diffusion in the white matter is anisotropic due to the existence of various barriers such as myelin sheaths surrounding the axons or axonal membranes. In this case, a single diffusion coefficient does not suffice to describe the directional dependence of the diffusion process. Instead, a 3×3 matrix, which is called a diffusion tensor $\mathbf{D}$, is used to describe diffusion:

$$
\mathbf{D} = \begin{bmatrix} D_{xx} & D_{xy} & D_{xz} \\ D_{xy} & D_{yy} & D_{yz} \\ D_{xz} & D_{yz} & D_{zz} \end{bmatrix}.
\tag{2.5}
$$

And the MR signal in the presence of diffusion can be proven to be (Basser et al., 1994)

$$
S = S_0 e^{-\sum\limits_{i=x,y,z} \sum\limits_{j=x,y,z} b_{i,j} D_{i,j}},
\tag{2.6}
$$

where the 3×3 matrix $\mathbf{b}$ depends on the direction and magnitude of the applied diffusion encoding as in

$$
\mathbf{b} = b(\mathbf{n} \cdot \mathbf{n}^{\mathrm{T}}),
\tag{2.7}
$$

where $\mathbf{n}$ is the unit vector pointing in the direction of the diffusion encoding and b is given in Equation 2.3.

In Equation 2.5, the diagonal elements represent the diffusion along the principal axes, and the off-diagonal elements describe the correlation of diffusion between these principle axes. A better way to visualize the diffusion tensor is by remembering that the main purpose of diffusion imaging is to describe the random motion of water molecules. In the case of DTI, the random motion of molecules is mathematically modeled using a Gaussian diffusion pdf:

$$
p_\Delta(\mathbf{r}) = \frac{1}{\sqrt{|\mathbf{D}|(4\pi\Delta)^3}} e^{-\frac{\mathbf{r}^{\mathrm{T}} \mathbf{D}^{-1} \mathbf{r}}{4\Delta}}.
\tag{2.8}
$$

Here, $\mathbf{r}$ is the displacement of a single spin, $\mathbf{D}$ is the diffusion tensor as shown in Equation 2.5, and Δ is the diffusion time as shown in Figure 2.1. p describes the probability that a single spin starting at position 0 will be at position $\mathbf{r}$ at time Δ. If we have spins (i.e., water protons), all of which start at position $\mathbf{r} = 0$, after time Δ, these spins will be displaced by a certain amount, as described by Equation 2.8. For the isotropic (i.e., same diffusion in all directions) case, the diagonal elements of the matrix $\mathbf{D}$ (D_{xx}, D_{yy}, D_{zz}) will be the same, and the off-diagonal elements (D_{xy}, D_{xz}, D_{yz}) will be 0. As shown in Figure 2.3a, the distribution of the spins in the case of isotropic diffusion

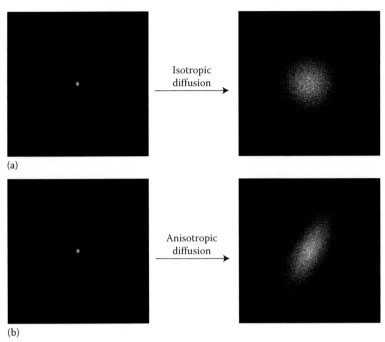

FIGURE 2.3
Illustration of particle distributions after time Δ in the case with isotropic (a) and anisotropic (b) diffusion. In the presence of isotropic diffusion, the spin distribution after diffusion time Δ looks like a sphere. In the case with anisotropic diffusion (i.e., the diffusion is hindered by barriers such as axons), the spin distribution after time Δ looks like an ellipsoid.

after time Δ will look like a sphere. However, if the diffusion is anisotropic, the spin distribution after time Δ results in an ellipsoid as in Figure 2.3b.

Thus, the ellipsoidal spin distribution shown in Figure 2.3b is a graphical representation of the 3 × 3 diffusion tensor. In fact, the shape of the ellipsoidal spin distribution (aka *diffusion ellipsoid*) can be mathematically extracted from the diffusion tensor using eigendecomposition:

$$
\mathbf{D} = \begin{bmatrix} D_{xx} & D_{xy} & D_{xz} \\ D_{xy} & D_{yy} & D_{yz} \\ D_{xz} & D_{yz} & D_{zz} \end{bmatrix}
$$

$$
= \begin{bmatrix} e_{1x} & e_{2x} & e_{3x} \\ e_{1y} & e_{2y} & e_{3y} \\ e_{1z} & e_{2z} & e_{3z} \end{bmatrix} \begin{bmatrix} \lambda_1 & & \\ & \lambda_2 & \\ & & \lambda_3 \end{bmatrix} \begin{bmatrix} e_{1x} & e_{1y} & e_{1z} \\ e_{2x} & e_{2y} & e_{3z} \\ e_{3x} & e_{3y} & e_{3z} \end{bmatrix},
$$

(2.9)

where e_i (i.e., eigenvectors) define the major axes of diffusion. The root mean square (rms) displacement of the spins along each major axis of the ellipsoid

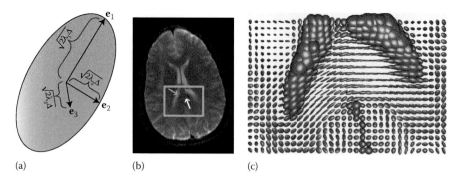

(a)   (b)   (c)

FIGURE 2.4
(Color version available from crcpress.com; see Preface.) (a) Graphical representation of the diffusion tensor as a diffusion ellipsoid. The eigenvectors of the diffusion tensor form the main axis of the ellipsoid, and the eigenvalues represent the diffusion-induced displacement in the direction of the corresponding eigenvector. (b–c) In the brain, the regions that contain densely packed white matter axons (such as the corpus callosum) possess high diffusion anisotropy. In such regions, the diffusion ellipsoids are elongated (b, light arrow). In other regions where the anisotropy is lower such as the CSF, the diffusion ellipsoids are more spherical (b, dark arrow).

is given by $\sqrt{2\lambda_i \Delta}$, where λ_i are the eigenvalues. These are graphically shown in Figure 2.4.

The eigenvector with the largest eigenvalue points to the direction with the largest rms displacement and is called the "major eigenvector" (e_1 in Figure 2.4a). To achieve high diffusion sensitivity, the conventional Stejskal–Tanner diffusion encoding usually uses long diffusion times (>20 ms) during which all the spins have had multiple interactions with the barriers; hence, the shape of diffusion tensor is determined only by the geometry of the barriers (i.e., tissue microarchitecture), and the term Δ is usually dropped while defining the ellipsoid in DTI (Figure 2.4a). However, when the diffusion time Δ is short (<10 ms), the frequencies of spin–barrier interaction are relatively small, causing the results of diffusion measurements look more and more like the isotropic diffusion as Δ decreases. In this case, the diffusion characteristics of a specific voxel will also be a function of the diffusion time Δ.

2.4 DTI Acquisition

The diffusion tensor describes both the direction and the magnitude of diffusion per voxel. Therefore, a single acquisition with a single diffusion encoding is not sufficient to infer the diffusion tensor **D**. Rather, multiple acquisitions (at least six) along noncollinear diffusion-encoding directions are necessary to determine the diffusion tensor. The *b*-value for

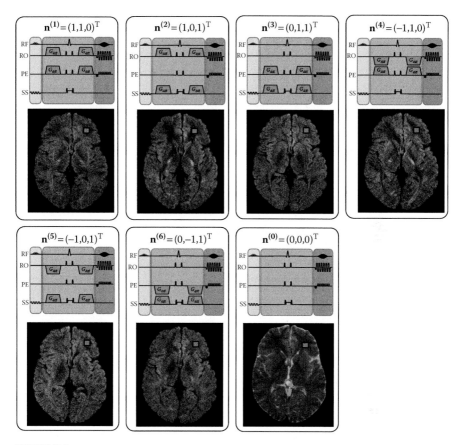

FIGURE 2.5
(Color version available from crcpress.com; see Preface.) An example DTI acquisition with six directions (same *b*-value) and one $b = 0$ image. The diffusion-encoding direction (i.e., the vector $\mathbf{n}_i$) is changed by applying diffusion encoding along different directions. An additional image is also acquired after turning off the diffusion-encoding gradients (n_0, also called the $b = 0$ image), thereby giving a T_2-weighted image. Thereafter, for each pixel, the diffusion-encoded and T_2-weighted image amplitudes are combined to get the 3×3 diffusion tensor $\mathbf{D}$.

diffusion-encoded data is adjusted by setting the gradient duration, amplitude, and separation (Equation 2.3), while the direction of the diffusion encoding is controlled by distributing the diffusion-encoding gradient on the *x*-, *y*-, and *z*-axes (Figures 2.1a and 2.5). Additional acquisitions with no diffusion weighting (i.e., $b = 0$ acquisitions) are also obtained, which correspond to the S_0 term in Equation 2.6. An example of DTI acquisition is shown in Figure 2.5, where six diffusion-weighted images with different diffusion-encoding directions and one $b = 0$ image are shown. It can be seen that if the applied diffusion-encoding direction $\mathbf{n}_i$ is aligned with the fiber orientation inside a voxel, the voxel has lower intensity because of high diffusion attenuation along the fiber. On the other hand, if n_i is perpendicular to the fiber orientation, the voxel has higher signal intensity because of lower diffusion perpendicular to the fiber. Next, we describe how we process and combine these images to determine the diffusion tensor **D**.

In order to get the diffusion tensor from the diffusion-weighted images in Figure 2.5, we start by augmenting the definition of b in Equation 2.7 to a 3×3 symmetric matrix in order to account for the different diffusion-encoding directions and/or strengths:

$$\mathbf{b}^{(k)} = b^{(k)}\left[\mathbf{n}^{(k)} \cdot \left(\mathbf{n}^{(k)}\right)^{\mathrm{T}}\right]. \tag{2.10}$$

Here, the index k was used to account for the different diffusion-encoding steps. Thus, for each k, there is one corresponding diffusion-weighted image that is acquired with the diffusion-encoding magnitude and direction defined by $\mathbf{b}^{(k)}$ (Figure 2.5). Writing Equation 2.6 for all different diffusion encodings and taking the logarithm, one can get

$$
\begin{pmatrix} \ln S_1 \\ \ln S_1 \\ \vdots \\ \ln S_n \end{pmatrix}
=
\begin{pmatrix}
1 & -\mathbf{b}_{xx}^{(1)} & -\mathbf{b}_{yy}^{(1)} & -\mathbf{b}_{zz}^{(1)} & -2\mathbf{b}_{xy}^{(1)} & -2\mathbf{b}_{xz}^{(1)} & -2\mathbf{b}_{yz}^{(1)} \\
1 & -\mathbf{b}_{xx}^{(2)} & -\mathbf{b}_{yy}^{(2)} & -\mathbf{b}_{zz}^{(2)} & -2\mathbf{b}_{xy}^{(2)} & -2\mathbf{b}_{xz}^{(2)} & -2\mathbf{b}_{yz}^{(2)} \\
 & & & \vdots & & & \\
1 & -\mathbf{b}_{xx}^{(n)} & -\mathbf{b}_{yy}^{(n)} & -\mathbf{b}_{zz}^{(n)} & -2\mathbf{b}_{xy}^{(n)} & -2\mathbf{b}_{xz}^{(n)} & -2\mathbf{b}_{yz}^{(n)}
\end{pmatrix}
\begin{pmatrix} \ln S_0 \\ \mathbf{D}_{xx} \\ \mathbf{D}_{yy} \\ \mathbf{D}_{zz} \\ \mathbf{D}_{xy} \\ \mathbf{D}_{xz} \\ \mathbf{D}_{yz} \end{pmatrix}
\tag{2.11}
$$

$$\mathbf{S} \qquad = \qquad\qquad \mathbf{M} \qquad\qquad\qquad \mathbf{d}.$$

Then, the unknown **d** (vector containing the elements of the diffusion tensor and the signal intensities of the pixels of the non-diffusion-weighted image) vector can be found in the least-squared sense using pseudo-inverse:

$$\mathbf{d} = (\mathbf{M}^{\mathrm{T}}\mathbf{M})^{-1}\mathbf{M}^{\mathrm{T}}\mathbf{S}. \tag{2.12}$$

This calculation is done on a per-pixel basis. It can be seen that the unknown vector **d** has seven parameters, which means at least seven diffusion acquisitions (including the $b = 0$ acquisition) need to be performed to get the diffusion tensor. However, in practice, it is customary to perform 20–25 diffusion acquisitions to get high SNR DTI maps and to avoid directional bias (Jones, 2004).

2.5 Quantitative Metrics from DTI

There are number of quantitative DTI metrics that are used to describe the shape of the diffusion tensors. The most common of those define the size (i.e., the magnitude of diffusion) and elongatedness (i.e., the directionality of diffusion) of the ellipsoid defining the diffusion tensor.

The first of these quantitative metrics is called the MD and is defined as the trace of the diffusion tensor:

$$MD = \bar{\lambda} = \frac{\lambda_1 + \lambda_2 + \lambda_3}{3}, \tag{2.13}$$

where λ_i are the eigenvalues of the diffusion tensor (Equation 2.9). The MD represents the isotropic portion of the diffusion tensor, that is, average diffusion restriction over all directions in a particular voxel. The higher the MD, the less restricted environment on average is. In terms of the shape of the diffusion tensor, a higher MD corresponds to a larger diffusion tensor ellipsoid volume, as seen in Figure 2.6.

The second frequently used quantitative metric is the FA (Basser and Pierpaoli, 1996):

$$FA = \frac{\sqrt{3}}{\sqrt{2}} \frac{\sqrt{(\lambda_1 - \bar{\lambda})^2 + (\lambda_2 - \bar{\lambda})^2 + (\lambda_3 - \bar{\lambda})^2}}{\sqrt{\lambda_1^2 + \lambda_2^2 + \lambda_3^2}}, \tag{2.14}$$

where $\bar{\lambda}$ is the mean of the eigenvalues (MD). FA is a measure of the anisotropic portion of the diffusion tensor (Figure 2.6). For an isotropic medium where diffusion is similar in all directions, FA approaches 0, whereas for

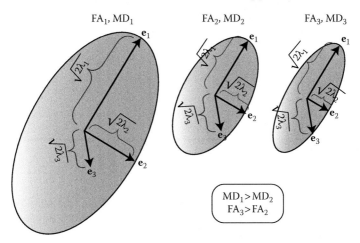

FIGURE 2.6
FA and MD are the most commonly used DTI-derived quantitative metrics. FA is related to how "elongated" and MD is related to how large the volume of the diffusion tensor ellipsoid is.

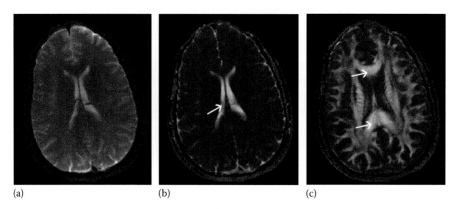

(a)   (b)   (c)

FIGURE 2.7
A DTI acquisition using single-shot EPI and 15 diffusion-encoding directions showing (a) the T_2 image (i.e., diffusion-encoding gradients turned off), (b) the MD image, and (c) the FA image. The FA is high in areas such as the corpus callosum due to the high number of axons, all pointing in the same direction [arrows in (c)]. CSF has a high MD and low FA due to less restriction and therefore isotropic diffusion, respectively [arrow in (b)].

a medium where diffusion has one primary orientation (i.e., anisotropic, $\lambda_1 \gg \lambda_2 = \lambda_3$), FA approaches 1. Figure 2.7 shows an example of *in vivo* FA and MD maps. FA is generally higher in areas like corpus callosum (Figures 2.4b, c and 2.7c) due to the high-density and unidirectional organization of the axons in these areas. It can also be seen that the cerebrospinal fluid (CSF) has high MD owing to the high mobility of water protons along all directions, but it has low FA because the diffusion is isotropic (Figures 2.4b, c and 2.7b).

Relative anisotropy (RA) is another metric that is derived from diffusion tensor MRI (Basser and Pierpaoli, 1996):

$$\mathrm{RA} = \sqrt{3}\,\frac{\sqrt{\left(\lambda_1 - \overline{\lambda}\right)^2 + \left(\lambda_2 - \overline{\lambda}\right)^2 + \left(\lambda_3 - \overline{\lambda}\right)^2}}{\lambda_1 + \lambda_2 + \lambda_3}. \tag{2.15}$$

RA defines the ratio of the anisotropic part of the tensor to the isotropic part. Like FA, RA = 0 for a complete isotropic medium.

Another diffusion tensor-derived metric is called volume ratio (VR) and is defined as (Pierpaoli and Basser, 1996):

$$\mathrm{VR} = \frac{\lambda_1 \lambda_2 \lambda_3}{\overline{\lambda}^3}. \tag{2.16}$$

VR is the volume of the ellipsoid defined by the diffusion tensor **D** divided by the volume of the sphere defined by the isotropic component of **D**, that is, the sphere whose radius is $\overline{\lambda}$. VR is 0 for highest anisotropy and 1 for complete isotropy.

Axial diffusivity (AD) and radial diffusivity (RD) are also frequently used as quantitative metrics to describe tissue architecture (Song et al., 2002).

As the name implies, AD describes the diffusion parallel to the axon and RD describes the diffusion perpendicular to the axon:

$$AD = \lambda_{\parallel} = \lambda_1$$

$$RD = \lambda_{\perp} = \frac{\lambda_2 + \lambda_3}{2}. \tag{2.17}$$

AD is believed to be an indicator of axonal damage, whereas RD is used to determine demyelination (Song et al., 2002). However, AD and RD must be interpreted with care, especially in regions of low anisotropy and crossing fibers (Wheeler-Kingshott and Cercignani, 2009).

2.6 High-Resolution Diffusion Imaging Techniques

As mentioned in Section 2.1, DTI is a unique imaging modality, thanks to its sensitivity to the tissue architecture at the cellular level. This ability of DTI comes from the fact that within biological tissues the diffusion of water molecules is modulated by structures such as cell membranes and microtubules in neuronal axons. However, the DTI model assumes that these bounding structures have homogeneous architecture (e.g., oriented in the same direction) inside the MRI voxel. In reality, since the voxel size is always multiple times larger than the cell size, different structures can coexist within a voxel—invalidating the DTI assumption and reducing the ability of DTI to accurately probe cellular microstructure. A classical example of heterogeneity is the case of crossing fibers: if two or more fibers with different orientation are present in a voxel, the diffusion tensor model will be invalid. There are three possible ways to address this shortcoming:

1. *High spatial resolution*—increasing spatial resolution so that heterogeneous voxels can be split up into smaller voxels, each of which are more homogeneous

2. *HARDI with enhanced signal model* (multitensor, ball-and-sticks, spherical harmonic decomposition), which use diffusion measurements with a large (>50) number of directions coupled with more complicated signal models to account for the structural heterogeneity in each voxel

3. *Model-free techniques,* which aim to represent the heterogeneous structure of the voxel without any assumption about the diffusion pdf by exploiting the Fourier relationship between the displacement probability space and the diffusion-encoding space, or more formally, *q*-space

The most common of these model-free techniques are q-ball and diffusion spectrum imaging (DSI). Further description of these approaches and their accompanying limitations will be discussed in the following sections.

2.6.1 High Spatial Resolution

In MRI, high spatial resolution corresponds to reading out further in k-space. However, this increases the EPI readout time and poses additional challenges. Specifically, there are two important challenges with long readouts:

1. The observable signal decays due to T_2^* relaxation, which causes blurring in the image, thus actually reducing the target resolution.
2. High sensitivity to inhomogeneities in the magnetic field (i.e., B_0 inhomogeneities). B_0 inhomogeneities cause pixel displacements and signal pile-ups, especially in regions where the local susceptibility changes rapidly, such as air–tissue interfaces. The pixel displacements caused by B_0 inhomogeneities are also called "geometric distortions" and will be discussed in more detail in Section 2.7.1.

The effect of B_0 inhomogeneities in EPI can be seen in Figure 2.8, shown in comparison to a fast spin-echo (FSE) acquisition, which is intrinsically distortion-free regardless of the resolution. In order to increase the resolution in single-shot EPI, one needs to acquire more data, which in turn increases the duration of the data acquisition window. Thus, the distortions and blurring increase as one increases the resolution from 128×128 (Figure 2.8b) to 288×288 (Figure 2.8d). The degree of distortion is visible through the mismatch between the contours obtained from the distortion-free FSE acquisition and the individual EPI acquisitions (Figure 2.8b–d, white contours).

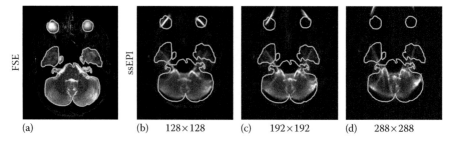

(a)   (b) 128×128 (c) 192×192 (d) 288×288

FIGURE 2.8
Geometric distortions that arise due to B_0 inhomogeneities in an EPI acquisition (only the T_2-weighted image of a DWI scan is shown; however, these same distortion artifacts will also be present on the diffusion-weighted volumes). (a) The FSE acquisition is a technique that is intrinsically distortion-free regardless of the resolution. In order to increase the resolution while using a single-shot EPI (ssEPI) readout, one needs to read out further in k-space. This increases the readout time, which in turn increases geometric distortions and T_2 blurring (b–d).

Both of these effects can be minimized by reducing the readout time and increasing the readout speed in the phase-encoding direction of an EPI trajectory (see Section 2.7.1). There are several techniques that can be used to increase the readout speed in EPI: (1) maximizing the readout bandwidth, (2) sampling on the ramps of the EPI trajectory, and (3) reducing the field-of-view (FOV) in the phase-encoding direction. While (1) and (2) are typically enabled for vendor-supplied EPI techniques, the readout speed remains limited by the gradient hardware specifications and by neurostimulation that is caused by quickly changing gradients (i.e., high slew rates). For (3), any significant reduction of FOV and thus distortion is only viable if one were interested in a specific region of interest.

One of the remaining techniques to reduce readout time is to use parallel imaging (Bammer et al., 2001). In this technique, a coil with multiple receiver channels is used to receive the MR signal such that each receiver channel is sensitive to a certain portion of the anatomy. The additional encoding information obtained from different coils complements the encoding that is obtained from the EPI readout. Thus, one can shorten the EPI readout and still be able to reconstruct the full image using this complementary encoding from different coils. This is called "parallel imaging reconstruction," and the two main methods under this category are Sensitivity Encoding (SENSE)-(Pruessmann et al., 1999, 2001) or generalized autocalibrating partially parallel acquisitions (GRAPPA)-based (Griswold et al., 2002) parallel imaging reconstruction. Although shortening the EPI readout time reduces geometric distortions and blurring, a disadvantage of parallel imaging is that the SNR of the image is reduced depending on both the reduction in readout duration and the coil configuration (Pruessmann et al., 1999). SNR loss limits the maximum loss in readout length that can be achieved.

The remaining technique that can be used to reduce the readout time of the trajectory in EPI is to use multishot (aka "interleaved" or "segmented") EPI acquisitions (Figure 2.9, first row; Figure 2.10b). In this technique, the image is acquired through multiple excitations and only a portion of k-space is acquired after each excitation. Thus, the readout is shorter, which reduces the blurring and distortions. However, in the presence of diffusion-weighting gradients, multishot acquisitions come with a challenge: the image has a different and random phase every time a k-space shot is acquired (Figure 2.9e). This random phase exists because the diffusion-weighting gradients are very sensitive to very small motion: even head motions with very small magnitude, such as those that can arise due to scanner vibrations (Gallichan et al., 2010) or cardiac pulsation, have detrimental effects on the MR data in the presence of diffusion-encoding gradients. Specifically, during the diffusion preparation period, small bulk head motion results in a constant and linearly changing phase term in the image (Anderson and Gore, 1994), while cardiac motion causes a nonlinear phase term and signal loss. In Figure 2.9, non-diffusion-weighted (i.e., $b = 0$) and diffusion-weighted images from an EPI sequence with three shots are shown. Each shot has been reconstructed separately using parallel

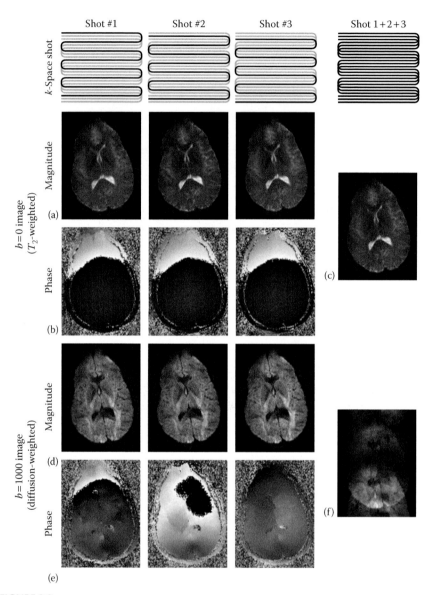

FIGURE 2.9

A multishot EPI acquisition with three shots and 192 × 192 resolution. (a, b, d, e) For each shot, a separate image was reconstructed using parallel imaging to demonstrate how the phase changes from one shot to the other. (c, f) The three shots were also combined without using parallel imaging. (b) Without diffusion-weighting gradients, the phase remains constant from one shot to the other, and the combined image is free of any phase-related artifacts (c). (e) When diffusion weighting is applied, small motion during acquisition, such as scanner vibrations or cardiac pulsation, results in a phase that varies from one shot to the other. (f) Combining the three shots in this case results in a corrupted image with threefold aliasing.

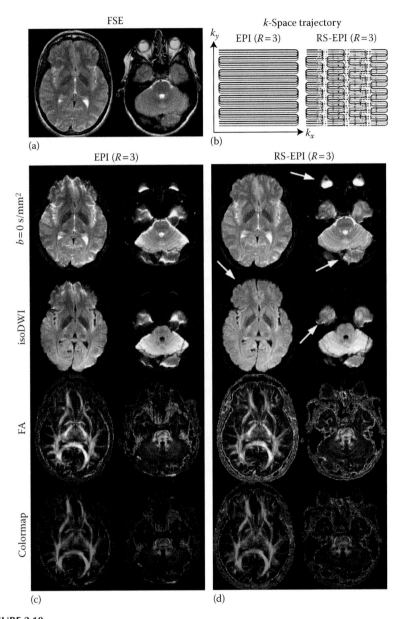

FIGURE 2.10
(See color insert.) High-resolution DTI using multishot EPI and RS-EPI with a resolution of
288 × 288. (b) The RS-EPI readout had seven "blinds," each of which had an acquisition matrix
size of 64 × 288. For both EPI and RS-EPI, three shots in the k_y direction were used and each
shot was reconstructed independently using parallel imaging. (c, d) The use of parallel imaging
reduced the distortions for both single-shot EPI (SS-EPI) and RS-EPI. However, RS-EPI has considerably lower geometric distortions compared to EPI due to the multishot readout used in
RS-EPI acquisition (white arrows). The non-diffusion-weighted FSE acquisition is free of any
geometric distortions and is shown in (a) for comparison.

imaging to demonstrate the phase changes between each shot (Figure 2.9a, b, d, and e). In a non-diffusion-weighted MRI acquisition, table vibration or cardiac pulsation causes negligible changes in the image phase (Figure 2.9b), yielding a good image after combining the three shots (Figure 2.9c). However, in the presence of diffusion weighting, the image phase changes between shots (Figure 2.9e), and combining the three shots yields an image with three-fold aliasing because of this phase mismatch (Figure 2.9f).

There have been numerous studies that have performed high-resolution DWI using multishot trajectories. Most of these studies acquire additional data before or after the actual imaging readout. This additional data acquisition is also called the navigator acquisition. The navigator acquisition contains the same phase as the imaging acquisition and can be used to subtract the motion-induced nonlinear phase that changes from one shot to the other. Some of these navigator-based methods use a 1D acquisition that passes through the center of k-space (Anderson and Gore, 1994; Bammer et al., 1999; de Crespigny et al., 1995; Jiang et al., 2002; Mori and van Zijl, 1998; Ordidge et al., 1994; Trouard et al., 1996; Williams et al., 1999). However, a 1D navigator acquisition is not sufficient to detect all of the phase terms. In particular, the linear phase components perpendicular to the 1D readout direction and the nonlinear phase terms due to cardiac pulsation are not detectable by a 1D readout. To overcome the limitation of 1D navigators, 2D navigator echoes can be used (Atkinson et al., 2000, 2006; Butts et al., 1997; Holdsworth et al., 2008; Miller and Pauly, 2003; Porter and Heidemann, 2009) to detect all terms of the phase. To remove the need for a separate 2D navigator and improve the scan efficiency, special trajectories can be used in which the phase could be extracted from the imaging acquisition without the need of a separate navigator acquisition (Liu et al., 2004b, 2005; Pipe et al., 2002; Skare et al., 2006). One such method is called periodically rotated overlapping parallel lines with enhanced reconstruction (PROPELLER), which collects diffusion data using rotating blades (Pipe et al., 2002). Each blade is acquired using an FSE train so that the final images are not affected by B_0 inhomogeneity distortions. Another method is called self-navigated interleaved spirals (SNAILS), where multiple variable density spirals are used to cover k-space (Liu et al., 2004b). Each spiral fully samples the center of k-space, from which the phase can be subtracted using iterative reconstruction (Liu et al., 2005).

An example of high-resolution DTI acquisition which uses a navigator-based multishot readout called readout-segmented EPI (RS-EPI) is shown in Figure 2.10 (Holdsworth et al., 2008; Porter and Heidemann, 2009). The multishot EPI and RS-EPI k-space trajectories used in this example are shown in Figure 2.10b. For both cases, three shots were used in the k_y direction, and each shot (or blind) was reconstructed independently using parallel imaging (Holdsworth et al., 2009). RS-EPI traverses k-space in "blinds" (Figure 2.10b), and for each blind, the central blind is reacquired to correct phase errors (Holdsworth et al., 2009). The use of blinds speeds up the traversal in k-space, thus reducing distortions. With standard EPI, despite the use of parallel imaging, the high-resolution DTI maps are degraded by geometric distortions and T_2 blurring (Figure 2.10c). The application of RS-EPI reduces distortions considerably (at the expense of SNR), as shown by the white arrows in Figure 2.10d.

The high-resolution techniques described above use 2D acquisitions to perform high-resolution DTI, which means that a slice-selective radio frequency (RF) pulse is used to excite a slice, followed by a 2D readout of the k_x–k_y plane. However, 2D acquisitions have an important shortcoming: due to the limitations in achieving sharp and narrow slice profiles, the resolution in the slice-select direction is usually lower than the in-plane resolution. For the purposes of mapping white matter connectivity using fiber tracking (Basser et al., 2000), isotropic resolution is usually desired. In order to achieve isotropic resolution, recently there have been some studies that employ a 3D acquisition of the k_x–k_y–k_z volume, which use a short repetition time (T_R) (Buxton, 1993; Kaiser et al., 1974; Wu and Buxton, 1990). However, 3D DWI/DTI has a lot of shortcomings in itself, and further improvement is warranted.

2.6.2 HARDI with Enhanced Signal Model

It is worth noticing that even at submillimeter resolution, depending on the complexity of the anatomy under investigation, structural heterogeneity can still exist in a voxel, and a simple signal model such as DTI renders estimated parameters inaccurate.

A straightforward extension from the DTI model is the multitensor model (Tuch et al., 2002). As its name implies, the multitensor model assumes the existence of multiple fiber populations in a voxel, each of which can be represented by a unique tensor. The MR signal equation in this case is a generalization of Equation 2.6:

$$S = S_0 \exp\left(-\sum_{t=1}^{N_t} \sum_i \sum_j f_t b_{ij} \mathbf{D}_{ij}^{(t)} \right), \tag{2.18}$$

where N_t is the total number of fiber populations/tensors assumed for that voxel and f_t is the apparent volume fractional of the diffusion tensor $\mathbf{D}^{(t)}$. Figure 2.11 shows the higher accuracy of the two-tensor model as compared to that of the one-tensor model (DTI) in depicting the region with crossing fibers.

Along the same line of the multitensor model is the ball-and-stick model (Behrens et al., 2003, 2007; Hosey et al., 2005), where the ball represents free diffusion population and the stick (or sticks) represents the restricted diffusion populations. To reduce the number of parameters that need estimation, each stick is assumed to have only one nonzero eigenvalue, meaning molecules move only in the fiber direction.

CHARMED is another beyond-DTI modeling method in which diffusion in each white matter voxel is assumed to follow two modes: hindered and

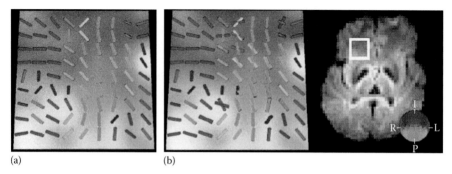

(a)   (b)

FIGURE 2.11
(See color insert.) Comparison of the principal eigenvectors (i.e., fiber orientation) estimated with single-tensor model (a) and two-tensor model in the ROI specified in the reference image (b). The single-tensor model fails to represent fiber crossings that are better represented by a two-tensor model. (From Tuch, D.S., Reese, T.G., Wiegell, M.R., Makris, N., Belliveau, J.W., Wedeen, V.J.: High angular resolution diffusion imaging reveals intravoxel white matter fiber heterogeneity. *Magn Reson Med.* 2002. 48. 577–82. Copyright Wiley-VCH Verlag GmbH & Co. KGaA. Reproduced with permission.)

restricted (Assaf and Basser, 2005; Assaf et al., 2004). Restricted diffusion is the diffusion within axons and is modeled as diffusion within impermeable cylinders. Hindered diffusion, modeled by a single tensor, comes from elsewhere outside axons. The MR signal equation for CHARMED is

$$E(b) = f_h E_h(b) + f_r E_r(b), \tag{2.19}$$

where $E(b)$ is the total signal decay, f_h and $E_h(b)$ are volume fraction and signal decay of the hindered part, respectively, and f_r and $E_r(b)$ are volume fraction and signal decay of the restricted part, respectively. E_h is the exponential part of Equation 2.6, and E_r has a closed-form solution under certain assumptions of the diffusion gradient pulses (Callaghan, 1991; Neuman, 1974). The diffusion acquisition of CHARMED requires a high number of directions and several b-values to ensure stability in this non-linear estimation problem. Apart from the capability to tease apart multiple fiber orientations, CHARMED also offers some other physically meaningful parameters such as diffusivity of extra-axonal medium and the axonal density (f_r) as shown in Figure 2.12.

When the orientation of the fiber population under question is known, a modified CHARMED framework, called AxCaliber, can yield axon diameter maps (Assaf et al., 2008; Barazany et al., 2009). In terms of acquisition, AxCaliber requires the CHARMED acquisition at multiple diffusion times. In terms of modeling, a gamma distribution of axon diameter is usually assumed.

Even though HARDI acquisitions with enhanced signal model methods have become quite popular, there are several major limitations to these methods including extended acquisition times, instability of the parameter estimation in what is usually a nonlinear optimization problem, and deviation of

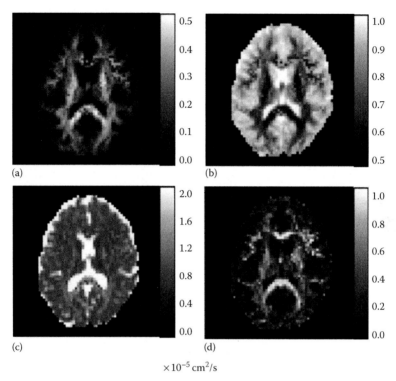

$\times 10^{-5}$ cm²/s

FIGURE 2.12
CHARMED analysis outputs: (a) Fraction of the restricted components (f_r), (b) fraction of the hindered component (f_h), (c) MD map calculated from the diffusion tensor eigenvalues of the hindered part of CHARMED, and (d) FA calculated for the diffusion tensor eigenvalues of the hindered part of CHARMED. (Reprinted from *Neuroimage*, 27, Assaf, Y., Basser, P.J., Composite hindered and restricted model of diffusion (CHARMED) MR imaging of the human brain, 48–58, Copyright 2005, with permission from Elsevier.)

the model from the actual underlying structure of investigation (e.g., a two-fiber model would fit badly for a single-fiber voxel).

2.6.3 Model-Free Techniques

In the previous section, it was shown how high spatial resolution can be used to represent heterogeneous structures by splitting up the voxel into smaller and more homogeneous voxels or how enhanced signal models can take into account (to a certain extent) the structural heterogeneity in a voxel. In this section, we will show how model-free techniques can be used to better represent the heterogeneous microarchitecture inside the voxel. Model-free techniques are mainly used to distinguish between different fiber populations in the same voxel that have different orientations. In order to understand how fibers with different directions can be distinguished, the concept of q-space (Callaghan, 1991) will be discussed first.

2.6.3.1 q-Space

The main purpose of diffusion imaging is to describe the Brownian motion of water molecules. Since diffusion is a random process, it can be described by a pdf. The pdf has the form $p_\Delta(\mathbf{r})$, and it describes the infinitesimal probability of a spin to move a distance of $\mathbf{r}$ during the diffusion time Δ. Here, Δ is the separation between two gradients as shown in Figure 2.1. So far, as a part of DTI formalism, $p_\Delta(\mathbf{r})$ was assumed to be Gaussian, which means that the spins displace according to the pdf given by Equation 2.8. A Gaussian pdf is defined by its mean (assumed to be 0) and covariance matrix (i.e., the 3×3 diffusion tensor $\mathbf{D}$) (Equation 2.8). Since $\mathbf{D}$ is defined by six elements, only six measurements (and one $b = 0$ acquisition) were enough to determine the diffusion pdf in this case. This method is called DTI since $\mathbf{D}$ is a 3×3 tensor.

One shortcoming of the Gaussian assumption is that it does not sufficiently describe the diffusion process in biological systems where the tissue architecture can be much more complicated due to restriction imposed by various barriers and due to the presence of crossing, kissing, or merging fibers within a voxel. In areas where two or more fibers with different orientations exist, standard DTI will give an averaged direction of diffusion, which will be inaccurate. To resolve complex fiber structures, one must relax the Gaussian assumption and therefore also replace DTI with a more general diffusion-encoding protocol.

In the presence of bipolar gradients as in Figure 2.1, each individual spin acquires a phase that is proportional to its net displacement during time Δ:

$$\varphi_\Delta(\mathbf{r}) = \gamma \delta \mathbf{G}_{\text{diff}} \cdot \mathbf{r}. \tag{2.20}$$

Δ, δ, and $\mathbf{G}_{\text{diff}}$ are defined as in Figure 2.1. For each voxel, the received MR signal is the sum of the signal from all spins within that voxel. Since each spin undergoes a random displacement according to the diffusion pdf $p_\Delta(\mathbf{r})$, the sum of signals from all spins can be expressed by the sum of all possible displacements $\mathbf{r}$ weighted by the number of spins which underwent that particular displacement (i.e., the probability $p_\Delta(\mathbf{r})$):

$$E_\Delta = S_0 \int_\mathbf{r} p_\Delta(\mathbf{r}) e^{j\varphi_\Delta(\mathbf{r})} d\mathbf{r} = S_0 \int_\mathbf{r} p_\Delta(\mathbf{r}) e^{j2\pi\gamma\delta\mathbf{G}_{\text{diff}} \cdot \mathbf{r}} d\mathbf{r}. \tag{2.21}$$

Defining $\mathbf{q} = \dfrac{1}{2\pi} \gamma \delta \mathbf{G}_{\text{diff}}$, one gets

$$E_\Delta(\mathbf{q}) = S_0 \int_\mathbf{r} p_\Delta(\mathbf{r}) e^{j2\pi\mathbf{q} \cdot \mathbf{r}} d\mathbf{r}. \tag{2.22}$$

It must be noted that Equation 2.22 is similar to the Fourier transform relation between the k-space and the MR image (Figure 2.13). Just the same way

Image ($m(\mathbf{r})$) (spatial domain) k-Space ($d(\mathbf{k})$)

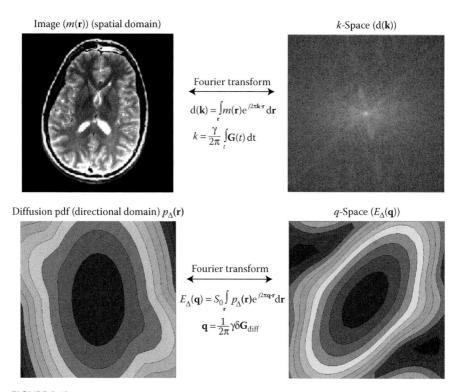

Fourier transform

$$d(\mathbf{k}) = \int_\mathbf{r} m(\mathbf{r}) e^{j2\pi\mathbf{k}\cdot\mathbf{r}} \, d\mathbf{r}$$

$$k = \frac{\gamma}{2\pi} \int_t \mathbf{G}(t) \, dt$$

Diffusion pdf (directional domain) $p_\Delta(\mathbf{r})$ q-Space ($E_\Delta(\mathbf{q})$)

Fourier transform

$$E_\Delta(\mathbf{q}) = S_0 \int_\mathbf{r} p_\Delta(\mathbf{r}) e^{j2\pi\mathbf{q}\cdot\mathbf{r}} \, d\mathbf{r}$$

$$\mathbf{q} = \frac{1}{2\pi} \gamma \delta \mathbf{G}_{\text{diff}}$$

FIGURE 2.13

(Color version available from crcpress.com; see Preface.) The Fourier transform relationship between image and k-space (upper row) is analogous to the relation between diffusion pdf and q-space (bottom row). Just the same way that k-space is traversed using readout gradients with changing amplitudes, q-space is acquired by changing diffusion-encoding gradients.

as one samples the k-space to get an MR image, one needs to sample q-space to fully describe the diffusion pdf.

There are two well-known methods that rely on q-space formalism: diffusion spectrum imaging (Wedeen et al., 2005) and q-ball imaging (QBI) (Tuch, 2004). Next, we discuss these two methods.

2.6.3.2 Diffusion Spectrum Imaging

In DSI, the diffusion pdf is determined by taking inverse Fourier transform of the sampled q-space according to Equation 2.22 (Wedeen et al., 2005). This is similar to standard structural imaging wherein the whole k-space is sampled and the acquired k-space data are inverse Fourier transformed to get the MRI image. A comparison between DTI and DSI is shown in Figure 2.14. In DTI, a single b-value is used in general, which corresponds to sampling the q-space on a sphere (Figure 2.14a; only q_x and q_y are shown for simplicity). Thereafter, using the diffusion-weighted signal values $E_\Delta(\mathbf{q})$

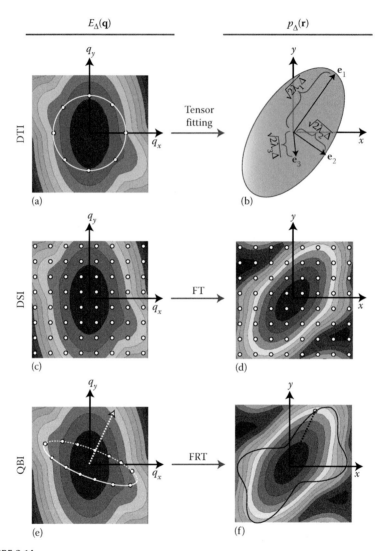

FIGURE 2.14

(Color version available from crcpress.com; see Preface.) Comparison between DTI, DSI, and QBI. The left column shows the different sampling strategies of the q-space corresponding to a diffusion pdf where two fibers cross at 30° (a, c, e). For clarity of representation, a 2D pdf is displayed using isocontours. In DTI, the diffusion pdf is usually sampled on a sphere at a limited number of directions, usually 6–25 (a), and a Gaussian pdf is generated (b). Due to the Gaussian assumption, the DTI model gives an "averaged" diffusion tensor and is not able to distinguish between the two crossing fibers. In DSI, the Gaussian assumption is removed, and the q-space is sampled in a rectilinear fashion using different b-values and diffusion-encoding directions (c). (d) The generalized pdf is then obtained using Fourier transformation (FT). For QBI, the diffusion pdf is approximated using an ODF. The q-space is sampled on a sphere using a much higher number of directions compared to DTI (e) and the ODF is determined through Funk-Radon transformation (FRT) of q-space samples (f).

acquired on the sphere, the diffusion pdf $p_\Delta(\mathbf{r})$ is modeled as a Gaussian pdf, and the 3×3 diffusion tensor $\mathbf{D}$ is determined by tensor fitting. Tensor fitting gives a continuous representation of the pdf (Figure 2.14b). However, DTI cannot resolve multiple fiber orientations inside the same voxel due to the Gaussian assumption, and this is clearly visible in Figure 2.14b. Although the voxel contains two fibers that are oriented $30°$ with respect to each other, DTI yields one fiber with an averaged orientation (Figure 2.14b). Conversely, DSI samples the whole q-space on a rectilinear grid (Figure 2.14c) and yields a discrete representation of the pdf that resolves both fibers (Figure 2.14d).

A disadvantage of DSI is that in order to characterize the diffusion pdf completely, a very large number of points in q-space are required. As an example, in order to represent a diffusion pdf with 8^3 resolution on a regular grid, 512 different diffusion-weighted scans are needed. Given an approximate T_R of 5 s, this will require 45 min, which is prohibitively long for clinical applications. In order to overcome this issue, QBI is proposed where sampling is performed on a spherical shell.

2.6.3.3 q-Ball Imaging

The idea behind QBI lies in the fact that we are not always interested in a complete 3D description of the pdf. What one usually needs is the direction and anisotropy of diffusion. In order to quantify direction and anisotropy of diffusion, a measure called "orientation distribution function" (ODF) is used (Tuch, 2004). As the name implies, ODF describes the number of fibers oriented in a specific direction. Mathematically, for a particular voxel, the ODF is defined as the summation of the diffusion pdf along radial lines:

$$\psi_\Delta(\mathbf{u}) = \int_0^\infty p_\Delta(\alpha\mathbf{u})d\alpha. \tag{2.23}$$

$\psi_\Delta(\mathbf{u})$ gives the probability of diffusion along direction $\mathbf{u}$. In QBI, the ODF $\psi_\Delta(\mathbf{u})$ is calculated without necessitating a complete sampling of the q-space. This is done by approximating $\psi_\Delta(\mathbf{u})$ by the Funk–Radon transform (FRT) of the acquired diffusion signal (Tuch et al., 2003):

$$\psi_\Delta(\mathbf{u}) \approx \frac{1}{Z}\mathcal{G}_{q'}\left[E(\mathbf{q})\right], \tag{2.24}$$

where $\mathcal{G}_{q'}$ is the FRT and $E(\mathbf{q})$ is the acquired diffusion signal. By definition, FRT is the integral of a function over the equator at a specific radius q'. Thus, Equation 2.24 means that we can obtain the ODF $\psi_\Delta(\mathbf{u})$ by going through the following steps:

1. Acquire multiple diffusion encodings at a fixed b-value over a sphere to get $E(\mathbf{q})$ in Equation 2.24 (Figure 2.14c).
2. Interpolate and calculate the sum of $E(\mathbf{q})$ over multiple equators on the sphere to get $\psi_\Delta(\mathbf{u})$.

In practice, QBI may require interpolation of the acquired signal to get a continuous representation of the ODF (Tuch, 2004).

Similar to the FA in DTI, a generalized FA (GFA) measure can be derived from QBI data. There, the GFA is defined as (Tuch, 2004)

$$GFA = \frac{\text{std}\left(\psi_\Delta(\mathbf{u})\right)}{\text{rms}\left(\psi_\Delta(\mathbf{u})\right)} = \left(\frac{\int\left(\psi_\Delta(\mathbf{u})-\overline{\psi}_\Delta\right)^2 d\mathbf{u}}{\int\left(\psi_\Delta(\mathbf{u})\right)^2 d\mathbf{u}} \right)^{\frac{1}{2}}. \tag{2.25}$$

2.6.3.4 Generalized DTI

In this technique, the standard 3×3 diffusion tensor $\mathbf{D}$ is expanded to include higher order terms (Liu et al., 2004a, 2010). Without going into the details of the mathematical description, the diffusion signal in this case is given by

$$S_k = S_0 \exp\left\{\sum_{n=2}^{\infty}\left[\sum_{i_1,i_2,\ldots i_n=\{x,y,z\}} j^n b_{i_1 i_2 \ldots i_n} D_{i_1 i_2 \ldots i_n}\right]\right\}. \tag{2.26}$$

In Equation 2.26, the coefficients $D_{i_1 i_2 \ldots i_n}$ are called the higher order tensors (HOTs) and represent the higher order moments of the diffusion pdf. The second-order tensor (i.e., $D_{i_1 i_2}$) represents the covariance matrix of the pdf. This covariance matrix is also the 3×3 diffusion tensor in standard DTI (Equation 2.6). In the same manner, the third-order tensor $D_{i_1 i_2 i_3}$ represents the skewness tensor and the fourth-order tensor $D_{i_1 i_2 i_3 i_4}$ represents the kurtosis tensor (Liu et al., 2004a). Figure 2.15 shows an *in vivo* example of generalized DTI (GDTI) in a brain region around the splenium of corpus callosum. The higher order tensors of order 4 are able to resolve multiple fibers merging at the same voxel (Figure 2.15b, white box). Due to its limitations in representing non-Gaussian diffusion, the standard DTI approach only gives an averaged diffusion tensor that does not describe the true underlying fiber structure (Figure 2.15c).

2.7 Artifacts in DWI and DTI

In this section, we will discuss common artifacts that appear during the acquisition and processing of diffusion data and methods to overcome them.

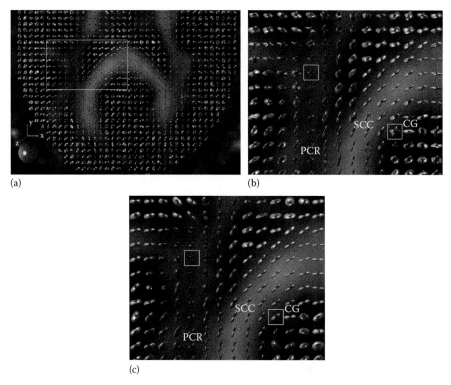

(a)   (b)

(c)

FIGURE 2.15
(Color version available from crcpress.com; see Preface.) (a) *In vivo* GDTI demonstrated around the splenium of corpus callosum. In this experiment, even order tensors up to the order 4 are measured and visualization of the HOTs is performed using pdf glyphs. pdf glyphs are closed surfaces such that the distance of each point on the surface of the glyph from the center is proportional to the value of the pdf evaluated at a radius R. (b) The white square contains regions where fibers with different orientations merge. The white box highlights a region where two major fiber bundles, namely the posterior corona radiata and the superior longitudinal fasciculus, cross. The pdf glyph visualization obtained from second- and fourth-order tensors using GDTI clearly depicts multiple fiber orientations in this voxel. (c) pdf glyph visualization obtained from standard DTI (i.e., second-order tensor only) shows an averaged orientation of the two merging fiber branches. (Image courtesy of Chunlei Liu, PhD, Brain Imaging and Analysis Center, Duke University, Durham, NC.)

2.7.1 B_0 Inhomogeneities and Geometric Distortion

The effects of B_0 inhomogeneities in high-resolution diffusion imaging were discussed in Section 2.6.1. In this section, a mathematical description of the geometric distortions will be provided.

MRI fundamentally relies on the assumption that the spins resonate at the Larmor frequency ($\omega_0 = \gamma B_0$). However, in practice, this requirement cannot be fulfilled, in which case the spins are said to be "off-resonance." Spins can be off-resonance due to inhomogeneities of the main magnetic field (i.e., B_0) caused by the imperfections in the manufacturing of the magnet.

Another reason for the effective B_0 to differ from desired B_0 arises from the ability of different materials (i.e., tissue or air) to get magnetized to varying extents. This magnetization ability is also defined as susceptibility and denoted by χ. Here, magnetic field gradients that occur in the presence of tissue interfaces with different χ can lead to detrimental off-resonant effects.

For conventional Cartesian readout methods which acquire one line in k-space per excitation, the B_0 inhomogeneities do not have significant effects on the final image. This is because one line of k-space can be acquired very quickly (i.e., the signal readout has a high bandwidth) before the effects of off-resonance become dominant. However, the conventional Cartesian readout is not commonly used in diffusion imaging due to two main reasons: (1) diffusion-sensitizing gradients cause a different random phase for each T_R, which is challenging for methods that need multiple excitations to acquire the whole k-space data (Section 2.6.1; Figure 2.9), and (2) diffusion acquisitions require long T_R that are on the order of seconds, and acquiring one line per T_R will prohibitively increase the scan time, particularly for DTI. For these reasons, single-shot EPI is the readout of choice for DWI/DTI in clinics (Figure 2.1b). However, the single-shot EPI readout trajectory is significantly longer than a conventional Cartesian readout, and this gives the B_0 inhomogeneities enough time to develop and distort the image.

In the case of single-shot EPI readouts, off-resonance causes pixel shifts in the phase-encoding direction, which is given by

$$\Delta y = \frac{\gamma(\Delta B(\mathbf{r}))}{2\pi\mathrm{BW}_{\mathrm{PE}}}. \tag{2.27}$$

Here, Δy is the shift in the phase-encoding direction (in units of mm), γ is the gyromagnetic ratio, $\Delta B(\mathbf{r})$ is the difference between the actual and nominal B_0 fields, and $\mathrm{BW}_{\mathrm{PE}}$ is the readout bandwidth (in units of 1/mm/sec) and is proportional to the speed of traversal of k-space in the phase-encoding direction.

One can reduce the field inhomogeneity caused by the patient with the use of "shimming gradients" that undo the effects of B_0 field alterations in the imaging volume. Typical MR systems have a linear shim that is applied over the selected imaging volume during the prescan period and corrects for the linear portion of the B_0 field alteration. Some systems have "higher order" shimming capability whereby a separate (and short) calibration scan is used to correct the field inhomogeneity of a selected volume. However, with the limited available shim channels, one will not overcome sudden field perturbations near tissue–air interfaces in whole-brain imaging, while shimming on a confined volume will cause detrimental shim currents in other parts of the brain.

Equation 2.27 suggests that the faster the traversal of k-space in the phase-encoding direction, that is, the higher $\mathrm{BW}_{\mathrm{PE}}$ is, the lower the distortions will be. Thus, in order to reduce geometric distortions, one must traverse the k-space faster in the phase-encoding direction. Some methods to reduce geometric distortions were described in Section 2.6.1. One of these methods is to use parallel imaging to speed k-space traversal by skipping lines in the phase-encoding direction. Another approach is to use multishot acquisitions together with a navigated phase correction (Figure 2.10d).

Another way to correct geometric distortions is with postprocessing. Some distortion correction strategies include coregistration to an undistorted reference image (Andersson et al., 2008; Ardekani and Sinha, 2005; Kybic et al., 2000; Maes et al., 1997; Studholme et al., 2000), the use of multireference methods (Robson et al., 1997), the field map method (Cusack et al., 2003; Jezzard and Balaban, 1995; Reber et al., 1998), and the reversed gradient polarity method (RGPM) (Andersson et al., 2003; Bowtell et al., 1994; Chang and Fitzpatrick, 1992; Kannengiesser et al., 1999; Skare and Andersson, 2005). For coregistration methods, geometric distortion is corrected using nonlinear registration procedures, by warping distorted EPI images onto a corresponding undistorted (anatomical) image so that their global shape would tend to be similar to their theoretical undistorted one. The task of eliminating distortions in this way is relatively cumbersome as the contrast between the reference and the diffusion-weighted image is often substantially different, and the transformation and considerably more degrees of freedom than a simple rigid body transform (Andersson and Skare, 2010).

The remaining methods must incorporate extra measurements in the acquisition. The "field mapping" approach obtains $\Delta B(\mathbf{r})$ with the use of a gradient echo sequence with a slightly shifted echo time. The "distorted" pixels can then be shifted to their original locations using Equation 2.27. This approach directly relies on phase information; however, phase wraps often occur in regions of severe susceptibility, and pixels often end up in the wrong locations.

Multireference methods use a separate reference scan that introduces an additional distortion-free dimension in k-space that can be used to undistort the image. In many multireference approaches, the correction is applied directly to the k-space data, which is difficult with DTI data due to the large phase shifts introduced by the diffusion-encoding period (Andersson and Skare, 2010). The RGPM method uses two "oppositely" distorted EPI images acquired using a positive and negative phase-encoding gradient to estimate $\Delta B(\mathbf{r})$. Since this approach does not involve extensive calibration or field map measurements, it is relatively practical to use as it has only a small scan time increase, and the additional acquisition can be used to improve the SNR of the total data set.

Note that for the RGPM (and for any of the aforementioned techniques) to work, the images must not be "too" distorted. In areas where the brain anatomy has been compressed so much that many voxels are compressed into one, the RGPM method will fail, in terms of both how well the $\Delta B(\mathbf{r})$ is estimated and the ability to resample the distorted EPI data back to its true voxel locations. Even with a hypothetically known displacement field, with image resampling alone, one cannot recreate this lost anatomic information from these areas, such as frequently seen at regions near tissue–air interfaces.

2.7.2 Ghost Correction

Another challenge with EPI trajectories is the mismatch between even and odd lines of the EPI train, which occurs due to hardware imperfections such as a delay of the gradients (Figure 2.16b). Even and odd line mismatch results in one primary and one "ghost" image that is shifted by half the field of view along the phase-encoding direction (Figure 2.16d). These aliasing artifacts are also called "FOV/2 ghosts." An odd/even phase error of zeroth order (just constant phase) gives discrete ghosts, whereas linear phase errors will lead to the banded replicas where the banding frequency depends on the slope of the first-order phase. For multishot EPI readouts, the number of ghosts will increase according to the number of shots. Also note that while gradient delays in EPI cause FOV/2 ghosts in EPI, they cause spiral data to be rotated and blurred.

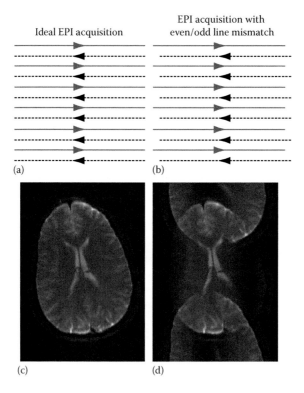

Ideal EPI acquisition

EPI acquisition with even/odd line mismatch

(a)   (b)

(c)   (d)

FIGURE 2.16
Even and odd line mismatch and ghosts in EPI. For a single-shot EPI scan, mismatch between even and odd lines (b) results in two separate images, each of which are undersampled by a factor of 2 (d). These are called "FOV/2 ghosts." An odd/even phase error of zeroth order (just constant phase) gives discrete ghosts, whereas linear phase errors will lead to the banded replicas where the banding frequency depends on the slope of the first-order phase. For EPI scans with more than one shot, the number of ghosts will increase accordingly. *k*-space trajectory without mismatch (a) and the corresponding image (c) are shown for comparison.

There is a large amount of literature focused on correction of ghost artifacts in EPI and a thorough discussion of all these methods is beyond the scope of this chapter. The most commonly used method is to perform an additional EPI "reference scan" without the phase-encoding gradients turned on. This results in a trajectory that traverses the k_x axis multiple times, and the mismatch between even and odd lines can be extracted out from this data (Bruder et al., 1992). Thereafter, ghosts are eliminated by retrospectively applying the mismatch parameters to the imaging data. If such reference data are not available, entropy-based autofocusing can also be used to remove ghost artifacts (Clare, 2003). Here, image entropy is used in the optimizer as a cost term because lower entropy is associated with the accumulation of image energy in as few pixels as possible (i.e., as small FOV/2 ghost energy as possible).

2.7.3 Eddy Currents

Another source of artifact that causes pixel displacements in EPI-based diffusion imaging is eddy currents. Eddy currents are caused by rapidly changing magnetic fields (i.e., gradients) in MRI. According to Faraday's law, changing magnetic fields induces electric (or "eddy") currents in the conducting structures of the magnet, such as the main magnet coil, shim coils, isolation, and gradient coils (Boesch et al., 1991). Thereafter, eddy currents in turn create their own magnetic field as described by Ampere's law. Thus, eddy currents can generate their own B_0 field variations (ΔB_0, ΔG, and higher order terms) and interfere with the actual imaging sequence. In MRI, eddy currents are created when gradient waveforms change amplitude, also called "gradient switching" (e.g., ramps in EPI). However, since the EPI trajectory has both positive and negative gradient switching (with short timescales) that are accompanied by opposite currents, EPI itself is largely unaffected by eddy currents. However, the diffusion-weighting gradients typically have long ramp times with more temporal separation between the "ramp-up" and "ramp-down" sections. In addition, for a standard Stejskal–Tanner sequence, the diffusion gradients are both positive, and thus, there are no opposite currents to compensate each other (Figure 2.17). The effects of these eddy currents vary with the direction of the applied diffusion weighting. Figure 2.17 demonstrates the three main effects that the diffusion-weighting gradients have on the image (Haselgrove and Moore, 1996):

1. Diffusion encoding along the readout direction results in a shear of the image (Figure 2.17, first row).
2. Diffusion encoding along the phase-encoding direction results in compression and stretching of the image (Figure 2.17, middle row).
3. Diffusion encoding along the slice-select direction results in a shift of the image in the phase-encoding direction (Figure 2.17, third row).

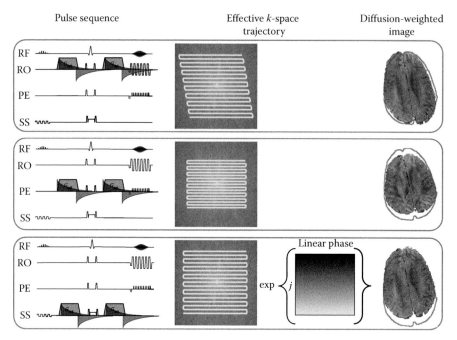

FIGURE 2.17
(Color version available from crcpress.com; see Preface.) Effect of eddy currents on diffusion-weighted images shown for all three diffusion-weighting directions. The eddy currents are created during the "ramp-up" and "ramp-down" sections of the diffusion-weighting gradients, as shown by the red and blue decaying gradients, respectively. The eddy currents arising from the ramp-down section of the last diffusion gradient is uncompensated and causes the effects seen on the right side. The x diffusion gradient results in an approximately constant eddy current gradient on the x-axis, which skews the k-space trajectory, and hence skews the image as well (first row). Likewise, the eddy currents resulting from the y diffusion gradient stretches or compresses the image (second row), whereas the eddy currents on the z-axis result in a linear phase in k-space and a shift in the image (third row). Both the eddy current plots and the corresponding k-space and image space representations are exaggerated compared to the real case for demonstration.

Apart from these three main distortions, there are also higher order eddy current distortions (Rohde et al., 2004). These distortions end up as a spatial mismatch between individual diffusion-weighted images. Thus, a characteristic effect of eddy currents is an artificially elevated FA value, which is a result of mismatched combined diffusion-weighted images (Figure 2.18c).

There are several methods used to either correct for or reduce eddy current-related distortions. These can roughly be divided into three methods: the first is to optimize the scanner hardware so that eddy currents do not affect the image quality in the first place. One example is to actively shield the gradient coils to minimize the generation of eddy currents in the other conductors. Another such method is to apply additional compensatory magnetic fields to balance the effect of eddy current-generating gradients. This is also called

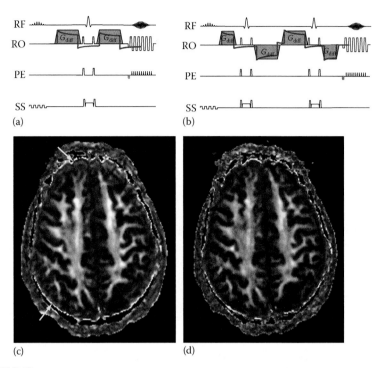

FIGURE 2.18
(Color version available from crcpress.com; see Preface.) Eddy current correction using a TRSE sequence. The exponential curves overlaid on RO gradient waveforms in (a) and (b) demonstrate the evolution of the eddy currents (a) When a regular diffusion-weighted spin-echo sequence is used, eddy currents result in mismatch between individual diffusion-weighted images. These mismatches manifest as elevated FA values (c, white arrows). (b) When a TRSE sequence is used, the eddy currents are nulled just before the readout. Thus, the FA maps are free of elevated FA values (d).

gradient preemphasis (Boesch et al., 1991; Papadakis et al., 2000; Schmithorst and Dardzinski, 2002). Fortunately, most modern scanners have built-in active shielding and gradient preemphasis. However, hardware correction may not be sufficient to compensate for the spatially and temporally varying magnetic fields due to eddy currents generated by diffusion gradients.

The second method to correct for eddy current distortions is to use post-processing algorithms. One way to perform this is to register the images in Figure 2.17 to a reference image (e.g., the T_2-weighted image) using an affine transformation (Andersson and Skare, 2002; Bastin, 1999; Bodammer et al., 2004; Haselgrove and Moore, 1996; Rohde et al., 2004). Additional calibration acquisitions can also be used to better infer the time-dependent characteristics of eddy currents for postprocessing (Calamante et al., 1999; Jezzard et al., 1998).

The third method is to change the pulse sequence. Among such techniques is the use of bipolar diffusion gradients (Alexander et al., 1997; Cotts et al., 1989; Hong and Thomas Dixon, 1992). Another approach is to use two reference

scans with opposite phase/frequency-encoding axes to allow the calculation of the long-term eddy currents along these axes (Jezzard et al., 1998).
Perhaps the most widely used sequence-based approach for eddy current
compensation is the twice-refocused spin-echo (TRSE) sequence of Wider
and Reese (Reese et al., 2003; Wider et al., 1994) (Figure 2.18b). Instead of a
single refocusing pulse, this sequence uses two refocusing pulses, separated
by bipolar diffusion-weighting gradients. Due to the existence of neighboring
bipolar diffusion gradients, the eddy current buildup becomes insignificant
prior to readout. While the twice-refocused approach has a longer T_E for a
given b-value compared with the Stejskal–Tanner diffusion preparation, it
remains the most efficient sequence-based approach for the correction of
long-term eddy currents thus far.

Note that short-term eddy currents that cannot be viewed as constant
during the EPI train are not guaranteed to be nulled with a sequence-based
approach. The nature of the artifacts due to short-term eddy currents is more
complicated because of the involvement of several decaying currents of different time constants and amplitudes that do not necessarily create a simple
affine image domain transformation.

2.7.4 Patient Motion

Patient motion is an important consideration in MRI. For MR examinations of the elderly, children, or patients with certain medical conditions
(i.e., Parkinson's disease, stroke), motion correction becomes pertinent for
obtaining images of diagnostic quality. Even for healthy subjects, head motion
can be observed for long scan protocols, such as functional MRI (fMRI) or
DTI. In DWI/DTI, it is important to make a distinction between three different types of motion depending on their effects on diffusion-weighted images:

1. *Microscopic Brownian motion of the spins*—this is also called diffusion,
   and this is the effect that is measured by DWI;
2. *Small bulk head motion*—this type of motion is caused by table
   vibrations or cardiac pulsation and causes random phase changes
   between shots, as described in Section 2.6.1; and
3. *Large bulk head motion*—this type of motion is caused by involuntary
   head rotation of the patient. It is usually on the scale of millimeters and
   causes pixel misregistration between diffusion-weighted volumes.

It is important to note that the second and third types of motion refer to the
"coherent" motion of all spins inside a voxel that occurs in the same direction with the same velocity. On the other hand, the first type of motion is
the random Brownian motion (i.e., diffusion) of spins inside the same voxel.

In this section, the second and third types of motion are discussed. As
discussed previously, bipolar gradients used in DWI have high motion

sensitivity. During this diffusion-weighted time, moving spins will accrue a phase depending on their velocity and location within the voxel, resulting in a varying phase within the same voxel (Figure 2.9). Thus, in the presence of either rigid body motion or large brain pulsation, signal cancellation can occur within a voxel—and in some instances, this signal loss can be irrecoverable. For large bulk head motion, in addition to large phase errors and misregistration of anatomical locations between images, the effective diffusion-encoding direction changes (i.e., the *b*-matrix)—and this must be corrected to achieve accurate fiber tracking (Aksoy et al., 2008, 2010; Leemans and Jones, 2009).

One method to correct motion artifacts in DTI is to retrospectively register all diffusion-weighted volumes to a reference volume after the scan is complete (Rohde et al., 2004). Volume-to-volume registration can also be performed prospectively by registering the diffusion-weighted volumes and updating the scan plane in real time while the scan is going on (Benner et al., 2011; Kober et al., 2012; Thesen et al., 2000). In addition, real-time correction enables reacquisition of volumes that are corrupted due to motion. One disadvantage of volume-to-volume registration is that if motion occurs *during* the acquisition of one volume, this particular volume will not be useful for registration. External motion-tracking solutions, such as those using cameras, can be used for prospective correction of patient motion to overcome the limitations of volume-to-volume registration (Aksoy et al., 2011; Herbst et al., 2012; Zaitsev et al., 2006).

2.8 Fiber Tractography

Diffusion-derived brain connectivity parameters, such as white matter tract volumes, axonal density, and regional connectivity, are pseudo-quantitative parameters widely used in neuroscience and psychology applications. These parameters are also used for surgical planning. The reconstruction of a continuous streamline representation of white matter architecture using the diffusion data is called "fiber tractography" (Basser et al., 2000; Conturo et al., 1999; Mori et al., 1999; Xue et al., 1999). The simplest way to perform fiber tractography includes selecting a region of interest, also called a "seed region," and tracing the path of an imaginary particle through the vector field defined by the major eigenvector (Figure 2.19b). Specifically, starting with the seed point, the new position of the particle is determined by moving in the direction of the major eigenvector orientation at the current location. The amount of motion is controlled by a parameter called "step size" and is usually smaller than a voxel size (Figure 2.19b). Since diffusion is bidirectional (unlike flow), tracking is initiated in both directions from the seed point, and it is stopped when the curvature of the tract exceeds, or if the FA goes below, a certain threshold. In combination with visualization

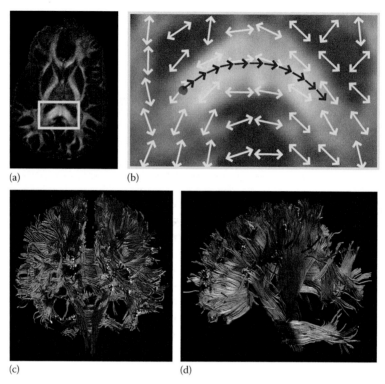

(a)   (b)

(c)   (d)

FIGURE 2.19
(Color version available from crcpress.com; see Preface.) Diffusion tensor data can be used
to reconstruct the white matter connections in the brain. This technique is called fiber tractography. A schematic description of fiber tractography using first-order interpolation in corpus
callosum is shown (a, rectangle). In this method, a seed point is selected [the circle in (b)].
Thereafter, starting from the seed point, one advances by a predefined step size parallel to the
direction of the local vector direction. The local vector direction is defined by the direction of
major eigenvector of the tensor. By selecting multiple seed points, a whole-brain tractography
or tractography of specific fibers can be performed. (c) Coronal view of a whole brain tractography. (d) Oblique sagittal view of corpus callosum, bilateral motor tracts, and transverse pon-
tine fibers. Fiber tractography was performed using ExploreDTI (http://www.exploredti.com).
(Image courtesy of Sjoerd Vos, Image Sciences Institute, University Medical Center Utrecht,
Utrecht, the Netherlands.)

software, constructing the fibers in this way can produce aesthetically pleas-
ing images of the white matter tracts in the brain (Figure 2.19c, d). The fiber-
tracking method just described is a linear interpolation technique called
Euler's method. Other methods, such as Runge–Kutta interpolation (Basser
et al., 2000), front propagation using fast marching (Parker et al., 2002), or
probabilistic tracking (Parker et al., 2003), also exist.

It is important to mention that the fiber tractography techniques mentioned above only apply to DTI where streamline interpolation is performed
only along the main direction of diffusion (i.e., the major eigenvector of the
diffusion tensor). With the use of more diffusion-encoding directions and

model-free techniques such as q-ball and DSI, streamline reconstruction of crossing fibers has also become possible (Campbell et al., 2005; Iturria-Medina et al., 2007; Wedeen et al., 2008).

Despite the promising potential of fiber tracking in understanding human brain connectivity, it is important to emphasize that the fibers obtained from fiber tracking do not correspond to real white matter pathways; these simulated fibers are just computed based on the diffusion anisotropy the real fibers impose in each voxel. The true underlying fiber architecture may look entirely different.

Fiber tracking can be used for quantification of white matter tracts. One way of tractography-based quantification is to use fiber tracts as a segmentation tool. In this way, quantitative metrics such as FA and MD can be measured on brain regions that have a specific function (Ciccarelli et al., 2003). Employment of these tract-specific metrics also reduces human errors that may arise during manual segmentation. These tractography-based segmentations can also be used to quantify other parameters, such as T_2 in the volume spanned by the specific tracts (Stieltjes et al., 2001). Another way of performing tractography-based quantification is to use the tract volume (Basser et al., 2000; Ciccarelli et al., 2003) or tract density (Calamante et al., 2010). It has been shown that the normalized tract volumes of the optic radiation and pyramidal tract are consistent with the known values from postmortem studies (Ciccarelli et al., 2003).

2.9 Conclusion

DWI allows one to investigate minuscule random motion in biological tissue. This very high sensitivity is achieved by powerful motion-probing gradients. Standard DWI, as applied in the clinics, typically employs three diffusion-encoding gradients to get a measure of isotropic diffusion, and measures such as the MD and ADC, can be extracted. DTI requires six (or more) gradient directions to model anisotropy of a cell, and metrics such as FA and directional (or "color") FA can additionally be extracted. With the limitations of the tensor model, alternatives such as HARDI-based methods or model-free techniques (such as q-ball and DSI) can be used—however, these often require several more directions (and thus long scan times). For all models, fiber tractography can be used, which gives a continuous streamline representation of white matter tract pathways.

There are a number of limitations of diffusion imaging that perhaps can be fundamentally attributed to the poor SNR of the diffusion acquisition. Due to its speed, EPI has typically been the acquisition of choice for diffusion imaging. However, despite the use of parallel imaging to speed the traversal of k-space, EPI remains plagued by artifacts such as distortion and blurring.

Alternative approaches, such as FSE, multishot, or 3D-based methods, can be used to enable high-resolution diffusion imaging with reduced distortion. However, these methods come with other problems, such as increased sensitivity to motion, the requirement for advanced reconstruction to correct for phase errors, and increased scan time. Additionally, for most acquisition methods used for diffusion imaging, other problems arise, such as motion (despite the use of EPI) and eddy current artifacts.

Several technical innovations have been developed to help substantially improve the quality of diffusion images. Despite this, there are still many unsolved problems (e.g., scan time efficiency, large-scale motion sensitivity, and isotropic voxel sizes) that remain to be solved. However, continual advances in MR scanner hardware, postprocessing, and acquisition strategies are paving the way for the acquisition of diffusion images with improved detail and more reliable quantitative parameters.

References

Aksoy, M., Forman, C., Straka, M., Skare, S., Holdsworth, S., Hornegger, J., Bammer, R., 2011. Real-time optical motion correction for diffusion tensor imaging. *Magn Reson Med.* 66, 366–78.

Aksoy, M., Liu, C., Moseley, M.E., Bammer, R., 2008. Single-step nonlinear diffusion tensor estimation in the presence of microscopic and macroscopic motion. *Magn Reson Med.* 59, 1138–50.

Aksoy, M., Skare, S., Holdsworth, S., Bammer, R., 2010. Effects of motion and b-matrix correction for high resolution DTI with short-axis PROPELLER-EPI. *NMR Biomed.* 23, 794–802.

Alexander, A.L., Tsuruda, J.S., Parker, D.L., 1997. Elimination of eddy current artifacts in diffusion-weighted echo-planar images: The use of bipolar gradients. *Magn Reson Med.* 38, 1016–21.

Anderson, A.W., Gore, J.C., 1994. Analysis and correction of motion artifacts in diffusion weighted imaging. *Magn Reson Med.* 32, 379–87.

Andersson, J., Skare, S., 2010. Image distortion and its correction in diffusion MRI. In: *Diffusion MRI: Theory, Methods, and Applications.* D.K. Jones, ed. Oxford: Oxford University Press, pp. 285–302.

Andersson, J., Smith, S., Jenkinson, M., 2008. FNIRT – FMRIB's non-linear image registration tool. In: *14th Annual Meeting of the Organization for Human Brain Mapping – HBM*, 15–19 June, Melbourne, Australia.

Andersson, J.L., Skare, S., 2002. A model-based method for retrospective correction of geometric distortions in diffusion-weighted EPI. *Neuroimage.* 16, 177–99.

Andersson, J.L.R., Skare, S., Ashburner, J., 2003. How to correct susceptibility distortions in spin-echo echo-planar images: Application to diffusion tensor imaging. *Neuroimage.* 20, 870–88.

Ardekani, S., Sinha, U., 2005. Geometric distortion correction of high-resolution 3 T diffusion tensor brain images. *Magn Reson Med.* 54, 1163–71.

Assaf, Y., Basser, P.J., 2005. Composite hindered and restricted model of diffusion (CHARMED) MR imaging of the human brain. *Neuroimage*. 27, 48–58.

Assaf, Y., Blumenfeld-Katzir, T., Yovel, Y., Basser, P.J., 2008. AxCaliber: A method for measuring axon diameter distribution from diffusion MRI. *Magn Reson Med*. 59, 1347–54.

Assaf, Y., Freidlin, R.Z., Rohde, G.K., Basser, P.J., 2004. New modeling and experimental framework to characterize hindered and restricted water diffusion in brain white matter. *Magn Reson Med*. 52, 965–78.

Atkinson, D., Counsell, S., Hajnal, J.V., Batchelor, P.G., Hill, D.L., Larkman, D.J., 2006. Nonlinear phase correction of navigated multi-coil diffusion images. *Magn Reson Med*. 56, 1135–9.

Atkinson, D., Porter, D.A., Hill, D.L., Calamante, F., Connelly, A., 2000. Sampling and reconstruction effects due to motion in diffusion-weighted interleaved echo planar imaging. *Magn Reson Med*. 44, 101–9.

Bammer, R., Keeling, S.L., Augustin, M., Pruessmann, K.P., Wolf, R., Stollberger, R., Hartung, H.P., Fazekas, F., 2001. Improved diffusion-weighted single-shot echo-planar imaging (EPI) in stroke using sensitivity encoding (SENSE). *Magn Reson Med*. 46, 548–54.

Bammer, R., Stollberger, R., Augustin, M., Simbrunner, J., Offenbacher, H., Kooijman, H., Ropele, S., et al., 1999. Diffusion-weighted imaging with navigated interleaved echo-planar imaging and a conventional gradient system. *Radiology*. 211, 799–806.

Barazany, D., Basser, P.J., Assaf, Y., 2009. In vivo measurement of axon diameter distribution in the corpus callosum of rat brain. *Brain*. 132, 1210–20.

Basser, P.J., Mattiello, J., LeBihan, D., 1994. MR diffusion tensor spectroscopy and imaging. *Biophys J*. 66, 259–67.

Basser, P.J., Pajevic, S., Pierpaoli, C., Duda, J., Aldroubi, A., 2000. In vivo fiber tractography using DT-MRI data. *Magn Reson Med*. 44, 625–32.

Basser, P.J., Pierpaoli, C., 1996. Microstructural and physiological features of tissues elucidated by quantitative-diffusion-tensor MRI. *J Magn Reson B*. 111, 209–19.

Bastin, M.E., 1999. Correction of eddy current-induced artefacts in diffusion tensor imaging using iterative cross-correlation. *Magn Reson Imaging*. 17, 1011–24.

Behrens, T.E., Berg, H.J., Jbabdi, S., Rushworth, M.F., Woolrich, M.W., 2007. Probabilistic diffusion tractography with multiple fibre orientations: What can we gain? *Neuroimage*. 34, 144–55.

Behrens, T.E., Woolrich, M.W., Jenkinson, M., Johansen-Berg, H., Nunes, R.G., Clare, S., Matthews, P.M., Brady, J.M., Smith, S.M., 2003. Characterization and propagation of uncertainty in diffusion-weighted MR imaging. *Magn Reson Med*. 50, 1077–88.

Benner, T., van der Kouwe, A.J., Sorensen, A.G., 2011. Diffusion imaging with prospective motion correction and reacquisition. *Magn Reson Med*. 66, 154–67.

Bodammer, N., Kaufmann, J., Kanowski, M., Tempelmann, C., 2004. Eddy current correction in diffusion-weighted imaging using pairs of images acquired with opposite diffusion gradient polarity. *Magn Reson Med*. 51, 188–93.

Boesch, C., Gruetter, R., Martin, E., 1991. Temporal and spatial analysis of fields generated by eddy currents in superconducting magnets: Optimization of corrections and quantitative characterization of magnet/gradient systems. *Magn Reson Med*. 20, 268–84.

Bowtell, R., McIntyre, D.J.O., Commandre, M.J., Glover, P.M., Suetens, P., 1994. Correction of geometric distortions in echo planar images. *Proceedings of the 2nd Annual Meeting of Society of Magnetic Resonance (SMR)*, 6–12 August, San Francisco, CA, p. 411.

Bruder, H., Fischer, H., Reinfelder, H.E., Schmitt, F., 1992. Image reconstruction for echo planar imaging with nonequidistant k-space sampling. *Magn Reson Med.* 23, 311–23.

Butts, K., Pauly, J., de Crespigny, A., Moseley, M., 1997. Isotropic diffusion-weighted and spiral-navigated interleaved EPI for routine imaging of acute stroke. *Magn Reson Med.* 38, 741–9.

Buxton, R.B., 1993. The diffusion sensitivity of fast steady-state free precession imaging. *Magn Reson Med.* 29, 235–43.

Calamante, F., Porter, D.A., Gadian, D.G., Connelly, A., 1999. Correction for eddy current induced Bo shifts in diffusion-weighted echo-planar imaging. *Magn Reson Med.* 41, 95–102.

Calamante, F., Tournier, J.D., Jackson, G.D., Connelly, A., 2010. Track-density imaging (TDI): Super-resolution white matter imaging using whole-brain track-density mapping. *Neuroimage.* 53, 1233–43.

Callaghan, P.T., 1991. *Principles of Nuclear Magnetic Resonance Microscopy.* Oxford: Clarendon Press.

Campbell, J.S., Siddiqi, K., Rymar, V.V., Sadikot, A.F., Pike, G.B., 2005. Flow-based fiber tracking with diffusion tensor and q-ball data: Validation and comparison to principal diffusion direction techniques. *Neuroimage.* 27, 725–36.

Chang, H., Fitzpatrick, J.M., 1992. A technique for accurate magnetic resonance imaging in the presence of field inhomogeneities. *IEEE Trans Med Imaging.* 11, 319–29.

Ciccarelli, O., Parker, G.J., Toosy, A.T., Wheeler-Kingshott, C.A., Barker, G.J., Boulby, P.A., Miller, D.H., Thompson, A.J., 2003. From diffusion tractography to quantitative white matter tract measures: A reproducibility study. *Neuroimage.* 18, 348–59.

Clare, S., 2003. Iterative Nyquist ghost correction for single and multi-shot EPI using an entropy measure. *Proceedings of the 11th Annual Meeting, International Society for Magnetic Resonance in Medicine*, 10–16 July, Toronto, Canada, p. 1041.

Conturo, T.E., Lori, N.F., Cull, T.S., Akbudak, E., Snyder, A.Z., Shimony, J.S., McKinstry, R.C., Burton, H., Raichle, M.E., 1999. Tracking neuronal fiber pathways in the living human brain. *Proc Natl Acad Sci USA.* 96, 10422–7.

Cotts, R., Hoch, M., Sun, T., Markert, J., 1989. Pulsed field gradient stimulated echo methods for improved NMR diffusion measurements in heterogeneous systems. *J Magn Reson.* 83, 252–66.

Cusack, R., Brett, M., Osswald, K., 2003. An evaluation of the use of magnetic field maps to undistort echo-planar images. *Neuroimage.* 18, 127–42.

de Crespigny, A.J., Marks, M.P., Enzmann, D.R., Moseley, M.E., 1995. Navigated diffusion imaging of normal and ischemic human brain. *Magn Reson Med.* 33, 720–8.

Einstein, A., 1905. Über die von der molekularkinetischen Theorie der Wärme geforderte Bewegung von in ruhenden Flüssigkeiten suspendierten Teilchen. *Ann Phys.* 322, 549–60.

Gallichan, D., Scholz, J., Bartsch, A., Behrens, T.E., Robson, M.D., Miller, K.L., 2010. Addressing a systematic vibration artifact in diffusion-weighted MRI. *Hum Brain Mapp.* 31, 193–202.

González, R.G., Schaefer, P.W., Buonanno, F.S., Schwamm, L.H., Budzik, R.F., Rordorf, G., Wang, B., Sorensen, A.G., Koroshetz, W.J., 1999. Diffusion-weighted MR imaging: Diagnostic accuracy in patients imaged within 6 hours of stroke symptom onset. *Radiology.* 210, 155–62.

Griswold, M.A., Jakob, P.M., Heidemann, R.M., Nittka, M., Jellus, V., Wang, J., Kiefer, B., Haase, A., 2002. Generalized autocalibrating partially parallel acquisitions (GRAPPA). *Magn Reson Med.* 47, 1202–10.

Haselgrove, J.C., Moore, J.R., 1996. Correction for distortion of echo-planar images used to calculate the apparent diffusion coefficient. *Magn Reson Med.* 36, 960–4.

Herbst, M., Maclaren, J., Weigel, M., Korvink, J., Hennig, J., Zaitsev, M., 2012. Prospective motion correction with continuous gradient updates in diffusion weighted imaging. *Magn Reson Med.* 67, 326–38.

Holdsworth, S.J., Skare, S., Newbould, R.D., Bammer, R., 2009. Robust GRAPPA-accelerated diffusion-weighted readout-segmented (RS)-EPI. *Magn Reson Med.* 62, 1629–40.

Holdsworth, S.J., Skare, S., Newbould, R.D., Guzmann, R., Blevins, N.H., Bammer, R., 2008. Readout-segmented EPI for rapid high resolution diffusion imaging at 3 T. *Eur J Radiol.* 65, 36–46.

Hong, X., Thomas Dixon, W., 1992. Measuring diffusion in inhomogeneous systems in imaging mode using antisymmetric sensitizing gradients. *J Magn Reson.* 99, 561–70.

Hosey, T., Williams, G., Ansorge, R., 2005. Inference of multiple fiber orientations in high angular resolution diffusion imaging. *Magn Reson Med.* 54, 1480–9.

Iturria-Medina, Y., Canales-Rodriguez, E.J., Melie-Garcia, L., Valdes-Hernandez, P.A., Martinez-Montes, E., Aleman-Gomez, Y., Sanchez-Bornot, J.M., 2007. Characterizing brain anatomical connections using diffusion weighted MRI and graph theory. *Neuroimage.* 36, 645–60.

Jezzard, P., Balaban, R.S., 1995. Correction for geometric distortion in echo planar images from B0 field variations. *Magn Reson Med.* 34, 65–73.

Jezzard, P., Barnett, A.S., Pierpaoli, C., 1998. Characterization of and correction for eddy current artifacts in echo planar diffusion imaging. *Magn Reson Med.* 39, 801–12.

Jiang, H., Golay, X., van Zijl, P.C., Mori, S., 2002. Origin and minimization of residual motion-related artifacts in navigator-corrected segmented diffusion-weighted EPI of the human brain. *Magn Reson Med.* 47, 818–22.

Jones, D.K., 2004. The effect of gradient sampling schemes on measures derived from diffusion tensor MRI: A Monte Carlo study. *Magn Reson Med.* 51, 807–15.

Kaiser, R., Bartholdi, E., Ernst, R., 1974. Diffusion and field-gradient effects in NMR Fourier spectroscopy. *J Chem Phys.* 60, 2966.

Kannengiesser, S.A., Wang, Y., Haacke, E.M., 1999. Geometric distortion correction in gradient-echo imaging by use of dynamic time warping. *Magn Reson Med.* 42, 585–90.

Kober, T., Gruetter, R., Krueger, G., 2012. Prospective and retrospective motion correction in diffusion magnetic resonance imaging of the human brain. *Neuroimage.* 59, 389–98.

Kybic, J., Thevenaz, P., Nirkko, A., Unser, M., 2000. Unwarping of unidirectionally distorted EPI images. *IEEE Trans Med Imaging.* 19, 80–93.

Le Bihan, D., Breton, E., Lallemand, D., Grenier, P., Cabanis, E., Laval-Jeantet, M., 1986. MR imaging of intravoxel incoherent motions: Application to diffusion and perfusion in neurologic disorders. *Radiology.* 161, 401–7.

Leemans, A., Jones, D.K., 2009. The B-matrix must be rotated when correcting for subject motion in DTI data. *Magn Reson Med*. 61, 1336–49.

Liu, C., Bammer, R., Acar, B., Moseley, M.E., 2004a. Characterizing non-Gaussian diffusion by using generalized diffusion tensors. *Magn Reson Med*. 51, 924–37.

Liu, C., Bammer, R., Kim, D.H., Moseley, M.E., 2004b. Self-navigated interleaved spiral (SNAILS): Application to high-resolution diffusion tensor imaging. *Magn Reson Med*. 52, 1388–96.

Liu, C., Mang, S.C., Moseley, M.E., 2010. In vivo generalized diffusion tensor imaging (GDTI) using higher-order tensors (HOT). *Magn Reson Med*. 63, 243–52.

Liu, C., Moseley, M.E., Bammer, R., 2005. Simultaneous phase correction and SENSE reconstruction for navigated multi-shot DWI with non-cartesian k-space sampling. *Magn Reson Med*. 54, 1412–22.

Lovblad, K.O., Laubach, H.J., Baird, A.E., Curtin, F., Schlaug, G., Edelman, R.R., Warach, S., 1998. Clinical experience with diffusion-weighted MR in patients with acute stroke. *AJNR Am J Neuroradiol*. 19, 1061–6.

Maes, F., Collignon, A., Vandermeulen, D., Marchal, G., Suetens, P., 1997. Multimodality image registration by maximization of mutual information. *IEEE Trans Med Imaging*. 16, 187–98.

Marks, M.P., de Crespigny, A., Lentz, D., Enzmann, D.R., Albers, G.W., Moseley, M.E., 1996. Acute and chronic stroke: Navigated spin-echo diffusion-weighted MR imaging. *Radiology*. 199, 403–8.

Miller, K.L., Pauly, J.M., 2003. Nonlinear phase correction for navigated diffusion imaging. *Magn Reson Med*. 50, 343–53.

Mori, S., Crain, B.J., Chacko, V.P., van Zijl, P.C., 1999. Three-dimensional tracking of axonal projections in the brain by magnetic resonance imaging. *Ann Neurol*. 45, 265–9.

Mori, S., van Zijl, P.C., 1998. A motion correction scheme by twin-echo navigation for diffusion-weighted magnetic resonance imaging with multiple RF echo acquisition. *Magn Reson Med*. 40, 511–16.

Neuman, C.H., 1974. Spin-echo of spins diffusing in a bounded medium. *J Chem Phys*. 60, 4508–11.

Ordidge, R.J., Helpern, J.A., Qing, Z.X., Knight, R.A., Nagesh, V., 1994. Correction of motional artifacts in diffusion-weighted MR images using navigator echoes. *Magn Reson Imaging*. 12, 455–60.

Papadakis, N.G., Martin, K.M., Pickard, J.D., Hall, L.D., Carpenter, T.A., Huang, C.L., 2000. Gradient preemphasis calibration in diffusion-weighted echo-planar imaging. *Magn Reson Med*. 44, 616–24.

Parker, G.J., Haroon, H.A., Wheeler-Kingshott, C.A., 2003. A framework for a streamline-based probabilistic index of connectivity (PICo) using a structural interpretation of MRI diffusion measurements. *J Magn Reson Imaging*. 18, 242–54.

Parker, G.J.M., Wheeler-Kingshott, C.A.M., Barker, G.J., 2002. Estimating distributed anatomical connectivity using fast marching methods and diffusion tensor imaging. *IEEE Trans Med Imaging* 21, 505–12.

Pierpaoli, C., Basser, P.J., 1996. Toward a quantitative assessment of diffusion anisotropy. *Magn Reson Med*. 36, 893–906.

Pipe, J.G., Farthing, V.G., Forbes, K.P., 2002. Multishot diffusion-weighted FSE using PROPELLER MRI. *Magn Reson Med*. 47, 42–52.

Porter, D.A., Heidemann, R.M., 2009. High resolution diffusion-weighted imaging using readout-segmented echo-planar imaging, parallel imaging and a two-dimensional navigator-based reacquisition. *Magn Reson Med.* 62, 468–75.

Pruessmann, K.P., Weiger, M., Bornert, P., Boesiger, P., 2001. Advances in sensitivity encoding with arbitrary k-space trajectories. *Magn Reson Med.* 46, 638–51.

Pruessmann, K.P., Weiger, M., Scheidegger, M.B., Boesiger, P., 1999. SENSE: Sensitivity encoding for fast MRI. *Magn Reson Med.* 42, 952–62.

Reber, P.J., Wong, E.C., Buxton, R.B., Frank, L.R., 1998. Correction of off resonance-related distortion in echo-planar imaging using EPI-based field maps. *Magn Reson Med.* 39, 328–30.

Reese, T.G., Heid, O., Weisskoff, R.M., Wedeen, V.J., 2003. Reduction of eddy-current-induced distortion in diffusion MRI using a twice-refocused spin echo. *Magn Reson Med.* 49, 177–82.

Robson, M.D., Gore, J.C., Constable, R.T., 1997. Measurement of the point spread function in MRI using constant time imaging. *Magn Reson Med.* 38, 733–40.

Rohde, G.K., Barnett, A.S., Basser, P.J., Marenco, S., Pierpaoli, C., 2004. Comprehensive approach for correction of motion and distortion in diffusion-weighted MRI. *Magn Reson Med.* 51, 103–14.

Schmithorst, V.J., Dardzinski, B.J., 2002. Automatic gradient preemphasis adjustment: A 15-minute journey to improved diffusion-weighted echo-planar imaging. *Magn Reson Med.* 47, 208–12.

Skare, S., Andersson, J.L., 2005. Correction of MR image distortions induced by metallic objects using a 3D cubic B-spline basis set: Application to stereotactic surgical planning. *Magn Reson Med.* 54, 169–81.

Skare, S., Newbould, R.D., Clayton, D.B., Bammer, R., 2006. Propeller EPI in the other direction. *Magn Reson Med.* 55, 1298–307.

Song, S.K., Sun, S.W., Ramsbottom, M.J., Chang, C., Russell, J., Cross, A.H., 2002. Dysmyelination revealed through MRI as increased radial (but unchanged axial) diffusion of water. *Neuroimage.* 17, 1429–36.

Stieltjes, B., Kaufmann, W.E., van Zijl, P.C., Fredericksen, K., Pearlson, G.D., Solaiyappan, M., Mori, S., 2001. Diffusion tensor imaging and axonal tracking in the human brainstem. *Neuroimage.* 14, 723–35.

Studholme, C., Constable, R.T., Duncan, J.S., 2000. Accurate alignment of functional EPI data to anatomical MRI using a physics-based distortion model. *IEEE Trans Med Imaging.* 19, 1115–27.

Thesen, S., Heid, O., Mueller, E., Schad, L.R., 2000. Prospective acquisition correction for head motion with image-based tracking for real-time fMRI. *Magn Reson Med.* 44, 457–65.

Trouard, T.P., Sabharwal, Y., Altbach, M.I., Gmitro, A.F., 1996. Analysis and comparison of motion-correction techniques in diffusion-weighted imaging. *J Magn Reson Imaging.* 6, 925–35.

Tuch, D.S., 2004. Q-ball imaging. *Magn Reson Med.* 52, 1358–72.

Tuch, D.S., Reese, T.G., Wiegell, M.R., Makris, N., Belliveau, J.W., Wedeen, V.J., 2002. High angular resolution diffusion imaging reveals intravoxel white matter fiber heterogeneity. *Magn Reson Med.* 48, 577–82.

Tuch, D.S., Reese, T.G., Wiegell, M.R., Wedeen, V.J., 2003. Diffusion MRI of complex neural architecture. *Neuron.* 40, 885–95.

Turner, R., Le Bihan, D., Maier, J., Vavrek, R., Hedges, L.K., Pekar, J., 1990. Echo-planar imaging of intravoxel incoherent motion. *Radiology.* 177, 407–14.

Wedeen, V.J., Hagmann, P., Tseng, W.Y., Reese, T.G., Weisskoff, R.M., 2005. Mapping complex tissue architecture with diffusion spectrum magnetic resonance imaging. *Magn Reson Med*. 54, 1377–86.

Wedeen, V.J., Wang, R.P., Schmahmann, J.D., Benner, T., Tseng, W.Y., Dai, G., Pandya, D.N., Hagmann, P., D'Arceuil, H., de Crespigny, A.J., 2008. Diffusion spectrum magnetic resonance imaging (DSI) tractography of crossing fibers. *Neuroimage*. 41, 1267–77.

Wheeler-Kingshott, C.A.M., Cercignani, M., 2009. About "axial" and "radial" diffusivities. *Magn Reson Med*. 61, 1255–60.

Wider, G., Dötsch, V., Wüthrich, K., 1994. Self-compensating pulsed magnetic-field gradients for short recovery times. *J Magn Reson A*. 108, 255–8.

Williams, C.F., Redpath, T.W., Norris, D.G., 1999. A novel fast split-echo multi-shot diffusion-weighted MRI method using navigator echoes. *Magn Reson Med*. 41, 734–42.

Witwer, B.P., Moftakhar, R., Hasan, K.M., Deshmukh, P., Haughton, V., Field, A., Arfanakis, K., et al., 2002. Diffusion-tensor imaging of white matter tracts in patients with cerebral neoplasm. *J Neurosurg*. 97, 568–75.

Wu, E.X., Buxton, R.B., 1990. Effect of diffusion on the steady-state magnetization with pulsed field gradients. *J Magn Reson*. 90, 243–53.

Xue, R., van Zijl, P.C., Crain, B.J., Solaiyappan, M., Mori, S., 1999. In vivo three-dimensional reconstruction of rat brain axonal projections by diffusion tensor imaging. *Magn Reson Med*. 42, 1123–7.

Zaitsev, M., Dold, C., Sakas, G., Hennig, J., Speck, O., 2006. Magnetic resonance imaging of freely moving objects: Prospective real-time motion correction using an external optical motion tracking system. *Neuroimage*. 31, 1038–50.

3

CMR$_{O2}$ Mapping by Calibrated fMRI

Fahmeed Hyder, Christina Y. Shu, Peter Herman, Basavaraju G. Sanganahalli, Daniel Coman, and Douglas L. Rothman

Yale University

CONTENTS

3.1 Introduction

The impact of functional magnetic resonance imaging (fMRI) on neuroscience has been vast since the earliest human experiments conducted at Medical College of Wisconsin, Yale University, Max Planck Institute, Massachusetts General Hospital, and University of Minnesota (Bandettini et al., 1992; Blamire et al., 1992; Frahm et al., 1992; Kwong et al., 1992; Ogawa et al., 1992). But shortly after these 1992 fMRI studies, there were concerns about how well the blood oxygenation level-dependent (BOLD) signal reflected underlying changes in neural activity (Barinaga, 1997). As described below, the component of the BOLD contrast that best correlates with neural activity is the change in oxygen consumption (ΔCMR$_{O2}$). A field of quantitative MRI emerged as a consequence to convert the BOLD response into a quantitative measure of the oxidative energy consumed during brain function [for a historical perspective, see Hyder and Rothman (2012)].

Human brain mapping became possible about three decades ago with positron emission tomography (PET). Within a decade, however, PET functional measurements were replaced by fMRI. But the basic concepts for mapping brain function are similar [for a historical perspective, see Raichle (2009)]. PET measures changes in blood flow or volume (CBF, CBV) and/or changes in glucose or oxygen metabolism (CMR_{glc}, CMR_{O2}) induced by external stimuli, whereas fMRI uses the BOLD contrast, which is a consequence of a complex combination of changes in CBV, CBF, and CMR_{O2} (Ogawa et al., 1990). The main advantage of fMRI over PET is that the BOLD signal can be mapped repeatedly in the same session because no external tracer is needed as the changing concentrations of oxyhemoglobin and deoxyhemoglobin inside red blood cells (i.e., diamagnetic and paramagnetic, respectively) during functional hyperemia provide the contrast.

Unfortunately, the BOLD signal itself is only a partial measure of neural activity (Barinaga, 1997; Fitzpatrick and Rothman, 1999). To convert the BOLD response into a reliable component of changes in CMR_{O2} demanded by neural activity, calibrated MRI attempts to detach changes in CMR_{O2} from concurrently measured changes in BOLD signal, CBF, and/or CBV with multimodal MRI techniques. Because animal experiments at high magnetic fields provide an opportunity to test assumptions of calibrated fMRI, here we focus on rat studies and examine how human calibrated fMRI studies differ. We first discuss some historical and fundamental aspects of calibrated fMRI. Then we discuss the accuracy of ΔCMR_{O2} predicted by calibrated fMRI in relation to changes in CBV and neural activity. The results will show that calibrated fMRI provides quantitative information on changes in neural activity. We finally discuss future challenges to advance calibrated fMRI for improved quantitative neuroimaging.

3.2 Historical Perspective on Oxidative Demand

It had long been established that glucose is the main energy substrate in the brain and its oxidation efficiently produces energy to support brain function (Siesjo, 1978). However in the late 1980s, Fox et al. published some PET data that revealed reduced glucose oxidation during sensory processing in awake human brain, where CMR_{glc} and CBF changed significantly and CMR_{O2} changed very little (Fox and Raichle, 1986; Fox et al., 1988). These PET results, along with calculations by Creutzfeld, supported a model of brain function in which neuronal signaling had minimal oxidative energy costs (1975). This conclusion contradicted with the generally accepted view of disproportionately high energy costs of human brain function at rest, especially given the brain's size relative to the body (Aiello and Wheeler, 1995; Sokoloff, 1991). Subsequent studies have shown that there is several-fold larger increase in CMR_{O2} with functional activation than originally measured by Fox et al., especially in relation to change in CBF (i.e., CMR_{O2}–CBF coupling underlying the BOLD contrast) as discussed below and in Sections 3.3 and 3.4.

While the Fox et al. results suggested some level of nonoxidative glycolysis during function, the amount of lactate produced as measured by ^{1}H magnetic resonance spectroscopy (MRS) at Yale was quantitatively much smaller than predicted by the PET studies (Chen et al., 1993; Prichard et al., 1991). While these early ^{1}H MRS results from Yale are in good agreement with independent measurements from other laboratories using ^{1}H MRS and PET (Frahm et al., 1996; Mangia et al., 2007; Merboldt et al., 1992; Mintun et al., 2002; Roland et al., 1987; Sappey-Marinier et al., 1992; Seitz and Roland, 1992; Vafaee et al., 1998, 1999), questions about the extent of oxidative demand during brain function could be tested with ^{13}C MRS and a related ^{1}H MRS method called proton observe carbon edit (POCE).

^{13}C MRS and POCE provide detailed measurements of metabolic fluxes [for historical and technological perspectives, see de Graaf et al. (2011) and Hyder et al. (2006)]. Following studies in cell suspensions and perfused organs in the late 1970s, ^{13}C MRS and POCE were both applied to the rodent and human brain in the 1980s and 1990s (Behar et al., 1986; Cerdan et al., 1990; Fitzpatrick et al., 1990; Gruetter et al., 1992; Petroff et al., 1986; Rothman et al., 1985, 1992). Based on the ability to detect *in vivo* turnover of ^{13}C label from substrates (e.g., glucose or acetate) into various metabolites (e.g., glutamate, glutamine, lactate, or aspartate) in real time, several metabolic fluxes (e.g., tri-carboxylic acid cycle, glutamine synthesis, neurotransmitter cycle, etc.) could be extracted with compartmental modeling of neuronal-glial trafficking of metabolites (Gruetter et al., 1996; Mason et al., 1992, 1995). The rates determined for glucose oxidation (i.e., which is stoichiometrically related to CMR_{O2}) were in good agreement with CMR_{glc} measured by other methods—for example, arteriovenous difference, autoradiography, and PET (Boumezbeur et al., 2005; Hyder et al., 2000a; Rothman et al., 2011).

To test the oxidative energy demand of functional brain activation (i.e., ΔCMR_{O2}), POCE studies were performed in animal and human. At Yale, POCE was used in an anesthetized rat model to demonstrate a large increase in neuronal glucose oxidation in the somatosensory cortex during sensory stimulation (Hyder et al., 1996, 1997). Because the earliest interpretations of BOLD signal increase were interpreted to reflect minimal rise in CMR_{O2} (Kim and Ugurbil, 1997; Ogawa et al., 1993), this finding was controversial at the time because a large increase in CMR_{O2} in the presence of a positive BOLD signal was considered to be an artifact. However concomitant large changes in CBF could explain the presence of large rise in CMR_{O2} with a positive BOLD response (Hyder et al., 2000a; Silva et al., 1999). This issue will be discussed more in Section 3.4 in context of CMR_{O2}–CBF coupling with calibrated fMRI, but POCE studies of the human brain conducted independently at the Universities of Minnesota and Nottingham also showed a substantial increase in CMR_{O2} with visual stimulation in the primary visual cortex (Chen et al., 2001; Chhina et al.,

2001). These MRS studies along with parallel studies using calibrated fMRI (see Sections 3.3 and 3.4) have led to the general acceptance that increases in brain activity are primarily fueled by ATP for oxidative glucose consumption, which in primary sensory cortices can be metabolically quite significant (Shulman et al., 2001a). However, the mechanism and function of enhanced nonoxidative metabolism remains an intense area of research, although several models have been proposed (Chih et al., 2001; Magistretti et al., 1999; Shulman et al., 2001b).

3.3 Principles of Calibrated fMRI

Because the BOLD signal is sensitive to functional hyperemic events (Ogawa et al., 1990), even the earliest biophysical descriptions hinted that CMR_{O2} could be extracted provided that CBF and CBV were measured with the baseline BOLD signal (Boxerman et al., 1995; Kennan et al., 1994; Ogawa et al., 1993). The BOLD contrast is generated by changing concentrations of oxyhemoglobin (diamagnetic) versus deoxyhemoglobin (paramagnetic) in the venous end of the vasculature, that is, the BOLD effect is weighted less toward arterial blood where hemoglobin oxygenation is nearly fully saturated. The magnetic properties of blood oxygenation changing (in the venous end) affect the tissue water MRI signal due to intravoxel spin dephasing (i.e., less paramagnetic deoxyhemoglobin, higher MRI signal), which can be captured with the transverse relaxation rate measured by either by gradient echo (R_2^*) or spin echo (R_2)

$$R_2^* = R_2'(Y) + R_2(Y) + \cdots \tag{3.1a}$$

$$R_2 = R_2(Y) + \cdots, \tag{3.1b}$$

where $R_2'(Y)$ and $R_2(Y)$ represent the reversible and nonreversible relaxation components due to blood oxygenation (Y) effects on tissue water relaxation rate (Hyder et al., 2000a; Kida et al., 1999, 2000). Since each of these relaxation terms in gradient-echo or spin-echo MRI varies as a function of Y, Fick's principle (i.e., which relates Y to CBF and CMR_{O2}) can be used to relate the BOLD signal change for calibrated fMRI [for a theoretical description, see Hyder et al. (2001)].

MRI contrast enhancement by deoxyhemoglobin for BOLD should focus on small vessel effects and exclude large vessel effects that are termed as intravascular contamination artifacts. Although the blood volume fraction for a typical MRI voxel tends to be small (2–5%), larger vessels (e.g., arteries and veins) can have significant partial volume when the transverse relaxation rates of water in blood and tissue are comparable. This is specifically an issue at lower magnetic field strengths, whereas generally at higher magnetic field strengths this issue can be reduced because the measured R_2^* or R_2 is largely dominated by the extravascular component. Contribution of the intravascular component on BOLD can be examined by using graded diffusion-weighted spin-echo MRI (Boxerman et al., 1995; Song et al., 1996). Results show negligible intravascular contamination of the BOLD effect at higher magnetic fields, although at fields below 7 T, there can be significant intravascular contributions.

The term "calibrated fMRI" was first used in the literature in 1998 where Davis et al. exploited the CMR$_{O2}$–CBF and CBV–CBF couplings embedded within the BOLD response ($\Delta S/S$) to extract ΔCMR$_{O2}$/CMR$_{O2}$ (1998):

$$\frac{\Delta S}{S_0} = M \left[1 - \left(\frac{CMR_{O2}}{CMR_{O2_0}} \right)^\beta \left(\frac{CBF}{CBF_0} \right)^{\alpha - \beta} \right], \tag{3.2}$$

where the subscript "0" represents the baseline state for each parameter and α describes the CBV–CBF coupling and is assumed to be 0.4 in human studies based on PET primate studies (Grubb et al., 1974). But as discussed in Section 3.3.2, α is measurable in animals (Hyder et al., 2001) and β describes the CMR$_{O2}$–CBF coupling and is assumed to be 1.5 in human studies based on simulations of BOLD effect at 1.5 T (Boxerman et al., 1995); and as discussed in Section 3.3.1, β approaches 1 at higher magnetic fields (Ogawa et al., 1993), and M represents the baseline BOLD signal which, under the approximation that the change in transverse relaxation component due to BOLD is small, is hypothesized to represent the product of $R_2'(Y)$ and echo time (TE).

$$M = R_2'(Y) \cdot TE. \tag{3.3a}$$

M is directly measurable by the difference between R_2^* and R_2 under well-shimmed conditions (Hyder et al., 2000a, 2001; Kida et al., 1999, 2000):

$$R_2'(Y) = R_2^* - R_2, \tag{3.3b}$$

where the first and second terms on the right-hand side are measured by gradient-echo and spin-echo MRI, respectively, under well-shimmed conditions.

The early development of calibrated fMRI studies, both in animals and humans, greatly benefited from MRI methods for CBV and CBF imaging. In the 1990s, several groups focused on novel MRI methods for CBF and CBV imaging, which proved to be seminal for the development of calibrated fMRI (Detre et al., 1992; Kennan et al., 1998; Kim, 1995; Williams et al., 1992). Just as hemoglobin is nature's endogenous MRI contrast agent to measure blood oxygenation changes, exogenous MRI contrast agents that are specifically blood-borne can reflect alterations in CBV with almost the same assumptions (i.e., more paramagnetic agent, lower MRI signal due to increase in R_2^* or R_2). Details of this CBV method are discussed in Section 3.3.2, but the method is used primarily in animals due to toxicity concerns about the contrast agent. Principles of CBF measured by MRI and PET are similar. MRI requires magnetic labeling of arterial blood water entering the brain and detecting subsequent MRI data from the brain. Measurements reflect the degree to which the label has decayed by mixing with unlabeled brain tissue water. This CBF technique is based on arterial spin labeling (ASL) using the sensitivity of the longitudinal relaxation rate of tissue water to reflect blood perfusion. Because magnetic labeling of arterial blood with ASL proves to be a sensitive enough perfusion tracer in brain tissue, the CBF method does not require exogenous contrast agents, and thus can be imaged reliably in both humans and animals.

3.3.1 Accuracy of ΔCMR_{O2} by Calibrated fMRI

Since POCE could measure CMR_{O2} directly (see Section 3.2), an opportunity presented itself at Yale to test the validity of Equation 3.2 in relation to Equation 3.3 by conducting multimodal measurements of CMR_{O2}, CBF, CBV, and BOLD signal with both spin echo and gradient echo, under various levels of brain activity in rats at 7.0 T (Hyder et al., 2001; Kida et al., 2000). A summary of the main results is shown in Figure 3.1a. From a baseline condition of morphine, variations were observed for CBF (measured by ASL, first column), CMR_{O2} (measured by POCE, second column), BOLD signal (measured by R_2 or R_2^*, third and fourth columns), and CBV (measured by intravascular MRI contrast agent, data not shown) with pentobarbital and nicotine. Cortical values of CBF, CMR_{O2}, and CBV increased with nicotine, whereas these parameters decreased with pentobarbital while R_2 or R_2^* changed in the opposite direction. At 7.0 T, the R_2^* range of 32–40 s^{-1} in the rat reported by Kida et al. is comparable to recently reported values in the human cerebral cortex at 7 T (Donahue et al., 2011). Similarly, the observed R_2 range of 21–25 s^{-1} at 7.0 T in the rat reported by Kida et al. is within expectations of other results in the rat (i.e., 15–24 s^{-1}) for magnetic fields spanning from 4.0 T to 9.4 T (de Graaf et al., 2006). The CBF and CMR_{O2} results in the Kida et al. study are also in accord with prior studies using alternate methods [see Hyder et al. (2000a) for details].

Since the POCE voxel (for CMR_{O2}) was localized in the somatosensory cortex, similar regions were interrogated in the MRI data to establish the relationship between $R_2'(Y)$, CBF, and CMR_{O2}, as shown in Figure 3.1b. It should be noted, however, that similar reliable relationships were also established with R_2^* or R_2, CBF, and CMR_{O2} (data not shown). Good agreement between measured ΔCMR_{O2} (by POCE) and predicted ΔCMR_{O2} (by calibrated fMRI) was consistent with an M value of about 25–35 at 7.0 T for the awake condition with a TE of 16–26 ms, as shown in the inset of Figure 3.1b.

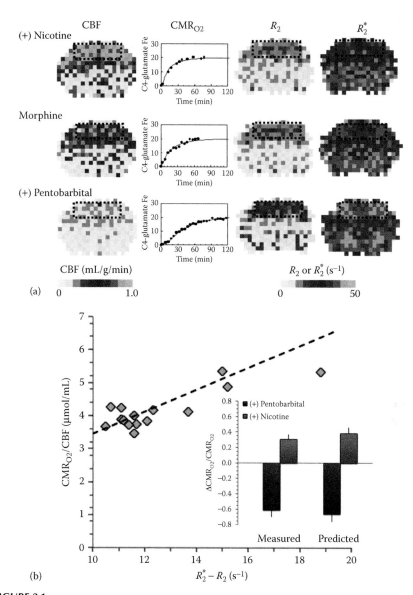

FIGURE 3.1
(See color insert.) Summary of multimodal investigations of calibrated fMRI in rat brain at 7.0 T: (a) Examples of CBF, CMR_{O2}, R_2, and R_2^* data from baseline condition of morphine. Cortical values of CBF, CMR_{O2}, and CBV increased with nicotine, whereas these parameters decreased with pentobarbital. The reverse was observed for by R_2 or R_2^*. (b) Relationship between CMR_{O2}/CBF and $R_2^*-R_2$ for all conditions. Inset: comparison between $\Delta CMR_{O2}/CMR_{O2}$ measured by ^{13}C MRS and predicted by calibrated fMRI. (Modified from Kida, I., et al., *J Cereb Blood Flow Metab.* 20, 847–60, 2000.)

The M value in awake human studies is significantly lower than animal studies, possibly due to differences in magnetic field strengths (e.g., 1.5–4.0 T in humans vs. >7.0 T in animals) and/or species differences, as discussed in more detail in Section 3.4. However another possibility, which needs further study, is the fact that the M value could be smaller under high brain activity levels (e.g., awake) than at low brain activity levels (e.g., deep anesthesia), as shown in Figure 3.1b, because the oxygen extraction fraction (OEF) is given by

$$OEF = \frac{CMR_{O2}}{C_a \cdot CBF} = 1 - Y, \tag{3.4}$$

where C_a is the arterial oxygen content and $1 - Y$ is the capillary arteriovenous oxygen difference (Hyder et al., 1998).

While the approach in the rat calibrated fMRI studies at 7.0 T differs from human calibrated fMRI studies at 1.5 T—as discussed below—some important observations have relevance for studies conducted at magnetic fields higher than 1.5 T. It was observed that the parameter β, which designates the CMR_{O2}–CBF coupling, is closer to 1 at 7.0 T (and higher). This β value is within expectations from simulations of the BOLD effect extrapolated to fields beyond 1.5 T (Boxerman et al., 1995; Kennan et al., 1994; Ogawa et al., 1993). The value of the parameter α, which describes the CBV–CBF coupling, is probably less than 0.4 as it was determined by PET studies in primate brain during CO_2 challenges (Grubb et al., 1974). This issue will be discussed further in Section 3.3.2. Another important issue deals with the inherent problem in calibrated fMRI to assess the so-called M parameter, which is affected by a variety of experimental settings including the static magnetic field homogeneity. An important component of the early Yale work by Kida et al. was that M could be directly measured under optimally shimmed conditions for both spin-echo and gradient-echo calibrated fMRI studies (Hyder et al., 2001; Kida et al., 2000). Recently, this view of directly measuring M has been raised in the literature (Blockley et al., 2012; Gauthier and Hoge, 2012).

The main pitfall with the direct M measurement is that residual components, which are unaccounted for in the relaxation rate terms (i.e., see Equation 3.1), may contaminate the calibration [see Kida et al. (2000) for details]. In human calibrated studies, the use of gas challenges—with O_2 (i.e., hyperoxia), CO_2 (i.e., hypercapnia), or combination of both (i.e., carbogen)—is the preferred approach to derive M [for a recent review, see Hoge (2012)]. If CMR_{O2} remains unperturbed with the gas perturbations while other factors (e.g., BOLD, CBF, CBV) change, then parameter M derived with the gas perturbations can be used for the functional activation. These gas perturbations are assumed to be physiologically benign. The oxygen saturation curve of hemoglobin is believed to be unaffected from the normal situation. No measurable consequences on neural activity are expected from the gases. Both the CBV–CBF and BOLD–CBF couplings are presumed to be the same as in functional activation. Various aspects of these assumptions remain intense areas of research (Chen and Pike, 2010; Donahue et al., 2009; Dutka et al., 2002; Hall et al., 2011; Jones et al., 2005; Liu et al., 2007; Mark et al., 2010; Martin et al., 2006; Wibral et al., 2007; Xu et al., 2011; Zappe et al., 2008).

3.3.2 Impact of ΔCBV on Calibrated fMRI

The earliest fMRI study of human brain function in 1991 was with CBV contrast using repeated bolus injections of a paramagnetic MRI contrast agent that has a relatively short half-life in blood circulation (Belliveau et al., 1990). A few years later, however, the CBV method in animals was based on intravascular-borne exogenous super-paramagnetic MRI contrast agents that maintained a steady blood concentration over several hours (Kennan et al., 1998; Mandeville et al., 1998). The susceptibility effect of the agent was so strong that it could drown out the diamagnetic BOLD effect during functional hyperemia. Recently, some new CBV methods have been developed for human studies that do not require exogenous MRI contrast agents (Lu et al., 2003; Stefanovic and Pike, 2004). However, early human calibrated fMRI studies combined only BOLD and CBF measurements to estimate CMR$_{O2}$ changes during functional activation (Davis et al., 1998; Hoge et al., 1999; Kastrup et al., 1999; Kim and Ugurbil, 1997; Kim et al., 1999), whereas animal calibrated fMRI studies also included CBV measurements because the contrast agents were limited to animals (Hyder et al., 2001; Kida et al., 1999, 2000; Mandeville et al., 1998).

Most early human calibrated fMRI studies assumed a specific value of α (i.e., 0.4) because CBV could not be independently measured, and thus it had to be derived from changes in CBF as indicated from hypercapnia PET studies in primates (Grubb et al., 1974):

$$\left(\frac{CBV}{CBV_0} \right) = \left(\frac{CBF}{CBF_0} \right)^{\alpha}, \tag{3.5}$$

where the subscript "0" represents the baseline state for each parameter. Animal studies at Harvard and Yale suggested that the α value could range from 0.1 to 0.4 because CBV and CBF dynamics were not fixed during the time course of the functional activation (Hyder et al., 2001; Mandeville et al., 1999b). These suggestions have now been independently supported by subsequent findings from studies in rat, cat, and human brain where CBV and CBF have been measured in the same session (Chen and Pike, 2009; Jin and Kim, 2008; Kida et al., 2007; Shen et al., 2008), where during a functional challenge the α value is found to range from 0.1 to 0.2. Closer inspection of dynamic changes in CBV and CBF over the course of the BOLD response during sensory stimulation in the rat, as shown in Figure 3.2a, suggests that

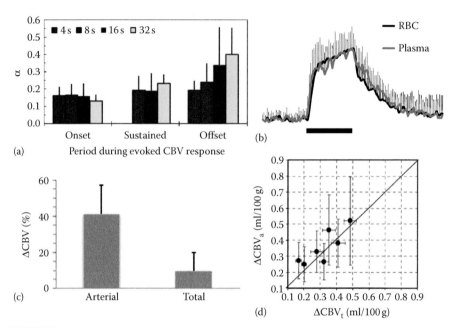

FIGURE 3.2

Summary of CBV studies in rat brain for calibrated fMRI: (a) Comparison of α in three periods of onset, sustained, and offset portions of the BOLD response for different durations of forepaw stimulation in rats anesthetized with α-chloralose. Note there were no α values reported for sustained response for the 4 s duration stimulation because of rapid rise and fall of the signal. (Modified from Kida, I., et al., *J Cereb Blood Flow Metab.* 27, 690–6, 2007.) (b) Comparison of CBV measured for plasma and erythrocyte (RBC) compartments in rat brain during forepaw stimulation under α-chloralose anesthesia. (Modified from Herman, P., et al., *J Cereb Blood Flow Metab.* 29, 19–24, 2009b.) (c) Comparison of fractional changes in CBV arterial and CBV total measured by Kim et al. in rat brain during forepaw stimulation under isoflurane anesthesia. (From Kim, T., et al., *J Cereb Blood Flow Metab.* 27, 1235–47, 2007. With permission.) (d) Comparison of normalized changes in CBV arterial (CBV$_a$) and CBV total (CBV$_t$) in rat brain during forepaw stimulation under isoflurane anesthesia. (Reprinted by permission from Macmillan Publishers Ltd. *J Cereb Blood Flow Metab.*, Kim, T. et al., 27, 1235–47, © 2007.)

α is initially quite small (i.e., ~0.1) and gradually become larger (i.e., >0.2), suggesting that calibrated fMRI should consider a dynamically varying α to accurately reflect ΔCMR_{O2} (Kida et al., 2007).

Since the MRI contrast agents used in animals are solely plasma-borne, concerns remained about whether these signals reflect changes in total CBV or only plasma CBV. Recently, a Yale study in anesthetized rats combined an MRI-based CBV measurement with a laser-Doppler backscattered light (805 nm) measurement to reflect the dynamic changes in plasma and erythrocyte compartments, respectively (Herman et al., 2009b). As shown in Figure 3.2b, dynamic comparisons of the two CBV signals during functional hyperemia revealed excellent correlations. These results suggest that CBV measurement from either compartment could be used to reflect dynamic changes in total CBV. Furthermore, if we assume steady-state mass balance and negligible counter flow, then these results indicate that volume hematocrit is not appreciably affected during functional activation—verifying a two-decade-old assumption about the BOLD effect (Ogawa et al., 1990).

It should be noted, however, that these CBV measurements reflect the entire blood volume in the vascular branch, whereas the BOLD effect is sensitive to mainly the venous compartment (see Section 3.3). Recently, some novel CBV methods developed at the University of Pittsburgh (Kim and Kim, 2005), which allows separation of the arterial and venous compartments, suggest that the contribution of CBV changes incorporated into calibrated fMRI may be too large because it is the arterial compartment that contributes most during functional hyperemia (Kim et al., 2007), as shown in Figure 3.2c. Since these CBV changes in the arterial compartment are well correlated with total CBV changes as shown in Figure 3.2d, these results have serious consequences for calibrated fMRI because of the potential to underestimate changes in CMR_{O2} by exaggerated contributions from CBV into Equation 3.2.

3.3.3 Neural Basis of Calibrated fMRI

The use of calibrated fMRI in block-design paradigms has advanced considerably in many different laboratories [for recent reviews in humans and animals, see Hoge (2012); Hyder et al. (2010)]. However the neural basis of ΔCMR_{O2} derived from calibrated fMRI remained unresolved until recently (see below). A problem encountered with dynamic calibrated fMRI for event-related paradigms is uncertainty about whether the conventional BOLD model can be applied transiently. With event-related paradigms, there is the issue of lack of a validation for the predicted transient energetics, whereas an advantage of steady-state calibrated fMRI is that calculated ΔCMR_{O2} can be validated by comparison with results from MRS or PET (see Section 3.3.1). A possible solution for both dynamic and steady-state calibrated fMRI is to test the linearity of the multimodal signals (i.e., BOLD, CBF, CBV) with short-lived and steady-state stimuli in relation to the neural response. If each BOLD-related component in Equation 3.2 is demonstrated to be linear across various stimulus durations, then the respective transfer functions generated by deconvolution with the neural signal should be linear and time-invariant and thus enable the possibility for calculating CMR_{O2} dynamics.

The Yale group has compared neural recordings of local field potential (LFP) and/or multiunit activity (MUA) with calibrated fMRI results in an animal model to show tight neurovascular and neurometabolic couplings with various stimulation paradigms in normal brain—that is, block-design (Maandag et al., 2007; Smith et al., 2002) and event-related paradigms (Herman et al., 2009a; Sanganahalli et al., 2009)—and various types of epileptic seizure models (Englot et al., 2008; Mishra et al., 2011; Schridde et al., 2008). The main results pertaining to comparison of dynamic changes in neural activity with

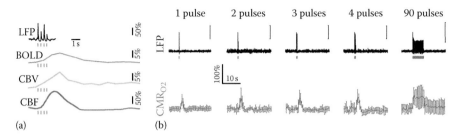

FIGURE 3.3
Summary of neural recordings with calibrated fMRI studies: (a) Single trial multimodal data of LFP, BOLD, CBV, and CBF recording during forepaw stimulation under α-chloralose anesthesia. (b) Measured LFP and calculated CMR_{O2} for short-lived and steady-state stimuli in α-chloralose anesthetized rats. (Modified from Herman, P., et al., *J Neurochem.* 109 Suppl 1, 73–9, 2009a; Sanganahalli, B.G., et al., *J Neurosci.* 29, 1707–18, 2009.)

CMR_{O2} dynamics are discussed below, where various stimulation paradigms in normal brain were examined by varying stimulus number and frequency within a narrow temporal regime.

Figure 3.3a shows a representative single-trial data set with four pulses of forepaw stimuli in the rat (Herman et al., 2009a). The evoked LFP response was immediate and short-lived in comparison to the imaging signals (i.e., BOLD, CBF, CBV), which lasted about 4 s for the 3 Hz forepaw stimuli. The neural response had two phases: a positive phase was initiated immediately after each stimulus pulse, which lasted about 150 ms; after the positive peak was a negative phase which lasted about 200 ms, but its amplitude was less than 5% of the initial positive peak which was below the standard deviation of the measurement. The dominant positive peaks, which represented more than 95% of the evoked neural response, were used as input signals for the convolution analysis. Evoked neural responses to multiple stimulus pulses demonstrated a unique pattern where the alternate responses were attenuated more with stimuli of higher frequencies (Sanganahalli et al., 2009). While MUA data are not shown, their behaviors were similar in amplitude and dynamics of the LFP response. Each imaging signal (i.e., BOLD, CBF, CBV) showed slightly different dynamic patterns, all of which was important for the determination of the transfer function for each modality.

Because each neural (i.e., LFP or MUA) and imaging (i.e., BOLD, CBF, CBV) signal is a function of varying amplitude over time, it was rationalized that signal strength would better capture the dynamic nuances of the evoked responses given that in the case of nonelectrical components both signal intensity and width varied, whereas for neural activity both the number and intensity of evoked signals differed. Negative parts of an evoked signal (i.e., below the baseline level) did not contribute significantly to the signal strength assessment. Although relationships between neural activity and stimulus features ranged from linear to nonlinear, associations between hyperemic components (i.e., BOLD, CBF, CBV) and neural activity (i.e., LFP or MUA) are linear. Given that all transfer functions were universal for all stimulus parameters examined, the linearity of the BOLD effect from event-related to block-design paradigms were confirmed. These results in the anesthetized rat showed that CMR_{O2} changes are correlated with LFP (and MUA) in the rat cerebral cortex, as shown in Figure 3.3b.

3.4 CMR$_{O2}$–CBF Coupling Underlying BOLD

The CMR_{O2}–CBF coupling underlying the BOLD contrast is a more dominant component than the CBV–CBF coupling, as discussed earlier in Section 3.3. Many levels of CMR_{O2}–CBF coupling have been reported for the BOLD response from oxygen delivery models—as small as 0.05 with nonlinear coupling (Buxton and Frank, 1997) to as large as 0.8 with pseudolinear coupling (Hyder et al., 2000a). Based on application of calibrated fMRI in a wide range of studies, across many different magnetic field strengths and several different types of species, it is possible to survey the CMR_{O2}–CBF couplings.

Figure 3.4 shows the latest comparison of CMR_{O2}–CBF couplings from calibrated fMRI studies, where the relationship ranges from as small as 0.08 at magnetic fields less than 1.5 T in humans (Fujita et al., 2006; Hoge et al., 1999; Kastrup et al., 1999, 2002; Kim et al., 1999; Lin et al., 2011; St Lawrence et al., 2003; Stefanovic et al., 2004) to as large as 0.70 at magnetic fields higher than 4.7 T in animals (Herman et al., 2009a; Hyder et al., 2000b; Jin and Kim, 2008; Kida et al., 2006; Kim et al., 2010; Lee et al., 2002; Maandag et al., 2007; Mandeville et al., 1999a; Sanganahalli et al., 2009; Silva et al., 1999, 2000; Smith et al., 2002; Wey et al., 2011; Wu et al., 2002). It would be tempting to conclude that this difference is due to magnetic field and/or species difference(s). However, in humans at magnetic field greater than 1.5 T but less than 4.0 T, the coupling is 0.36 (Ances et al., 2008, 2009; Bangen et al., 2009; Bulte et al., 2012; Chiarelli et al., 2007a, 2007b; Donahue et al., 2009; Feng et al., 2003; Gauthier et al., 2011; Goodwin et al., 2009; Leontiev et al., 2007; Lin et al., 2009; Lu et al., 2006; Mark et al., 2010; Mohtasib et al., 2012; Moradi et al., 2012; Pasley et al., 2007; Perthen et al., 2008; Restom et al., 2007, 2008; Stefanovic et al., 2006; Uludag et al., 2004), which is roughly 4.5 times greater than the coupling found at 1.5 T. Since the CMR_{O2}–CBF coupling is based on fundamental physiology, the different CMR_{O2}–CBF couplings found at higher fields in humans might only be explained on the basis of superior specificity and sensitivity for BOLD. But is there CMR_{O2}–CBF coupling difference between humans and animals? Although past meta-analysis of the CMR_{O2}–CBF couplings do not suggest species difference (Hyder et al., 1998), future human calibrated fMRI studies at higher magnetic fields are needed to assess if the improved BOLD specificity/sensitivity can lead to different CMR_{O2}–CBF couplings than currently observed at lower magnetic fields.

Another contentious issue is estimation of M, which is also related to CMR_{O2}–CBF coupling, but for resting values only. How variable is M across

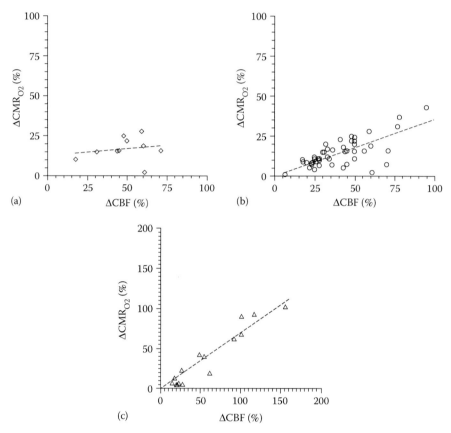

FIGURE 3.4
Summary of CMR_{O2}–CBF couplings with calibrated fMRI studies in animals and humans across a wide range of magnetic fields. (a) CMR_{O2}–CBF couplings for human studies at magnetic field less than 1.5 T (Adapted from Fujita et al., 2006; Hoge et al., 1999; Kastrup et al., 1999, 2002; Kim et al., 1999; Lin et al., 2011; St Lawrence et al., 2003; Stefanovic et al., 2004). (b) CMR_{O2}–CBF couplings for human studies at magnetic field greater than 1.5 T but less than 4.0 T (Adapted from Ances et al., 2008, 2009; Bulte et al., 2012; Chiarelli et al., 2007a, 2007b; Davis et al., 1998; Gauthier et al., 2011; Goodwin et al., 2009; Hoge et al., 1999; Kastrup et al., 1999, 2002; Kim and Ugurbil, 1997; Kim et al., 1999; Leontiev et al., 2007; Lin et al., 2009; Mark et al., 2010; Mohtasib et al., 2012; Moradi et al., 2012; Perthen et al., 2008; Restom et al., 2007, 2008; St Lawrence et al., 2003; Stefanovic et al., 2004, 2006; Uludag et al., 2004). (c) CMR_{O2}–CBF couplings for animal studies at magnetic field higher than 4.7 T (Adapted from Herman et al., 2009a; Hyder et al., 2000b; Jin and Kim, 2008; Kida et al., 2006; Kim et al., 2010; Lee et al., 2002; Maandag et al., 2007; Mandeville et al., 1999a; Sanganahalli et al., 2009; Silva et al., 1999, 2000; Smith et al., 2002; Wey et al., 2011; Wu et al., 2002). The slopes of the data shown are 0.08, 0.36, and 0.70, respectively.

magnetic fields, regions, and/or species? The issue of species differences may be difficult to answer given that quite different approaches are used to derive M in rats and humans, as discussed in Section 3.3.1. Figure 3.5 shows the latest survey of M values from calibrated fMRI studies in humans for various fields and regions (Ances et al., 2008, 2009; Bulte et al., 2012; Chiarelli

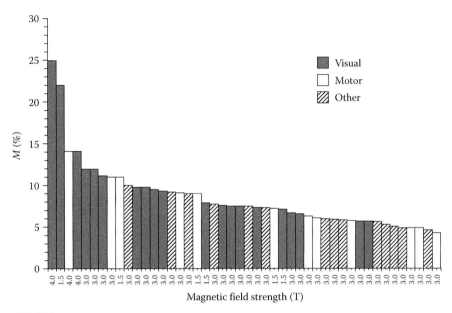

FIGURE 3.5
Summary of M values with calibrated fMRI studies in humans across a wide range of magnetic fields and across various brain regions (Adapted from Ances et al., 2008, 2009; Bulte et al., 2012; Chiarelli et al., 2007a, 2007b; Davis et al., 1998; Gauthier et al., 2011; Goodwin et al., 2009; Hoge et al., 1999; Kastrup et al., 1999, 2002; Kim and Ugurbil, 1997; Kim et al., 1999; Leontiev et al., 2007; Lin et al., 2009; Mark et al., 2010; Mohtasib et al., 2012; Moradi et al., 2012; Perthen et al., 2008; Restom et al., 2007, 2008; St Lawrence et al., 2003; Stefanovic et al., 2004, 2006; Uludag et al., 2004).

et al., 2007a, 2007b; Davis et al., 1998; Gauthier et al., 2011; Goodwin et al., 2009; Hoge et al., 1999; Kastrup et al., 1999, 2002; Kim and Ugurbil, 1997; Kim et al., 1999; Leontiev et al., 2007; Lin et al., 2009; Mark et al., 2010; Mohtasib et al., 2012; Moradi et al., 2012; Perthen et al., 2008; Restom et al., 2007, 2008; St Lawrence et al., 2003; Stefanovic et al., 2004, 2006; Uludag et al., 2004). While the analysis does not reveal any regional or field dependencies, the range of M values is quite significant and worthy of some discussion.

At 4.0 T, the M values between motor and visual areas differ by nearly a factor of 2; whereas at 3.0 T, the M values in those regions for some studies are quite similar to each other. However, there are other 3.0 T studies that vary by a factor of 2 for those same regions, indicating experimental variations across laboratories and sometimes for studies even from the same laboratory. Could these differences be due to actual differences across regions and/or subjects for resting values of CMR$_{O2}$ and/or CBF? Does the effect of variable static magnetic field shimming from subject to subject contribute to the variation of M? Further studies are needed to unify the M measurement, and because this parameter is also dependent on the TE value used in the BOLD experiment, some standardization is needed for the reporting of M.

3.5 Concluding Remarks

Overall, the calibrated fMRI studies in animal models quantitatively explained the presence of large CMR_{O2} changes observed during sensory stimulation with a positive BOLD signal (Hyder et al., 1996, 1997) with concomitant large changes in CBF (Hyder et al., 2000a; Silva et al., 1999). Notably, both in human and animal studies, the measured $\Delta CBF/CBF$ and calculated $\Delta CMR_{O2}/CMR_{O2}$ from calibrated fMRI studies (i.e., the CMR_{O2}–CBF coupling) show a linear trend suggesting rapid oxygen equilibration between blood and tissue pools within the physiological range (Hyder et al., 1998). In support of this hypothesis, independent studies from the Universities of Pennsylvania and Pittsburgh show that tissue pO_2 dynamics are just as fast as the CBF transients during functional activation (Ances et al., 2001; Vazquez et al., 2008). Together, these results support that calibrated fMRI at high magnetic fields can provide high spatio-temporal mapping of CMR_{O2} changes. In summary, the underlying theme in these calibrated fMRI studies reviewed, across animals and humans, is that functional brain imaging studies can examine neurometabolic and neuro-vascular components, both in health and disease. Further improvements in calibrated fMRI will require continued developments and innovations in multimodal neuroimaging, especially at high magnetic field strengths where animal studies will continue to play a significant role.

Acknowledgments

The authors thank their colleagues at Yale University for their insightful comments. This work was supported by National Institutes of Health Grants (R01 MH-067528 to FH, P30 NS-052519 to FH, and R01 AG-034953 to DLR).

References

Aiello, L.C., Wheeler, P., 1995. The expensive-tissue hypothesis—The brain and the digestive system in human and primate evolution. *Curr Anthropol.* 36, 199–221.

Ances, B.M., Buerk, D.G., Greenberg, J.H., Detre, J.A., 2001. Temporal dynamics of the partial pressure of brain tissue oxygen during functional forepaw stimulation in rats. *Neurosci Lett.* 306, 106–10.

Ances, B.M., Leontiev, O., Perthen, J.E., Liang, C., Lansing, A.E., Buxton, R.B., 2008. Regional differences in the coupling of cerebral blood flow and oxygen metabolism changes in response to activation: Implications for BOLD-fMRI. *Neuroimage.* 39, 1510–21.

Ances, B.M., Liang, C.L., Leontiev, O., Perthen, J.E., Fleisher, A.S., Lansing, A.E., Buxton, R.B., 2009. Effects of aging on cerebral blood flow, oxygen metabolism, and blood oxygenation level dependent responses to visual stimulation. *Hum Brain Mapp.* 30, 1120–32.

Bandettini, P.A., Wong, E.C., Hinks, R.S., Tikofsky, R.S., Hyde, J.S., 1992. Time course EPI of human brain function during task activation. *Magn Reson Med.* 25, 390–7.

Bangen, K.J., Restom, K., Liu, T.T., Jak, A.J., Wierenga, C.E., Salmon, D.P., Bondi, M.W., 2009. Differential age effects on cerebral blood flow and BOLD response to encoding: Associations with cognition and stroke risk. *Neurobiol Aging.* 30, 1276–87.

Barinaga, M., 1997. What makes brain neurons run? *Science.* 276, 196–8.

Behar, K.L., Petroff, O.A., Prichard, J.W., Alger, J.R., Shulman, R.G., 1986. Detection of metabolites in rabbit brain by 13C NMR spectroscopy following administration of [1-13C]glucose. *Magn Reson Med.* 3, 911–20.

Belliveau, J.W., Rosen, B.R., Kantor, H.L., Rzedzian, R.R., Kennedy, D.N., McKinstry, R.C., Vevea, J.M., Cohen, M.S., Pykett, I.L., Brady, T.J., 1990. Functional cerebral imaging by susceptibility-contrast NMR. *Magn Reson Med.* 14, 538–46.

Blamire, A.M., Ogawa, S., Ugurbil, K., Rothman, D., McCarthy, G., Ellermann, J.M., Hyder, F., Rattner, Z., Shulman, R.G., 1992. Dynamic mapping of the human visual cortex by high-speed magnetic resonance imaging. *Proc Natl Acad Sci USA.* 89, 11069–73.

Blockley, N.P., Griffeth, V.E., Buxton, R.B., 2012. A general analysis of calibrated BOLD methodology for measuring CMRO(2) responses: Comparison of a new approach with existing methods. *Neuroimage.* 60, 279–89.

Boumezbeur, F., Besret, L., Valette, J., Gregoire, M.C., Delzescaux, T., Maroy, R., Vaufrey, F., et al., 2005. Glycolysis versus TCA cycle in the primate brain as measured by combining 18F-FDG PET and 13C-NMR. *J Cereb Blood Flow Metab.* 25, 1418–23.

Boxerman, J.L., Hamberg, L.M., Rosen, B.R., Weisskoff, R.M., 1995. MR contrast due to intravascular magnetic susceptibility perturbations. *Magn Reson Med.* 34, 555–66.

Bulte, D.P., Kelly, M., Germuska, M., Xie, J., Chappell, M.A., Okell, T.W., Bright, M.G., Jezzard, P., 2012. Quantitative measurement of cerebral physiology using respiratory-calibrated MRI. *Neuroimage.* 60, 582–91.

Buxton, R.B., Frank, L.R., 1997. A model for the coupling between cerebral blood flow and oxygen metabolism during neural stimulation. *J Cereb Blood Flow Metab.* 17, 64–72.

Cerdan, S., Kunnecke, B., Seelig, J., 1990. Cerebral metabolism of [1,2-13C2]acetate as detected by in vivo and in vitro 13C NMR. *J Biol Chem.* 265, 12916–26.

Chen, J.J., Pike, G.B., 2009. BOLD-specific cerebral blood volume and blood flow changes during neuronal activation in humans. *NMR Biomed.* 22, 1054–62.

Chen, J.J., Pike, G.B., 2010. Global cerebral oxidative metabolism during hypercapnia and hypocapnia in humans: Implications for BOLD fMRI. *J Cereb Blood Flow Metab.* 30, 1094–9.

Chen, W., Novotny, E.J., Zhu, X.H., Rothman, D.L., Shulman, R.G., 1993. Localized 1H NMR measurement of glucose consumption in the human brain during visual stimulation. *Proc Natl Acad Sci USA.* 90, 9896–900.

Chen, W., Zhu, X.H., Gruetter, R., Seaquist, E.R., Adriany, G., Ugurbil, K., 2001. Study of tricarboxylic acid cycle flux changes in human visual cortex during hemifield visual stimulation using (1)H-[(13)C] MRS and fMRI. *Magn Reson Med.* 45, 349–55.

Chhina, N., Kuestermann, E., Halliday, J., Simpson, L.J., Macdonald, I.A., Bachelard, H.S., Morris, P.G., 2001. Measurement of human tricarboxylic acid cycle rates during visual activation by (13)C magnetic resonance spectroscopy. *J Neurosci Res.* 66, 737–46.

Chiarelli, P.A., Bulte, D.P., Gallichan, D., Piechnik, S.K., Wise, R., Jezzard, P., 2007a. Flow-metabolism coupling in human visual, motor, and supplementary motor areas assessed by magnetic resonance imaging. *Magn Reson Med.* 57, 538–47.

Chiarelli, P.A., Bulte, D.P., Wise, R., Gallichan, D., Jezzard, P., 2007b. A calibration method for quantitative BOLD fMRI based on hyperoxia. *Neuroimage.* 37, 808–20.

Chih, C.P., Lipton, P., Roberts, E.L., Jr., 2001. Do active cerebral neurons really use lactate rather than glucose? *Trends Neurosci.* 24, 573–8.

Creutzfeldt, O., 1975. Neurophysiological correlates of different functional states of the brain. In: *Brain Work. The Coupling of Function, Metabolism and Blood Flow in the Brain.* Alfred Benzon Symposium VIII. D. Ingvar, N. Lassen, eds. Academic Press, New York, pp. 21–46.

Davis, T.L., Kwong, K.K., Weisskoff, R.M., Rosen, B.R., 1998. Calibrated functional MRI: Mapping the dynamics of oxidative metabolism. *Proc Natl Acad Sci USA.* 95, 1834–9.

de Graaf, R.A., Brown, P.B., McIntyre, S., Nixon, T.W., Behar, K.L., Rothman, D.L., 2006. High magnetic field water and metabolite proton T1 and T2 relaxation in rat brain in vivo. *Magn Reson Med.* 56, 386–94.

de Graaf, R.A., Rothman, D.L., Behar, K.L., 2011. State of the art direct 13C and indirect 1H-[13C] NMR spectroscopy in vivo. A practical guide. *NMR Biomed.* 24, 958–72.

Detre, J.A., Leigh, J.S., Williams, D.S., Koretsky, A.P., 1992. Perfusion imaging. *Magn Reson Med.* 23, 37–45.

Donahue, M.J., Hoogduin, H., van Zijl, P.C., Jezzard, P., Luijten, P.R., Hendrikse, J., 2011. Blood oxygenation level-dependent (BOLD) total and extravascular signal changes and ΔR2* in human visual cortex at 1.5, 3.0 and 7.0 T. *NMR Biomed.* 24, 25–34.

Donahue, M.J., Stevens, R.D., de Boorder, M., Pekar, J.J., Hendrikse, J., van Zijl, P.C., 2009. Hemodynamic changes after visual stimulation and breath holding provide evidence for an uncoupling of cerebral blood flow and volume from oxygen metabolism. *J Cereb Blood Flow Metab.* 29, 176–85.

Dutka, M.V., Scanley, B.E., Does, M.D., Gore, J.C., 2002. Changes in CBF–BOLD coupling detected by MRI during and after repeated transient hypercapnia in rat. *Magn Reson Med.* 48, 262–70.

Englot, D.J., Mishra, A.M., Mansuripur, P.K., Herman, P., Hyder, F., Blumenfeld, H., 2008. Remote effects of focal hippocampal seizures on the rat neocortex. *J Neurosci.* 28, 9066–81.

Feng, C.M., Liu, H.L., Fox, P.T., Gao, J.H., 2003. Dynamic changes in the cerebral metabolic rate of O2 and oxygen extraction ratio in event-related functional MRI. *Neuroimage.* 18, 257–62.

Fitzpatrick, S.M., Hetherington, H.P., Behar, K.L., Shulman, R.G., 1990. The flux from glucose to glutamate in the rat brain in vivo as determined by 1H-observed, 13C-edited NMR spectroscopy. *J Cereb Blood Flow Metab.* 10, 170–9.

Fitzpatrick, S.M., Rothman, D., 1999. New approaches to functional neuroenergetics. *J Cogn Neurosci.* 11, 467–71.

Fox, P.T., Raichle, M.E., 1986. Focal physiological uncoupling of cerebral blood flow and oxidative metabolism during somatosensory stimulation in human subjects. *Proc Natl Acad Sci USA.* 83, 1140–4.

Fox, P.T., Raichle, M.E., Mintun, M.A., Dence, C., 1988. Nonoxidative glucose consumption during focal physiologic neural activity. *Science*. 241, 462–4.

Frahm, J., Bruhn, H., Merboldt, K.D., Hanicke, W., 1992. Dynamic MR imaging of human brain oxygenation during rest and photic stimulation. *J Magn Reson Imaging*. 2, 501–5.

Frahm, J., Kruger, G., Merboldt, K.D., Kleinschmidt, A., 1996. Dynamic uncoupling and recoupling of perfusion and oxidative metabolism during focal brain activation in man. *Magn Reson Med*. 35, 143–8.

Fujita, N., Matsumoto, K., Tanaka, H., Watanabe, Y., Murase, K., 2006. Quantitative study of changes in oxidative metabolism during visual stimulation using absolute relaxation rates. *NMR Biomed*. 19, 60–8.

Gauthier, C.J., Hoge, R.D., 2012. Magnetic resonance imaging of resting OEF and CMRO using a generalized calibration model for hypercapnia and hyperoxia. *Neuroimage*. 60, 1212–25.

Gauthier, C.J., Madjar, C., Tancredi, F.B., Stefanovic, B., Hoge, R.D., 2011. Elimination of visually evoked BOLD responses during carbogen inhalation: Implications for calibrated MRI. *Neuroimage*. 54, 1001–11.

Goodwin, J.A., Vidyasagar, R., Balanos, G.M., Bulte, D., Parkes, L.M., 2009. Quantitative fMRI using hyperoxia calibration: Reproducibility during a cognitive Stroop task. *Neuroimage*. 47, 573–80.

Grubb, R.L., Jr., Raichle, M.E., Eichling, J.O., Ter-Pogossian, M.M., 1974. The effects of changes in PaCO2 on cerebral blood volume, blood flow, and vascular mean transit time. *Stroke*. 5, 630–9.

Gruetter, R., Novotny, E.J., Boulware, S.D., Rothman, D.L., Shulman, R.G., 1996. 1H NMR studies of glucose transport in the human brain. *J Cereb Blood Flow Metab*. 16, 427–38.

Gruetter, R., Rothman, D.L., Novotny, E.J., Shulman, R.G., 1992. Localized 13C NMR spectroscopy of myo-inositol in the human brain in vivo. *Magn Reson Med*. 25, 204–10.

Hall, E.L., Driver, I.D., Croal, P.L., Francis, S.T., Gowland, P.A., Morris, P.G., Brookes, M.J., 2011. The effect of hypercapnia on resting and stimulus induced MEG signals. *Neuroimage*. 58, 1034–43.

Herman, P., Sanganahalli, B.G., Blumenfeld, H., Hyder, F., 2009a. Cerebral oxygen demand for short-lived and steady-state events. *J Neurochem*. 109 Suppl 1, 73–9.

Herman, P., Sanganahalli, B.G., Hyder, F., 2009b. Multimodal measurements of blood plasma and red blood cell volumes during functional brain activation. *J Cereb Blood Flow Metab*. 29, 19–24.

Hoge, R.D., 2012. Calibrated fMRI. *Neuroimage*. 62, 930–7.

Hoge, R.D., Atkinson, J., Gill, B., Crelier, G.R., Marrett, S., Pike, G.B., 1999. Linear coupling between cerebral blood flow and oxygen consumption in activated human cortex. *Proc Natl Acad Sci USA*. 96, 9403–8.

Hyder, F., Chase, J.R., Behar, K.L., Mason, G.F., Siddeek, M., Rothman, D.L., Shulman, R.G., 1996. Increased tricarboxylic acid cycle flux in rat brain during forepaw stimulation detected with ^{1}H[^{13}C]NMR. *Proc Natl Acad Sci USA*. 93, 7612–7.

Hyder, F., Kennan, R.P., Kida, I., Mason, G.F., Behar, K.L., Rothman, D., 2000a. Dependence of oxygen delivery on blood flow in rat brain: A 7 Tesla nuclear magnetic resonance study. *J Cereb Blood Flow Metab*. 20, 485–98.

Hyder, F., Kida, I., Behar, K.L., Kennan, R.P., Maciejewski, P.K., Rothman, D.L., 2001. Quantitative functional imaging of the brain: Towards mapping neuronal activity by BOLD fMRI. *NMR Biomed*. 14, 413–31.

Hyder, F., Patel, A.B., Gjedde, A., Rothman, D.L., Behar, K.L., Shulman, R.G., 2006. Neuronal-glial glucose oxidation and glutamatergic-GABAergic function. *J Cereb Blood Flow Metab*. 26, 865–77.

Hyder, F., Renken, R., Kennan, R.P., Rothman, D.L., 2000b. Quantitative multi-modal functional MRI with blood oxygenation level dependent exponential decays adjusted for flow attenuated inversion recovery (BOLDED AFFAIR). *Magn Reson Imaging*. 18, 227–35.

Hyder, F., Rothman, D.L., 2012. Quantitative fMRI and oxidative neuroenergetics. *Neuroimage*. 62, 985–94.

Hyder, F., Rothman, D.L., Mason, G.F., Rangarajan, A., Behar, K.L., Shulman, R.G., 1997. Oxidative glucose metabolism in rat brain during single forepaw stimulation: A spatially localized 1H[13C] nuclear magnetic resonance study. *J Cereb Blood Flow Metab*. 17, 1040–7.

Hyder, F., Sanganahalli, B.G., Herman, P., Coman, D., Maandag, N.J., Behar, K.L., Blumenfeld, H., Rothman, D.L., 2010. Neurovascular and neurometabolic couplings in dynamic calibrated fMRI: Transient oxidative neuroenergetics for block-design and event-related paradigms. *Front Neuroenergetics*. 2, 18.

Hyder, F., Shulman, R.G., Rothman, D.L., 1998. A model for the regulation of cerebral oxygen delivery. *J Appl Physiol*. 85, 554–64.

Jin, T., Kim, S.G., 2008. Cortical layer-dependent dynamic blood oxygenation, cerebral blood flow and cerebral blood volume responses during visual stimulation. *Neuroimage*. 43, 1–9.

Jones, M., Berwick, J., Hewson-Stoate, N., Gias, C., Mayhew, J., 2005. The effect of hypercapnia on the neural and hemodynamic responses to somatosensory stimulation. *Neuroimage*. 27, 609–23.

Kastrup, A., Kruger, G., Glover, G.H., Moseley, M.E., 1999. Assessment of cerebral oxidative metabolism with breath holding and fMRI. *Magn Reson Med*. 42, 608–11.

Kastrup, A., Kruger, G., Neumann-Haefelin, T., Glover, G.H., Moseley, M.E., 2002. Changes of cerebral blood flow, oxygenation, and oxidative metabolism during graded motor activation. *Neuroimage*. 15, 74–82.

Kennan, R.P., Scanley, B.E., Innis, R.B., Gore, J.C., 1998. Physiological basis for BOLD MR signal changes due to neuronal stimulation: Separation of blood volume and magnetic susceptibility effects. *Magn Reson Med*. 40, 840–6.

Kennan, R.P., Zhong, J., Gore, J.C., 1994. Intravascular susceptibility contrast mechanisms in tissues. *Magn Reson Med*. 31, 9–21.

Kida, I., Hyder, F., Kennan, R.P., Behar, K.L., 1999. Toward absolute quantitation of bold functional MRI. *Adv Exp Med Biol*. 471, 681–9.

Kida, I., Kennan, R.P., Rothman, D.L., Behar, K.L., Hyder, F., 2000. High-resolution $CMR(O_2)$ mapping in rat cortex: A multiparametric approach to calibration of BOLD image contrast at 7 Tesla. *J Cereb Blood Flow Metab*. 20, 847–60.

Kida, I., Rothman, D.L., Hyder, F., 2007. Dynamics of changes in blood flow, volume, and oxygenation: Implications for dynamic functional magnetic resonance imaging calibration. *J Cereb Blood Flow Metab*. 27, 690–6.

Kida, I., Smith, A.J., Blumenfeld, H., Behar, K.L., Hyder, F., 2006. Lamotrigine suppresses neurophysiological responses to somatosensory stimulation in the rodent. *Neuroimage*. 29, 216–24.

Kim, S.G., 1995. Quantification of relative cerebral blood flow change by flow-sensitive alternating inversion recovery (FAIR) technique: Application to functional mapping. *Magn Reson Med*. 34, 293–301.

Kim, S.G., Rostrup, E., Larsson, H.B., Ogawa, S., Paulson, O.B., 1999. Determination of relative CMRO2 from CBF and BOLD changes: Significant increase of oxygen consumption rate during visual stimulation. *Magn Reson Med*. 41, 1152–61.

Kim, S.G., Ugurbil, K., 1997. Comparison of blood oxygenation and cerebral blood flow effects in fMRI: Estimation of relative oxygen consumption change. *Magn Reson Med*. 38, 59–65.

Kim, T., Hendrich, K.S., Masamoto, K., Kim, S.G., 2007. Arterial versus total blood volume changes during neural activity-induced cerebral blood flow change: Implication for BOLD fMRI. *J Cereb Blood Flow Metab*. 27, 1235–47.

Kim, T., Kim, S.G., 2005. Quantification of cerebral arterial blood volume and cerebral blood flow using MRI with modulation of tissue and vessel (MOTIVE) signals. *Magn Reson Med*. 54, 333–42.

Kim, T., Masamoto, K., Fukuda, M., Vazquez, A., Kim, S.G., 2010. Frequency-dependent neural activity, CBF, and BOLD fMRI to somatosensory stimuli in isoflurane-anesthetized rats. *Neuroimage*. 52, 224–33.

Kwong, K.K., Belliveau, J.W., Chesler, D.A., Goldberg, I.E., Weisskoff, R.M., Poncelet, B.P., Kennedy, D.N., et al., 1992. Dynamic magnetic resonance imaging of human brain activity during primary sensory stimulation. *Proc Natl Acad Sci USA*. 89, 5675–9.

Lee, S.P., Silva, A.C., Kim, S.G., 2002. Comparison of diffusion-weighted high-resolution CBF and spin-echo BOLD fMRI at 9.4 T. *Magn Reson Med*. 47, 736–41.

Leontiev, O., Dubowitz, D.J., Buxton, R.B., 2007. CBF/CMRO$_2$ coupling measured with calibrated BOLD fMRI: Sources of bias. *Neuroimage*. 36, 1110–22.

Lin, A.L., Fox, P.T., Yang, Y., Lu, H., Tan, L.H., Gao, J.H., 2009. Time-dependent correlation of cerebral blood flow with oxygen metabolism in activated human visual cortex as measured by fMRI. *Neuroimage*. 44, 16–22.

Lin, P., Hasson, U., Jovicich, J., Robinson, S., 2011. A neuronal basis for task-negative responses in the human brain. *Cereb Cortex*. 21, 821–30.

Liu, Y.J., Juan, C.J., Chen, C.Y., Wang, C.Y., Wu, M.L., Lo, C.P., Chou, M.C., et al., 2007. Are the local blood oxygen level-dependent (BOLD) signals caused by neural stimulation response dependent on global BOLD signals induced by hypercapnia in the functional MR imaging experiment? Experiments of long-duration hypercapnia and multilevel carbon dioxide concentration. *AJNR Am J Neuroradiol*. 28, 1009–14.

Lu, H., Donahue, M.J., van Zijl, P.C., 2006. Detrimental effects of BOLD signal in arterial spin labeling fMRI at high field strength. *Magn Reson Med*. 56, 546–52.

Lu, H., Golay, X., Pekar, J.J., Van Zijl, P.C., 2003. Functional magnetic resonance imaging based on changes in vascular space occupancy. *Magn Reson Med*. 50, 263–74.

Maandag, N.J., Coman, D., Sanganahalli, B.G., Herman, P., Smith, A.J., Blumenfeld, H., Shulman, R.G., Hyder, F., 2007. Energetics of neuronal signaling and fMRI activity. *Proc Natl Acad Sci USA*. 104, 20546–51.

Magistretti, P.J., Pellerin, L., Rothman, D.L., Shulman, R.G., 1999. Energy on demand. *Science*. 283, 496–7.

Mandeville, J.B., Marota, J.J., Ayata, C., Moskowitz, M.A., Weisskoff, R.M., Rosen, B.R., 1999a. MRI measurement of the temporal evolution of relative CMRO(2) during rat forepaw stimulation. *Magn Reson Med.* 42, 944–51.

Mandeville, J.B., Marota, J.J., Ayata, C., Zaharchuk, G., Moskowitz, M.A., Rosen, B.R., Weisskoff, R.M., 1999b. Evidence of a cerebrovascular postarteriole windkessel with delayed compliance. *J Cereb Blood Flow Metab.* 19, 679–89.

Mandeville, J.B., Marota, J.J., Kosofsky, B.E., Keltner, J.R., Weissleder, R., Rosen, B.R., Weisskoff, R.M., 1998. Dynamic functional imaging of relative cerebral blood volume during rat forepaw stimulation. *Magn Reson Med.* 39, 615–24.

Mangia, S., Tkac, I., Gruetter, R., Van de Moortele, P.F., Maraviglia, B., Ugurbil, K., 2007. Sustained neuronal activation raises oxidative metabolism to a new steady-state level: Evidence from 1H NMR spectroscopy in the human visual cortex. *J Cereb Blood Flow Metab.* 27, 1055–63.

Mark, C.I., Fisher, J.A., Pike, G.B., 2010. Improved fMRI calibration: Precisely controlled hyperoxic versus hypercapnic stimuli. *Neuroimage.* 54, 1102–11.

Martin, C., Jones, M., Martindale, J., Mayhew, J., 2006. Haemodynamic and neural responses to hypercapnia in the awake rat. *Eur J Neurosci.* 24, 2601–10.

Mason, G.F., Gruetter, R., Rothman, D.L., Behar, K.L., Shulman, R.G., Novotny, E.J., 1995. Simultaneous determination of the rates of the TCA cycle, glucose utilization, alpha-ketoglutarate/glutamate exchange, and glutamine synthesis in human brain by NMR. *J Cereb Blood Flow Metab.* 15, 12–25.

Mason, G.F., Rothman, D.L., Behar, K.L., Shulman, R.G., 1992. NMR determination of the TCA cycle rate and alpha-ketoglutarate/glutamate exchange rate in rat brain. *J Cereb Blood Flow Metab.* 12, 434–47.

Merboldt, K.D., Bruhn, H., Hanicke, W., Michaelis, T., Frahm, J., 1992. Decrease of glucose in the human visual cortex during photic stimulation. *Magn Reson Med.* 25, 187–94.

Mintun, M.A., Vlassenko, A.G., Shulman, G.L., Snyder, A.Z., 2002. Time-related increase of oxygen utilization in continuously activated human visual cortex. *Neuroimage.* 16, 531–7.

Mishra, A.M., Ellens, D.J., Schridde, U., Motelow, J.E., Purcaro, M.J., DeSalvo, M.N., Enev, M., Sanganahalli, B.G., Hyder, F., Blumenfeld, H., 2011. Where fMRI and electrophysiology agree to disagree: Corticothalamic and striatal activity patterns in the WAG/Rij rat. *J Neurosci.* 31, 15053–64.

Mohtasib, R.S., Lumley, G., Goodwin, J.A., Emsley, H.C., Sluming, V., Parkes, L.M., 2012. Calibrated fMRI during a cognitive Stroop task reveals reduced metabolic response with increasing age. *Neuroimage.* 59, 1143–51.

Moradi, F., Buracas, G.T., Buxton, R.B., 2012. Attention strongly increases oxygen metabolic response to stimulus in primary visual cortex. *Neuroimage.* 59, 601–7.

Ogawa, S., Lee, T.M., Kay, A.R., Tank, D.W., 1990. Brain magnetic resonance imaging with contrast dependent on blood oxygenation. *Proc Natl Acad Sci USA.* 87, 9868–72.

Ogawa, S., Menon, R.S., Tank, D.W., Kim, S.G., Merkle, H., Ellermann, J.M., Ugurbil, K., 1993. Functional brain mapping by blood oxygenation level-dependent contrast magnetic resonance imaging. A comparison of signal characteristics with a biophysical model. *Biophys J.* 64, 803–12.

Ogawa, S., Tank, D.W., Menon, R., Ellermann, J.M., Kim, S.G., Merkle, H., Ugurbil, K., 1992. Intrinsic signal changes accompanying sensory stimulation: Functional brain mapping with magnetic resonance imaging. *Proc Natl Acad Sci USA.* 89, 5951–5.

Pasley, B.N., Inglis, B.A., Freeman, R.D., 2007. Analysis of oxygen metabolism implies a neural origin for the negative BOLD response in human visual cortex. *Neuroimage.* 36, 269–76.

Perthen, J.E., Lansing, A.E., Liau, J., Liu, T.T., Buxton, R.B., 2008. Caffeine-induced uncoupling of cerebral blood flow and oxygen metabolism: A calibrated BOLD fMRI study. *Neuroimage.* 40, 237–47.

Petroff, O.A., Prichard, J.W., Ogino, T., Avison, M., Alger, J.R., Shulman, R.G., 1986. Combined 1H and 31P nuclear magnetic resonance spectroscopic studies of bicuculline-induced seizures in vivo. *Ann Neurol.* 20, 185–93.

Prichard, J., Rothman, D., Novotny, E., Petroff, O., Kuwabara, T., Avison, M., Howseman, A., Hanstock, C., Shulman, R., 1991. Lactate rise detected by 1H NMR in human visual cortex during physiologic stimulation. *Proc Natl Acad Sci USA.* 88, 5829–31.

Raichle, M.E., 2009. A brief history of human brain mapping. *Trends Neurosci.* 32, 118–26.

Restom, K., Bangen, K.J., Bondi, M.W., Perthen, J.E., Liu, T.T., 2007. Cerebral blood flow and BOLD responses to a memory encoding task: A comparison between healthy young and elderly adults. *Neuroimage.* 37, 430–9.

Restom, K., Perthen, J.E., Liu, T.T., 2008. Calibrated fMRI in the medial temporal lobe during a memory-encoding task. *Neuroimage.* 40, 1495–502.

Roland, P.E., Eriksson, L., Stone-Elander, S., Widen, L., 1987. Does mental activity change the oxidative metabolism of the brain? *J Neurosci.* 7, 2373–89.

Rothman, D.L., Behar, K.L., Hetherington, H.P., den Hollander, J.A., Bendall, M.R., Petroff, O.A., Shulman, R.G., 1985. 1H-Observe/13C-decouple spectroscopic measurements of lactate and glutamate in the rat brain in vivo. *Proc Natl Acad Sci USA.* 82, 1633–7.

Rothman, D.L., De Feyter, H.M., de Graaf, R.A., Mason, G.F., Behar, K.L., 2011. 13C MRS studies of neuroenergetics and neurotransmitter cycling in humans. *NMR Biomed.* 24, 943–57.

Rothman, D.L., Novotny, E.J., Shulman, G.I., Howseman, A.M., Petroff, O.A., Mason, G., Nixon, T., Hanstock, C.C., Prichard, J.W., Shulman, R.G., 1992. 1H-[13C] NMR measurements of [4-13C]glutamate turnover in human brain. *Proc Natl Acad Sci USA.* 89, 9603–6.

Sanganahalli, B.G., Herman, P., Blumenfeld, H., Hyder, F., 2009. Oxidative neuroenergetics in event-related paradigms. *J Neurosci.* 29, 1707–18.

Sappey-Marinier, D., Calabrese, G., Fein, G., Hugg, J.W., Biggins, C., Weiner, M.W., 1992. Effect of photic stimulation on human visual cortex lactate and phosphates using 1H and 31P magnetic resonance spectroscopy. *J Cereb Blood Flow Metab.* 12, 584–92.

Schridde, U., Khubchandani, M., Motelow, J.E., Sanganahalli, B.G., Hyder, F., Blumenfeld, H., 2008. Negative BOLD with large increases in neuronal activity. *Cereb Cortex.* 18, 1814–27.

Seitz, R.J., Roland, P.E., 1992. Vibratory stimulation increases and decreases the regional cerebral blood flow and oxidative metabolism: A positron emission tomography (PET) study. *Acta Neurol Scand.* 86, 60–7.

Shen, Q., Ren, H., Duong, T.Q., 2008. CBF, BOLD, CBV, and CMRO$_2$ fMRI signal temporal dynamics at 500-msec resolution. *J Magn Reson Imaging.* 27, 599–606.

Shulman, R.G., Hyder, F., Rothman, D.L., 2001a. Lactate efflux and the neuroenergetic basis of brain function. *NMR Biomed.* 14, 389–96.

Shulman, R.G., Hyder, F., Rothman, D.L., 2001b. Cerebral energetics and the glycogen shunt: Neurochemical basis of functional imaging. *Proc Natl Acad Sci USA*. 98, 6417–22.

Siesjo, B.K., 1978. *Brain Energy Metabolism*. Wiley and Sons, New York.

Silva, A.C., Lee, S.P., Iadecola, C., Kim, S.G., 2000. Early temporal characteristics of cerebral blood flow and deoxyhemoglobin changes during somatosensory stimulation. *J Cereb Blood Flow Metab*. 20, 201–6.

Silva, A.C., Lee, S.P., Yang, G., Iadecola, C., Kim, S.G., 1999. Simultaneous blood oxygenation level-dependent and cerebral blood flow functional magnetic resonance imaging during forepaw stimulation in the rat. *J Cereb Blood Flow Metab*. 19, 871–9.

Smith, A.J., Blumenfeld, H., Behar, K.L., Rothman, D.L., Shulman, R.G., Hyder, F., 2002. Cerebral energetics and spiking frequency: The neurophysiological basis of fMRI. *Proc Natl Acad Sci USA*. 99, 10765–70.

Sokoloff, L., 1991. Relationship between functional activity and energy metabolism in the nervous system: Whether, where and why? In: *Brain Work and Mental Activity*. N.A. Lassen, D.H. Ingvar, M.E. Raichle, L. Friberg, eds. Munksgaard, Copenhagen, pp. 52–64.

Song, A.W., Wong, E.C., Tan, S.G., Hyde, J.S., 1996. Diffusion weighted fMRI at 1.5 T. *Magn Reson Med*. 35, 155–8.

St Lawrence, K.S., Ye, F.Q., Lewis, B.K., Frank, J.A., McLaughlin, A.C., 2003. Measuring the effects of indomethacin on changes in cerebral oxidative metabolism and cerebral blood flow during sensorimotor activation. *Magn Reson Med*. 50, 99–106.

Stefanovic, B., Pike, G.B., 2004. Human whole-blood relaxometry at 1.5 T: Assessment of diffusion and exchange models. *Magn Reson Med*. 52, 716–23.

Stefanovic, B., Warnking, J.M., Pike, G.B., 2004. Hemodynamic and metabolic responses to neuronal inhibition. *Neuroimage*. 22, 771–8.

Stefanovic, B., Warnking, J.M., Rylander, K.M., Pike, G.B., 2006. The effect of global cerebral vasodilation on focal activation hemodynamics. *Neuroimage*. 30, 726–34.

Uludag, K., Dubowitz, D.J., Yoder, E.J., Restom, K., Liu, T.T., Buxton, R.B., 2004. Coupling of cerebral blood flow and oxygen consumption during physiological activation and deactivation measured with fMRI. *Neuroimage*. 23, 148–55.

Vafaee, M.S., Marrett, S., Meyer, E., Evans, A.C., Gjedde, A., 1998. Increased oxygen consumption in human visual cortex: Response to visual stimulation. *Acta Neurol Scand*. 98, 85–9.

Vafaee, M.S., Meyer, E., Marrett, S., Paus, T., Evans, A.C., Gjedde, A., 1999. Frequency-dependent changes in cerebral metabolic rate of oxygen during activation of human visual cortex. *J Cereb Blood Flow Metab*. 19, 272–7.

Vazquez, A.L., Masamoto, K., Kim, S.G., 2008. Dynamics of oxygen delivery and consumption during evoked neural stimulation using a compartment model and CBF and tissue P_{O2} measurements. *Neuroimage*. 42, 49–59.

Wey, H.Y., Wang, D.J., Duong, T.Q., 2011. Baseline CBF, and BOLD, CBF, and CMRO2 fMRI of visual and vibrotactile stimulations in baboons. *J Cereb Blood Flow Metab*. 31, 715–24.

Wibral, M., Muckli, L., Melnikovic, K., Scheller, B., Alink, A., Singer, W., Munk, M.H., 2007. Time-dependent effects of hyperoxia on the BOLD fMRI signal in primate visual cortex and LGN. *Neuroimage*. 35, 1044–63.

Williams, D.S., Detre, J.A., Leigh, J.S., Koretsky, A.P., 1992. Magnetic resonance imaging of perfusion using spin inversion of arterial water. *Proc Natl Acad Sci USA.* 89, 212–6.

Wu, G., Luo, F., Li, Z., Zhao, X., Li, S.J., 2002. Transient relationships among BOLD, CBV, and CBF changes in rat brain as detected by functional MRI. *Magn Reson Med.* 48, 987–93.

Xu, F., Uh, J., Brier, M.R., Hart, J., Jr., Yezhuvath, U.S., Gu, H., Yang, Y., Lu, H., 2011. The influence of carbon dioxide on brain activity and metabolism in conscious humans. *J Cereb Blood Flow Metab.* 31, 58–67.

Zappe, A.C., Uludag, K., Oeltermann, A., Ugurbil, K., Logothetis, N.K., 2008. The influence of moderate hypercapnia on neural activity in the anesthetized non-human primate. *Cereb Cortex.* 18, 2666–73.

4

Quantifying Functional Connectivity in the Brain

Gang Chen and Shi-Jiang Li

Medical College of Wisconsin

CONTENTS

4.1 Introduction

A series of snapshots of the resting brain using functional magnetic resonance imaging (fMRI) reveal signals that are temporally correlated across functionally related areas (Biswal et al., 1995). These correlations are known as functional connectivity (Biswal et al., 1995; Fox and Raichle, 2007). Analysis of the correlation patterns reveals the normal brain's functional architecture (Biswal et al., 1995, 2010; Cordes et al., 2000; Lowe et al., 1998; Smith et al., 2009) and its deteriorations in a variety of neurological and psychiatric disorders (Chen et al., 2011, 2012a; Fox and Greicius, 2010; Fox and Raichle, 2007; Li et al., 2002, 2012; Rosen and Napadow, 2011; Xie et al., 2011a, 2011b).

In this chapter, we outline the techniques that are commonly performed to measure functional connectivity in the human brain. Then we discuss a method that can quantify functional connectivity-based biomarkers for disease status classification.

4.2 Special Issues in the Acquisition and Processing of Functional Connectivity MRI Data

4.2.1 Data Acquisition

To obtain functional connectivity information in the human brain, a resting-state fMRI data set, a high-resolution T_1-weighted anatomical MRI, a pulse oximeter to monitor and record the heart rate, and a respiratory belt to record the respiratory cycles are need.

A typical resting-state fMRI data set can be acquired, for instance, using a standard whole-body 3 T MR imager (Signa; GE Medical Systems, Milwaukee, WI) with a standard head coil. No specific cognitive tasks are performed during the resting-state acquisitions, and usually study participants are instructed to close their eyes and relax inside the scanner. A series of sagittal scans that cover the whole brain are acquired every 2–3 s with a single-shot gradient echo-planar imaging (EPI) pulse sequence. A typical resting-state fMRI data set is obtained in about 6 min. The acquisition parameters that are typically used would be repetition time (TR) of 2000 ms, echo time (TE) of 25 ms, 90° flip angle, 4 mm-thick slices acquired without a gap (usually 35–40 slices cover the whole brain), 64 × 64 acquisition matrix, and field of view (FOV) of 24 × 24 cm (Chen et al., 2011).

It should be noted that several scan parameters used in acquiring the resting-state fMRI data set influence the outcome. Functional connectivity strength has a noticeable dependence on parameters such as the duration of the scan, TR, and voxel size (Van Dijk et al., 2010).

A high-resolution T_1-weighted anatomical MRI data set is usually acquired, in addition to the resting-state fMRI, using a three-dimensional magnetization-prepared rapid gradient-echo (MPRAGE) pulse sequence for anatomic reference. The commonly used acquisition parameters are TR = 10 ms, TE = 4 ms, inversion time (TI) = 450 ms, flip angle = 12°, 144 slices with a thickness of 1 mm, and acquisition matrix of 256 × 192 (Chen et al., 2011).

Researchers should acquire the anatomical MRI data set for several reasons. The anatomical images can be segmented to identify the gray matter regions so that the functional connectivity analysis can be con-fined within the gray matter regions of interest. They can also be used to identify the white matter and cerebrospinal fluid (CSF) regions. The signals in the white matter and CSF regions correlate with the physiological noise embedded in the resting-state fMRI data (Chang and Glover, 2009). Therefore, they should be considered "nuisance" regressors and excluded from the analysis. Finally, the anatomical image can be used to map the data from a native space to a standard space so that the analysis can be compared among subjects with different brain shapes and sizes.

As we mentioned earlier, the pulse oximeter and the respiratory belt are used to record the cardiac and respiratory cycles. These physiological processes have been shown to induce signal modulations in the resting-state fMRI data (Glover et al., 2000) and must be removed as "nuisance" regressors as well.

4.2.2 Preprocessing of Functional Connectivity MRI Data

Resting-state fMRI data analysis involves the assessment of temporal coherences across different brain regions. These data sets are usually contaminated by various fluctuations that are not related to neural activity. These include subject's head motion, physiological artifacts (caused by respiration and cardiac functions), and hardware instabilities and magnetic field drifts. These signal fluctuations from nonneural processes might introduce false-positive coherences and cause an overestimation of functional connectivity strengths. Therefore, sophisticated preprocessing techniques are commonly employed to remove such confounding effects prior to the connectivity assessment.

The common fMRI preprocessing steps can be conducted using free software tools such as Analysis of Functional NeuroImages (AFNI) software (Medical College of Wisconsin, Milwaukee, WI); (http://afni.nimh.nih.gov/afni/), SPM8 (Wellcome Trust, London, UK), or Matlab (Mathworks, Natick, MA). The preprocessing usually begins with the removal of the first five volumes of fMRI data to allow for T_1 equilibration effects. This is followed by the removal of cardiac and respiratory artifacts (e.g., using 3dretroicor; Birn et al., 2006; Glover et al., 2000), slice acquisition-dependent time shift correction (3dTshift, AFNI), motion correction (3dvolreg, AFNI), detrending (3dDetrend, AFNI), and despiking (3dDespike, AFNI). Then, each subject's anatomical image is segmented to identify white matter and CSF compartments (SPM8) from which white matter and CSF masks are created (3dcalc and 3dfractionize, AFNI). The averaged white matter and CSF signals are obtained (3dROIstats, AFNI). A program, such as 3dDeconvolve of AFNI, is used to remove white matter and CSF signals and motion effects from the resting-state fMRI data sets. Finally, low-frequency band-pass filtering (3dFourier, AFNI) is applied to remove unwanted high-frequency noise, and images are spatially normalized to Talairach space. (It is important to note that the anatomical images are first transformed from their original spaces to Talairach spaces using @auto_tlrc. The process also generates the information that can transform any other images such as the filtered fMRI images from original space to Talairach space. Then, using adwarp tool of

AFNI and the coordinate transformation information, the filtered fMRI images in the original space are transformed to Talairach space.) Once the resting-state fMRI data set is processed through this preprocessing pipeline, the data can be analyzed to quantify the strength of functional connectivity between different brain regions.

In addition to these common fMRI preprocessing steps, a "global signal regression" technique has been introduced (Macey et al., 2004). The global signal regression technique is intended to eliminate undesired global noise, that is, nonneural noise that has a global effect on the resting-state fMRI signal. It was hypothesized that the resting-state fMRI signals may cancel each other out in the global average, while the global noise may be additive and dominate the global-averaged signal. The global signal is obtained by averaging all brain voxel time courses. A number of resting-state fMRI studies have used this technique (Fair et al., 2008; Fox et al., 2005; Fransson, 2005, 2006; Kelly et al., 2008; Tian et al., 2007; Uddin et al., 2009; Wang et al., 2007). However, it was found that global signal regression forces a bell-shaped distribution of functional connectivity measures centered on zero. Therefore, this technique introduces spurious negative measures (Murphy et al., 2009; Weissenbacher et al., 2009) and might lead to underestimation of true-positive measures.

Several groups have explored techniques to monitor the sources that may induce nonneural noise in the resting-state fMRI signal (Anderson et al., 2011; Birn et al., 2006; Chang and Glover, 2009). When no measurement of these sources is made, they can be estimated by data-driven methods (Beall and Lowe, 2007). It has been shown that the low-frequency signals related to respiratory and cardiac cycles had significant covariance with the global signal (Chang and Glover, 2009). Therefore, these nonneural noise correction techniques may reduce the global noise level and diminish the motivation for adopting global signal regression in data preprocessing.

Until recently, it was not clear to what degree the global noise levels could be reduced, or if it was truly necessary to use global signal regression in further reducing the global noise following other nonneural noise removal procedures. If the global noise level is high, the global signal may represent the global noise. Therefore, the global signal regression might be beneficial. However, if the global noise level is low, the global signal may contain a higher portion of useful signal (probably from the dominant network component), rather than the global noise. Taking into account that the default mode network (a network of brain regions that is active when the individual is at wakeful rest) has shown high glucose metabolism during rest (Buckner et al., 2008; Raichle et al., 2001), it can be hypothesized that the global signal may be most strongly correlated with the default mode network signal. For this reason, removing the global signal may generate more errors in the functional connectivity measurements, instead of reducing them.

A quantitative method has been introduced to determine the need for global signal regression on a single subject level (Chen, 2012). The accuracy of the global signal regression is determined by the level of global noise. There is a monotonic relationship between the signal-to-global noise ratio and the global negative index:

$$\text{Global negative index} = \frac{\text{Number of voxels negatively correlated with global signal}}{\text{Total number of voxels}}. \qquad (4.1)$$

The number of voxels, which are negatively correlated with the global signal and survive a threshold of $p < .05$ (uncorrected for multiple comparisons), is used in Equation 4.1. There is a criterial global negative index (3.0) associated with a criterial signal-to-global noise ratio (7.03). Below this criterial global negative index, performing global signal regression induces less errors. Therefore, it is highly suggested that the global regression be performed. Above this criterial global negative index, performing global signal regression induces more errors. Therefore, global signal regression is not recommended. One can decide whether to apply this technique for each individual data set by comparing the global negative index of the data set to the criterial global negative index of 3.0.

4.3 Brain Functional Network and Quantifying Functional Connectivity

4.3.1 Brain Functional Networks in Normative Subjects

Several approaches have been proposed to obtain information about brain functional networks, which have been reviewed in the paper published by Margulies et al. (2010). The seed-based analysis method is the simplest method. In a seed-based functional connectivity analysis, the seed region of interest (ROI) in the brain is commonly defined using the knowledge about the neuroanatomy. The reason we call it "seed" ROI is that the corresponding brain functional network is defined as the regions that have strong functional connectivity to the seed ROI. Therefore, we use the term h to indicate that the brain functional network "grows" from the selected "seed" ROI. The functional connectivity between the seed region and the brain can be assessed with the Pearson product moment correlation coefficient (r) (Rodgers and Nicewander, 1988) between the preprocessed resting-state fMRI time courses averaged within the seed region and those of the other brain voxels.

4.3.2 Changes in Functional Connectivity in Cognitive and Psychiatric Diseases

In this section, we discuss a method that detects the functional connectivity changes in cognitive and psychiatric disorders and obtains the functional connectivity-based biomarkers for disease status classification.

To reduce redundancy and computational costs, the preprocessed resting-state fMRI time courses within predefined ROIs are usually averaged in order to generate the functional connectivity of the ROIs. As a result, each subject will have an $n \times n$ functional connectivity matrix (r-matrix), where n is the number of ROIs in the predefined regions.

To explore group differences in functional connectivity, a parametric two-sample t-test or a nonparametric two-sample Wilcoxon rank-sum test (Noether, 1991) can be employed. In the parametric t-test, the Fisher transformation $\{m = 0.5\ln[(1 + r)/(1 - r)]\}$ (Zar, 1996), which yields variants of approximately normal distribution, is applied to the individual r-matrix to generate the m-matrix. The functional connectivity difference is calculated using the m-matrices and two-tailed two-sample t-test. The p-values are transformed to Z-values using the inverse of the normal cumulative distribution function. The result is an $n \times n$ Z-matrix. To avoid making any assumption about the distribution of r values, the nonparametric two-sample Wilcoxon rank-sum test can be employed. The Wilcoxon rank-sum test also yields an $n \times n$ Z-matrix. The absolute Z-value indicates statistical significance, while the sign of the Z-value indicates which group has the higher mean r value.

Previously, we discussed the method to detect functional connectivity differences between groups. Each of these differences is a feature. To reduce the number of features to manageable two values per subject, one can define a decreased connectivity index (DCI) by averaging the r-values in the decreased connection set (connections that are significantly decreased in the diseased group). Similarly, one can define an increased connectivity index (ICI) by averaging the r-values in the increased connection set (connections that are significantly increased in the diseased group). Next, the DCI and ICI indices can classify the subjects with different disease statuses using Fisher's linear discriminant analysis, as shown in Figure 4.1 (Theodoridis and Koutroumbas, 2006). The Fisher's linear discriminant function (diagonal line) is used to identify the subjects with Alzheimer's disease (AD) and those without AD (non-AD).

4.3.3 Cross-Validation

In the previous section, we discussed a method to obtain biomarkers for disease status classification by quantifying functional connectivity in the brain. However, in this example (Figure 4.2), the same subjects were used for training and evaluation. As a result, the method would give an overly optimistic estimate. Therefore, an independent validation cohort cross-validates

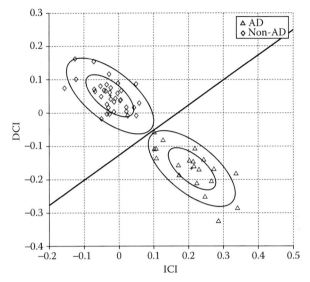

FIGURE 4.1

Classifying subjects as AD or non-AD using Fisher's linear discriminant analysis. The Fisher's linear discriminant function (diagonal line) is used to identify the subjects with AD and those without AD (non-AD). Concentric ellipses represent 50% and 85% probability containment for the AD and non-AD groups, respectively.

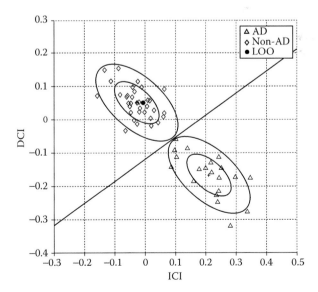

FIGURE 4.2

Classifying subjects using the leave-one-out (LOO) method. In the LOO procedure, one subject without AD (non-AD) is left out; the remaining 20 AD subjects and 34 non-AD subjects are used for training. Thus, the classification criteria are determined by using all subjects except the one to be evaluated. As a result, Fisher's linear discriminant function (diagonal line) is used to classify the subjects as AD or non-AD.

the accuracy of the classification. Alternatively, with limited data set size, one can employ a leave-one-out (LOO) method in order to make up for the lack of independence between the training and testing sets (Theodoridis and Koutroumbas, 2006). The LOO method uses one subject at a time for evaluation, while the remaining subjects are used for training. The classification criteria are determined using all subjects, except the one subject to be evaluated. This entire process is repeated for each subject, thereby providing a relative unbiased estimate of the classification error rate.

4.3.4 Areas That Need Improvement

To reduce redundancy and computational costs, the preprocessed resting-state fMRI time courses within predefined ROIs are usually averaged in order to produce the functional connectivity of the ROIs. However, these predefined regions have relied on atlases derived from anatomical or cytoarchitectonic boundaries. It is not clear whether these ROIs are suitable for resting-state functional connectivity studies. It has been found that the anatomical atlases exhibit poor ROI homogeneity (i.e., the ROI contains

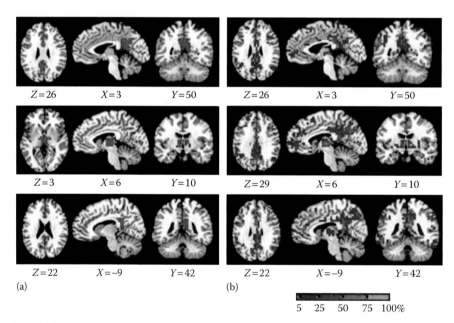

FIGURE 4.3
(See color insert.) An example result using the *x*-guided clustering method. (a) The clusters identified using group difference information between amnestic mild cognitively impaired and age-matched cognitively normal subjects. (b) The decreased connectivity pattern to the clusters shown on the left. Color bar indicates the percentage of disconnection. For example, light blue indicates that a voxel is disconnected to 75–100% of the voxels in the corresponding cluster.

voxels with dissimilar functional properties, such as voxel-wise resting-state signals and voxel-wise functional connectivity maps) and do not accurately reproduce functional connectivity patterns present at the voxel scale (Craddock et al., 2011; Margulies et al., 2007). To solve the problem, data-driven methods have been introduced to generate functionally homogeneous ROIs. Clustering and independent component analysis (ICA) are among the most popular methodologies (Margulies et al., 2010).

Recently, an x-guided clustering method was introduced to detect functional connectivity differences (Chen et al., 2012a, 2012b, 2012c). Using the x-guided clustering-based method, one can clearly define the reference clusters (ROIs) for detecting functional connectivity differences between patients and control subjects. The functional connectivity changes (i.e., changes in connectivity between controls and patients) of all voxels within the reference clusters are guaranteed to be homogeneous above a threshold level, which is not possible in previously available methods. Since the x-guided clustering method can provide better functional homogeneity, it may further improve the functional connectivity MRI-based biomarkers in the classification of subject's disease state. Figure 4.3 illustrates an example of the x-guided clustering method that investigates the percentage change in brain connectivity between amnestic mild cognitively impaired and age-matched cognitively normal subjects.

References

Anderson, J.S., Druzgal, T.J., Lopez-Larson, M., Jeong, E.K., Desai, K., Yurgelun-Todd, D., 2011. Network anticorrelations, global regression, and phase-shifted soft tissue correction. *Hum Brain Mapp*. 32, 919–34.

Beall, E.B., Lowe, M.J., 2007. Isolating physiologic noise sources with independently determined spatial measures. *Neuroimage*. 37, 1286–300.

Birn, R.M., Diamond, J.B., Smith, M.A., Bandettini, P.A., 2006. Separating respiratory-variation-related fluctuations from neuronal-activity-related fluctuations in fMRI. *Neuroimage*. 31, 1536–48.

Biswal, B., Yetkin, F.Z., Haughton, V.M., Hyde, J.S., 1995. Functional connectivity in the motor cortex of resting human brain using echo-planar MRI. *Magn Reson Med*. 34, 537–41.

Biswal, B.B., Mennes, M., Zuo, X.N., Gohel, S., Kelly, C., Smith, S.M., Beckmann, C.F., et al., 2010. Toward discovery science of human brain function. *Proc Natl Acad Sci USA*. 107, 4734–9.

Buckner, R.L., Andrews-Hanna, J.R., Schacter, D.L., 2008. The brain's default network: Anatomy, function, and relevance to disease. *Ann NY Acad Sci*. 1124, 1–38.

Chang, C., Glover, G.H., 2009. Effects of model-based physiological noise correction on default mode network anti-correlations and correlations. *Neuroimage*. 47, 1448–59.

Chen, G., 2012. A method to determine the necessity for global signal regression in resting-state fMRI studies. *Magn Reson Med*. 68, 1828–35.

Chen, G., Ward, B.D., Xie, C., Li, W., Chen, G., Antuono, P., Li, S.J., 2012a. A clustering based method to detect functional connectivity differences. *Neuroimage*. 61, 56–61.

Chen, G., Ward, B.D., Xie, C., Li, W., Chen, G., Antuono, P., Li, S.J., 2012b. The x-guided clustering method and its application to unbiased detection of differences in functional connectivity networks. *Proceedings of the 20th Annual Meeting of the International Society of Magnetic Resonance in Medicine (ISMRM)*, 5–11 May, Melbourne, Australia. p. 2828.

Chen, G., Ward, B.D., Xie, C., Li, W., Chen, G., Antuono, P., Li, S.J., 2012c. The x-guided clustering method to detect functional connectivity differences. *Proceedings of the 18th Annual Meeting of the Organization for Human Brain Mapping*, 10–14 June, Beijing, China. p. 488.

Chen, G., Ward, B.D., Xie, C., Li, W., Wu, Z., Jones, J.L., Franczak, M., Antuono, P., Li, S.J., 2011. Classification of Alzheimer disease, mild cognitive impairment, and normal cognitive status with large-scale network analysis based on resting-state functional MR imaging. *Radiology*. 259, 213–21.

Cordes, D., Haughton, V.M., Arfanakis, K., Wendt, G.J., Turski, P.A., Moritz, C.H., Quigley, M.A., Meyerand, M.E., 2000. Mapping functionally related regions of brain with functional connectivity MR imaging. *AJNR Am J Neuroradiol*. 21, 1636–44.

Craddock, R.C., James, G.A., Holtzheimer, P.E., III, Hu, X.P., Mayberg, H.S., 2011. A whole brain fMRI atlas generated via spatially constrained spectral clustering. *Hum Brain Mapp*. 33, 1914–28.

Fair, D.A., Cohen, A.L., Dosenbach, N.U., Church, J.A., Miezin, F.M., Barch, D.M., Raichle, M.E., Petersen, S.E., Schlaggar, B.L., 2008. The maturing architecture of the brain's default network. *Proc Natl Acad Sci USA*. 105, 4028–32.

Fox, M.D., Greicius, M., 2010. Clinical applications of resting state functional connectivity. *Front Syst Neurosci*. 4, 19.

Fox, M.D., Raichle, M.E., 2007. Spontaneous fluctuations in brain activity observed with functional magnetic resonance imaging. *Nat Rev Neurosci*. 8, 700–11.

Fox, M.D., Snyder, A.Z., Vincent, J.L., Corbetta, M., Van Essen, D.C., Raichle, M.E., 2005. The human brain is intrinsically organized into dynamic, anticorrelated functional networks. *Proc Natl Acad Sci USA*. 102, 9673–8.

Fransson, P., 2005. Spontaneous low-frequency BOLD signal fluctuations: An fMRI investigation of the resting-state default mode of brain function hypothesis. *Hum Brain Mapp*. 26, 15–29.

Fransson, P., 2006. How default is the default mode of brain function? Further evidence from intrinsic BOLD signal fluctuations. *Neuropsychologia*. 44, 2836–45.

Glover, G.H., Li, T.Q., Ress, D., 2000. Image-based method for retrospective correction of physiological motion effects in fMRI: RETROICOR. *Magn Reson Med*. 44, 162–7.

Kelly, A.M., Uddin, L.Q., Biswal, B.B., Castellanos, F.X., Milham, M.P., 2008. Competition between functional brain networks mediates behavioral variability. *Neuroimage*. 39, 527–37.

Li, S.J., Li, Z., Wu, G., Zhang, M.J., Franczak, M., Antuono, P.G., 2002. Alzheimer disease: Evaluation of a functional MR imaging index as a marker. *Radiology*. 225, 253–9.

Li, W., Antuono, P.G., Xie, C., Chen, G., Jones, J.L., Ward, B.D., Franczak, M.B., Goveas, J.S., Li, S.J., 2012. Changes in regional cerebral blood flow and functional connectivity in the cholinergic pathway associated with cognitive performance in subjects with mild Alzheimer's disease after 12-week donepezil treatment. *Neuroimage*. 60, 1083–91.

Lowe, M.J., Mock, B.J., Sorenson, J.A., 1998. Functional connectivity in single and multislice echoplanar imaging using resting-state fluctuations. *Neuroimage*. 7, 119–32.

Macey, P.M., Macey, K.E., Kumar, R., Harper, R.M., 2004. A method for removal of global effects from fMRI time series. *Neuroimage*. 22, 360–6.

Margulies, D.S., Bottger, J., Long, X., Lv, Y., Kelly, C., Schafer, A., Goldhahn, D., et al., 2010. Resting developments: A review of fMRI post-processing methodologies for spontaneous brain activity. *MAGMA*. 23, 289–307.

Margulies, D.S., Kelly, A.M., Uddin, L.Q., Biswal, B.B., Castellanos, F.X., Milham, M.P., 2007. Mapping the functional connectivity of anterior cingulate cortex. *Neuroimage*. 37, 579–88.

Murphy, K., Birn, R.M., Handwerker, D.A., Jones, T.B., Bandettini, P.A., 2009. The impact of global signal regression on resting state correlations: Are anticorrelated networks introduced? *Neuroimage*. 44, 893–905.

Noether, G., ed., 1991. The Wilcoxon two-sample test. In: *Introduction to Statistics: The Nonparametric Way*. New York: Springer-Verlag, 103–12.

Raichle, M.E., MacLeod, A.M., Snyder, A.Z., Powers, W.J., Gusnard, D.A., Shulman, G.L., 2001. A default mode of brain function. *Proc Natl Acad Sci USA*. 98, 676–82.

Rodgers, J.L., Nicewander, W.A., 1988. Thirteen ways to look at the correlation coefficient. *Am Stat*. 42, 59–66.

Rosen, B.R., Napadow, V., 2011. Quantitative markers for neuropsychiatric disease: Give it a rest. *Radiology*. 259, 17–19.

Smith, S.M., Fox, P.T., Miller, K.L., Glahn, D.C., Fox, P.M., Mackay, C.E., Filippini, N., et al., 2009. Correspondence of the brain's functional architecture during activation and rest. *Proc Natl Acad Sci USA*. 106, 13040–5.

Theodoridis, S., Koutroumbas, K., eds., 2006. Linear classifiers. In: *Pattern Recognition*. 3rd ed. San Diego, CA: Academic Press, 69–78, 474–5.

Tian, L., Jiang, T., Liu, Y., Yu, C., Wang, K., Zhou, Y., Song, M., Li, K., 2007. The relationship within and between the extrinsic and intrinsic systems indicated by resting state correlational patterns of sensory cortices. *Neuroimage*. 36, 684–90.

Uddin, L.Q., Kelly, A.M., Biswal, B.B., Xavier Castellanos, F., Milham, M.P., 2009. Functional connectivity of default mode network components: Correlation, anticorrelation, and causality. *Hum Brain Mapp*. 30, 625–37.

Van Dijk, K.R.A., Hedden, T., Venkataraman, A., Evans, K.C., Lazar, S.W., Buckner, R.L., 2010. Intrinsic functional connectivity as a tool for human connectomics: Theory, properties, and optimization. *J Neurophysiol*. 103, 297–321.

Wang, K., Liang, M., Wang, L., Tian, L., Zhang, X., Li, K., Jiang, T., 2007. Altered functional connectivity in early Alzheimer's disease: A resting-state fMRI study. *Hum Brain Mapp*. 28, 967–78.

Weissenbacher, A., Kasess, C., Gerstl, F., Lanzenberger, R., Moser, E., Windischberger, C., 2009. Correlations and anticorrelations in resting-state functional connectivity MRI: A quantitative comparison of preprocessing strategies. *Neuroimage*. 47, 1408–16.

Xie, C., Goveas, J., Wu, Z., Li, W., Chen, G., Franczak, M., Antuono, P.G., Jones, J.L., Zhang, Z., Li, S.J., 2011a. Neural basis of the association between depressive symptoms and memory deficits in nondemented subjects: Resting-state fMRI study. *Hum Brain Mapp*. 33, 1352–63.

Xie, C., Li, S.J., Shao, Y., Fu, L., Goveas, J., Ye, E., Li, W., et al., 2011b. Identification of hyperactive intrinsic amygdala network connectivity associated with impulsivity in abstinent heroin addicts. *Behav Brain Res*. 216, 639–46.

Zar, J.H., 1996. *Biostatistical Analysis*. Englewood Cliffs, NJ: Prentice-Hall.

5

Brain MR Spectroscopy In Vivo: Basics and Quantitation of Metabolites

Manoj K. Sarma, Rajakumar Nagarajan, and M. Albert Thomas
University of California, Los Angeles

CONTENTS

5.1 Introduction

Nuclear magnetic resonance (NMR) spectroscopy is a powerful noninvasive technique that allows the study of properties of different nuclei such as proton (^{1}H), phosphorus (^{31}P), carbon (^{13}C), fluorine (^{19}F), and so on. It is a diverse research tool widely used by biochemists to investigate pathophysiological processes *in vitro* and by physicians to determine disease abnormalities *in vivo*. NMR was first described and measured in molecular beams by Isidor Rabi in 1938. Eight years later, in 1946, Felix Bloch and Edward Mills Purcell refined the technique for use on liquids and solids, for which they shared the Nobel Prize in physics in 1952. The conception of Fourier transform (FT) spectroscopy by Ernst and Anderson (1966) opened a new page in modern NMR spectroscopy and completely revolutionized the field. Whole body magnetic resonance imaging (MRI) scanners were introduced in the early 1980s and since then multinuclear MR Spectroscopy (MRS) has been investigated in living tissues. During the last three decades ^{1}H MR spectra have been recorded in the range of 64–300 MHz using the superconducting magnets operating at the field strengths of 1.5–7 T.

There are two main sources of protons in living systems, namely water and fat. Due to the high abundance of water and fat they become hindrance in acquiring signals from low concentration metabolites present in tissues, and initially this made the use of ^{1}H MRS limited to a small number of centers possessing a specific technological expertise. With the development of frequency-selective excitation pulse schemes such as chemical shift-selective suppression (CHESS; Haase et al., 1985), global water suppression enhanced through T_1 effect (WET) (Ogg et al., 1994) and frequency-selective 180° inversion (Shen and Saunders, 1993), excellent water/fat suppression has been achieved and ^{1}H MRS has become a routine technique in almost all clinical MR systems. Excellent water-suppressed ^{1}H MR spectra have been recorded in different organs during the last two decades (Fan et al., 2004; Hanstock et al., 1988; Prichard, 1997; Ross, 1991). Earlier MRS attempts included topical localization of region of interest (ROI) using surface coil (Ackerman et al., 1980; Weiner et al., 1989), which introduced inhomogeneous B_1 field. By the end of the 1980s, they were almost completely replaced by volume coils with more homogeneous transmission/reception in combination with several localization sequences namely, volume selective excitation (VSE), stimulated echo acquisition mode (STEAM), point-resolved spectroscopic sequence (PRESS), and image selected *in vivo* spectroscopy (ISIS) (Aue et al., 1984; Bottomley, 1987; Frahm et al., 1987; Ordidge et al., 1986).

The success of MRS as a clinical analysis tool depends on the way in which the MRS data can be analyzed. The analyzed spectral parameters are then translated into relevant biochemical or medical parameters for the diagnosis. In this chapter, we have presented the basics of one-dimensional (1D) MRS, two-dimensional (2D) MR spectroscopic imaging (MRSI), single-voxel-based 2D MRS, and the techniques for the accurate analysis of MRS signals. Some postprocessing algorithms that can improve the analysis of MRS data are also discussed. The signals are modeled by an appropriate model function, and mathematical techniques are subsequently applied to determine the model parameters.

5.2 MRS Localization

For *in vivo* biochemical and biomedical applications, it is of great importance to restrict signal detection to a well-defined ROI such that spatial origin of the collected signals is accurately known. As the biological tissues are spatially inhomogeneous in terms of metabolic compositions as well as magnetic field strengths, it is quite difficult to interpret lines from nonlocalized spectrum. The poor sensitivity makes spatial localization complicated in MRS as the concentrations of metabolites of interests are usually lower than the concentration of water in tissues by four orders of magnitude. Also the size of the selected volume needs to be large in MRS, which may lead to partial volume effects and signal contamination from adjacent tissue. Therefore, in order to acquire high-quality spectra for more genuine tissue representation, a good control of the spatial origin is crucial.

Over the past 25 years, a wide range of spatial localization techniques has been developed utilizing various combinations of pulse sequence design, static magnetic field gradients, and radio frequency (RF) coil properties. These techniques can be divided into two categories: single voxel and multivoxel. In single-voxel spectroscopy (SVS), a spectrum is obtained from a certain volume element called volume of interest (VOI) or ROI, whereas in multivoxel multiple spectra are recorded simultaneously from different regions in one, two, and three spatial dimensions.

5.2.1 Single-Voxel Spectroscopy

The basic principle underlying single-voxel localization techniques typically is to acquire a spectrum from a small cubic or rectangular-based volume element or voxel of tissue generated by the intersection of three orthogonal planes. To select a voxel, selective pulses are applied one after the other, in the presence of mutually orthogonal field gradients. The two basic spectroscopy sequences commonly used in SVS are STEAM (Frahm et al., 1987, 1989) and PRESS (Bottomley, 1987; Keevil, 2006). Both localization techniques use a series of three slice-selective RF pulses, but have different flip angles, sequence timing and placement of spoiling gradient pulses.

5.2.1.1 STEAM

The STEAM technique, which is mostly used in ^{1}H MRS, is based on the stimulated echo (STE) generated by three consecutive slice-selective 90° RF pulses applied with slice-selection gradients each along one of the orthogonal directions x, y, or z (Figure 5.1). The first two pulses are separated by a delay of half the echo time, that is, TE/2 and the last two pulses are separated by a delay TM, the mixing time. The three RF pulses produce three free-induction decays (FIDs) and five echoes. The STE is formed at position TE + TM (or after a delay TE/2 following the last 90° pulse). As STE contains the signal of interest, in order to sample the STE only during acquisition time, the FID and spin-echo signals are eliminated by proper design of slice-selection gradient. The main disadvantage of this method is that the echo comes with a 50% signal loss. This technique is best suited for short TE acquisitions.

5.2.1.2 PRESS

In the PRESS technique, a slice-selective 90° RF pulse is followed by two slice-selective 180° refocusing RF pulses. Each pulse is applied in the presence of a gradient magnetic field pulse directed along one of the three orthogonal axes. Figure 5.2 explains the PRESS localization sequence. First spin-echo is formed at time TE_1 after the 90° pulse and a second spin-echo, which only contains signal from the intersection of the three planes selected by all the three pulses, is formed at time $TE = TE_1 + TE_2$, where TE_2 is the delay during which the first spin-echo is refocused again by the second 180°

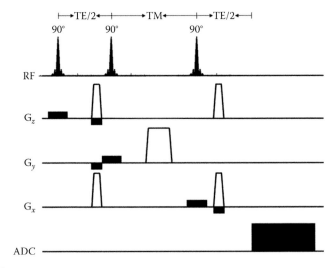

FIGURE 5.1
Scheme of the STEAM sequence (90°–90°–90°). The three 90° RF pulses are slice selective along three planes. ADC, analog-to-digital conversion.

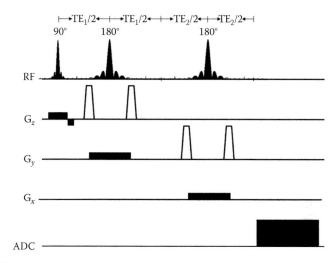

FIGURE 5.2
Scheme of the PRESS (90°–180°–180°) sequence which is based on two Hahn echoes. ADC, analog-to-digital conversion.

pulse. TE crushers are applied surrounding each 180° pulse so that only the final spin-echo is sampled and collected. PRESS gives two times better signal-to-noise ratio (SNR) over STEAM. This technique is most used in ^{1}H MRS for detecting metabolites with long T_2-relaxation times (>30 ms) and for small volume.

5.2.2 2D Chemical Shift Imaging

Chemical shift imaging (CSI), also known as magnetic resonance spectroscopic imaging (MRSI) or spectroscopic imaging (SI), is a technique that combines features of both imaging and spectroscopy (De Graaf, 1998; Duijn et al., 1992; Moonen et al., 1992; Vigneron et al., 1990). Just like the MRI sequence, it uses phase-encoding gradients in order to encode spatial distribution of metabolite concentrations in tissue. A spectrum is recorded at each phase-encoding step in the absence of any gradient to allow the spectral acquisition in a number of volumes covering the whole sample. Multidimensional FT is used to obtain the spectral data and can be represented or displayed either as spectral map (parametric images) analog to single spectra related to individual voxels, or as metabolite images.

In 2D CSI, phase-encoding gradients are applied in two dimensions. In clinical applications, 2D CSI are usually combined with VSE. In general, this is achieved by combining two phase-encoding gradients to a PRESS or STEAM sequence (Duijn et al., 1992; Moonen et al., 1992). Figure 5.3 illustrates a 2D CSI sequence based on a PRESS sequence. The second spin-echo signal is collected after switching the gradient off and the experiment is repeated at intervals of repetition time (TR) with an incremental value of

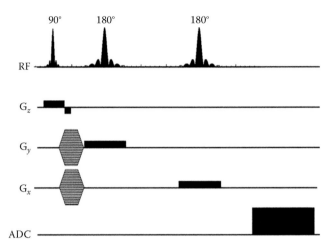

FIGURE 5.3
A pulse sequence diagram of 2D CSI with VOI selection by PRESS. ADC, analog-to-digital conversion.

gradient amplitude. The STEAM sequence can also be modified similarly to acquire 2D CSI. Total measurement time for a 2D CSI with N_x, N_y, phase-encoding steps along the x and y directions and with $N_{av,CSI}$ representing the number of averages is given by

$$T_{CSI} = TR \cdot N_x \cdot N_y \cdot N_{av,CSI}. \tag{5.1}$$

Apart from the elimination of spurious signals, another advantage of volume-selective CSI is that a better spatial resolution can be achieved by reducing the field of view (FOV) as only a restricted area of the sample gets excited. But due to the long required acquisition times, conventional CSI is not clinically feasible. To overcome this problem, variants of the basic CSI sequence have been proposed. Instead of two phase-encoding gradients, echo planar images (EPI) (Mansfield, 1977) or fast or turbo spin echo (FSE/TSE) can be used for spatial encoding of one of the two dimensions (Duyn and Moonen, 1993). This will result in an acceleration of N_x or N_y as shown in Equation 5.1.

5.2.3 2D MRS Techniques

Conventional 1D MRS cannot fully characterize a molecular system, especially of large molecules. It suffers from severe signal overlap leading to unreliable quantitation of metabolites and lipids. Through 2D MR spectroscopy, more detailed information about the system under investigation can be obtained. Jeener (1971) first introduced the concept of 2D NMR in 1971, experimentally first realized by Ernst and coworkers in 1976, and since then an astonishing number of powerful 2D techniques have been invented and applied. The power of 2D MRS lies in its ability to separate overlapping spectral lines and correlates interactions along independent frequency domains, providing enhancement in spectral dispersion and information not available in the 1D counterparts. Correlation spectroscopy (COSY), J-resolved spectroscopy, exchange spectroscopy (EXSY), and nuclear Overhauser effect spectroscopy (NOESY) are some of selected types of 2D NMR spectroscopy. In 2D J-resolved spectroscopy, chemical shift versus J-coupling pattern of the same metabolite system is plotted; whereas in COSY, correlation of chemical shifts of J-coupled partner as well as uncoupled spins is plotted.

5.2.3.1 J-Resolved Spectroscopy

Two-dimensional (2D) J-resolved spectroscopy, first demonstrated by Ernst and coworkers almost two decades ago (Aue et al., 1976; Ernst et al., 1987) on a high-resolution NMR spectrometer using a conventional Hahn spin-echo pulse scheme, is one of the simplest 2D sequences well suited for *in vivo* applications. The *in vivo* MRS version of this sequence is termed as JPRESS as it uses PRESS for volume localization and is often compared as 2D analog of the PRESS sequence. It encodes chemical shift and J-coupling information along the first (direct, t_2) dimension and indirect spin–spin (J) coupling along the second (indirect, t_1) dimension. After a 45° rotation of the 2D data matrix, chemical shift along the direct and J-coupling along the indirect dimensions can be presented. As a result, it reduces the spectral crowding often seen in 1D spectrum introduced by multiplicity of various peaks in the presence of coupled spins, and a better separation of the resonance lines are achieved.

Ryner et al. proposed a 2D JPRESS (Figure 5.4) sequence based on the product PRESS sequence (Ryner et al., 1995a, 1995b; Thomas et al., 1996).

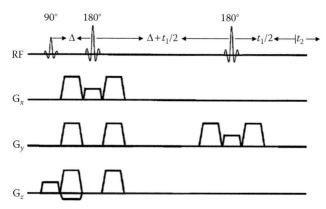

FIGURE 5.4
Schematic diagram of a 2D JPRESS sequence.

They modified the PRESS sequence to allow simultaneous incrementation of the two periods before and after the last 180° RF pulse and came up with the following sequence:

$$90^{\circ}_{sl} - \Delta - 180^{\circ}_{sl} - \Delta - \frac{t_1}{2} - 180^{\circ}_{sl} - \frac{t_1}{2} - t_2 \left(\text{Acquisition} \right),$$

where t_1 refers to the evolution time (indirect) and the subscript "sl" refers to a slice-selective RF pulse along one of the three orthogonal axes; 2Δ was the minimum time necessary to play out RF pulses and gradient pulses for the first echo to occur. An incremental delay $(t_1/2)$ was added to the intervals before and after the last 180° RF pulse which monitored the J-evolution. Apart from reducing the spectral overlap, another major advantage of the JPRESS sequence is that 100% of the available magnetization is retained, neglecting losses caused by T_2 relaxation and spin diffusion. The 2D JPRESS sequence is versatile in the sense that it can be used for the acquisition of a standard 1D PRESS spectrum by simply eliminating the incremental increases of the t_1-period. JPRESS has been applied to different human organs *in vivo*, including brain (Hurd et al., 2004, 1998; Ke et al., 2000, 2004; Levy and Hallett, 2002; Ryner et al., 1995a, 1995b; Thomas et al., 1996, 2003), prostate (Swanson et al., 2001; Yue et al., 2002), breast (Bolan et al., 2002), and muscle (Bolan et al., 2002; Kreis and Boesch, 1994, 1996). There are other variants of the JPRESS sequence also (Adalsteinsson and Spielman, 1999; Dreher and Leibfritz, 1999; Kuhn et al., 1999; Schulte et al., 2006).

5.2.3.2 2D Correlated MR Spectroscopy

2D COSY experiments enable identification of the coupled spins in a metabolite. During the mixing period of COSY, the last 90° RF pulse causes a transfer of coherences between J-coupled spins or protons. Different versions of the localized COSY sequence have been proposed. In 2001, Thomas et al. proposed a new version of localized COSY (L-COSY) sequence (Binesh et al., 2002; Thomas et al., 2001) for a 1.5 T MRI scanner. They subsequently implemented its extension on a conventional 3 T MRI scanner and evaluated in healthy human brains (Thomas et al., 2003).

The 2D L-COSY sequence, proposed by Thomas et al. consisted of three slice-selective RF pulses (90°–180°–90°) (Figure 5.5) for the volume localization. The last 90° RF pulse also enabled the coherence transfer necessary for correlating the metabolite peaks in the second dimension. Immediately after the formation of the Hahn spin-echo, an incremental period for the second dimension was inserted. In addition to the slice-selective 180° RF pulse, there were refocusing B_0 gradient crusher pulses before and after the last 90° RF pulses (Mcrobbie et al., 2006). There was also an eight-step phase cycling, two steps superimposed on each RF pulse in order to minimize the effect

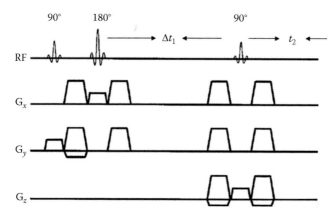

FIGURE 5.5
Localized 2D shift correlated MR spectroscopic sequence (L-COSY). The RF pulse scheme consisted of three RF pulses (90°, 180°, and 90°) which were slice-selective along three orthogonal axes. (From Thomas, M.A., et al., *NMR in Biomedicine.* 16, 245–251, 2003. With permission.)

of the RF pulse imperfections. The 2D L-COSY improved the spectral dispersion due to increased spectral width along the second dimension, and consequently, this enabled the detection of several brain metabolites at low concentrations.

5.2.3.3 Multivoxel 2D MRS

Multivoxel 2D MRS can be obtained by combining the classic phase-encoding SI with 2D spectroscopy, and different variants of these techniques have been introduced to date (Herigault et al., 2004; Jensen et al., 2005; Lipnick et al., 2010; Mayer et al., 2000, 2003; von Kienlin et al., 2000). Through these multivoxel techniques, spatial distributions of better-resolved spectrum of metabolites are obtained. One of the main drawbacks of the multivoxel 2D MRS is the excessively long scan duration. Because of this problem, it has not found wide acceptance in the clinical settings despite the prospects of high sensitivity in the evaluation of different diseases.

Lipnick et al. recently implemented the multivoxel 2D COSY, called echo planar correlated spectroscopic imaging (EP-COSI), a 4D MRSI technique combining two spectral with two spatial-encoding dimensions on a whole body 3 T MRI scanner (2010). The sequence (Figure 5.6) consists of a 90°–180°–90° scheme for the VOI localization with crusher gradients used before and after the refocusing 180° and coherence transfer 90° RF pulses so that magnetization outside of the VOI undergoes dephasing and will not significantly contribute to the acquired signal. The evolution time in between the 180° and final 90° RF pulses was incremented between scans (Δt_1) to facilitate acquisition of the second (indirect) spectral dimension (t_1). The bipolar

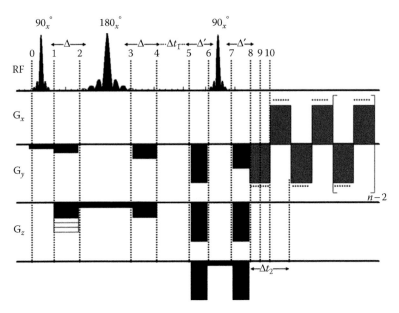

FIGURE 5.6
An illustration of the pulse sequence diagram for the EP-COSI sequence. (From Lipnick, S., et al., *Magnetic Resonance in Medicine*. 64, 947–956, 2010. With permission.)

read-out gradient train facilitates frequency encoding for one of the spatial dimensions as well the spectral encoding along the t_2 dimension. The phase-encoding gradient was applied after the first slice-selective 90° RF pulse to encode the remaining spatial dimension.

5.3 MRS Data Processing

Success of 1D and 2D MRS among other things depends on the way in which the data can be processed and analyzed by the computer. Measured MRS signals are required to interpret in terms of relevant physical parameters which, in turn, can be translated into relevant biochemical or medical parameters such as ratios of metabolite concentrations or the pH value of the tissue concerned. MRS data processing involves preprocessing of the data either in the time domain before performing FT or in the frequency domain and quantitation of the resultant data. Processing of MRS data can either be performed by programs available on commercial MR systems or by stand-alone programs (see Figure 5.7). The raw time domain data serve as input for most of the commonly available software packages, which report estimations of spectral

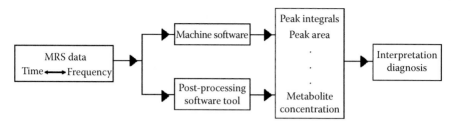

FIGURE 5.7
Flow chart of processing steps from raw time domain data, produced by an MRS/MRI machine, to biomedical or physical interpretation.

parameters or results on a higher level of sophistication. In the next two sections, an overview of existing preprocessing and quantification methods will be discussed.

5.3.1 Preprocessing of MRS Signals

MRS signals are affected by the presence of artifacts, instrumental errors, noise, and other unwanted components such as residual water and lipid signals. Though most pulse sequences have incorporated pulses that suppress the water and lipid signals such as CHESS (Rosen et al., 1984) and BASING (Star-Lack et al., 1997) sequences, there is an inability of accomplishing complete water suppression. Because of all these reasons, preprocessing should be performed in order to enable a better quantification of the signals. The aim of the preprocessing step is to remove the irrelevant information, while enhancing the key features and spectral quality. Some frequently used preprocessing steps (shown in Figure 5.8) are discussed below.

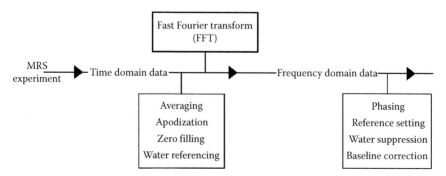

FIGURE 5.8
Flow chart of common preprocessing steps either in time or frequency domain.

5.3.1.1 Apodization

If an MRS data set is directly subjected to fast Fourier transform (FFT), a fast version of FT, it can suffer from truncation artifacts, low SNR ratio, or spectral resolution, and/or broad spectral components in the frequency domain. These can be removed through a process called apodization by convoluting the spectrum $F(v')$ with a different lineshape function $G(v')$:

$$F'(v') = F(v') * G(v') = \int_{-\infty}^{\infty} F(\tau)G(v' - \tau)d\tau. \tag{5.2}$$

In practice, this is performed by multiplying the raw MR data in the time domain utilizing the property that multiplication in the time domain is equivalent to convolution in the frequency domain:

$$F'(v') = \mathfrak{F}\big(f(t) \cdot g(t)\big). \tag{5.3}$$

The time domain data are multiplied by an appropriate filter or weighting function characterized by a starting value of 1 and that decays regularly to 0 at the end of the acquisition time. There are numerous such mathematical functions used in the time domain, but the selection of proper one depends on the application. Some of the most used filters are exponential function, Gaussian function, Lorentzian/Gaussian function, Sinebell function, Hamming function, and Hanning function (Hoch and Stern, 1996; Lindon and Ferrige, 1980).

5.3.1.2 Zero Filling

This process, also known as zero padding, extends the experimental time domain data set by appending a number of zeros. This implies that the part of the signal that is not acquired has relaxed completely, thus it has a value of zero. This has a similar effect as increasing the sampling points after the signal has decayed. It results in increased spectral resolution, equivalent to a smooth interpolation between spectral data points and improves spectral visualization after Fourier transformation (Lindon and Ferrige, 1980). However, it is important to note that zero filling has no effect on the SNR. Zero filling will induce a discontinuity if the signal has not decayed to zero at the end of the acquisition which will result in a distorted spectrum. Also the FFT requires that the number of data points, N, is a power of 2. Thus, if $2\,m-1 < N < 2\,m$, then zero filling is required to bring the number of points to $2\,m$. In theory, zero filling by at least a factor of 2 is highly recommended because it enforces causality (Hoch and Stern, 1996). But beyond that, the gain in resolution is purely cosmetic, producing only more data points per hertz without adding any new information.

5.3.1.3 Lineshape Correction

Due to field inhomogeneity caused by imperfect shim, small sample size, proximity of the sample and RF coil, nonspinning samples, or samples with intrinsic spatial inhomogeneity, measured MRS signal may have insufficient spectral resolution and actual lineshape may deviate from the model lineshapes. Reference deconvolution, which uses a reference signal, is a resolution enhancement technique that has been applied by a number of investigators. The reference signal that is generally used is the unsuppressed water signal. One of the general applicable methods for lineshape correction is QUALITY (de Graaf et al., 1990), which can remove all deviations from the pure Lorentzian lineshape in the experimental spectrum via time domain deconvolution. It improves the accuracy of spectral quantification through a model-based algorithm and also preserves peak areas.

Another utility of water reference signal is for eddy current correction (Klose, 1990). Eddy currents are induced by the switching magnetic field gradients in localized spectroscopy sequences. The eddy currents produce time-dependent shifts of the resonance frequency in the selected volume, which results in a distortion of the spectrum after Fourier transformation. The correction is performed in the time domain by dividing the water-suppressed signal by the phase factor of the water signal for each data point. It should be noted that the acquisition of a reference signal itself is sensitive to errors and success of reference deconvolution depends on choosing a suitable reference signal. If the reference signal is acquired separately, both the spectral data and the reference signal should have the same digital resolution. Another drawback is that for clinical applications of MRS, the acquisition of additional reference signals can be too time consuming in the CSI experiments.

5.3.1.4 Phase Correction

Due to quadrature acquisition, time delay between transmit and receive periods introduces a phase shift in the frequency domain. This can be explained by the "time shifting" theorem for an FT, which states that

$$\mathfrak{F}\{s(t-\tau)\} = e^{-i2\pi v' \tau} s(v').$$
(5.4)

This may lead to signal loss, since addition of spectra with different phases during signal averaging will result in phase cancellation. There are two kinds of phase correction involved: zero-order phase correction, which is frequency independent and first-order phase correction, which is frequency dependent. Phase correction routines are based on optimizing these two parameters.

5.3.1.5 Water Suppression

In MRS, it is desirable to minimize dominating unsuppressed tails from water or lipids, which hamper accurate quantification of other nearby small-intensity resonances. Even though ^{1}H signals are normally acquired with some water-suppression technique, some water signals still remain, which produce a relatively high peak and can obscure signals of interest. Removal of water peak can reduce the computational burden as well as improve the accuracy of the parameter estimates. Many methods have been proposed in the literature for the suppression of water and unwanted large solvent peaks. These methods can be classified in two groups: the pulse sequence and hardware-based methods and the postprocessing methods (Price, 1999).

The earliest postprocessing techniques for residual water removal in ^{1}H NMR were based on filtering. In Kuroda et al. (1989), a simple high-pass finite impulse response (FIR) filter was used to suppress the water peak. Marion et al. (1989) proposed an extrapolation method that uses a low-pass FIR filter of relatively high order to suppress all metabolite resonances at higher frequencies. There are other filters such as extraction and reduction (ER) filter (Cavassila et al., 1997), the filter proposed by Dologlou et al. (1998), or the FIR filter (Sundin et al., 1999). In the time domain, the contribution of unwanted peaks can be reduced by introducing an appropriate weighting vector in the nonlinear least-squares (NLLS) cost function of the parameter estimation techniques (Knijn et al., 1992). Another approach is to model the nuisance peaks by a black-box method and subtract the reconstructed time domain signal from the original signal prior to parameter estimation. A black-box method that can be used in this context is the Hankel singular value decomposition (HSVD) method (Leclerc, 1994; van den Boogaart et al., 1994). Later a fast version of HSVD called Hankel-Lanczos singular value decomposition (HLSVD), which is based on the Lanczos procedure (Golub and Van Loan, 1996) was proposed by Pijnappel et al. (1992).

5.3.1.6 Baseline Correction

NMR spectrum is generally composed of broad and slowly varying background underlying sharp peaks; the latter are referred to as spectral lines and the former is termed baseline. The spectral lines arise from the tissue under investigation and contain information about the metabolite of interest, while the baseline often originates from various sources and in most of the cases interferes with the desired signal (Dong et al., 2006). Nearly all type of experimental NMR spectra contains baseline distortion artifacts. One of the main sources of baseline distortion is the large, less mobile macromolecules and lipids. Signals of macromolecules and lipids have much shorter spin–spin relaxation time (T_2) than the low molecular weight metabolites making their signals decay very rapidly in the time domain. At long TE their contribution to the spectra become smaller, but at short TE they contribute signals with broad features in the frequency domain causing distorted baseline.

Some of the other sources of baseline distortion are instrumental effects such as residual water signal, eddy current induced by pulsed gradient signals, "dead time" between signal excitation and detection and postprocessing procedures such as water suppression or discarding the first few data points in the FID. These distortions may obscure chemical information and affect spectral analysis. To ensure reliable and accurate spectral analysis, especially for spectral quantification, it is important to have a flat baseline. Therefore an effective baseline correction algorithm is essential.

During the past three decades, numerous baseline correction algorithms have been developed (Carlos Cobas et al., 2006; Chang et al., 2007; Golotvin and Williams, 2000; Pearson, 1977; Ratiney et al., 2004; Soher et al., 2001; Stanley and Pettegrew, 2001; Young et al., 1998). Most of these fall into three categories: (1) correction of baseline during actual data acquisition; (2) postprocessing method to improve the time domain data; and (3) postprocessing method to improve the frequency domain data.

5.3.2 Data Quantification

In vivo MRS data quantitation can be done either in time or frequency domain. These two ways of analysis are different in principle: time domain methods operate directly on the measured FID signal, whereas frequency domain methods use the Fourier transformed spectrum of the original FID. The time domain signal and its frequency domain counterpart contain exactly the same information and are in this sense equivalent. If performed correctly, both frequency and time domain spectral analytic methods could produce the same results (Joliot et al., 1991; van den Boogaart et al., 1994).

5.3.2.1 Frequency Domain Methods

Frequency domain methods can be divided into two classes: nonparametric and parametric quantification. Nonparametric quantification can be done by peak integration, a method rarely used nowadays. It is based on integration of the peak area of the frequency-domain signal (Meyer et al., 1988). The advantages are that no assumptions have to be made concerning the signal and it is very robust. But the method has low estimation accuracy. The accuracy of the integration-based methods is highly dependent on proper phase corrections which is quite difficult to perform.

The *parametric* frequency domain methods are based on a model function to model the spectrum and some fitting method. They are more complicated than the nonparametric approach and NLSS minimization algorithm are used to obtain the estimates of the parameters in the frequency domain. Theoretically, this approach is equivalent to the minimization problem in the time domain. Several spectral fitting methods have been developed recently, such as TLS (Laatikainen et al., 1996), second derivative method (Sokol, 2001), deconvolution method (Swanson et al., 2006), or the methods proposed by Stanley et al. (1995), Van de Veen et al. (2000), or Umesh and Tufts (1996).

A semiparametric method called LCModel has been proposed by Provencher (1993, 2001) and is the most widely used commercial metabolite quantification method for *in vivo* proton MRS. Although we have mentioned several frequency domain methods, most of those are not successful or appropriate for quantification of short TE *in vivo* ¹H MRS due to low SNR or overlapping peaks and significant artifacts. LCModel excel in these cases compared to other methods.

5.3.2.1.1 LCModel

The LCModel method fits an *in vivo* spectrum as a linear combination of a basis set of complete model spectra of metabolite solutions *in vitro*. The model used can be written in the time domain (Pels, 2005),

$$S(t_n) = \sum_{k=1}^{K} \Delta a_k \cdot M_k(t_n) \cdot e^{i\Delta f_k} \cdot e^{(-\Delta d_k + i2\pi\Delta v_k)t_k}, \tag{5.5}$$

where K is the number of measured metabolite spectra M_k and Δa_k, Δd_k, $\Delta \varphi_k$, Δv_k represent the difference in amplitude, damping, phase and frequency between the acquired metabolite solutions and the signal to be analyzed, respectively. A maximum likelihood solution is found through a NLLS minimization problem. It has the advantage that a maximal amount of prior knowledge, in terms of model spectra, is already provided. Moreover, model distortions due to acquisition effects can also be taken into account. Major drawback of the method is that it is necessary to measure model spectra of all metabolites *in vitro*, with the same experimental conditions and parameters as those of *in vivo* experiments. Whenever experimental conditions or parameters change, new model spectra must be measured and calibrated. This can be tedious and rather time consuming. The phantoms of the model solutions should be specially treated and well preserved from contamination for long term use. Alternatively, synthetic 1D MR spectra of different metabolites can be simulated using GAMMA library (Smith et al., 1994).

5.3.2.1.2 Prior-Knowledge Fitting

Schulte and Boesiger first proposed the 2D fitting algorithm called Prior knowledge Fitting (ProFit) (Schulte et al., 2006) and showed an increase in the number of quantifiable metabolites using maximum echo sampling 2D JPRESS compared with 1D PRESS. A modified version of the algorithm was used for the processing of 2D COSY spectra by Sarma et al. (2011). In contrast to commonly applied methods based on simple line fitting and peak integration for metabolite quantification in 2D MRS of brain, ProFit yields metabolite concentration ratios; thus absolute concentrations can be calculated using the reported concentration of Creatine (Cr). It fits spectra as linear combinations of 2D basis spectra using an NLLS algorithm in combination with a linear least-squares algorithm to minimize a cost function made up of the sum of squares residual plus penalty factors for regularization. At the start of the fitting process, the global phase correction and frequency shifts are determined using the dominant singlets. Then it enters the first of the three iterations, where relative frequency shifts and phases are fitted nonlinearly for the dominant metabolites and subsequently fitting concentrations linearly. In the second iteration, a line-shape distortion is added to the nonlinear fit. Finally, in the third iteration all the basis metabolites are included. In this way, by dividing the algorithm into a linear and nonlinear part, the number of parameters is reduced, leading to faster overall convergence. At the same time, ProFit incorporates maximum prior knowledge available. For basis-set creation synthetic 2D JPRESS and L-COSY spectra of different metabolites are simulated using GAMMA library in combination with the chemical shift and J-coupling values reported in the literature (Govindaraju et al., 2000; Smith et al., 1994).

5.3.2.2 Time Domain Methods

Time domain methods are usually divided into two categories: interactive and black-box methods. Interactive methods have the flexibility that all types of mathematical shape of the model functions can be considered and prior knowledge can be employed easily. But these methods require user involvement, they are time consuming and there is no guarantee that global optimum can be obtained. In the first iteration step, starting values for the nonlinear parameters need to be provided. On the contrary, black box methods need minimal user interaction, model parameters are calculated in one single step and there is no need for initial values. However, prior knowledge can be applied in a limited scale and model function is limited to the exponential decay model.

5.3.2.2.1 Black-Box Methods

Black-box methods can further be subdivided into two classes: linear prediction methods based on the linear prediction principle (Koehl, 1999; Stephenson, 1988) and state-space methods (Barkhuijsen et al., 1987; de Beer and van Ormondt, 1992; Kung et al., 1983) based on state-space formalism. One of the best known state-space methods is HSVD (Barkhuijsen et al., 1987; Kung et al., 1983). Many variants of the HSVD algorithm exist. An improved version is the HTLS method, which computes the solution of the model equation in a total least squares (TLS) sense which leads to a more accurate parameter estimation (Vanhuffel et al., 1994). HLSVD is a computationally more efficient version (Pijnappel et al., 1992). In this method, only part of the SVD is computed using the Lanczos algorithm. Apart from using for MRS data quantification, HSVD can also be applied for filtering of unwanted components in the signals, such as residual water. The HLSVD method has also been used to quantify 2D spectra (de Beer et al., 1992).

5.3.2.2.2 *Interactive Methods*

The first well-known interactive method for 1D MRS data quantification was VARPRO (van der Veen et al., 1988). It uses a simple Levenberg–Marquardt algorithm to minimize the variable projection functional (Osborne, 1972). VARPRO was further improved to the AMARES algorithm (Vanhamme et al., 1997) in terms of robustness and flexibility and minimizes the general functional using a version of NL2SOL (Dennis and Schnabel, 1983), a sophisticated NLLS algorithm. AMARES allows the imposition of more types of prior knowledge. Other features of AMARES are the possibility of choosing a Lorentzian as well as a Gaussian lineshape for each peak and imposing lower and upper bounds on the parameters.

5.4 ^{1}H MR Spectroscopy

A representative ^{1}H MRS spectrum of the human brain acquired at 3 T with identifiable major metabolite peaks assigned is shown in Figure 5.9. When water- and fat-suppression techniques are used, ^{1}H MRS detects a number of

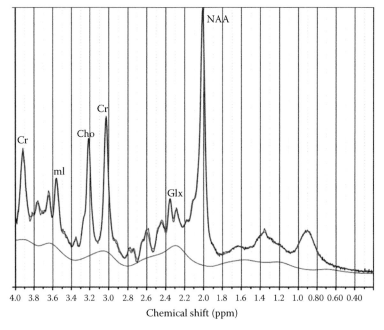

FIGURE 5.9

(Color version available from crcpress.com; see Preface.) A PRESS-localized 1D spectrum of healthy control at 3 T and processed using LCModel. It shows the fitted curve (solid line) overlaid on the original data with a smooth fitted baseline (lower curve).

metabolites present in relatively low concentrations (<10 mM). Major brain metabolites that can be analyzed with ^{1}H MRS are discussed below.

5.4.1 N-Acetylaspartate

In normal spectra, *N*-acetylaspartate (NAA) is the largest peak observed in human brain. The presence of NAA is attributable to its *N*-acetyl methyl group, which resonates at 2.0 ppm. This peak also contains contributions from less important *N*-acetyl groups (Frahm et al., 1991). In adult brain, NAA has been shown to be predominantly localized in neurons, axons, and dendrites within the central nervous system (Simmons et al., 1991). The exact role of NAA in the brain is not known. NAA is labeled as a neuronal marker. It acts as neurotransmitter, precursor of brain lipids, involved in coenzyme A interaction and balance anion deficit in neural tissue. Decreased NAA was reported in developmental delay, infancy, hypoxia, anoxia, ischemia, herpes encephalitis, neoplasms, epilepsy, and so on (Miller, 1991). On the other hand, a marked increase in NAA is observed in Canavan disease.

5.4.2 Creatine

Creatine (Cr) has two groups of peaks, one at 3.0 ppm and the other at 3.9 ppm. The peak at 3 ppm contains contributions from creatine, creatine phosphate, and, to a lesser degree, γ-aminobutyric acid, lysine, and glutathione. Therefore, the Cr peak is sometimes referred to as "total Cr." It is an important mediator in energy metabolism (Kreis et al., 1993). Creatine remains fairly stable under many pathological conditions in the brain and is used as internal reference for absolute quantification. Creatine is increased in hyperosmolar states and decreased in hypoosmolar states.

5.4.3 Choline

The peak for choline (Cho) is seen at 3.2 ppm and contains contributions from glycerophosphocholine, phosphocholine, and free choline. Choline reflects the structural components of membrane metabolism-myelin sheaths (Miller, 1991). It is increased in tumor, demyelinization, multiple sclerosis (MS), trauma, diabetes mellitus (DM), neonates, postliver transplant, chronic hypoxia, and hyperosmolar states and is decreased in asymptomatic liver disease, hypoxic encephalopathy, stroke, and nonspecific dementias.

5.4.4 Myoinositol

One of the largest peaks from (mI) occurs at 3.56 ppm. The amino acid glycine may also contribute to the mI resonance. Myoinositol is a glial marker and an osmolite important for cell growth. It is involved in hormone-sensitive neuroreception, detoxification (van der Knaap et al., 1994). Its levels have been found to be reduced in chronic hypoxic encephalopathy (Ross et al., 1994), stroke, and tumor and increased in neonate, Alzheimer's disease (Shonk et al., 1995), DM, recovered hypoxia, and hyperosmolar states.

5.4.5 Lactate

Lactate (Lac) doublet occurs at 1.32 ppm. A second multiplet (quartet) for lactate occurs at 4.1 ppm. In normal human brain, lactate levels are low and generally not detectable. Any detectable increase of lactate except perhaps in CSF indicates that the normal cellular oxidative respiration mechanism is no longer in effect, and that carbohydrate catabolism is taking place (Sanders, 1995).

5.4.6 Glutamate/Glutamine

Glutamate/Glutamine (Glu/Gln), also known as Glx, peaks are located between 1.9 and 2.4 ppm and also between 3.6 and 3.9 ppm. It is not possible to separate one from the other using 1D MRS. Glutamine is a precursor of glutamate. Glutamate is an excitatory neurotransmitter important in neurotransmission. Gamma-aminobutyric acid (GABA), also present but in lower concentrations during normal physiological conditions, may overlap with the Glu/Gln resonance at 1.5 T field strength. Glx is found to be elevated in liver disease such as hepatic encephalopathy.

5.5 ^{31}P MR Spectroscopy

Next to ^{1}H MRS, ^{31}P MRS is the most widely used clinical spectroscopy modality. Compared to hydrogen (^{1}H), phosphorus (^{31}P) is less MR sensitive, but naturally present in abundance. Also, it has a larger chemical shift range (~30 ppm), making it possible to acquire simple and excellent spectrum even at low magnetic field strengths. ^{31}P MRS is extensively used for the assessment of tissue energy metabolism (Pettegrew et al., 1986, 1987), brain membrane phospholipid (Pettegrew et al., 1986, 1987), and phosphorus-containing markers of disease. It provides a noninvasive and dynamic method for determining biologically relevant parameters such as intracellular pH (Petroff et al., 1985; Pettegrew et al., 1988), intracellular free magnesium concentration [Mg^{2+}], and rate constant of creatine kinase (CK) reaction. Figure 5.10 shows a representative ^{31}P MRS spectrum of the human brain at 3 T (Martin, 2007) where peaks of major metabolites observed are labeled. The principal high energy phosphate metabolites that can be identified in an *in vivo* human ^{31}P spectrum are phosphocreatine (PCr), three separate signal from adenosine triphosphate (ATP) labeled as α-, β-, and γ-ATP, inorganic phosphate (Pi), phosphodiesters (PDE), and phosphomonoesters

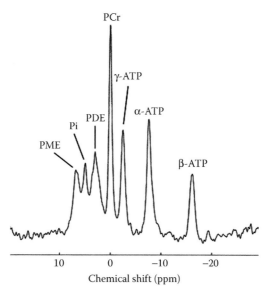

FIGURE 5.10

A representative ^{31}P MRS spectrum from the temporo-occipital cortex in a normal volunteer at 3 T. (From Martin, W.R., *Molecular Imaging and Biology.* 9, 196–203, 2007. With permission.)

(PME) such as phosphocholine, phosphoethanolamine, and sugar phosphates. Conventionally, PCr peak is used as internal chemical shift reference and a ppm value of 0 is assigned. At lower field strengths, PME peaks may overlap with Pi resonances.

Though ^{31}P NMR has been used for the study of animal brains only since the introduction of surface coils (Ackerman et al., 1980), investigations on humans only started in 1980s. In 1983, Cady et al. investigated the intracellular metabolism in the brains of infants born at 33–40 weeks' gestation using a surface RF coil (1983). They observed that the PCr/Pi ratio was low in infants who had severe birth asphyxia and increased as their clinical condition improved (or on infusion of a solution of mannitol). During the last two decades, ^{31}P MRS has been used extensively for clinical diagnosis. [31]P has been found to be useful for studying neurological diseases, such as brain ischemia and seizure (Gadian et al., 1993; Welch et al., 1992), epilepsy (Hetherington et al., 2002; Weiner, 1987), Alzheimer disease (Brown et al., 1989), Parkinson (Hoang et al., 1998), and schizophrenia (Pettegrew and Minshew, 1992). *In vivo* studies using ^{31}P MRS have shown that tumors are often characterized by elevated PME and/or PDE (de Certaines et al., 1993).

Owing to the short T_2 relaxation times, ^{31}P MRS requires pulse sequences with short TE. The spectrum may be acquired during the FID or a short TE spin-echo. Spatial localization is often executed by a surface coil or through use of the ISIS technique where the z-magnetization is selected using slice-selective inversion pulses.

References

Ackerman, J.J., Grove, T.H., Wong, G.G., Gadian, D.G., Radda, G.K., 1980. Mapping of metabolites in whole animals by 31P NMR using surface coils. *Nature.* 283, 167–170.

Adalsteinsson, E., Spielman, D.M., 1999. Spatially resolved two-dimensional spectroscopy. *Magnetic Resonance in Medicine.* 41, 8–12.

Aue, W.P., Karhan, J., Ernst, R.R., 1976. Homonuclear broad band decoupling and two-dimensional J-resolved NMR spectroscopy. *The Journal of Chemical Physics.* 64, 4226–4227.

Aue, W.P., Muller, S., Cross, T.A., Seelig, J., 1984. Volume-selective excitation. A novel approach to topical NMR. *Journal of Magnetic Resonance.* 56, 350–354.

Barkhuijsen, H., de Beer, R., van Ormondt, D., 1987. Improved algorithm for noniterative time-domain model fitting to exponentially damped magnetic resonance signals. *Journal of Magnetic Resonance.* 73, 553–557.

Binesh, N., Yue, K., Fairbanks, L., Thomas, M.A., 2002. Reproducibility of localized 2D correlated MR spectroscopy. *Magnetic Resonance in Medicine.* 48, 942–948.

Bolan, P.J., DelaBarre, L., Baker, E.H., Merkle, H., Everson, L.I., Yee, D., Garwood, M., 2002. Eliminating spurious lipid sidebands in 1H MRS of breast lesions. *Magnetic Resonance in Medicine.* 48, 215–222.

Bottomley, P.A., 1987. Spatial localization in NMR spectroscopy in vivo. *Annals of the New York Academy of Sciences.* 508, 333–348.

Brown, G.G., Levine, S.R., Gorell, J.M., Pettegrew, J.W., Gdowski, J.W., Bueri, J.A., Helpern, J.A., Welch, K.M., 1989. In vivo 31P NMR profiles of Alzheimer's disease and multiple subcortical infarct dementia. *Neurology.* 39, 1423–1427.

Cady, E.B., Costello, A.M., Dawson, M.J., Delpy, D.T., Hope, P.L., Reynolds, E.O., Tofts, P.S., Wilkie, D.R., 1983. Non-invasive investigation of cerebral metabolism in newborn infants by phosphorus nuclear magnetic resonance spectroscopy. *Lancet.* 1, 1059–1062.

Carlos Cobas, J., Bernstein, M.A., Martin-Pastor, M., Tahoces, P.G., 2006. A new general-purpose fully automatic baseline-correction procedure for 1D and 2D NMR data. *Journal of Magnetic Resonance.* 183, 145–151.

Cavassila, S., Fenet, B., van den Boogaart, A., Remy, C., Briguet, A., Graveron-Demilly, D., 1997. ER-Filter: A preprocessing technique for frequency-selective time-domain analysis. *Magnetic Resonance Analysis.* 3, 87–92.

Chang, D., Banack, C.D., Shah, S.L., 2007. Robust baseline correction algorithm for signal dense NMR spectra. *Journal of Magnetic Resonance.* 187, 288–292.

de Beer, R., van Ormondt, D., 1992. Analysis of NMR data using time-domain fitting procedures. In: *NMR, Basic Principles and Progress.* Rudin, M., Seelig, J., eds., vol. 26. Springer-Verlag, Berlin, pp. 201–248.

de Beer, R., van Ormondt, D., Pijnappel, W.W.F., 1992. Quantification of 1-D and 2-D magnetic resonance time domain signals. *Pure and Applied Chemistry.* 64, 815–823.

de Certaines, J.D., Larsen, V.A., Podo, F., Carpinelli, G., Briot, O., Henriksen, O., 1993. In vivo 31P MRS of experimental tumours. *NMR in Biomedicine.* 6, 345–365.

de Graaf, A.A., van Dijk, J.E., Bovee, W.M., 1990. QUALITY: Quantification improvement by converting lineshapes to the Lorentzian type. *Magnetic Resonance in Medicine.* 13, 343–357.

De Graaf, R.A., 1998. *In vivo NMR Spectroscopy: Principles and Techniques.* Wiley, Chichester.

Dennis, J.E., Schnabel, R.B., 1983. *Numerical Methods for Unconstrained Optimization and Nonlinear Equations.* Prentice-Hall, Englewood Cliffs.

Dologlou, I., Van Huffel, S., Van Ormondt, D., 1998. Frequency-selective MRS data quantification with frequency prior knowledge. *Journal of Magnetic Resonance.* 130, 238–243.

Dong, Z., Dreher, W., Leibfritz, D., 2006. Toward quantitative short-echo-time in vivo proton MR spectroscopy without water suppression. *Magnetic Resonance in Medicine.* 55, 1441–1446.

Dreher, W., Leibfritz, D., 1999. Detection of homonuclear decoupled in vivo proton NMR spectra using constant time chemical shift encoding: CT-PRESS. *Magnetic Resonance Imaging.* 17, 141–150.

Duijn, J.H., Matson, G.B., Maudsley, A.A., Weiner, M.W., 1992. 3D phase encoding 1H spectroscopic imaging of human brain. *Magnetic Resonance Imaging.* 10, 315–319.

Duyn, J.H., Moonen, C.T., 1993. Fast proton spectroscopic imaging of human brain using multiple spin-echoes. *Magnetic Resonance in Medicine.* 30, 409–414.

Emst, R., Bodenhausen, G., Wokaun, A., 1987. *Principles of NMR in One and Two Dimensions.* Oxford Publications, Oxford.

Ernst, R.R., Anderson, W.A., 1966. Applications of Fourier transform spectroscopy to magnetic resonance. *Review of Scientific Instruments* 37, 93–102.

Fan, G., Sun, B., Wu, Z., Guo, Q., Guo, Y., 2004. In vivo single-voxel proton MR spectroscopy in the differentiation of high-grade gliomas and solitary metastases. *Clinical Radiology.* 59, 77–85.

Frahm, J., Bruhn, H., Gyngell, M.L., Merboldt, K.D., Hanicke, W., Sauter, R., 1989. Localized high-resolution proton NMR spectroscopy using stimulated echoes: Initial applications to human brain in vivo. *Magnetic Resonance in Medicine.* 9, 79–93.

Frahm, J., Merboldt, K.-D., Hanicke, W., 1987. Localized proton spectroscopy using stimulated echoes. *Journal of Magnetic Resonance.* 72, 502–508.

Frahm, J., Michaelis, T., Merboldt, K.D., Hanicke, W., Gyngell, M.L., Bruhn, H., 1991. On the N-acetyl methyl resonance in localized 1H NMR spectra of human brain in vivo. *NMR in Biomedicine.* 4, 201–204.

Gadian, D.G., Williams, S.R., Bates, T.E., Kauppinen, R.A., 1993. NMR spectroscopy: Current status and future possibilities. *Acta Neurochirurgica: Supplementum.* 57, 1–8.

Golotvin, S., Williams, A., 2000. Improved baseline recognition and modeling of FT NMR spectra. *Journal of Magnetic Resonance.* 146, 122–125.

Golub, G.H., Van Loan, C.F., 1996. *Matrix Computations.* Johns Hopkins University Press, Baltimore.

Govindaraju, V., Young, K., Maudsley, A.A., 2000. Proton NMR chemical shifts and coupling constants for brain metabolites. *NMR in Biomedicine.* 13, 129–153.

Haase, A., Frahm, J., Hanicke, W., Matthaei, D., 1985. 1H NMR chemical shift selective (CHESS) imaging. *Physics in Medicine and Biology.* 30, 341–344.

Hanstock, C.C., Rothman, D.L., Prichard, J.W., Jue, T., Shulman, R.G., 1988. Spatially localized 1H NMR spectra of metabolites in the human brain. *Proceedings of the National Academy of Sciences of the United States of America.* 85, 1821–1825.

Herigault, G., Zoula, S., Remy, C., Decorps, M., Ziegler, A., 2004. Multi-spin-echo J-resolved spectroscopic imaging without water suppression: Application to a rat glioma at 7T. *MAGMA.* 17, 140–148.

Hetherington, H.P., Pan, J.W., Spencer, D.D., 2002. 1H and 31P spectroscopy and bio-energetics in the lateralization of seizures in temporal lobe epilepsy. *Journal of Magnetic Resonance Imaging.* 16, 477–483.

Hoang, T.Q., Bluml, S., Dubowitz, D.J., Moats, R., Kopyov, O., Jacques, D., Ross, B.D., 1998. Quantitative proton-decoupled 31P MRS and 1H MRS in the evaluation of Huntington's and Parkinson's diseases. *Neurology.* 50, 1033–1040.

Hoch, J.C., Stern, A.S., 1996. *NMR Data Processing.* Wiley, New York.

Hurd, R., Sailasuta, N., Srinivasan, R., Vigneron, D.B., Pelletier, D., Nelson, S.J., 2004. Measurement of brain glutamate using TE-averaged PRESS at 3T. *Magnetic Resonance in Medicine.* 51, 435–440.

Hurd, R.E., Gurr, D., Sailasuta, N., 1998. Proton spectroscopy without water suppression: The oversampled J-resolved experiment. *Magnetic Resonance in Medicine.* 40, 343–347.

Jeener, J., 1971. Unpublished Lecture. Ampere International Summer School, Basko Polje, Yugoslavia.

Jensen, J.E., Frederick, B.D., Wang, L., Brown, J., Renshaw, P.F., 2005. Two-dimensional, J-resolved spectroscopic imaging of GABA at 4 Tesla in the human brain. *Magnetic Resonance in Medicine.* 54, 783–788.

Joliot, M., Mazoyer, B.M., Huesman, R.H., 1991. In vivo NMR spectral parameter estimation: A comparison between time and frequency domain methods. *Magnetic Resonance in Medicine.* 18, 358–370.

Ke, Y., Cohen, B.M., Bang, J.Y., Yang, M., Renshaw, P.F., 2000. Assessment of GABA concentration in human brain using two-dimensional proton magnetic resonance spectroscopy. *Psychiatry Research.* 100, 169–178.

Ke, Y., Streeter, C.C., Nassar, L.E., Sarid-Segal, O., Hennen, J., Yurgelun-Todd, D.A., Awad, L.A., et al., 2004. Frontal lobe GABA levels in cocaine dependence: A two-dimensional, J-resolved magnetic resonance spectroscopy study. *Psychiatry Research.* 130, 283–293.

Keevil, S.F., 2006. Spatial localization in nuclear magnetic resonance spectroscopy. *Physics in Medicine and Biology.* 51, R579–R636.

Klose, U., 1990. In vivo proton spectroscopy in presence of eddy currents. *Magnetic Resonance in Medicine.* 14, 26–30.

Knijn, A., De Beer, R., Van Ormondt, D., 1992. Frequency-selective quantification in the time domain. *Journal of Magnetic Resonance.* 97, 444–450.

Koehl, P., 1999. Linear prediction spectral analysis of NMR data. *Progress in Nuclear Magnetic Resonance Spectroscopy.* 34, 257–299.

Kreis, R., Boesch, C., 1994. Liquid-crystal-like structures of human muscle demonstrated by in vivo observation of direct dipolar coupling in localized proton magnetic resonance spectroscopy. *Journal of Magnetic Resonance Series B.* 104, 189–192.

Kreis, R., Boesch, C., 1996. Spatially localized, one- and two-dimensional NMR spectroscopy and in vivo application to human muscle. *Journal of Magnetic Resonance Series B.* 113, 103–118.

Kreis, R., Ernst, T., Ross, B.D., 1993. Development of the human brain: In vivo quantification of metabolite and water content with proton magnetic resonance spectroscopy. *Magnetic Resonance in Medicine.* 30, 424–437.

Kuhn, B., Dreher, W., Leibfritz, D., Heller, M., 1999. Homonuclear uncoupled 1H-spectroscopy of the human brain using weighted accumulation schemes. *Magnectic Resonance Imaging.* 17, 1193–1201.

Kung, S.Y., Arun, K.S., Rao, D.V.B., 1983. State-space and singular-value decomposition-based approximation methods for the harmonic retrieval problem. *Journal of the Optical Society of America.* 73, 1799–1811.

Kuroda, Y., Wada, A., Yamazaki, T., Nagayama, K., 1989. Postacquisition data processing method for suppression of the solvent signal. *Journal of Magnetic Resonance.* 84, 604–610.

Laatikainen, R., Niemitz, M., Malaisse, W.J., Biesemans, M., Willem, R., 1996. A computational strategy for the deconvolution of NMR spectra with multiplet structures and constraints: Analysis of overlapping 13C-2H multiplets of 13C enriched metabolites from cell suspensions incubated in deuterated media. *Magnetic Resonance in Medicine.* 36, 359–365.

Leclerc, J.H.J., 1994. Distortion-free suppression of the residual water peak in proton spectra by postprocessing. *Journal of Magnetic Resonance, Series B.* 103, 64–67.

Levy, L.M., Hallett, M., 2002. Impaired brain GABA in focal dystonia. *Annals of Neurology.* 51, 93–101.

Lindon, J.C., Ferrige, A.G., 1980. Digitisation and data processing in Fourier transform NMR. *Progress in Nuclear Magnetic Resonance Spectroscopy.* 14, 27–66.

Lipnick, S., Verma, G., Ramadan, S., Furuyama, J., Thomas, M.A., 2010. Echo planar correlated spectroscopic imaging: Implementation and pilot evaluation in human calf in vivo. *Magnetic Resonance in Medicine.* 64, 947–956.

Mansfield, P., 1977. Multi-planar image formation using NMR spin echoes. *Journal of Physics C: Solid State Physics.* 10, L55.

Marion, D., Ikura, M., Bax, A., 1989. Improved solvent suppression in one- and two-dimensional NMR spectra by convolution of time-domain data. *Journal of Magnetic Resonance.* 84, 425–430.

Martin, W.R., 2007. MR spectroscopy in neurodegenerative disease. *Molecular Imaging and Biology.* 9, 196–203.

Mayer, D., Dreher, W., Leibfritz, D., 2000. Fast echo planar based correlation-peak imaging: Demonstration on the rat brain in vivo. *Magnetic Resonance in Medicine.* 44, 23–28.

Mayer, D., Dreher, W., Leibfritz, D., 2003. Fast U-FLARE-based correlation-peak imaging with complete effective homonuclear decoupling. *Magnetic Resonance in Medicine.* 49, 810–816.

Mcrobbie, D., Moore, E., Graves, M., Prince, M., 2006. *MRI from Picture to Proton* (2nd Ed.). Cambridge University Press, Cambridge.

Meyer, R.A., Fisher, M.J., Nelson, S.J., Brown, T.R., 1988. Evaluation of manual methods for integration of in vivo phosphorus NMR spectra. *NMR in Biomedicine.* 1, 131–135.

Miller, B.L., 1991. A review of chemical issues in 1H NMR spectroscopy: N-acetyl-L-aspartate, creatine and choline. *NMR in Biomedicine.* 4, 47–52.

Moonen, C.T.W., Sobering, G., Van Zijl, P.C.M., Gillen, J., Von Kienlin, M., Bizzi, A., 1992. Proton spectroscopic imaging of human brain. *Journal of Magnetic Resonance.* 98, 556–575.

Ogg, R.J., Kingsley, R.B., Taylor, J.S., 1994. WET, a T1- and B1-insensitive water-suppression method for in vivo localized 1H NMR spectroscopy. *Journal of Magnetic Resonance, Series B.* 104, 1–10.

Ordidge, R.J., Connelly, A., Lohman, J.A.B., 1986. Image-selected in vivo spectroscopy (ISIS). A new technique for spatially selective NMR spectroscopy. *Journal of Magnetic Resonance.* 66, 283–294.

Osborne, M., 1972. Some aspects of non-linear least squares calculations. In: *Numerical Methods for Nonlinear Optimization: Conference Sponsored by the Science Research Council*, Lootsma, F.A., ed. [and held at the] University of Dundee, Scotland, 1971. Academic Press, London, pp. 171–189.

Pearson, G.A., 1977. A general baseline-recognition and baseline-flattening algorithm. *Journal of Magnetic Resonance.* 27, 265–272.

Pels, P., 2005. *Analysis and Improvement of Quantification Algorithms for Magnetic Resonance Spectroscopy*, Katholieke Universiteit Leuven, Leuven, p. 140.

Petroff, O.A., Prichard, J.W., Behar, K.L., Rothman, D.L., Alger, J.R., Shulman, R.G., 1985. Cerebral metabolism in hyper- and hypocarbia: 31P and 1H nuclear magnetic resonance studies. *Neurology.* 35, 1681–1688.

Pettegrew, J.W., Kopp, S.J., Dadok, J., Minshew, N.J., Feliksik, J.M., Glonek, T., Cohen, M.M., 1986. Chemical characterization of a prominent phosphomono-ester resonance from mammalian brain. 31P and 1H NMR analysis at 4.7 and 14.1 Tesla. *Journal of Magnetic Resonance.* 67, 443–450.

Pettegrew, J.W., Minshew, N.J., 1992. Molecular insights into schizophrenia. *Journal of Neural Transmission. Supplementum.* 36, 23–40.

Pettegrew, J.W., Withers, G., Panchalingam, K., Post, J.F., 1987. 31P nuclear magnetic resonance (NMR) spectroscopy of brain in aging and Alzheimer's disease. *Journal of Neural Transmission. Supplementa.* 24, 261–268.

Pettegrew, J.W., Withers, G., Panchalingam, K., Post, J.F., 1988. Considerations for brain pH assessment by 31P NMR. *Magnectic Resonance Imaging.* 6, 135–142.

Pijnappel, W.W.F., van den Boogaart, A., de Beer, R., van Ormondt, D., 1992. SVD-based quantification of magnetic resonance signals. *Journal of Magnetic Resonance.* 97, 122–134.

Price, W.S., 1999. Water signal suppression in NMR spectroscopy. *Annual Reports on NMR Spectroscopy.* 38, 289–354.

Prichard, J.W., 1997. New nuclear magnetic resonance data in epilepsy. *Current Opinion in Neurology.* 10, 98–102.

Provencher, S.W., 1993. Estimation of metabolite concentrations from localized in vivo proton NMR spectra. *Magnetic Resonance in Medicine.* 30, 672–679.

Provencher, S.W., 2001. Automatic quantitation of localized in vivo 1H spectra with LCModel. *NMR in Biomedicine.* 14, 260–264.

Ratiney, H., Coenradie, Y., Cavassila, S., van Ormondt, D., Graveron-Demilly, D., 2004. Time-domain quantitation of 1H short echo-time signals: Background accommodation. *MAGMA.* 16, 284–96.

Rosen, B.R., Wedeen, V.J., Brady, T.J., 1984. Selective saturation NMR imaging. *Journal of Computer Assisted Tomography.* 8, 813–818.

Ross, B.D., 1991. Biochemical considerations in 1H spectroscopy. Glutamate and glutamine; myo-inositol and related metabolites. *NMR in Biomedicine.* 4, 59–63.

Ross, B.D., Jacobson, S., Villamil, F., Korula, J., Kreis, R., Ernst, T., Shonk, T., Moats, R.A., 1994. Subclinical hepatic encephalopathy: Proton MR spectroscopic abnormalities. *Radiology.* 193, 457–463.

Ryner, L.N., Sorenson, J.A., Thomas, M.A., 1995a. 3D localized 2D NMR spectroscopy on an MRI scanner. *Journal of Magnetic Resonance Series B.* 107, 126–137.

Ryner, L.N., Sorenson, J.A., Thomas, M.A., 1995b. Localized 2D J-resolved 1H MR spectroscopy: Strong coupling effects in vitro and in vivo. *Magnectic Resonance Imaging.* 13, 853–869.

Sanders, J.A., 1995. Magnetic resonance imaging. In: *Functional Brain Imaging.* Orrison, W.W., Lewine, J.D., Sanders, J.A., Hartshorne, M.F., eds. Moseby, St. Louis, MO, pp. 419–467.

Sarma, M.K., Huda, A., Nagarajan, R., Hinkin, C.H., Wilson, N., Gupta, R.K., Frias-Martinez, E., et al., 2011. Multi-dimensional MR spectroscopy: Towards a better understanding of hepatic encephalopathy. *Metabolic Brain Disease.* 26, 173–184.

Schulte, R.F., Lange, T., Beck, J., Meier, D., Boesiger, P., 2006. Improved two-dimensional J-resolved spectroscopy. *NMR in Biomedicine.* 19, 264–270.

Shen, J.F., Saunders, J.K., 1993. Double inversion recovery improves water suppression in vivo. *Magnetic Resonance in Medicine.* 29, 540–542.

Shonk, T.K., Moats, R.A., Gifford, P., Michaelis, T., Mandigo, J.C., Izumi, J., Ross, B.D., 1995. Probable Alzheimer disease: Diagnosis with proton MR spectroscopy. *Radiology.* 195, 65–72.

Simmons, M.L., Frondoza, C.G., Coyle, J.T., 1991. Immunocytochemical localization of N-acetyl-aspartate with monoclonal antibodies. *Neuroscience.* 45, 37–45.

Smith, S.A., Levante, T.O., Meier, B.H., Ernst, R.R., 1994. Computer simulations in magnetic resonance. An object-oriented programming approach. *Journal of Magnetic Resonance, Series A.* 106, 75–105.

Soher, B.J., Young, K., Maudsley, A.A., 2001. Representation of strong baseline contributions in 1H MR spectra. *Magnetic Resonance in Medicine.* 45, 966–972.

Sokol, M., 2001. In vivo 1H MR spectra analysis by means of second derivative method. *MAGMA.* 12, 177–183.

Stanley, J.A., Drost, D.J., Williamson, P.C., Thompson, R.T., 1995. The use of a priori knowledge to quantify short echo in vivo 1H MR spectra. *Magnetic Resonance in Medicine.* 34, 17–24.

Stanley, J.A., Pettegrew, J.W., 2001. Postprocessing method to segregate and quantify the broad components underlying the phosphodiester spectral region of in vivo (31)P brain spectra. *Magnetic Resonance in Medicine.* 45, 390–396.

Star-Lack, J., Nelson, S.J., Kurhanewicz, J., Huang, L.R., Vigneron, D.B., 1997. Improved water and lipid suppression for 3D PRESS CSI using RF band selective inversion with gradient dephasing (BASING). *Magnetic Resonance in Medicine.* 38, 311–321.

Stephenson, D.S., 1988. Linear prediction and maximum entropy methods in NMR spectroscopy. *Progress in Nuclear Magnetic Resonance Spectroscopy.* 20, 515–626.

Sundin, T., Vanhamme, L., Van Hecke, P., Dologlou, I., Van Huffel, S., 1999. Accurate quantification of (1)H spectra: From finite impulse response filter design for solvent suppression to parameter estimation. *Journal of Magnetic Resonance.* 139, 189–204.

Swanson, M.G., Vigneron, D.B., Tran, T.K., Sailasuta, N., Hurd, R.E., Kurhanewicz, J., 2001. Single-voxel oversampled J-resolved spectroscopy of in vivo human prostate tissue. *Magnetic Resonance in Medicine.* 45, 973–980.

Swanson, M.G., Zektzer, A.S., Tabatabai, Z.L., Simko, J., Jarso, S., Keshari, K.R., Schmitt, L., et al., 2006. Quantitative analysis of prostate metabolites using 1H HR-MAS spectroscopy. *Magnetic Resonance in Medicine.* 55, 1257–1264.

Thomas, M.A., Hattori, N., Umeda, M., Sawada, T., Naruse, S., 2003. Evaluation of two-dimensional L-COSY and JPRESS using a 3 T MRI scanner: From phantoms to human brain in vivo. *NMR in Biomedicine.* 16, 245–251.

Thomas, M.A., Ryner, L.N., Mehta, M.P., Turski, P.A., Sorenson, J.A., 1996. Localized 2D J-resolved 1H MR spectroscopy of human brain tumors in vivo. *Journal of Magnetic Resonance Imaging.* 6, 453–459.

Thomas, M.A., Yue, K., Binesh, N., Davanzo, P., Kumar, A., Siegel, B., Frye, M., et al., 2001. Localized two-dimensional shift correlated MR spectroscopy of human brain. *Magnetic Resonance in Medicine*. 46, 58–67.

Umesh, S., Tufts, D.W., 1996. Estimation of parameters of exponentially damped sinusoids using fast maximum likelihood estimation with application to NMR spectroscopy data. *IEEE Transactions on Signal Processing*, 44, 2245–2259.

van den Boogaart, A., Ala-Korpela, M., Jokisaari, J., Griffiths, J.R., 1994. Time and frequency domain analysis of NMR data compared: An application to 1D 1H spectra of lipoproteins. *Magnetic Resonance in Medicine*. 31, 347–358.

van der Knaap, M., Ross, B., Valk, J., 1994. Uses of MR in inborn errors of metabolism. In: *Magnetic Resonance Neuroimaging*. Kucharczyk, J., Moseley, M.E., Barkovich, A.J., eds. CRC Press, Boca Raton, FL, 485 pp.

van der Veen, J.W., de Beer, R., Luyten, P.R., van Ormondt, D., 1988. Accurate quantification of in vivo 31P NMR signals using the variable projection method and prior knowledge. *Magnetic Resonance in Medicine*. 6, 92–98.

van der Veen, J.W., Weinberger, D.R., Tedeschi, G., Frank, J.A., Duyn, J.H., 2000. Proton MR spectroscopic imaging without water suppression. *Radiology*. 217, 296–300.

Vanhamme, L., van den Boogaart, A., Van Huffel, S., 1997. Improved method for accurate and efficient quantification of MRS data with use of prior knowledge. *Journal of Magnetic Resonance*. 129, 35–43.

Vanhuffel, S., Chen, H., Decanniere, C., Vanhecke, P., 1994. Algorithm for time-domain NMR data fitting based on total least squares. *Journal of Magnetic Resonance, Series A*. 110, 228–237.

Vigneron, D.B., Nelson, S.J., Murphy-Boesch, J., Kelley, D.A., Kessler, H.B., Brown, T.R., Taylor, J.S., 1990. Chemical shift imaging of human brain: Axial, sagittal, and coronal P-31 metabolite images. *Radiology*. 177, 643–649.

von Kienlin, M., Ziegler, A., Le Fur, Y., Rubin, C., Decorps, M., Remy, C., 2000. 2D-spatial/2D-spectral spectroscopic imaging of intracerebral gliomas in rat brain. *Magnetic Resonance in Medicine*. 43, 211–219.

Weiner, M.W., 1987. NMR spectroscopy for clinical medicine animal models and clinical examplesa. *Annals of the New York Academy of Sciences*. 508, 287–299.

Weiner, M.W., Hetherington, H., Hubesch, B., Karczmar, G., Massie, B., Maudsley, A., Meyerhoff, D.J., Sappey-Marinier, D., Schaefer, S., Twieg, D.B., 1989. Clinical magnetic resonance spectroscopy of brain, heart, liver, kidney, and cancer. A quantitative approach. *NMR in Biomedicine*. 2, 290–297.

Welch, K.M., Levine, S.R., Martin, G., Ordidge, R., Vande Linde, A.M., Helpern, J.A., 1992. Magnetic resonance spectroscopy in cerebral ischemia. *Neurologic Clinics*. 10, 1–29.

Young, K., Soher, B.J., Maudsley, A.A., 1998. Automated spectral analysis II: Application of wavelet shrinkage for characterization of non-parameterized signals. *Magnetic Resonance in Medicine*. 40, 816–821.

Yue, K., Marumoto, A., Binesh, N., Thomas, M.A., 2002. 2D JPRESS of human prostates using an endorectal receiver coil. *Magnetic Resonance in Medicine*. 47, 1059–1064.

Part II

Body

6

Quantitative MRI of Articular Cartilage and Its Clinical Applications

Xiaojuan Li and Sharmila Majumdar

University of California, San Francisco

CONTENTS

6.1 Introduction

Cartilage is one of the most essential tissues for healthy joint function and is compromised in degenerative and traumatic joint diseases. There have been tremendous advances during the past decade using quantitative MRI technique as a noninvasive tool for evaluating cartilage, with a focus on assessing cartilage degeneration during osteoarthritis (OA). In this chapter, after a brief overview of cartilage composition and degeneration, we will discuss techniques that quantify morphologic changes as well as changes in the extracellular matrix (ECM), followed by a review of studies correlating quantitative MRI with biomechanics of the knee joint. The basic principles, *in vivo* applications, advantages, and challenges of each technique will be discussed.

6.2 Overview of Cartilage Composition and Degeneration

6.2.1 Normal Cartilage

Hyaline cartilage consists of a low density of chondrocytes and a large ECM. The ECM is composed primarily of water (~75% of cartilage by weight) and a mixture of amorphous and fibrous components. The amorphous component predominantly contains proteoglycans (PGs). The PGs consist of polysaccharide chains [glycosaminoglycans (GAGs)] such as keratin sulfate and chondroitin sulfate, which are covalently bound to a protein core. These core proteins are in turn noncovalently bound to a long filament of hyaluronic acid to form large PG aggregates (Jeffrey and Watt, 2003). The GAG is highly negatively charged. A key function of these aggregates is to provide a stable environment of high fixed charge density (FCD), essential for imbibing and retaining water in the tissue by the high osmotic swelling pressure created. The formed component of the ground substance is composed of collagen fibers (mainly type II) that interact electrostatically with the GAGs to form a cross-linked matrix. Biomechanically, the collagen network provides tensile stiffness to the tissue and the PG provides compressive stiffness. The distribution and orientation of collagen in cartilage demonstrates anatomical zones at microscopy (Verstraete et al., 2004). The collagen fibers are oriented parallel to the articular surface in the superficial zone, perpendicular in the radial zone, and arcade-like in the transitional zone. The superficial zone can be further divided into two sub-zones: (1) the lamina splenden, the more superficial layer with dense small fibrils and little polysaccharide and no cells, and (2) the cellular layer with flat-tened chondrocytes and collagen fibers tangentially to the articular surface. Below radial zone is a thin layer of calcified cartilage separated by the tidemark. The exact function of each of these zones is not known. The water concentration differs slightly between the zones, being 82% in the superficial zone and 76% in the radial zone. The superficial zone has a lower PG concentration than the other zones (Jeffrey and Watt, 2003). "Minor" amounts of other collagens, PGs, glycoproteins, and other structural and enzymatic molecules make up a small percentage of the tissue but may be of critical importance in the structure and function of the tissue, which is however beyond the scope of this chapter.

6.2.2 Cartilage Degeneration

Although the etiopathogenesis of OA is not fully understood, it is believed that OA results from an imbalance between predominantly chondrocyte-controlled anabolic and catabolic processes, and is characterized by progressive degrada-tion of the components of the ECM of the cartilage (Brandt et al., 1998; Dijkgraaf et al., 1995; Goldring, 2012). In this chapter, we focus on changes in two major components of the cartilage matrix: the collagen network and the PG aggregates.

During the early stage of OA, the increased synthesis of the ECM compo-nents (the attempt to repair by cartilage chondrocytes) is exceeded by their degradation due to increased synthesis and activity of proteases. Collagen breakdown is characterized by thinning of the collagen fibrils and loosen-ing of the tight weave of the collagen network. However, it has been sug-gested that neither the content nor the type of collagen is altered in early OA (Dijkgraaf et al., 1995). The PGs, which are no longer constrained (or less con-strained) by the tension of the collagen network, are able to bind increased amounts of water, resulting in increased water content in early OA. This accounts for the swelling observed histologically. There is also a loss of PGs, or fixed charges, during early OA. However, it is not clear yet which comes first: a loss of PGs or a loosening of the collagen network, as both affect each other. As shown by Maroudas and Venn (1977), a loosening of the collagen network leads to a loss of PGs, which creates mechanical overload resulting in further damage and loosening of the collagen network.

As the OA process progresses, the PGs, which were captured underneath the intact articular surface, diffuse into the synovial fluid. The swollen tissue observed histologically becomes ulcerated. This loss of PGs results in a decreased water content of the cartilage and a subsequent loss of its biomechanical char-acteristics, such as compressive stiffness. The collagen network becomes more disorganized. The synthesis of the ECM components may fail, and the synthesis

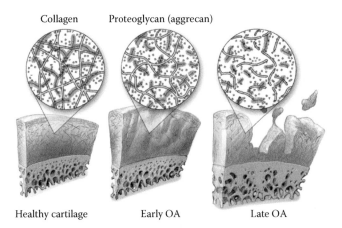

FIGURE 6.1
(Color version available from crcpress.com; see Preface.) Articular cartilage and degeneration during OA. Changes at early stages of OA include hydration, loss of proteoglycan, and thinning and loosening disruption of collagen. Changes at late stages include further loss of proteoglycan and collagen, dehydration, extensive fibrillation and cartilage thinning, and eventually denudation of the subchondral bone.

and activity of the proteases remain increased, resulting in progressive degradation and loss of the articular cartilage. This intermediate OA is characterized histologically by fibrillation (vertical splitting), detachments (horizontal splitting), and thinning of the cartilage due to mechanical wear. In the late stage of OA, there is a further loss of ECM components, including water, PGs, and collagen. The synthesis and activity of proteases may remain increased or may be reduced when the cartilage is extremely thin or nearly completely destroyed, resulting in residual OA. This late stage of OA is characterized histologically by extensive fibrillation of the cartilage and eventually denudation of the subchondral bone. Figure 6.1 illustrates the different stages of cartilage degeneration.

6.3 Morphological Changes in Cartilage in OA

Magnetic resonance provides excellent soft tissue contrast, and clinical evaluation of articular cartilage is typically based on proton density or intermediate-weighted fast spin-echo (FSE) images. Morphological grading of the cartilage to quantify cartilage changes in OA has also been developed. The oldest is one developed by Recht et al. (1996), which utilized the Noyes classification (Noyes and Stabler, 1989) modified to MR imaging. In this system, five grades (0–4) are assigned to the cartilage depending on the thickness of cartilage defects and signal inhomogeneity on high-resolution spoiled gradient-echo (SPGR) sequences. The Whole Organ Magnetic Resonance Imaging Score (WORMS) of the knee in OA was proposed by Peterfy et al. (2004) and is one

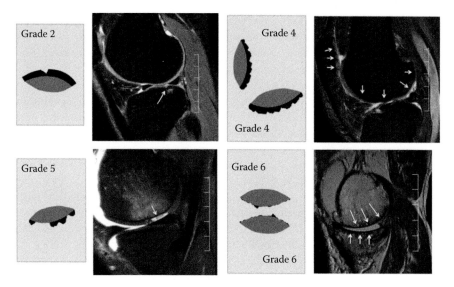

FIGURE 6.2
Representative MR images with different stages of cartilage lesions and the corresponding WORMSs.

that characterizes multiple features in cartilage, meniscus, ligaments, and bone marrow, all believed to be relevant to the pathophysiology of OA. Figure 6.2 shows representative MR images with corresponding WORMSs. The Boston-Leeds OA Knee Score or the BLOKS (Hunter et al., 2008) focuses on bone marrow edema-like lesions (BMELs) and is related to pain in OA.

Apart from the grading systems, high-resolution, typically fat-suppressed MR images have also been used to quantify cartilage thickness and volume. Apart from compartmental analysis (i.e., medial vs. lateral compartments), subregional analysis of cartilage volume and thickness has been undertaken especially by Eckstein et al. (1997). Pulse sequences that are used to obtain images for cartilage volume estimation have high signal-to-noise ratio (SNR), high contrast-to-noise ratio, and high resolution as the cartilage thickness of a few millimeters warrants that the image resolution be as high as possible. SPGR images and a number of variants on these sequences such as dual echo steady state (DESS) images (Hardy et al., 1996), with different methods of fat suppression as well as newly developed 3D FSE images, have been used for the assessment of cartilage morphology.

The steps involved in assessing cartilage volume and thickness include the segmentation of the cartilage. Several different image processing approaches have been proposed for this purpose. Stammberger and colleagues (1999b) applied a quadratic B-spline to guide the segmentation. An internal energy term controlled the rigidity of the contour, an external energy term attracted the contour to the cartilage edges, and a coupling force enforced smooth changes from one slice to another. In another two-dimensional (2D) approach, Lynch

et al. (2000) combined expert knowledge with cubic splines and image processing algorithms reporting better reproducibility and less user interaction than
region growing techniques. Grau et al. (2004) extended the watershed technique
to examining difference in class probability of neighboring pixels. Pakin et al.
(2002) used a region growing method with prior knowledge. Warfield and colleagues (2000) developed an adaptive, template-moderated, spatially varying statistical classification method, which consisted of an iterative technique between
classification and template registration. Manual editing is often required in
the context of semiautomated cartilage segmentation. After segmentation, the
simplest morphological measure to compute is the sum of the total number of
voxels representing cartilage, followed by corresponding scaling according to
the voxel dimensions. The most common approach to compute cartilage thickness is based on 2D or three-dimensional (3D) minimum Euclidean distances
(Carballido-Gamio et al., 2008a; Kauffmann et al., 2003; Stammberger et al.,
1999a). For each point on the bone–cartilage interface or articular surface, the
closest point on the opposite surface is computed. Another 3D approach consists of computing normal vectors on one surface (articular or bone–cartilage
interface) and finding the intersection of the vectors on the opposite surface
(Ateshian et al., 1991; Cohen et al., 1999). The average of the minimum distances
or the length of the vectors, respectively, is reported as the average cartilage
thickness. Figure 6.3 shows a thickness map of the knee cartilage.

To obtain a measure of the spatial distribution of change (between time
points) in cartilage volume and thickness, image registration is required.
For this purpose, some investigators have used the bone–cartilage interface
(Jaremko et al., 2006; Pelletier et al., 2007; Stammberger et al., 2000), whereas
others have used the total bone shape (Carballido-Gamio et al., 2008a, 2008b).
Once shapes have been aligned, matching of cartilage thickness patterns can
be performed at a local level. Because there is an increasing interest in performing regional comparisons of cartilage properties at specific anatomic
locations between different populations for a better understanding of OA,
techniques have also been developed for this purpose (Carballido-Gamio

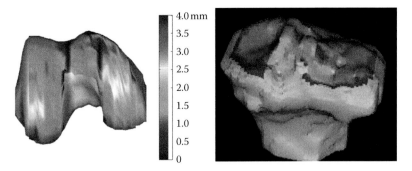

FIGURE 6.3
(Color version available from crcpress.com; see Preface.) Cartilage thickness maps of the knee
cartilage overlaying over the femur (left) and the tibia (right), respectively.

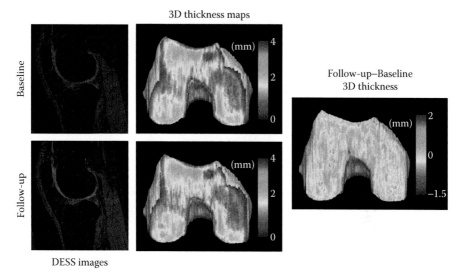

FIGURE 6.4
(See color insert.) An example of intrasubject longitudinal cartilage thickness matching using the technique described by Carballido-Gamio et al. (Modified from Carballido-Gamio, J., et al., *Med Image Anal.* 12, 120–35, 2008a.)

et al., 2008a; Cohen et al., 2003). For intersubject comparisons, shapes have to be registered prior to any comparison, but in this case, scale factors as well as nonlinear differences between shapes have to be considered. The bone shape, and not only the interface, has recently been proposed for this type of matching (Carballido-Gamio et al., 2008a). Figure 6.4 shows an intrasubject longitudinal example of cartilage thickness matching using the technique described by Carballido-Gamio and colleagues (2008a), which can also be applied for intersubject comparisons of cartilage properties.

6.3.1 Quantitative Morphology in the OAI

The OA Initiative (OAI; http://oai.epi-ucsf.org/datarelease/StudyOverview. asp) is a multicenter, longitudinal, prospective observational study of knee OA. The aim of the OAI is to develop a public domain research resource to facilitate the scientific evaluation of biomarkers for OA as potential surrogate end points for disease onset and progression. The OAI data include clinical evaluation data, radiological (x-ray and 3 T MR) images, and a biospecimen repository from 4796 men and women aged 45–79 years. The participants include those who are at high risk for developing symptomatic knee OA. Using the MR data from these studies, investigators have determined that the MR sequences used to estimate the cartilage thickness and volume have adequate reproducibility (Eckstein et al., 2007; Wirth et al., 2010b). Rates of change for cartilage morphological measures in different regions have been established as a function of radiographic disease extent and existence of pain (Benichou et al., 2010; Eckstein et al., 2009a, 2009b, 2009c, 2010a, 2010b, 2011; Wirth et al., 2009, 2010a, 2011).

6.4 Quantitative MRI for Cartilage Matrix Composition

As discussed previously, cartilage degeneration starts with changes in hydration and the macromolecular structures, specifically PG and collagen network, within the matrix. Techniques that are sensitive and specific to these changes are ideal for early detection of cartilage degeneration.

In the early days, water content in cartilage has been quantified by using proton density images (a measurement of the net magnetization vector) after corrections of T_1 and T_2 relaxation values (Shapiro et al., 2001; Xia et al., 1994). The signal in this absolute proton density images was calibrated to the bathing solution, or a known phantom (a test tube filled with water), to determine the water content. Although the measurement itself is straightforward and Shapiro et al. (2001) have reported that the measured water content agreed remarkably with the value obtained by gravimetric analysis, the challenge lies in the fact that OA-induced hydration changes are relatively small [up to 9% (Mankin and Thrasher, 1975)]. Lusse et al. (2000) performed T_2-mapping experiments on both *in vitro* and *in vivo* human knees and correlated these values to a predetermined calibration curve correlating T_2 versus water content. This method, however, may not be accurate if T_2 changes occur due to any mechanism other than water content changes, such as changes in collagen–PG matrix, as in the case of OA.

Later efforts have been made more toward measuring the changes associated with macromolecular changes, PG and collagen, within the matrix. However, both PG and collagen protons sustain very short T_2 relaxation time due to their macromolecular structures, and it is difficult to measure them directly with MRI, except for using *ex vivo* high-field NMR techniques. As a result, current MRI techniques are primarily focused on sensitizing the measurement of water proton signals to the macromolecular contents and structures in the matrix based on energy and magnetization exchange between bulk water protons and protons associated with macromolecules. Lattanzio et al. (2000) proposed a four-site relaxation and interspin group coupling model for articular cartilage (Figure 6.5). The quantitative MRI methods for cartilage matrix biochemistry were divided into four categories: (1) methods based on relaxometry including T_1 (with presence of contrast agent), T_2 and $T_{1\rho}$ relaxation time quantification, (2) methods based on diffusion measurement, (3) methods based on magnetization transfer (MT) measurement including conventional MT and chemical exchange saturation transfer (CEST), and (4) a nonproton method, sodium imaging, developed and applied to measure cartilage matrix changes, which will also be discussed in this chapter.

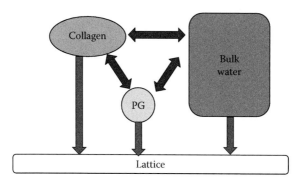

FIGURE 6.5
A four-site relaxation and interspin group coupling model for articular cartilage.

6.4.1 Postcontrast T_1 Relaxation Time by Delayed Gadolinium-Enhanced Proton MRI of Cartilage

6.4.1.1 Basic Principles

As T_1 reflects activities with relaxation correlation time in the range of $1/(\gamma B_0)$, native T_1 measurement is normally not sensitive to cartilage matrix changes due to the slow motion of the restricted protons in the matrix. However, delayed gadolinium (Gd)-enhanced proton MRI of cartilage (dGEMRIC) has been developed to quantify PG content in cartilage by quantifying T_1 in the presence of the contrast agent (Bashir et al., 1999; Gray et al., 2008; Tiderius et al., 2003). It is based on the fact that PG, or the associated GAG, has abundant negatively charged carboxyl and sulfate groups. In the study of dGEMRIC, contrast agent of Gd diethylene triamine pentaacetic acid, or Gd-DTPA^{2-} (Magnevist; Berlex Laboratories, Wayne, NJ), is injected intravenously (or intraarticularly) and distributed in the cartilage by diffusion. The diffusion time depends on the cartilage thickness and is approximately 2 h in femoral weight-bearing cartilage. Because Gd-DTPA^{2-} is negatively charged, it will be distributed in a relatively low concentration in areas that are rich in GAG (normal cartilage) and in higher concentrations in regions with depleted GAG (degenerated cartilage). Gd-DTPA^{2-} has a concentration-dependent effect on the MR parameter T_1. Therefore, T_1-weighted images in the presence of Gd-DTPA^{2-} reflect the Gd-DTPA^{2-} concentration, and hence tissue GAG concentration. By calculating T_1 relaxation time, the concentration of GAG can be quantified based on a modified electrochemical equilibrium theory, assuming that the Gd-DTPA^{2-} is equilibrated in the tissue (Bashir et al., 1996).

dGEMRIC measures have been validated by comparison with gold-standard biochemical and histologic measurements of GAG concentration in cartilage with *ex vivo* studies (Bashir et al., 1996, 1999; Nieminen et al., 2002; Trattnig et al., 1999). The *in vivo* validation of dGEMRIC techniques, however, was not straightforward especially for the conversion from T_1 quantification to GAG concentration; therefore, the direct T_1 measure was normally reported as a "dGEMRIC index" for clinical studies (Gray et al., 2008). Loss of GAG will result in a decreased T_1 and a decreased "dGEMRIC index."

6.4.1.2 In Vivo *Applications*

The most commonly used methods for *in vivo* T_1 quantification in dGEMRIC studies are 2D inversion recovery FSE (IR-FSE) sequences due to its widespread availability of the sequence and desirable contrast properties (cartilage vs. fluid) (Williams et al., 2004). However, this method has limited joint coverage (especially with single slice) and needs a long acquisition time. To obtain full joint coverage, 3D acquisitions based on IR-SPGR were developed at 1.5 T and 3 T (McKenzie et al., 2006), which however still require a long scan time. More recently, fast 3D methods based on Look-Locker (Kimelman et al., 2006; Li et al., 2008a) or two SPGR images with optimized flip angles (Mamisch et al., 2008a; Sur et al., 2009) were developed. Good correlations were reported between measures from the 3D methods and the gold-standard 2D IR-FSE technique (Li et al., 2008a; Mamisch et al., 2008a; McKenzie et al., 2006). The reproducibility of the 3D-variable flip angle method [root-mean-square coefficients of variation (RMS-CV): 9.3–15.2%] was reported to be inferior compared to 2D IR-FSE and 3D Look-Locker methods (RMS-CV: 5.8–8.4%) (Siversson et al., 2010). In particular, the investigators suggested that the positioning of the analyzed images is crucial to generate reliable repeatability results. The recommended clinical protocol was summarized by Burstein et al. (2001).

Decreased dGEMRIC index was observed in cartilage with early degeneration as detected by arthroscopic evaluation (Tiderius et al., 2003) and in compartment with radiographic joint space narrowing compared to other compartment (Williams et al., 2005). Large variation of dGEMRIC index was observed with the same Kellgren–Lawrence (KL) scores, an established scoring system for OA based on radiographs, including knees without joint space narrowing (Hart and Spector, 1998; Kellgren and Lawrence, 1957). In a longitudinal study with a relatively small cohort ($n = 17$), dGEMRIC index at baseline for those who developed radiographic OA after 6 years was lower than that in the knees without radiographic OA ($p = .03$) (Owman et al., 2008), suggesting potential capability of dGEMRIC index to predict OA development.

dGEMRIC has also been applied to the cohort with high risk of OA. In obese subjects, knee dGEMRIC index was associated with age, clinical knee OA, abnormal tibiofemoral alignment, and quadriceps strength (Anandacoomarasamy et al., 2009). Subjects with medial meniscectomy (1–6 years before imaging) showed significantly lower dGEMRIC index of the medial compartment after surgery (Ericsson et al., 2009). In subjects with acute anterior cruciate ligament (ACL) injuries, significantly decreased dGEMRIC index was observed in ACL-injured knees compared to either contralateral knees (Fleming et al., 2010) or knees from healthy control subjects (Neuman et al., 2011; Tiderius et al., 2005) in both medial and lateral compartments, suggesting that dGEMRIC index may be indicative of potential early degeneration in knees with risk of developing posttraumatic OA. dGEMRIC has also been used in cartilage repair as a useful tool to evaluate matrix composition of the repaired tissue, which was reviewed by Trattnig et al. (2009).

In the hip, lower dGEMRIC index was reported in subjects with femoroac-etabular impingement (FAI) (Bittersohl et al., 2009) and correlated with the magnitude of the deformity (Jessel et al., 2009; Pollard et al., 2010) and pain (Jessel et al., 2009). Further analysis of subregions revealed different pattern of subjects with cam and pincer FAI (Mamisch et al., 2011). In subjects with hip dysplasia, a developmental malformation of the hip associated with early OA, Kim et al. (2003) reported that dGEMRIC but not joint space measured with radiographs significantly correlated with severity of dysplasia and pain. Longitudinal studies in subjects with hip dysplasia and treated with Bernese periacetabular osteotomy suggested that among other factors including age, radiographic arthritis severity, and dysplasia severity, dGEMRIC index was the best predictor of failure of the osteotomy (Cunningham et al., 2006). This study suggested that the dGEMRIC index, as a potential marker for early OA, appears to be useful for identifying poor candidates for a pelvic osteotomy.

6.4.1.3 Advantages and Challenges

Previous studies support the dGEMRIC index as a clinically relevant measure of cartilage integrity, with specific measures related to GAG concentration. However, one should be aware that especially for *in vivo* studies, bias may be introduced during conversion from dGEMRIC measures to GAG concentration due to factors such as the possibility of altered contrast agent relaxivity under different tissue conditions (Gillis et al., 2002) and other factors such as the transport of contrast agent into cartilage with a variable blood concentration that might affect the equilibrating concentration of Gd-DTPA^{2-} *in vivo* (Burstein et al., 2001; Gray et al., 2008). Other challenges that hinder dGEM-RIC from widespread clinical applications include the necessity of contrast agent injection, joint exercise, and long delay of MR scan after Gd injection.

6.4.2 T_2 Relaxation Time

6.4.2.1 Basic Principles

T_2 relaxation time, or the spin–spin relaxation time, reflects the ability of free water proton molecules to move and exchange energy inside the cartilaginous matrix. Immobilization of water protons in cartilage by the collagen–PG matrix promotes T_2 decay and renders the cartilage low in signal intensity on long echo time (TE) (T_2-weighted) images, whereas mobile water protons in synovial fluid retain their high signal. It has been shown that in normal cartilage this trans-verse (T_2) relaxation is dominated by the anisotropic motion of water molecules in a fibrous collagen network. Damage to collagen–PG matrix and increase of water content in degenerating cartilage may increase T_2 relaxation times.

T_2 quantification is normally performed by fitting T_2-weighted images acquired with different TEs to mono- or multiexponential decay curve. Pai et al. (2008) compared the T_2-mapping techniques in phantoms and *in vivo* based on five different sequences with regard to SNR, reproducibility, and quantification: spin-echo (SE), FSE, multi-echo SE (MESE), magnetization-prepared 2D spiral, and magnetization-prepared 3D SPGR. Variation of T_2 quantification was observed, which may be due to different sensitivity of each sequence to system imperfections including stimulated echoes, off-resonance signals, and eddy currents. Different fitting methods will also introduce bias to T_2 quantification (Koff et al., 2008). Caution needs to be exercised when comparing results from different studies and when designing a multicenter study.

In vitro studies have reported that T_2 correlated well with water content (Lusse et al., 2000) but poorly with PG content (Regatte et al., 2002; Toffanin et al., 2001), and PG cleavage did not affect T_2 values significantly (Nieminen et al., 2000), whereas one study on rat patellar cartilage has shown significantly increased T_2 within PG-depleted cartilage (Watrin-Pinzano et al., 2005). Using high-field microscopic MRI (µMRI), Xia (1998) showed that T_2 spatial variation is dominated by the ultrastructure of collagen fibrils, and thus angular dependency of T_2 with respect to the external magnetic field B_0 can provide specific information about the collagen structure. T_2 variation was also correlated with polarized light microscopy (PLM) (Nieminen et al., 2001). This angular dependency of T_2, however, also results in the "magic angle" effect and commonly seen as a laminar appearance in cartilage imaging (David-Vaudey et al., 2004; Xia et al., 1994). Multiexponential T_2 relaxation components were identified by Reiter et al. (2009, 2011) using intact and enzyme-digested bovine nasal cartilage. Using bovine nasal cartilage, it was demonstrated that these multicomponent properties may depend on the orientation of collagen fibers with respect to B_0—at the magic angle (55°), T_2 (and $T_{1\rho}$) was a single component rather than multiple components (Wang and Xia, 2011). This multiexponential approach has the potential to improve specificity of cartilage matrix evaluation using T_2 relaxation parameters.

6.4.2.2 In Vivo *Applications*

In vivo T_2 values have been shown to vary from the subchondral bone to the cartilage surface (Dardzinski et al., 1997; Li et al., 2009). With the exception of the lamina splendens, T_2 decreased from the superficial layer to the deep layer. Cartilage T_2 has been correlated with age (Mosher et al., 2000, 2004a), but not with gender (Mosher et al., 2004b). Elevated T_2 relaxation times were observed in OA cartilage (Mosher et al., 2000) and associated with the severity of disease (Dunn et al., 2004; Li et al., 2007; Yao et al., 2009). In one study examining T_2 in patellar cartilage, however, no differences of T_2 values were found across the stages of OA based on radiographic KL scores

($p = .25$). The same study reported that the increased T_2 was associated with the increased body mass index (BMI), suggesting that BMI is a significant factor for increasing T_2.

The relationship between the cartilage T_2 and the underlying bone structure has been investigated (Blumenkrantz et al., 2004; Bolbos et al., 2008b; Lammentausta et al., 2007). In a cross-sectional study at 3 T, a negative relationship was observed between T_2 relaxation time, bone structural parameters including bone volume/total volume fraction (BV/TV), and trabecular number (Tb.N) (Bolbos et al., 2008b), highlighting the interplay between cartilage and bone structure in OA. In a pilot, 2-year follow-up study, cartilage T_2 increased and bone structural parameters decreased in the OA subjects (Blumenkrantz et al., 2004). Interestingly, a positive relationship was established between cartilage changes and localized bone changes closest to the joint line, whereas a negative relationship was established between cartilage changes and global bone changes farthest from the joint line (Blumenkrantz et al., 2004), implying bone degeneration and adaption in OA during the joint degeneration.

In vivo T_2 quantification using a multislice MESE sequence has been included in the OAI (Peterfy et al., 2008). Data from the monthly quality assurance procedure over a 3-year period showed excellent longitudinal reproducibility (RMS–CV: <5%) except for one site (Schneider et al., 2008).

Cartilage T_2 values have been correlated with physical activity levels in OAI cohort (Hovis et al., 2011; Stehling et al., 2010), which will be detailed in our later section on biomechanics and vastus lateralis/vastus medialis cross-sectional area (CSA) ratio (Pan et al., 2011b). More recently, a 2-year longitudinal analysis of OAI normal cohort showed a significant increase in T_2 in the tibiofemoral cartilage, but no changes in the patellofemoral cartilage (Pan et al., 2011a). A greater increase in cartilage T_2 was found to correlate with an increase in the progression of cartilage abnormalities (graded with WORMS), suggesting that characterizing and monitoring the cartilage matrix integrity with longitudinal T_2 measurements might enable identification of individuals at risk for the development of early OA before irreversible cartilage loss occurs.

6.4.2.3 Advantages and Challenges

Cartilage T_2 quantification is a promising marker for cartilage matrix biochemistry and has been applied in the largest multicenter studies of OA with the effort of the OAI. T_2 changes are dominated by the collagen fibers and may not be sensitive to subtle changes of PG loss. Caution is also needed during data interpretation due to the magic angle effect of T_2. In addition, due to the layer variation of cartilage T_2 as discussed earlier, laminar analysis is recommended to be used for analyzing either OA or cartilage repair data. It should be noted that sampling quantitative data in the superficial zone or lamina splendens is technically challenging, and normally two layers (with equal thickness) were used for *in vivo* T_2 laminar analysis due to the limited resolution of *in vivo* T_2 mapping.

6.4.3 $T_{1\rho}$ Relaxation Time Quantification

6.4.3.1 Basic Principles

The $T_{1\rho}$ parameter is defined as the time constant describing the spin-lattice relaxation in the rotating frame (Sepponen, 1992). It probes the slow motion interactions between motion-restricted water molecules and their local macromolecular environment. The macromolecules in articular cartilage ECM restrict the motion of water molecules. Changes to the ECM, such as PG loss, therefore, can be reflected in measurements of $T_{1\rho}$. The $T_{1\rho}$ is normally measured by the spin-lock (SL) technique (Redfield, 1969), which enabled the study of relaxation at very low magnetic fields without sacrificing the SNR afforded by higher field strengths. In an SL experiment, spins are flipped into the transverse plane along one axis, immediately followed by an SL pulse applied along the same axis. An SL pulse is an on-resonance, continuous wave (CW) radio frequency (RF) pulse, normally long duration and low energy. Since the magnetization and RF field are along the same direction, the magnetization is said to be "spin-locked," provided that the locking condition is satisfied, that is, the B_1 of locking pulses is much stronger than the local magnetic fields generated by, for example, magnetic moments of nuclei. The spins will relax with a time constant $T_{1\rho}$ along B_1 of locking pulses in the transverse plane (Sepponen, 1992). The amplitude of the SL pulse is commonly referenced in terms of the nutation frequency ($f = \gamma B_1$). The normal range of SL frequency is a few hundred hertz to a few kilohertz. $T_{1\rho}$ relaxation phenomena are sensitive to physicochemical processes with inverse correlation times on the order of the nutation frequency of the SL pulse. $T_{1\rho}$ increases as the strength of the SL field increases, a phenomenon termed $T_{1\rho}$ dispersion. $T_{1\rho}$ dispersions may also have tissue specificity (Sepponen, 1992).

The mechanism of $T_{1\rho}$ relaxation time in biological tissues, particularly in cartilage, is not fully understood yet. Using native and immobilized protein solution, Makela et al. (2001) suggested that proton exchange between the protein side chain groups and the bulk water contribute significantly to the $T_{1\rho}$ relaxation. Based on spectroscopy experiments with peptide solutions, GAG solutions, and bovine cartilage samples before and after PG degradation, Duvvuri et al. (2001b) further suggested that in cartilage hydrogen exchange from NH and OH groups to water may dominate the low-frequency (0–1.5 kHz) water $T_{1\rho}$ dispersion. They speculated that an increase of the low-frequency correlation rate with PG loss could be the result of increased proton exchange rates. Other evidence of a proton exchange pathway is the pH dependency of $T_{1\rho}$ values in the ischemic rat brain tissues (Kettunen et al., 2002). Mlynarik et al. (2004), on the other hand, have suggested that the dominant relaxation mechanism in the rotating frame in cartilage at $B_0 \leq 3$ T seems to be dipolar interaction. The contribution of scalar relaxation caused by proton exchange is only relevant at high fields such as 7 T. Clearly, further investigations are needed to better understand this relaxation mechanism.

The $T_{1\rho}$-weighted imaging sequences are composed of two parts: magnetization preparation with $T_{1\rho}$ weighting using SL pulse cluster and a following 2D

(Borthakur et al., 2006a; Li et al., 2005; Ristanis et al., 2006; Wheaton et al., 2004b) or 3D (Borthakur et al., 2003; Li et al., 2008b; Regatte et al., 2004; Witschey et al., 2008) data acquisition. Compared to 2D acquisition, 3D sequences have the advantage of higher image resolution, especially in the slice direction. Among 3D sequences, the method using transient signals immediately after $T_{1\rho}$ preparation based on either SPGR acquisition (Li et al., 2008b) [magnetization-prepared angle-modulated partitioned k-space spoiled gradient-echo snapshots (MAPSS)] or balanced GRE acquisition (Witschey et al., 2008) is more SNR-efficient and less-specific absorption rate (SAR) intensive compared to the method based on the steady-state GRE acquisition (Regatte et al., 2004). These sequences have been implemented at both 1.5 T and 3 T on scanners from different manufactures. Several modifications in the SL pulse cluster and phase cycling were performed in order to improve the robustness of spin locking to B_0 and B_1 inhomogeneities (Charagundla et al., 2003; Chen et al., 2011; Dixon et al., 1996; Li et al., 2005, 2008b; Witschey et al., 2007). $T_{1\rho}$ sequences have also been implemented using parallel imaging techniques (Pakin et al., 2006; Zuo et al., 2007). In these studies, no significant differences were found between parallel and nonparallel $T_{1\rho}$ quantification.

In vitro studies have evaluated the relationship between $T_{1\rho}$ relaxation time and the biochemical composition of cartilage. Akella et al. (2001) have demonstrated that over 50% depletion of PG from bovine articular cartilage resulted in average $T_{1\rho}$ increases from 110 ms to 170 ms. Regression analysis of the data showed a strong correlation ($R^2 = 0.987$) between changes in PG and $T_{1\rho}$. In another study on bovine cartilage at 4 T, $T_{1\rho}$ has been proposed as a more specific indicator of PG content than T_2 relaxation in trypsinized cartilage (Regatte et al., 2002). The authors reported that there was an excellent correlation ($R^2 = 0.89$) between $1/T_{1\rho}$ and GAG concentration, whereas the correlation between $1/T_2$ and GAG concentration was rather poor ($R^2 = 0.01$). In $T_{1\rho}$ quantification experiments, the SL techniques reduce dipolar interactions, and therefore reduce the dependence of the relaxation time constant on collagen fiber orientation (Akella et al., 2004). This may enable more sensitive and specific detection of changes in PG content using $T_{1\rho}$ quantification in cartilage. Regatte et al. (2006) demonstrated that $T_{1\rho}$ values in human cartilage specimens increased with increased clinical grades of degeneration. More recently, Li et al. (2011a) reported that in human OA cartilage obtained from patients who underwent total knee arthroplasty, $R_{1\rho}$ ($1/T_{1\rho}$) values had a significant, but moderate correlation with GAG contents ($R = 0.45, p = .002$). These studies form the experimental basis for using the $T_{1\rho}$-mapping techniques in studying cartilage pathology in OA.

The reduced dipolar interaction also results in less "magic angle effect" in $T_{1\rho}$ imaging compared to T_2 imaging. Less laminar appearance was observed in $T_{1\rho}$-weighted images compared to T_2-weighted images (Akella et al., 2004). Previous specimen studies reported that $T_{1\rho}$ values at the magic angle (54.7°) were significantly higher than at other angles, but the difference was smaller than that in T_2 values at the same angles (Li et al., 2010). The difference was decreased with increased $T_{1\rho}$ spin-lock frequencies and was diminished when $T_{1\rho}$ spin-lock frequency was equal to or higher than 2KHz (Akella et al., 2004).

Studies have demonstrated the regional variations of cartilage $T_{1\rho}$. $T_{1\rho}$ values were highest at the superficial zone, decreased gradually in the middle zone, and increased in the region near the subchondral bone in bovine cartilage (Akella et al., 2001) and human cartilage specimens (Regatte et al., 2006). Although the trend of $T_{1\rho}$ and T_2 values is similar, $T_{1\rho}$ shows a larger dynamic range from the bone–cartilage interface to the cartilage surface, which may allow a higher sensitivity of distinguishing between healthy and degenerated cartilage. Furthermore, less laminar appearance was observed in $T_{1\rho}$-weighted images compared to T_2-weighted images due to reduced dipolar interaction during $T_{1\rho}$ relaxation (Akella et al., 2004). Previous studies also investigated the correlation between T_2 and $T_{1\rho}$ values in cartilage. Mlynarik et al. (1999) showed that the $T_{1\rho}$ relaxation times obtained were slightly longer than the corresponding T_2 values, but both parameters showed almost identical spatial distributions. Menezes et al. (2004) have shown that $T_{1\rho}$ and T_2 changes in articular cartilage do not necessarily coincide and might provide complementary information. These studies suggested that $T_{1\rho}$ and T_2 reflect changes that may be associated with PG, collagen content, and hydration, and the true mechanism of $T_{1\rho}$ may arise from a weighted average of multiple biochemical changes occurring in cartilage in OA.

6.4.3.2 In Vivo *Applications*

In vivo $T_{1\rho}$ reproducibility was reported to range from 1.7% to 8.7% (Li et al., 2005, 2008b). A recent multicenter study reported reproducibility CVs for both T_2 (4–14%) and $T_{1\rho}$ (7–19%) (Mosher et al., 2011). Future studies are warranted to evaluate multisite, multivendor variation of $T_{1\rho}$ and T_2 quantification with a clinical trial setup. *In vivo* studies showed increased cartilage $T_{1\rho}$ values in OA subjects compared to controls (Bolbos et al., 2008b; Duvvuri et al., 2001a; Li et al., 2005, 2007; Regatte et al., 2004; Stahl et al., 2009; Wang et al., 2012) and were moderately age-dependent (Stahl et al., 2009). $T_{1\rho}$ measurements were reported to be superior to T_2 in differentiating OA patients from healthy subjects (Li et al., 2007; Stahl et al., 2009). Stahl et al. (2009) reported that $T_{1\rho}$ and T_2 values in asymptomatic active healthy subjects with focal cartilage abnormalities were significantly higher than those without focal lesions. The authors suggested that $T_{1\rho}$ and T_2 could be the parameters suited to identify active healthy subjects at higher risk for developing cartilage pathology. Further, elevated $T_{1\rho}$ was observed in subcompartments in OA subjects where no obvious morphologic changes were observed, suggesting the capability of $T_{1\rho}$ in detecting very early biochemical changes within the cartilage matrix (Li et al., 2007). Elevated $T_{1\rho}$ has also been correlated with lesions evaluated with arthroscopy (Lozano et al., 2006; Witschey et al., 2010), especially when a small region of interest (ROI) was used for $T_{1\rho}$ quantification (Witschey et al., 2010).

The relationship between meniscus and cartilage degeneration was studied in 44 OA patients and 19 controls using $T_{1\rho}$ and T_2. After stratifying the patients based on the posterior horn of medial meniscus (PHMED) status where the most lesions were observed, $T_{1\rho}$ and T_2 of the PHMED and the medial tibial (MT) cartilage were significantly higher in subjects having a meniscal tear (meniscal grade 2–4) compared to subjects with a meniscal grade of 0 or 1 ($p < .05$) (Zarins et al., 2010).

The spatial correlation between T_2 and $T_{1\rho}$ in cartilage has also been explored *in vivo* (Li et al., 2009). The investigators showed that although the average $T_{1\rho}$ and T_2 values correlated significantly, the pixel-by-pixel correlation between $T_{1\rho}$ and T_2 showed a large range in both controls and OA patients ($R^2 = 0.522 \pm 0.183$, ranging from 0.221 to 0.763 in OA patients, vs. $R^2 = 0.624 \pm 0.060$, ranging from 0.547 to 0.726 in controls) (Li et al., 2009). Figure 6.6 shows $T_{1\rho}$ (top row) and T_2 (bottom row) maps from a control subject, a subject with mild OA, and a subject with severe OA. The differences between the $T_{1\rho}$ and T_2 maps are evident. These results suggested that $T_{1\rho}$ and T_2 show different spatial distribution and may provide complementary information regarding cartilage degeneration in OA. Combining these two parameters may further improve our capability to diagnose early cartilage degeneration and injury.

Studies have also used $T_{1\rho}$ imaging to evaluate cartilage overlying BMEL in OA knees (Li et al., 2008c; Zhao et al., 2010). In a 1-year longitudinal study with 23 OA patients, Zhao et al. (2010) have found that patients with BMEL showed overall higher $T_{1\rho}$ values in cartilage compared with those who had

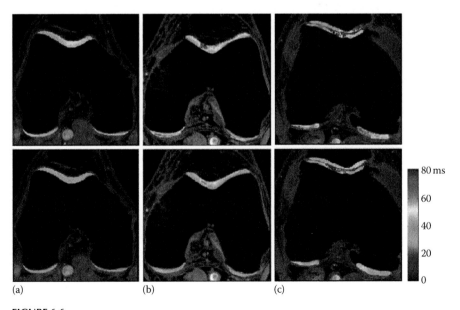

(a) (b) (c)

FIGURE 6.6
(See color insert.) $T_{1\rho}$ (top row) and T_2 maps (bottom row) of a healthy control (a), a subject with mild OA (b), and a subject with severe OA (c). Significant elevation of $T_{1\rho}$ and T_2 values was observed in subjects with OA. $T_{1\rho}$ and T_2 elevation had different spatial distribution and may provide complementary information associated with the cartilage degeneration.

no BMEL (42.0 ± 3.7 ms vs. 39.8 ± 1.4 ms, $p = .032$), suggesting that BMEL may be correlated with disease severity of OA. Furthermore, in patients with BMEL, both $T_{1\rho}$ values and clinical WORMS grading were elevated significantly ($p < .05$) in cartilage overlying BMEL, suggesting a local spatial correlation between BMEL and more advanced cartilage degeneration. At 1-year follow-up, cartilage overlying BMEL showed a significantly higher $T_{1\rho}$ value increase compared with surrounding cartilage, suggesting that BMEL is indicative of accelerated cartilage degeneration. Interestingly, no such difference was found using WORMS scoring. This result suggests that quantitative cartilage imaging, such as $T_{1\rho}$, may be a more sensitive indicator of cartilage degeneration than semiquantitative scoring systems.

In addition to patients with OA, $T_{1\rho}$ quantification techniques have been applied to patients with acutely injured knees, who have a high risk of developing OA later in life. In patients with acute ACL tears, significantly increased $T_{1\rho}$ values were found at baseline (after injury but prior to ACL reconstruction) in cartilage overlying BMEL compared with surrounding cartilage at the lateral tibia (Bolbos et al., 2008a; Li et al., 2008c). Follow-up exams at 1 year after ACL reconstruction of these ACL-injured knees (Li et al., 2011b; Theologis et al., 2011) showed that (1) in lateral sides, despite the resolution of BMEL, cartilage overlying the baseline BMEL still shows significantly higher $T_{1\rho}$ compared to the surrounding cartilage, suggesting potential irreversible damage of cartilage in these regions; (2) in medial sides, $T_{1\rho}$ values in MT and medial femoral condyles, especially the contact area (the contact area during supine unloaded MRI), show significant elevation at as early as 1-year after ACL reconstruction compared to healthy controls, probably due to abnormal kinematics even after ACL reconstruction. These results suggest that cartilage damage after acute knee injuries can be a risk factor for predisposing these knees to OA. Quantitative $T_{1\rho}$ can probe these degenerations at earlier time points than radiographs or conventional MRI. T_2 also showed an increasing trend in these regions of cartilage; however, it did not reach statistical significance. The authors speculated that $T_{1\rho}$ is more sensitive than T_2 in detecting cartilage damages and potential early degeneration in ACL-injured knees.

It should be noted that at the spin-lock frequencies that are commonly used for *in vivo* studies (400–500 Hz), $T_{1\rho}$ quantification also showed orientation dependence as discussed earlier. Therefore, $T_{1\rho}$ values of subregions along the femoral condyles cannot be compared directly. Rather, the values should be compared to match regions in control knees. To increase the spin-lock frequency will decrease the orientation dependence, but the highest spin-lock frequency will be limited by the energy deposited to the tissue as evaluated by SAR or by the maximum B_1 allowed by the hardware.

6.4.3.3 Advantages and Challenges

$T_{1\rho}$ quantification in cartilage can provide valuable information related to biochemical changes in cartilage matrix. In particular, compared to more established T_2 relaxation time, $T_{1\rho}$ provides more sensitive detection of PG loss at early stages of cartilage degeneration. Similar to T_2 quantification, $T_{1\rho}$ quantification requires no contrast agent injection and no special hardware. Technical challenges of $T_{1\rho}$ quantification include high energy deposited to tissue (high SAR), especially at high and ultrahigh field strengths, and relatively long acquisition time. However advanced acceleration techniques might help reduce acquisition time. Previous studies have suggested $T_{1\rho}$-mapping sequences with reduced SAR using partial *k*-space acquisition approach (Wheaton et al., 2004a) and "keyhole" acquisition (Wheaton et al., 2003).

6.4.4 MT and CEST MRI

MR imaging techniques based on MT are sensitive to macromolecular changes in tissues by exploring the exchange of magnetization between the pool of semisolid macromolecule-associated (bound) protons and the pool of unbound protons associated with bulk water (Henkelman et al., 2001). During MT experiments, the macromolecule-bound protons are preferentially saturated with selective off-resonance RF pulses. Off-resonance irradiation in MRI can be applied in a CW mode or in a pulsed mode (Graham and Henkelman, 1997; Pike, 1996; Pike et al., 1993). Off-resonance RF pulses are usually delivered with an amplitude modulation that varies smoothly in time. The frequency spectrum of this RF pulse does not contain any energy in the vicinity of the Larmor frequency of the liquid pool (Henkelman et al., 2001). The exchange of magnetization that transfers macromolecular saturation to the unbound proton pool produces an observable decrease in the longitudinal magnetization of the unbound spins, which forms the basis of MT experiments. The magnetization exchange can occur via dipolar coupling or chemical exchange. The most common parameter to quantify the amount of MT is by calculating the MT ratio (MTR) of the signal intensity before and after saturation.

The MT effect in cartilage has been attributed primarily to collagen (Kim et al., 1993; Laurent et al., 2001), while studies showed that changes in PG concentration and tissue structure also contribute to changes in MTR in cartilage (Gray et al., 1995; Wachsmuth et al., 1997). It is worthy of noting that although MTR provides valuable information regarding tissue composition, MTR is not an intrinsic MR property of tissues. This parameter is reflective of a complex combination of sequence (including MT pulse frequency and power) and the tissue relaxation (T_1 of both free and bound pool) parameters in addition to MT parameters (including bound pool fraction, the exchange rate between the bound and the free pool) (Pike, 1996). The reduction of the entire MT phenomenon to a single MTR value may prevent a physical interpretation and overlook potentially useful diagnostic information. In an effort to overcome these limitations and decouple the contribution of each of these parameters, quantitative MT (qMT) techniques have been proposed, earlier work with NMR (Henkelman et al., 1993; Morrison et al., 1995) and later with *in vivo* MRI (Chai et al., 1996; Gloor et al., 2008; Sled and Pike, 2001). These methods derive the fraction of exchanging protons that are bound to macromolecules and the exchange rate between the bound and the free pool based on the two-compartment model. Very recently, Stikov et al. (2011) applied qMT techniques in bovine cartilage and revealed that in the top layer of cartilage, the bound pool fraction was moderately correlated with PG content, whereas the cross-relaxation rate and the longitudinal relaxation time were moderately correlated with collagen. This study suggested that qMT may provide useful parameters that are indicative of different aspects of the macromolecular structures in cartilage matrix. More studies in the future are warranted to examine the feasibility and validity of qMT in assessing matrix composition of healthy and degenerated cartilage.

CEST imaging is a relatively new MRI contrast approach (van Zijl and Yadav, 2011; Ward et al., 2000). In CEST experiments, exogenous or endogenous compounds containing exchangeable protons from macromolecules are selectively saturated and, after transfer of this saturation upon chemical exchange to the bulk water, detected indirectly through the water signal with enhanced sensitivity. In CEST MRI, transfer of magnetization is studied in mobile compounds instead of semisolids in contrast to conventional MT. CEST compounds differ from the semisolid MT protons in transverse relaxation properties ($T_{2\text{mobile}}^{*} \gg T_{2\text{solid}}^{*}$) and average exchange rate ($k_{\text{CEST}} \gg k_{\text{solid}}$). These differences provide opportunities for alternative labeling techniques (other than saturation) that cannot be applied to conventional MT and for separating CEST and conventional MT effects (van Zijl and Yadav, 2011).

In order to account for direct saturation of water and background MT that is related to mechanism other than chemical exchange, such as nuclear Overhauser effect (NOE) in cartilage (Ling et al., 2008), two images are normally acquired in CEST experiments: one with a saturation pulse applied at the resonance frequency of interest ($-\delta$) and the other acquired with an equal frequency offset but applied on the other side of the bulk water peak (δ). The CEST effect was then quantified as the difference of these two images (Ling et al., 2008).

In cartilage, CEST exploits the exchangeable protons, including NH, OH, and NH_2 proton groups, on the GAG side chains of PG (Ling et al., 2008; Schmitt et al., 2011) and was termed as gagCEST. Ling et al. (2008) showed that –OH at $\delta = -1.0$ ppm, where δ is the frequency offset relative to the water, among other labile protons, can be used to monitor GAG concentration in cartilage *in vivo*. gagCEST revealed difference in GAG concentration between trypsin PG-depleted cartilage and controls, and low GAG concentration *in vivo* with focal cartilage lesion (Ling et al., 2008). At 7 T, a strong correlation ($r = 0.701$) was found between ratios of signal intensity from native cartilage to signal intensity from repair tissue obtained with gagCEST (asymmetries in gagCEST z-spectra summed over all offsets from 0 ppm to

1.3 ppm) or ^{23}Na imaging (Schmitt et al., 2011), which suggested the specificity of gagCEST for detecting GAG concentration.

gagCEST is a promising new method of evaluating cartilage matrix composition, with high specificity to GAG concentration and with no need of contrast agent injection. However, the sensitivity to multiple factors—including pH changes and changes in hydration and collagen that may also change the exchange rate of –OH protons—and to pulse sequence parameters complicates the interpretation of gagCEST and comparison studies between laboratories. In addition, the field inhomogeneity, susceptibility-induced artifacts, and motion may confound the results from gagCEST experiments. Further, at 3 T, the slow-to-intermediate exchange condition, $\Delta\omega > k$ (where $\Delta\omega$ is the chemical shift of the solute spin relative to the bulk water proton and k is the exchange rate of the solute spin), which is needed to efficiently observe CEST effect, is not fulfilled for –OH protons with exchange site at 1 ppm. The CEST effect is expected to increase at ultrahigh field strength such as 7 T. The major advantages are the larger frequency separation for better adherence to the slow-exchange condition and reduced interference of direct water saturation, but increased power deposition (SAR) might impose limitations.

6.4.5 Diffusion Imaging

Diffusion MRI measures Brownian motion of water molecules in tissues, which provides a noninvasive method for evaluating microscopic cellular structures of the tissue. Diffusion-weighted images provide a new contrast qualitatively related to tissue properties. Diffusion tensor imaging (DTI), where diffusion-weighted images are acquired in multiple directions, can provide quantitative information, including mean diffusivity (MD) or apparent diffusion coefficient (ADC), indicating the degree of water diffusion, and the fractional anisotropy (FA), indicating the anisotropy of water diffusion in different orientations.

Ex vivo studies have documented promising results that MRI diffusion parameters may be sensitive to both the PG content and the collagen architecture within cartilage matrix, and therefore can be promising indicators of early degeneration of the tissue (Azuma et al., 2009; Burstein et al., 1993; de Visser et al., 2008a; Deng et al., 2007; Filidoro et al., 2005; Meder et al., 2006; Raya et al., 2011a, 2011b; Xia et al., 1995). Zonal variation was observed where ADC continuously decreased and FA increased from the articular surface to the bone–cartilage interface (Meder et al., 2006; Raya et al., 2011a). The collagen fiber architecture has been shown to be the primary source of the anisotropy of diffusion (de Visser et al., 2008a; Deng et al., 2007; Filidoro et al., 2005; Meder et al., 2006; Raya et al., 2011b). The primary eigenvector of the diffusion tensor (i.e., the eigenvector associated with the largest eigenvalue) has been shown to correlate with the orientation of the collagen fibrils using PLM (de Visser et al., 2008a) and scanning electron microscopy (SEM) (Filidoro et al., 2005; Raya et al., 2011). Loss of PG caused significant increases in MD and minimal changes in FA (Deng et al., 2007; Meder et al., 2006; Raya et al., 2011b), whereas collagen depletion led to changes in ADC, FA, and primary eigenvector (Deng et al., 2007).

In bovine cartilage, compression resulted in a decrease in both the maximum and mean eigenvalues, particularly in the surface and transitional zones, while the change in orientation of the eigenvectors corresponding to maximum diffusion was greatest in the transitional region (de Visser et al., 2008b). Raya et al. (2011a) imaged human patella samples under indentation loading and observed that compression reduced ADC in the superficial 30% of cartilage, increased FA in the superficial 5% of cartilage, and caused reorientation of the eigenvectors. However, no significant correlation was found between ADC, FA, and compressive stiffness. DTI was also used in establishing a novel finite element simulation for modeling the complex morphology and response of articular cartilage under compression (Pierce et al., 2010). These studies suggested that DTI can be used to monitor the changes in the microstructure of cartilage matrix due to compression and provides valuable information of the tissue functional state.

Despite promising results from *ex vivo* studies, *in vivo* applications of diffusion MRI to cartilage are rather limited. Fast high-resolution (submillimeter in plane resolution) 3D diffusion-weighted images have been developed based on steady-state free precession (SSFP) sequences (Bieri et al., 2011; Miller et al., 2004). *In vivo* diffusion-weighted MRI has been applied in monitoring tissue integrity after cartilage repair (Mamisch et al., 2008b; Welsch et al., 2009) using the mean diffusion quotients (signal intensity without diffusion weighting divided by signal intensity with diffusion weighting). A study in five healthy controls showed the feasibility of performing *in vivo* DTI based on spin-echo echo-planar imaging (SE-EPI) sequences with GRAPPA (Azuma et al., 2009). The study reported that the MD was increased at the surface of cartilage after 60 min of bed resting compared to 3 min of bed resting, suggesting the loading effect on diffusion parameters in cartilage (Azuma et al., 2009).

Diffusion MRI, in particular DTI, has the advantage that multicomponents of the cartilage matrix including PG concentration and collagen network orientation and fraction can be quantified from one sequence. The major challenges of *in vivo* diffusion MRI of cartilage include the low SNR due to the short T_2 relaxation time of the tissue, the requirement of high resolution due to small dimension of cartilage and the sensitivity of diffusion MRI to motion, and the long acquisition time. Further studies are warranted to evaluate the reliability and reproducibility of *in vivo* diffusion parameter quantification and its capability in diagnosing and monitoring cartilage degeneration.

6.4.6 Sodium Imaging

In addition to the proton MRI methods, sodium (^{23}Na) MRI has also been developed to evaluate cartilage matrix composition, in particular PG concentration, which was reviewed in detail by Borthakur et al. (2006b).

As discussed earlier in this chapter, the GAG side chains of the PG are negatively charged, which concentrates cations including Na^+ and K^+ in the cartilaginous interstitial fluid. Based on the ideal Donnan theory, the FCD, which is correlated with GAG concentration (Maroudas et al., 1969), can be estimated based on sodium content (Lesperance et al., 1992; Shapiro et al., 2002): $FCD = [Na_s^+]^2/[Na_t^+] - [Na_s^+]$, where $[Na_s^+]$ is the sodium concentration in the surrounding synovial fluid and $[Na_t^+]$ is the sodium concentration in the tissue (cartilage). Therefore, sodium can serve as an attractive natural endogenous contrast agent for PG evaluation.

Compared to proton MRI, sodium MRI, however, suffers from inherent low SNR due to (1) low concentrations *in vivo* [~300 mM of ^{23}Na (Shapiro et al., 2002; Wang et al., 2009) vs. 50 M of ^1H in healthy cartilage]; (2) a four times lower gyromagnetic ratio (11.262 MHz/T of ^{23}Na vs. 42.575 MHz/T for ^1H); and (3) the ultrashort T_2 and T_2^* relaxation times. Sodium in cartilage showed bioexpential T_2 and T_2^* relaxation. T_{2short} and T_{2long} in human patellar cartilage was reported to be approximately 1 ms and 12 ms respectively using *in vivo* sodium multiple quantum spectra at 2 T (Reddy et al., 1997). T_{2long}^*/T_{2short}^* in different compartments of knee cartilage was recently measured using *in vivo* sodium MRI at 7 T and reported to range from 0.5 ms to 1.4 ms for T_{2short}^* and 11.4 ms to 14.8 ms for T_{2long}^* (Madelin et al., 2012). Thus, it is highly challenging to acquire *in vivo* sodium MR images with adequate SNR and spatial resolution under a clinically reasonable scan time. Higher static magnetic field strengths, dedicated coils, and optimal pulse sequences are essential for *in vivo* sodium MRI. At ultrahigh field strength such as 7 T, the low gyromagnetic ratio and consequently the low Larmor frequency of sodium allow for imaging without certain limitations that normally apply to proton imaging at high fields including hardware limitation and power deposition. Field inhomogeneity and off-resonance effects are also reduced for sodium imaging due to the low gyromagnetic ratio.

The 3D GRE sequences have demonstrated the feasibility of sodium MRI of the human knee *in vivo* (Granot, 1988; Reddy et al., 1998; Shapiro et al., 2002; Trattnig et al., 2010; Wheaton et al., 2004c). However, these approaches are suboptimal as they required long TE values of 2–6.4 ms. The 3D projection imaging permits extremely short TE and has been implemented for sodium knee imaging with twisted projection imaging (TPI; TE = 0.4 ms) (Boada et al., 1997b; Borthakur et al., 1999), straight radial projections (TE = 0.16 ms) (Wang et al., 2009), or 3D cones (TE = 0.6 ms) (Staroswiecki et al., 2010). Nonselective hard RF pulses are also essential to achieve short TE in these 3D projection sequences. Using these techniques at 3 T and/or 7 T, images can be acquired approximately within 15–30 min with a reasonable SNR and spatial resolution (SNR ~30 with nominal resolution 3.4 mm³, scan time ~14 min using a quadrature ^{23}Na coil at 7 T (Wang et al., 2009); SNR ~7 at 3 T using a sodium-only coil; and SNR ~17 at 7 T using a dual-tuned sodium/proton surface coil). Both 3 T and 7 T measures had TE = 0.6 ms, nominal resolution 2 mm³, scan time ~26 min (Staroswiecki et al., 2010).

In addition to the above efforts to reduce the TE to the lowest limit allowed by hardware, a recent study suggested an optimization of sodium SNR under SAR constraints with a shorter "steady-state" TR and smaller flip angles to take advantage of short sodium T_1 [~24 ms at 4 T (Wheaton et al., 2004c)] (Stobbe and Beaulieu, 2008; Watts et al., 2011). This shorter TR allowed more data averaging (although it meanwhile increased the TE due to SAR limit), which resulted in an approximately 30% SNR increase compared to "fully relaxed" parameters. At 4.7 T, using this projection acquisition in the steady-state (PASS) sequence, sodium images (TR/TE = 30/0.185 ms), with SNR 8–10 were acquired within 9 min (Watts et al., 2011).

Sodium content in cartilage measured by NMR and MRI was highly correlated with measures from inductively coupled plasma (ICP) emission spectroscopy (Lesperance et al., 1992) and standard dimethylmethylene blue PG assays (Shapiro et al., 2002), respectively. Using relaxation normalized calibration phantoms, sodium absolute concentration and then FCD were quantified (Shapiro et al., 2000). In trypsin-digested bovine cartilage, with over a 20% PG depletion, sodium image signal change correlated significantly with the observed PG loss, whereas changes in proton MRI (proton density-weighted, proton T_1 and T_2 mapping) did not exhibit a definite trend (Borthakur et al., 2000). Improved methods were suggested later to correct for B_1 inhomogeneity, as well as for T_1 and T_2^* weighting during sodium quantification, and showed the feasibility of using sodium MRI to distinguish early degenerated cartilage using *in vivo* data from controls at 4.7 T with a surface coil (Wheaton et al., 2004c). The FCD was reported to be 182 ± 9 mmol/L in healthy patella cartilage ($n = 9$) and 108 mmol/L to 144 mmol/L in OA patella cartilage ($n = 3$) (Wheaton et al., 2004c). At 7 T, sodium concentrations were reduced significantly in OA subjects ($n = 3$) compared to controls ($n = 5$) by approximately 30–60%, depending on the degree of cartilage degeneration (Wang et al., 2009). Sodium MRI has also been applied to monitor tissue after cartilage repair. Mean normalized sodium signal intensities were significantly lower in the cartilage transplant after matrix-associated autologous chondrocyte transplantation (MACT) compared to healthy cartilage, which was also correlated with dGEMRIC measures in the same cohort (Trattnig et al., 2010).

The sodium relaxation times measured using NMR spectrometer were reported to change with PG depletion such that T_1 and T_{2long}^* increased, and T_{2short}^* decreased with decreased PG concentration (Insko et al., 1999). In addition to the total sodium signal, the ordered sodium pool can be detected in MRI with the use of the double-quantum filter technique or methods based on detection via the central transition through the use of frequency-swept pulses or the nutation effect (Ling et al., 2006). NMR spectroscopy experiments showed the ordered sodium signals decreased, while the quadrupolar couplings increased as a function of PG depletion time in bovine cartilage (Ling et al., 2006). These studies suggested other promising sodium markers for detection of early changes in PG and matrix structure during cartilage degeneration.

The major advantage of sodium MR imaging of cartilage is its high specificity to PG content without the need for any exogenous contrast agent. Moreover, the low sodium content (<50 mmol/L) of surrounding structures in the joint enables visualization of the cartilage with very high tissue contrast. However, sodium MRI is primarily limited by special hardware requirements and inherently low sensitivity at the standard clinical field strength of 1.5 T. With the increasing availability of scanners with high field strength (3 T or higher), sodium MRI might be utilized more commonly in clinical applications.

6.4.7 Ultrashort Echo Imaging Methods

Sequences that acquire data with ultrashort or negligible time between excitation and data acquisition are designed to image tissue components with very short T_2 of a few milliseconds or less (Bergin et al., 1991; Gatehouse and Bydder, 2003; Idiyatullin et al., 2006), which are otherwise "invisible" with conventional MR sequences. In the musculoskeletal (MSK) system, the ultrashort echo time (UTE) methods have been applied to cortical bone, tendons, menisci, vertebral disc, and cartilage (Bae et al., 2010; Carl et al., 2011; Du et al., 2007, 2010a; Gold et al., 1998; Krug et al., 2011; Qian et al., 2012; Reichert et al., 2005; Techawiboonwong et al., 2008a; Williams et al., 2010). In cartilage, the T_2 relaxation times decrease from the superficial zone to the deep zone as discussed previously. In the calcified zone close to the bone–cartilage interface, the T_2 relaxation times can be 10 ms or less (Freeman et al., 1997; Xia et al., 1994). Changes within these regions may be important in early detection of cartilage degeneration.

The typical UTE sequences apply half excitation pulses, and the data are acquired on the gradient ramp to keep a minimal TE, normally from a few microseconds (μs) to less than 100 μs. The data acquisition was developed with both 2D and 3D implementations, including 2D radial, twisted radial, or spiral acquisition (Du et al., 2008; Jackson et al., 1992; Wansapura et al., 2001) and 3D projection reconstruction (3D PR), TPI, and hybrid methods such as acquisition-weighted stack of spirals (Boada et al., 1997a; Gurney et al., 2006; Rahmer et al., 2006; Techawiboonwong et al., 2008b; Wu et al., 2003). Multiple techniques for long T_2 component suppression in UTE were proposed in order to improve the contrast and visualization of short T_2 component, including dual-echo acquisition with echo or scaled echo subtraction (Du et al., 2011; Gatehouse and Bydder, 2003; Rahmer et al., 2007), methods based on RF pulses that selectively saturate or null long T_2 component for saturation and its modifications (Du et al., 2010c, 2011; Larson et al., 2006, 2007; Pauly et al., 1992; Rahmer et al., 2007; Sussman et al., 1998).

UTE sequences have been employed for investigating T_2^* (Du et al., 2009; Williams et al., 2010, 2011) and $T_{1\rho}$ (Du et al., 2010b) in cartilage and other MSK tissues with short T_2 such as Achilles tendon and meniscus. The root-mean-square average coefficients of variation (RMSA-CV) of *in vivo* UTE-T_2^* measures in cartilage was reported to be 8%, 6%, and 16% for full-thickness, superficial, and deep central medial femoral condyle (cMFC) ROIs, corresponding to absolute errors (SDs) of 1.2 ms, 1.5 ms, and 1.5 ms, respectively (Williams et al., 2011). These sequences also have potential to quantify the relaxation times and the fractions of short and long water components in cartilage and other tissues.

6.4.8 Advanced Image and Data Analysis Methods

Advanced image analysis methods have been applied in quantitative MRI of cartilage to explore parameters other than the mean values. In particular, texture analysis (Petrou and Sevilla, 2006) has been applied to examine the spatial distribution of pixel values and quantify the heterogeneity in cartilage $T_{1\rho}$ and T_2 maps (Blumenkrantz et al., 2008; Carballido-Gamio et al., 2009; Li et al., 2009). The most commonly used texture analysis parameters are those extracted from the gray-level co-occurrence matrix (GLCM) as proposed by Haralick et al. (1973). The GLCM determines the frequency that neighboring gray-level values occur in an image. Parameters derived from GLCM provide information on the variation between neighboring pixels and directly quantify the distribution of the image signal. Higher GLCM contrast and mean, lower angular second moment (ASM), higher entropy, higher variance, and higher correlation were reported for both $T_{1\rho}$ and T_2 in OA cartilage compared to controls (Blumenkrantz et al., 2008; Carballido-Gamio et al., 2009; Li et al., 2009). Using receiver operating curve (ROC) analysis, laminar and texture analyses improved subject discrimination between OA subjects and controls compared to the global mean values (Carballido-Gamio et al., 2009). Cartilage flattening techniques based on Bezier splines and warping were further developed for texture analysis parallel and perpendicular to the natural cartilage layers (Carballido-Gamio et al., 2008b). Cartilage flattening showed minimum distortion (~0.5 ms) of mean T_2 values between nonflattened and flattened T_2 maps (Carballido-Gamio et al., 2010). Using this flattening technique, Carballido-Gamio et al. (2010) revealed significant longitudinal T_2 texture changes reflecting subtle signs of a laminar disruption. The authors showed decreasing contrast, dissimilarity, and entropy; and increasing homogeneity, energy, and correlation, from baseline to 1- and 2-year follow-ups. Using similar techniques, Joseph et al. (2011) examined T_2 from subjects with risk factors for OA ($n = 92$) and healthy controls ($n = 53$), randomly selected from the OAI incidence and control cohorts, respectively. They demonstrated that while there was no significant difference in the prevalence of knee abnormalities (cartilage lesions, BME, and meniscus lesions) between controls and subjects at risk for OA, T_2 parameters (mean T_2, GLCM contrast, and GLCM variance) were significantly elevated in those at risk for OA. Further studies from the same group using subjects with risk factors for OA ($n = 289$) from OAI suggested that the baseline mean and heterogeneity of cartilage T_2 were significantly ($p < .05$) associated with morphologic joint degeneration in the cartilage, meniscus, and bone marrow over 3 years (Joseph et al., 2012).

The results from these studies indicate that $T_{1\rho}$ and T_2 relaxation time constants are not only increased but are also more heterogeneous in osteoarthritic cartilage. Feasibility of texture analysis of *in vivo* dGEMRIC in menisci was demonstrated recently, and significantly increased entropy and decreased ASM of dGEMRIC index were observed in menisci after 50% body-weight loading (Mayerhoefer et al., 2010).

While univariate analysis of each MR parameter discussed above demonstrated promising results, combining them using advanced statistical analysis method will provide multiparametric evaluation (Lin et al., 2009) and integrate the information related to different aspects of the cartilage matrix, which has the potential to further improve our capability to distinguish degenerated cartilage from controls, as well as to monitor subtle changes after interventions.

6.5 Repair and Regeneration

MR methods, especially those characterizing changes in the ECM, play an important role in assessing cartilage repair (Trattnig et al., 2011). Microfracture is a frequently used technique for the repair of articular cartilage lesions of the knee, which is performed by inducing multiple penetrating injuries into the subchondral bone. Pluripotent mesenchymal progenitor cells migrate into the injury area, which subsequently form reparative fibrocartilage. Such generated fibrocartilage typically lacks stratification of hyaline cartilage matrix structure. In contrast, other techniques, including mosaicplasty, autologous chondrocyte transplantation (ACT) or MACT, are expected to generate more hyaline cartilage with stratification of matrix structure.

In a prospective 2-year study in symptomatic patients with isolated full-thickness articular cartilage defects treated with the microfracture technique outcomes, clinical rating and MRI were combined. MRI showed good repair tissue fill in the defect in 13 patients (54%), moderate fill in 7 (29%), and poor fill in 4 patients (17%), and correlated with the knee function scores (Mithoefer et al., 2006). A multimodal approach using T_2, diffusion-weighted imaging and grading of MR images have been used for assessing postoperative cartilage. While grading did not show differences between the two repair techniques, microfracture therapy and MACT, T_2 mapping showed lower T_2 values after microfracture, and diffusion-weighted imaging between healthy cartilage and cartilage repair tissue in both procedures (Welsch et al., 2009). Mamisch et al. (2010) prospectively used T_2 cartilage maps to study the effect of unloading during the MR scan in the postoperative follow-up of patients after MACT of the knee joint. They demonstrated that T_2 values change with the time of unloading during the MR scan, and this difference was more pronounced in repair tissue. The difference between the repair and the control tissue was also greater after longer unloading times, implying that assessment of cartilage repair is affected by the timing of the image acquisition relative to unloading the joint. A combined $T_{1\rho}$ and T_2 study examining repaired and the surrounding cartilage has demonstrated differences in cartilage after microfracture and mosaicplasty at 3–6 months and after a year (Holtzman et al., 2010).

Articular cartilage has very limited intrinsic regenerative capacity, making cell-based therapy a possible approach for cartilage repair. Tissue-engineered collagen matrix seeded with autogenous chondrocytes designed for the repair of hyaline articular cartilage has also been proposed, and early studies combining MR grading and quantitative T_2 mapping have been used to assess the impact of such repair (Crawford et al., 2009). In addition to fill efficacy, the layered appearance or partial stratification of T_2 as a result of collagen orientation was detected in this study for two patients at 12 months and four patients at 24 months. Cell tracking can be a major step toward unraveling and improving the repair process of some of these therapies. Micron-sized iron oxide particles (MPIOs) and superparamagnetic iron oxides (SPIOs) for labeling bone marrow-derived mesenchymal stem cells regarding effectivity, cell viability, long-term metabolic cell activity, and chondrogenic differentiation have been studied (Majumdar et al., 2006; Saldanha et al., 2011). MPIO labeling results in efficient contrast uptake and signal loss that can be visualized and quantitatively characterized via MRI. SPGR imaging of implanted cells results in *ex vivo* detection within native tissue, and $T_{1\rho}$ imaging is unaffected by the presence of labeled cells immediately following implantation. MPIO labeling does not affect quantitative GAG production during chondrogenesis, but iron aggregation hinders ECM visualization (Saldanha et al., 2011). Similarly, SPIO labeling was effective and did not impair any of the studied safety aspects, and injected and implanted SPIO-labeled cells can be accurately visualized by MRI in a clinically relevant sized joint (van Buul et al., 2011).

6.6 Quantitative Cartilage MRI and Biomechanics

It is well established that biomechanical factors play significant roles in cartilage development, maintenance, and degeneration (Carter et al., 2004). Conventionally, in the laboratory, cartilage response to load has been shown at scales ranging from the cellular to tissue to whole joint, as reviewed by Carter et al. (2004). During the past decades, evidence has emerged supporting a relationship between biomechanical factors and *in vivo* human cartilage health, thanks to the advances of MRI techniques that can quantify cartilage morphology and matrix composition noninvasively. Normal physiological loading and joint motion (kinematics) are critical for maintaining healthy cartilage, and abnormalities in these biomechanical factors may initiate and accelerate OA. Studies of correlating quantitative MRI with biomechanical measures normally fall into two categories: (1) at the tissue level, correlating quantitative MRI with biomechanical properties of cartilage in order to validate quantitative MRI as a predictor of functional state of cartilage, and (2) at the whole-body function level, correlating factors such as loading and joint kinematics with quantitative MRI in order to explore the relationship between biomechanical factors and cartilage health.

6.6.1 Correlation of *Ex Vivo* Cartilage MRI Measures to Biomechanical Properties

The mechanical behavior of articular cartilage is derived from the triphasic interactions between its primary constituents of collagen, PGs, and water (Kwan et al., 1984). The collagen network in cartilage is optimized to resist shear and tensile strains, and it also provides attachment for the negatively charged PG network (Bank et al., 1997) that can resist compressive loading (Kwan et al., 1984). The ability of advanced MRI techniques, as mentioned in previous sections, to map the biochemical composition of cartilage may provide a noninvasive method to estimate its mechanical properties.

Several studies have evaluated the relationship between quantitative MRI, including T_2, $T_{1\rho}$, and dGEMRIC, and mechanical properties (Juras et al., 2009; Kurkijarvi et al., 2004; Lammentausta et al., 2006; Nieminen et al., 2004; Nissi et al., 2007; Wheaton et al., 2005). In these studies, mechanical properties including equilibrium (Young's) modulus, dynamic modulus, aggregate modulus (HA), and hydraulic permeability of the samples were measured in unconfined or confined compression using a material testing device.

Using bovine cartilage, Nieminen et al. (2004) reported that both T_2 and dGEMRIC at 9.4 T are correlated with mechanical parameters and that quantitative MRI parameters were able to explain up to 87% of the variations in certain biomechanical properties. Average $T_{1\rho}$ relaxation rate was also shown to strongly correlate with HA and hydraulic permeability ($R^2 = 0.862$) in bovine cartilage (Wheaton et al., 2005).

Using human cadaver patellar cartilage, Lammentausta et al. (2006) observed both T_2 and dGEMRIC measurements at 1.5 T and 9.4 T were correlated with the static and dynamic compressive moduli, especially between T_2 and Young's modulus. The same group in later studies examined mature human (nonarthritis cadaver patellar cartilage), juvenile porcine, and juvenile bovine articular cartilage at 9.4 T, and revealed significant but complicated relationships between T_2 and mechanical properties depending on the group studied (Nissi et al., 2007). dGEMRIC however showed no significant associations with the mechanical properties of the tested patellar samples. The investigators discussed that variations in collagen architecture may provide the dominant mechanism for accounting for the differences in mechanical properties, thereby masking the positive effect of PGs on compressive stiffness, especially in juvenile tissue with immature matrix structure.

In a more recent study with human OA cartilage harvested during total knee arthroplasty, Juras et al. (2009) observed significant correlation between dGEMRIC index with mechanical parameters including equilibrium modulus; however, T_2 and ADC were unable to show significant correlation with mechanical measures.

Differences in cartilage preparations as well as differences in mechanical testing procedures may explain the discrepancy between these studies. Further investigations with larger samples are needed to determine the relationship between mechanical properties and quantitative MRI in healthy and diseased human cartilage. Combining MRI techniques that have different sensitivities for PG and collagen may improve the capability of predicting biomechanical properties of cartilage (Nieminen et al., 2004).

A few other correlation studies used indentation techniques, which allow nondestructive measures of the tissue biomechanical properties at different locations immediately before or after the quantitative MRI measures. Significant correlations were reported between dGEMRIC and site-matched indentation stiffness measurements in human tibial plateau (Baldassarri et al., 2007; Samosky et al., 2005). However, it was noted that differences in stiffness in different areas of the joint could not be explained by GAG measurements alone, suggesting the need for combined GAG/collagen studies. More recently, site-specific $T_{1\rho}$ relaxation times were correlated significantly with the phase angle ($p < .001$, $R = 0.908$; Figure 6.7), a viscoelastic mechanical behavior of the cartilage, in human OA tibial specimens (Tang et al., 2011). In these studies, the investigators discussed that the correlation was significant when quantitative MRI (dGEMRIC index or $T_{1\rho}$) at each test location was averaged over a depth of tissue comparable to that affected by the indentation. When quantitative MRI was averaged over larger depths, the correlations were generally lower. This result suggested the heterogeneous nature of the tissue biochemical and biomechanical properties in OA cartilage, and the coregistration between imaging and biomechanical measures is critical.

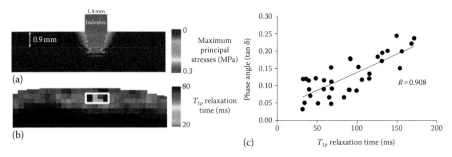

FIGURE 6.7
(Color version available from crcpress.com; see Preface.) MR $T_{1\rho}$ relaxation times significantly correlated with local cartilage biomechanics measured by indentation techniques in human OA cartilage. (a) A simple finite element contact analysis confirms that the maximum principal stresses due to indentation occur at a depth of 0.9 mm. (b) The $T_{1\rho}$ relaxation times were computed for each site of indentation by averaging the corresponding pixels as shown by the white-enclosed box. (c) The scatter plot showing the significant correlation between the phase angle and the $T_{1\rho}$ relaxation time. (Adapted from Tang, S.Y., et al., *J Orthop Res.* 29, 1312–19, 2011.)

6.6.2 Correlation of *In Vivo* Cartilage MRI Measures to Loading and Joint Kinematics

6.6.2.1 Loading

Loading plays a major role in the development of healthy cartilage, as well as in the progression of OA. Both excessive load and inadequate load have been shown to result in degenerative changes in articular cartilage (Hagiwara et al., 2009; Neidhart et al., 2000; Vanwanseele et al., 2003). In this section, we will focus more on recent studies using quantitative MRI for evaluating cartilage matrix changes with loading. Extensive reviews of morphologic changes of cartilage as a response to loading and different exercises can be found in Eckstein et al. (2006).

Using an MR compatible loading device, significant decrease in T_2 (Nag et al., 2004; Nishii et al., 2008) and $T_{1\rho}$ (Souza et al., 2010) was observed with static axial loading (50% body weight), suggesting a loss of water content and alteration to the matrix structure under acute loading. Interestingly, Souza et al. (2010) suggested that cartilage compartments with small focal lesions experienced greater $T_{1\rho}$ change with loading compared to cartilage without lesions or cartilage with larger defects. This greater change in $T_{1\rho}$ with loading may represent a deficit in cartilage mechanical properties (i.e., tissue elasticity, permeability, etc.) that can occur at early stages of the disease. Thus, the results of quantitative MRI with loading will provide more information than unloaded relaxation time mapping that may be related to cartilage biomechanical properties and functional states.

Cartilage response to cyclical and dynamic physiological loading such as running (Mosher et al., 2005, 2010) and knee bending (Liess et al., 2002) has been investigated using MRI T_2 quantification. Mosher et al. (2005) reported that after running, MRI T_2 values decreased in the superficial layer of femorotibial cartilage, along with a decrease in cartilage thickness. Smaller decrease and no change in cartilage T_2 values were observed in the middle and deep zone of cartilage, respectively. There were no differences in cartilage deformation or T_2 response to running as a function of age or level of physical activity, although there is a decreasing trend of smaller cartilage thickness changes with increased age, suggesting that older subjects have stiffer cartilage independent of their level of physical activity (Mosher et al., 2010). Interestingly, in that study, the investigators found no regional differences in cartilage T_2 relaxation, which suggests that in the normal joint, rather uniform tissue strains are generated within femorotibial cartilage during running exercise. Further studies in subjects with OA would be interesting to examine the loading pattern and cartilage response to running in degenerative joints.

Cartilage response to long-distance running has also been investigated using quantitative MRI recently because there is continuing controversy whether long-distance running results in irreversible articular cartilage damage. At the biochemical level, marathon runners have been demonstrated to have increased OA-associated biomarkers in serum, including tumor necrosis factor-α, soluble interleukin-6 receptor, and cartilage oligomeric matrix protein (COMP) (Neidhart et al., 2000). However, early studies using anatomical MRI reveal no significant structural changes in the knee and hip after marathon running except for some subtle and transient changes in intrameniscal signal, effusion, and bone marrow (Foley et al., 2007; Hohmann et al., 2004; Kessler et al., 2008; Schueller-Weidekamm et al., 2006). More recently, Luke et al. (2010) and Stehling et al. (2011) MRI $T_{1\rho}$ and T_2 to evaluate cartilage changes in asymptomatic marathon runners 2 weeks before, within 48 h after, and 10–12 weeks after running a marathon. They showed that although runners did not demonstrate any gross morphologic MRI changes after running a marathon, consistent with previous studies, significantly higher T_2 and $T_{1\rho}$ values in patella, trochlea, and medial femorotibial cartilage were found ($p < .01$). More interestingly, the T_2 values recovered to baseline in all compartments except in the medial femoral condyle after 3 months. Average $T_{1\rho}$ values increased after the marathon and remained increased at 3 months (Luke et al., 2010). A similar pattern was observed in meniscal $T_{1\rho}$ and T_2 in the same cohort, that is, both increased after running, and $T_{1\rho}$ stayed elevated, while T_2 recovered (Stehling et al., 2011). These results suggest that $T_{1\rho}$ reflects changes, such as early loss of PG, other than fluid shifts (the change and recovery of T_2 were probably caused by fluid shifts) in the articular cartilage and meniscus. Although whether these $T_{1\rho}$ and T_2 changes after running lead to detrimental structural articular cartilage and meniscal changes over time requires longer term studies, results from this study suggested that there is potential long-term effect on cartilage and meniscal matrix after long-distance running.

Using data from the OAI, the association between physical activity levels (determined by Physical Activity Scale for the Elderly), in particular knee-bending activities, and cartilage T_2 relaxation time has been explored. In 128 subjects with risk factors for knee OA, light exercisers had lower T_2 values compared with sedentary and moderate/strenuous exercisers, suggesting a potential beneficial effect of mild exercise on cartilage health, whereas subjects ($n = 33$) without risk factors displayed no significant differences in T_2 values according to exercise level (Hovis et al., 2011). There has been controversy over the effect of physical activities to cartilage health as some studies showed a beneficial effect (Rogers et al., 2002; Roos and Dahlberg, 2005), while other studies showed a detrimental effect (Cheng et al., 2000; Spector et al., 1996). The results from the OAI data suggested that there could be a "U" shape relationship between exercise levels and cartilage biochemistry—either too little or too much will be detrimental. The optimized appropriate level of exercise for each individual may be affected by many factors including BMI, genetic profile, and biomechanical characteristics such as alignment and kinematics. Nonetheless, as exercise is a modifiable factor, exploring the relationship of different types and levels of exercise to cartilage biochemistry as quantified by advanced MRI can be clinically valuable to allow intervention and training to prevent and slow down cartilage degeneration.

6.6.2.2 Kinematics

In addition to the volume of physical activity, joint, kinematics and kinetics also play critical roles in cartilage health. Many biomechanical studies in the literature suggested that increased external knee adduction moment (KAM), a kinetic variable that describes the frontal plane torque about the knee joint and provides important information regarding the load distribution between the medial and lateral compartments, is associated with OA to varying degrees (Astephen et al., 2008; Gok et al., 2002; Thorp et al., 2006).

Using 3D MRI cartilage morphology quantification sequences, several studies investigated the relationship of KAM and angular impulse to morphological characteristics of cartilage in healthy and OA subjects (Creaby et al., 2010; Jackson et al., 2004; Koo and Andriacchi, 2007; Vanwanseele et al., 2010). While Jackson and colleagues (2004) reported no relationship between peak KAM and cartilage thickness in healthy females, Koo and Andriacchi (2007) reported a significant positive correlation between peak KAM during gait and the medial–lateral cartilage thickness ratio (M/L ratio) in both the femur and the tibia in healthy young males. Two recent studies reported that KAM was not related to cartilage volume, thickness, or M/L ratio in subjects with OA (Creaby et al., 2010; Vanwanseele et al., 2010), while there was a significant relationship between the knee adduction angular impulse and the M/L ratio of femoral cartilage (Vanwanseele et al., 2010).

In a more recent study, MRI $T_{1\rho}$ and T_2 relaxation times, which are more sensitive imaging markers than morphologic quantification for cartilage matrix changes, were correlated to joint kinetics including parameters in frontal (such as KAM), transverse, and sagittal planes in 28 knees from 14 healthy young subjects (Souza et al., 2012). Increased flexion moments during hopping were associated with decreased $T_{1\rho}$ and T_2 values, suggesting that a higher loading in the sagittal plane is correlated with richer collagen–PG matrix in the cartilage in this cohort, which may be protective against cartilage degeneration. Additionally, the M/L ratio of $T_{1\rho}$ and T_2 times was reported to be related to the frontal (peak KAM) and transverse [internal rotation (IR) moment] plane joint mechanics, respectively. These data suggest that higher loads across the knee are significantly related to composition of the articular cartilage, which offer promising findings of potentially modifiable determinants of cartilage biochemistry.

Abnormal kinematics after acute injuries have been suggested as significant factors contributing to posttraumatic OA development. In particular, subjects with ACL injury have a high risk of developing posttraumatic OA despite surgical reconstruction (Daniel et al., 1994; Ferretti et al., 1991; Hanypsiak et al., 2008; Lohmander et al., 2004, 2007; van der Hart et al., 2008; von Porat et al., 2004). Abnormal biomechanics even after ACL reconstruction, along with biochemical changes caused by the initial injuries, are hypothesized as the contributing factors of the joint degeneration. Previous studies reported abnormal knee kinematics after ACL injury and reconstruction using high-speed radiography and fluoroscopic analysis (Kozanek et al., 2011; Li et al., 2006; Seon et al., 2009; Tashman et al., 2004, 2007; Van de Velde et al., 2009), kinematic MRI (Carpenter et al., 2009; Shefelbine et al., 2006) with static loading or upright MRI (Logan et al., 2004; Nicholson et al., 2012; Scarvell et al., 2005), and motion analysis (Andriacchi and Dyrby, 2005; Butler et al., 2009; Gao and Zheng, 2010; Georgoulis et al., 2003; Scanlan et al., 2010; Yamazaki et al., 2010). Interestingly, the increased KAM observed in patients with knee OA was also observed in patients after ACL reconstruction (Baliunas et al., 2002).

Despite the hypothesis that abnormal biomechanics were associated with OA development in ACL-injured and ACL-reconstructed knees, limited studies have provided direct evidence of such link. Using finite element analysis, the ACL-deficient knee with IR offset predicted a more rapid rate of cartilage thinning throughout the knee, especially in the medial compartment, compared to knees with normal kinematics (Andriacchi et al., 2006). Very recently, Haughom et al. (2011) documented a correlation between abnormal

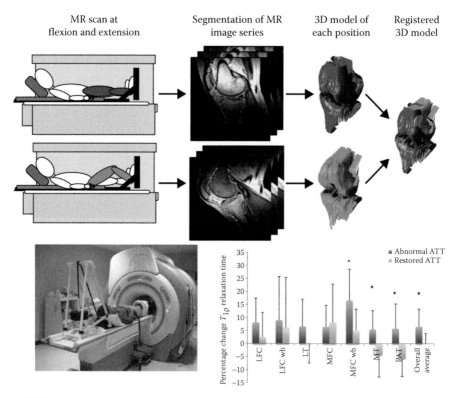

FIGURE 6.8
(**Color version available from crcpress.com; see Preface.**) Elevated $T_{1\rho}$ values in ACL-reconstructed knees are associated with abnormal kinematics quantified by 3D MRI. $^*p < .05$.

knee kinematics and early cartilage degeneration in ACL-reconstructed knees using kinematic MR and $T_{1\rho}$ quantification. Elevated $T_{1\rho}$ values in medial femoral condyle weight-bearing region, MT, and patellar cartilage were observed in patients with "abnormal" anterior tibial translation (ATT) compared to those with "restored" ATT (Figure 6.8). Similar trend was observed in the weight-bearing region of MFC in patients with "abnormal" IR compared to those with "restored" IR. These results suggested that the abnormal kinematics after ACL reconstruction might lead to accelerated cartilage degeneration in the joint. Quantitative MRI techniques such as $T_{1\rho}$ can be valuable tools in detecting early cartilage degeneration and in exploring the interrelationship between biomechanical abnormalities and cartilage health.

These studies revealed the close interrelationship between cartilage biochemical composition and kinematics of the joint. Kinematic changes may shift loads that cause both overloading to infrequently loaded regions and unloading to frequently loaded regions and may initiate and accelerate OA (Andriacchi et al., 2004). High sensitivity of quantitative MRI for detecting early biochemical changes in cartilage matrix offers a unique opportunity to explore the direct correlation between biochemical and biomechanical abnormalities in the joint. Kinematics can be modifiable variables, which implies potential effective interventions to prevent or slow down the cartilage degeneration.

6.7 Summary and Conclusions

Quantitative MRI provides noninvasive measures of changes in both morphology and matrix composition of cartilage during degeneration or injuries. In particular, quantification of cartilage matrix changes provides valuable information at the earliest stages of joint degeneration, which is essential for efforts toward prevention and early intervention in OA. Combining quantitative cartilage MRI with biomechanical measures will provide insights toward interaction and potential causal relationship between biomechanical abnormality and joint degeneration.

Acknowledgments

The authors would like to acknowledge research support by National Institutes of Health grants K25 AR053633, RO1 AR46905, RO1 AG17762, U01 AR055079, and R21 056773; the Aircast Foundation; Pfizer Inc.; and Glaxo Smith Kline Inc.

References

Akella, S.V., Regatte, R.R., Gougoutas, A.J., Borthakur, A., Shapiro, E.M., Kneeland, J.B., Leigh, J.S., Reddy, R., 2001. Proteoglycan-induced changes in T1rho-relaxation of articular cartilage at 4T. *Magn Reson Med.* 46, 419–23.

Akella, S.V., Regatte, R.R., Wheaton, A.J., Borthakur, A., Reddy, R., 2004. Reduction of residual dipolar interaction in cartilage by spin-lock technique. *Magn Reson Med.* 52, 1103–9.

Anandacoomarasamy, A., Giuffre, B.M., Leibman, S., Caterson, I.D., Smith, G.S., Fransen, M., Sambrook, P.N., March, L.M., 2009. Delayed gadolinium-enhanced magnetic resonance imaging of cartilage: Clinical associations in obese adults. *J Rheumatol.* 36, 1056–62.

Andriacchi, T., Dyrby, C., 2005. Interactions between kinematics and loading during walking for the normal and ACL deficient knee. *J Biomech.* 38, 293–8.

Andriacchi, T.P., Briant, P.L., Bevill, S.L., Koo, S., 2006. Rotational changes at the knee after ACL injury cause cartilage thinning. *Clin Orthop Relat Res.* 442, 39–44.

Andriacchi, T.P., Mundermann, A., Smith, R.L., Alexander, E.J., Dyrby, C.O., Koo, S., 2004. A framework for the in vivo pathomechanics of osteoarthritis at the knee. *Ann Biomed Eng.* 32, 447–57.

Astephen, J.L., Deluzio, K.J., Caldwell, G.E., Dunbar, M.J., 2008. Biomechanical changes at the hip, knee, and ankle joints during gait are associated with knee osteoarthritis severity. *J Orthop Res.* 26, 332–41.

Ateshian, G.A., Soslowsky, L.J., Mow, V.C., 1991. Quantitation of articular surface topography and cartilage thickness in knee joints using stereophotogrammetry. *J Biomech.* 24, 761–76.

Azuma, T., Nakai, R., Takizawa, O., Tsutsumi, S., 2009. In vivo structural analysis of articular cartilage using diffusion tensor magnetic resonance imaging. *Magn Reson Imaging.* 27, 1242–8.

Bae, W.C., Du, J., Bydder, G.M., Chung, C.B., 2010. Conventional and ultrashort time-to-echo magnetic resonance imaging of articular cartilage, meniscus, and intervertebral disk. *Top Magn Reson Imaging.* 21, 275–89.

Baldassarri, M., Goodwin, J.S., Farley, M.L., Bierbaum, B.E., Goldring, S.R., Goldring, M.B., Burstein, D., Gray, M.L., 2007. Relationship between cartilage stiffness and dGEMRIC index: Correlation and prediction. *J Orthop Res.* 25, 904–12.

Baliunas, A., Hurwitz, D., Ryals, A., Karrar, A., Case, J., Block, J., Andriacchi, T., 2002. Increased knee joint loads during walking are present in subjects with knee osteoarthritis. *Osteoarthritis Cartilage.* 10, 573–9.

Bank, R.A., Krikken, M., Beekman, B., Stoop, R., Maroudas, A., Lafeber, F.P., te Koppele, J.M., 1997. A simplified measurement of degraded collagen in tissues: Application in healthy, fibrillated and osteoarthritic cartilage. *Matrix Biol.* 16, 233–43.

Bashir, A., Gray, M., Burstein, D., 1996. Gd-DTPA2- as a measure of cartilage degradation. *Magn Reson Med.* 36, 665–73.

Bashir, A., Gray, M.L., Hartke, J., Burstein, D., 1999. Nondestructive imaging of human cartilage glycosaminoglycan concentration by MRI. *Magn Reson Med.* 41, 857–65.

Benichou, O.D., Hunter, D.J., Nelson, D.R., Guermazi, A., Eckstein, F., Kwoh, K., Myers, S.L., Wirth, W., Duryea, J., 2010. One-year change in radiographic joint space width in patients with unilateral joint space narrowing: Data from the Osteoarthritis Initiative. *Arthritis Care Res (Hoboken)*. 62, 924–31.

Bergin, C.J., Pauly, J.M., Macovski, A., 1991. Lung parenchyma: Projection reconstruction MR imaging. *Radiology*. 179, 777–81.

Bieri, O., Ganter, C., Welsch, G.H., Trattnig, S., Mamisch, T.C., Scheffler, K., 2011. Fast diffusion-weighted steady state free precession imaging of in vivo knee cartilage. *Magn Reson Med*. 67, 691–700.

Bittersohl, B., Steppacher, S., Haamberg, T., Kim, Y.J., Werlen, S., Beck, M., Siebenrock, K.A., Mamisch, T.C., 2009. Cartilage damage in femoroacetabular impingement (FAI): Preliminary results on comparison of standard diagnostic vs delayed gadolinium-enhanced magnetic resonance imaging of cartilage (dGEMRIC). *Osteoarthritis Cartilage*. 17, 1297–306.

Blumenkrantz, G., Lindsey, C.T., Dunn, T.C., Jin, H., Ries, M.D., Link, T.M., Steinbach, L.S., Majumdar, S., 2004. A pilot, two-year longitudinal study of the interrelationship between trabecular bone and articular cartilage in the osteoarthritic knee. *Osteoarthritis Cartilage*. 12, 997–1005.

Blumenkrantz, G., Stahl, R., Carballido-Gamio, J., Zhao, S., Lu, Y., Munoz, T., Hellio Le Graverand-Gastineau, M., Jain, S., Link, T., Majumdar, S., 2008. The feasibility of characterizing the spatial distribution of cartilage T(2) using texture analysis. *Osteoarthritis Cartilage*. 16, 584–90.

Boada, F.E., Christensen, J.D., Gillen, J.S., Thulborn, K.R., 1997a. Three-dimensional projection imaging with half the number of projections. *Magn Reson Med*. 37, 470–7.

Boada, F.E., Shen, G.X., Chang, S.Y., Thulborn, K.R., 1997b. Spectrally weighted twisted projection imaging: Reducing T2 signal attenuation effects in fast three-dimensional sodium imaging. *Magn Reson Med*. 38, 1022–8.

Bolbos, R., Ma, C., Link, T., Majumdar, S., Li, X., 2008a. In vivo T1rho quantitative assessment of knee cartilage after anterior cruciate ligament injury using 3 Tesla magnetic resonance imaging. *Invest Radiol*. 43, 782–8.

Bolbos, R.I., Zuo, J., Banerjee, S., Link, T.M., Ma, C.B., Li, X., Majumdar, S., 2008b. Relationship between trabecular bone structure and articular cartilage morphology and relaxation times in early OA of the knee joint using parallel MRI at 3 T. *Osteoarthritis Cartilage*. 16, 1150–9.

Borthakur, A., Hancu, I., Boada, F.E., Shen, G.X., Shapiro, E.M., Reddy, R., 1999. In vivo triple quantum filtered twisted projection sodium MRI of human articular cartilage. *J Magn Reson*. 141, 286–90.

Borthakur, A., Hulvershorn, J., Gualtieri, E., Wheaton, A., Charagundla, S., Elliott, M., Reddy, R., 2006a. A pulse sequence for rapid in vivo spin-locked MRI. *J Magn Reson Imaging*. 23, 591–6.

Borthakur, A., Mellon, E., Niyogi, S., Witschey, W., Kneeland, J., Reddy, R., 2006b. Sodium and T1rho MRI for molecular and diagnostic imaging of articular cartilage. *NMR Biomed*. 19, 781–821.

Borthakur, A., Shapiro, E.M., Beers, J., Kudchodkar, S., Kneeland, J.B., Reddy, R., 2000. Sensitivity of MRI to proteoglycan depletion in cartilage: Comparison of sodium and proton MRI. *Osteoarthritis Cartilage*. 8, 288–93.

Borthakur, A., Wheaton, A., Charagundla, S.R., Shapiro, E.M., Regatte, R.R., Akella, S.V., Kneeland, J.B., Reddy, R., 2003. Three-dimensional T1rho-weighted MRI at 1.5 Tesla. *J Magn Reson Imaging*. 17, 730–6.

Brandt, K.D., Doherty, M., Lohmander, L.S. Eds., 1998. *Osteoarthritis*. New York: Oxford University Press Inc.

Burstein, D., Gray, M.L., Hartman, A.L., Gipe, R., Foy, B.D., 1993. Diffusion of small solutes in cartilage as measured by nuclear magnetic resonance (NMR) spectroscopy and imaging. *J Orthop Res*. 11, 465–78.

Burstein, D., Velyvis, J., Scott, K.T., Stock, K.W., Kim, Y.J., Jaramillo, D., Boutin, R.D., Gray, M.L., 2001. Protocol issues for delayed Gd(DTPA)(2-)-enhanced MRI (dGEMRIC) for clinical evaluation of articular cartilage. *Magn Reson Med*. 45, 36–41.

Butler, R., Minick, K., Ferber, R., Underwood, F., 2009. Gait mechanics after ACL reconstruction: Implications for the early onset of knee osteoarthritis. *Br J Sports Med*. 43, 366–70.

Carballido-Gamio, J., Bauer, J.S., Stahl, R., Lee, K.Y., Krause, S., Link, T.M., Majumdar, S., 2008a. Inter-subject comparison of MRI knee cartilage thickness. *Med Image Anal*. 12, 120–35.

Carballido-Gamio, J., Blumenkrantz, G., Lynch, J., Link, T., Majumdar, S., 2010. Longitudinal analysis of MRI T(2) knee cartilage laminar organization in a subset of patients from the osteoarthritis initiative. *Magn Reson Med*. 63, 465–72.

Carballido-Gamio, J., Link, T.M., Majumdar, S., 2008b. New techniques for cartilage magnetic resonance imaging relaxation time analysis: Texture analysis of flattened cartilage and localized intra- and inter-subject comparisons. *Magn Reson Med*. 59, 1472–7.

Carballido-Gamio, J., Stahl, R., Blumenkrantz, G., Romero, A., Majumdar, S., Link, T., 2009. Spatial analysis of magnetic resonance T1rho and T2 relaxation times improves classification between subjects with and without osteoarthritis. *Med Phys*. 36, 4059–67.

Carl, M., Sanal, H.T., Diaz, E., Du, J., Girard, O., Statum, S., Znamirowski, R., Chung, C.B., 2011. Optimizing MR signal contrast of the temporomandibular joint disk. *J Magn Reson Imaging*. 34, 1458–64.

Carpenter, R., Majumdar, S., Ma, C., 2009. Magnetic resonance imaging of 3-dimensional in vivo tibiofemoral kinematics in anterior cruciate ligament-reconstructed knees. *Arthroscopy*. 25, 760–6.

Carter, D.R., Beaupre, G.S., Wong, M., Smith, R.L., Andriacchi, T.P., Schurman, D.J., 2004. The mechanobiology of articular cartilage development and degeneration. *Clin Orthop Relat Res*. 427 Suppl, S69–77.

Chai, J.W., Chen, C., Chen, J.H., Lee, S.K., Yeung, H.N., 1996. Estimation of in vivo proton intrinsic and cross-relaxation rate in human brain. *Magn Reson Med*. 36, 147–52.

Charagundla, S.R., Borthakur, A., Leigh, J.S., Reddy, R., 2003. Artifacts in T(1rho)-weighted imaging: Correction with a self-compensating spin-locking pulse. *J Magn Reson*. 162, 113–21.

Chen, W., Takahashi, A., Han, E., 2011. Quantitative T(1)(rho) imaging using phase cycling for B0 and B1 field inhomogeneity compensation. *Magn Reson Imaging*. 29, 608–19.

Cheng, Y., Macera, C.A., Davis, D.R., Ainsworth, B.E., Troped, P.J., Blair, S.N., 2000. Physical activity and self-reported, physician-diagnosed osteoarthritis: Is physical activity a risk factor? *J Clin Epidemiol*. 53, 315–22.

Cohen, Z.A., McCarthy, D.M., Kwak, S.D., Legrand, P., Fogarasi, F., Ciaccio, E.J., Ateshian, G.A., 1999. Knee cartilage topography, thickness, and contact areas from MRI: In-vitro calibration and in-vivo measurements. *Osteoarthritis Cartilage*. 7, 95–109.

Cohen, Z.A., Mow, V.C., Henry, J.H., Levine, W.N., Ateshian, G.A., 2003. Templates of the cartilage layers of the patellofemoral joint and their use in the assessment of osteoarthritic cartilage damage. *Osteoarthritis Cartilage*. 11, 569–79.

Crawford, D.C., Heveran, C.M., Cannon, W.D., Jr., Foo, L.F., Potter, H.G., 2009. An autologous cartilage tissue implant NeoCart for treatment of grade III chondral injury to the distal femur: Prospective clinical safety trial at 2 years. *Am J Sports Med*. 37, 1334–43.

Creaby, M.W., Wang, Y., Bennell, K.L., Hinman, R.S., Metcalf, B.R., Bowles, K.A., Cicuttini, F.M., 2010. Dynamic knee loading is related to cartilage defects and tibial plateau bone area in medial knee osteoarthritis. *Osteoarthritis Cartilage*. 18, 1380–5.

Cunningham, T., Jessel, R., Zurakowski, D., Millis, M.B., Kim, Y.J., 2006. Delayed gadolinium-enhanced magnetic resonance imaging of cartilage to predict early failure of Bernese periacetabular osteotomy for hip dysplasia. *J Bone Joint Surg Am*. 88, 1540–8.

Daniel, D., Stone, M., Dobson, B., Fithian, D., Rossman, D., Kaufman, K., 1994. Fate of the ACL-injured patient. A prospective outcome study. *Am J Sports Med*. 22, 632–44.

Dardzinski, B.J., Mosher, T.J., Li, S., Van Slyke, M.A., Smith, M.B., 1997. Spatial variation of T2 in human articular cartilage. *Radiology*. 205, 546–50.

David-Vaudey, E., Ghosh, S., Ries, M., Majumdar, S., 2004. T2 relaxation time measurements in osteoarthritis. *Magn Reson Imaging*. 22, 673–82.

de Visser, S.K., Bowden, J.C., Wentrup-Byrne, E., Rintoul, L., Bostrom, T., Pope, J.M., Momot, K.I., 2008a. Anisotropy of collagen fibre alignment in bovine cartilage: Comparison of polarised light microscopy and spatially resolved diffusion-tensor measurements. *Osteoarthritis Cartilage*. 16, 689–97.

de Visser, S.K., Crawford, R.W., Pope, J.M., 2008b. Structural adaptations in compressed articular cartilage measured by diffusion tensor imaging. *Osteoarthritis Cartilage*. 16, 83–9.

Deng, X., Farley, M., Nieminen, M.T., Gray, M., Burstein, D., 2007. Diffusion tensor imaging of native and degenerated human articular cartilage. *Magn Reson Imaging*. 25, 168–71.

Dijkgraaf, L.C., de Bont, L.G., Boering, G., Liem, R.S., 1995. The structure, biochemistry, and metabolism of osteoarthritic cartilage: A review of the literature. *J Oral Maxillofac Surg*. 53, 1182–92.

Dixon, W.T., Oshinski, J.N., Trudeau, J.D., Arnold, B.C., Pettigrew, R.I., 1996. Myocardial suppression in vivo by spin locking with composite pulses. *Magn Reson Med*. 36, 90–4.

Du, J., Bydder, M., Takahashi, A.M., Carl, M., Chung, C.B., Bydder, G.M., 2011. Short T2 contrast with three-dimensional ultrashort echo time imaging. *Magn Reson Imaging*. 29, 470–82.

Du, J., Bydder, M., Takahashi, A.M., Chung, C.B., 2008. Two-dimensional ultrashort echo time imaging using a spiral trajectory. *Magn Reson Imaging*. 26, 304–12.

Du, J., Carl, M., Bydder, M., Takahashi, A., Chung, C.B., Bydder, G.M., 2010a. Qualitative and quantitative ultrashort echo time (UTE) imaging of cortical bone. *J Magn Reson*. 207, 304–11.

Du, J., Carl, M., Diaz, E., Takahashi, A., Han, E., Szeverenyi, N.M., Chung, C.B., Bydder, G.M., 2010b. Ultrashort TE T1rho (UTE T1rho) imaging of the Achilles tendon and meniscus. *Magn Reson Med*. 64, 834–42.

Du, J., Hamilton, G., Takahashi, A., Bydder, M., Chung, C.B., 2007. Ultrashort echo time spectroscopic imaging (UTESI) of cortical bone. *Magn Reson Med*. 58, 1001–9.

Du, J., Takahashi, A.M., Bae, W.C., Chung, C.B., Bydder, G.M., 2010c. Dual inversion recovery, ultrashort echo time (DIR UTE) imaging: Creating high contrast for short-T(2) species. *Magn Reson Med*. 63, 447–55.

Du, J., Takahashi, A.M., Chung, C.B., 2009. Ultrashort TE spectroscopic imaging (UTESI): Application to the imaging of short T2 relaxation tissues in the musculoskeletal system. *J Magn Reson Imaging*. 29, 412–21.

Dunn, T.C., Lu, Y., Jin, H., Ries, M.D., Majumdar, S., 2004. T2 relaxation time of cartilage at MR imaging: Comparison with severity of knee osteoarthritis. *Radiology*. 232, 592–8.

Duvvuri, U., Charagundla, S.R., Kudchodkar, S.B., Kaufman, J.H., Kneeland, J.B., Rizi, R., Leigh, J.S., Reddy, R., 2001a. Human knee: In vivo T1(rho)-weighted MR imaging at 1.5 T—Preliminary experience. *Radiology*. 220, 822–6.

Duvvuri, U., Goldberg, A.D., Kranz, J.K., Hoang, L., Reddy, R., Wehrli, F.W., Wand, A.J., Englander, S.W., Leigh, J.S., 2001b. Water magnetic relaxation dispersion in biological systems: The contribution of proton exchange and implications for the noninvasive detection of cartilage degradation. *Proc Natl Acad Sci USA*. 98, 12479–84.

Eckstein, F., Adam, C., Sittek, H., Becker, C., Milz, S., Schulte, E., Reiser, M., Putz, R., 1997. Non-invasive determination of cartilage thickness throughout joint surfaces using magnetic resonance imaging. *J Biomech*. 30, 285–9.

Eckstein, F., Benichou, O., Wirth, W., Nelson, D.R., Maschek, S., Hudelmaier, M., Kwoh, C.K., Guermazi, A., Hunter, D., 2009a. Magnetic resonance imaging-based cartilage loss in painful contralateral knees with and without radiographic joint space narrowing: Data from the Osteoarthritis Initiative. *Arthritis Rheum*. 61, 1218–25.

Eckstein, F., Cotofana, S., Wirth, W., Nevitt, M., John, M.R., Dreher, D., Frobell, R., 2011. Greater rates of cartilage loss in painful knees than in pain-free knees after adjustment for radiographic disease stage: Data from the osteoarthritis initiative. *Arthritis Rheum*. 63, 2257–67.

Eckstein, F., Hudelmaier, M., Putz, R., 2006. The effects of exercise on human articular cartilage. *J Anat*. 208, 491–512.

Eckstein, F., Kunz, M., Schutzer, M., Hudelmaier, M., Jackson, R.D., Yu, J., Eaton, C.B., Schneider, E., 2007. Two year longitudinal change and test-retest-precision of knee cartilage morphology in a pilot study for the osteoarthritis initiative. *Osteoarthritis Cartilage*. 15, 1326–32.

Eckstein, F., Maschek, S., Wirth, W., Hudelmaier, M., Hitzl, W., Wyman, B., Nevitt, M., Le Graverand, M.P., 2009b. One year change of knee cartilage morphology in the first release of participants from the Osteoarthritis Initiative progression subcohort: Association with sex, body mass index, symptoms and radiographic osteoarthritis status. *Ann Rheum Dis*. 68, 674–9.

Eckstein, F., Wirth, W., Hudelmaier, M.I., Maschek, S., Hitzl, W., Wyman, B.T., Nevitt, M., Hellio Le Graverand, M.P., Hunter, D., 2009c. Relationship of compartment-specific structural knee status at baseline with change in cartilage morphology: A prospective observational study using data from the osteoarthritis initiative. *Arthritis Res Ther*. 11, R90.

Eckstein, F., Wirth, W., Hunter, D.J., Guermazi, A., Kwoh, C.K., Nelson, D.R., Benichou, O., 2010a. Magnitude and regional distribution of cartilage loss associated with grades of joint space narrowing in radiographic osteoarthritis—Data from the Osteoarthritis Initiative (OAI). *Osteoarthritis Cartilage*. 18, 760–8.

Eckstein, F., Yang, M., Guermazi, A., Roemer, F.W., Hudelmaier, M., Picha, K., Baribaud, F., Wirth, W., Felson, D.T., 2010b. Reference values and Z-scores for subregional femorotibial cartilage thickness—Results from a large population-based sample (Framingham) and comparison with the non-exposed Osteoarthritis Initiative reference cohort. *Osteoarthritis Cartilage*. 18, 1275–83.

Ericsson, Y.B., Tjornstrand, J., Tiderius, C.J., Dahlberg, L.E., 2009. Relationship between cartilage glycosaminoglycan content (assessed with dGEMRIC) and OA risk factors in meniscectomized patients. *Osteoarthritis Cartilage*. 17, 565–70.

Ferretti, A., Conteduca, F., De Carli, A., Fontana, M., Mariani, P.P., 1991. Osteoarthritis of the knee after ACL reconstruction. *Int Orthop*. 15, 367–71.

Filidoro, L., Dietrich, O., Weber, J., Rauch, E., Oerther, T., Wick, M., Reiser, M.F., Glaser, C., 2005. High-resolution diffusion tensor imaging of human patellar cartilage: Feasibility and preliminary findings. *Magn Reson Med*. 53, 993–8.

Fleming, B.C., Oksendahl, H.L., Mehan, W.A., Portnoy, R., Fadale, P.D., Hulstyn, M.J., Bowers, M.E., Machan, J.T., Tung, G.A., 2010. Delayed gadolinium-enhanced MR imaging of cartilage (dGEMRIC) following ACL injury. *Osteoarthritis Cartilage*. 18, 662–7.

Foley, S., Ding, C., Cicuttini, F., Jones, G., 2007. Physical activity and knee structural change: A longitudinal study using MRI. *Med Sci Sports Exerc*. 39, 426–34.

Freeman, D.M., Bergman, G., Glover, G., 1997. Short TE MR microscopy: Accurate measurement and zonal differentiation of normal hyaline cartilage. *Magn Reson Med*. 38, 72–81.

Gao, B., Zheng, N., 2010. Alterations in three-dimensional joint kinematics of anterior cruciate ligament-deficient and -reconstructed knees during walking. *Clin Biomech (Bristol, Avon)*. 25, 222–9.

Gatehouse, P.D., Bydder, G.M., 2003. Magnetic resonance imaging of short T2 components in tissue. *Clin Radiol*. 58, 1–19.

Georgoulis, A., Papadonikolakis, A., Papageorgiou, C., Mitsou, A., Stergiou, N., 2003. Three-dimensional tibiofemoral kinematics of the anterior cruciate ligament-deficient and reconstructed knee during walking. *Am J Sports Med*. 31, 75–9.

Gillis, A., Gray, M., Burstein, D., 2002. Relaxivity and diffusion of gadolinium agents in cartilage. *Magn Reson Med*. 48, 1068–71.

Gloor, M., Scheffler, K., Bieri, O., 2008. Quantitative magnetization transfer imaging using balanced SSFP. *Magn Reson Med*. 60, 691–700.

Gok, H., Ergin, S., Yavuzer, G., 2002. Kinetic and kinematic characteristics of gait in patients with medial knee arthrosis. *Acta Orthop Scand*. 73, 647–52.

Gold, G.E., Thedens, D.R., Pauly, J.M., Fechner, K.P., Bergman, G., Beaulieu, C.F., Macovski, A., 1998. MR imaging of articular cartilage of the knee: New methods using ultrashort TEs. *AJR Am J Roentgenol*. 170, 1223–6.

Goldring, M.B., 2012. Chondrogenesis, chondrocyte differentiation, and articular cartilage metabolism in health and osteoarthritis. *Ther Adv Musculoskelet Dis*. 4, 269–85.

Graham, S.J., Henkelman, R.M., 1997. Understanding pulsed magnetization transfer. *J Magn Reson Imaging*. 7, 903–12.

Granot, J., 1988. Sodium imaging of human body organs and extremities in vivo. *Radiology*. 167, 547–50.

Grau, V., Mewes, A.U., Alcaniz, M., Kikinis, R., Warfield, S.K., 2004. Improved watershed transform for medical image segmentation using prior information. *IEEE Trans Med Imaging*. 23, 447–58.

Gray, M.L., Burstein, D., Kim, Y.J., Maroudas, A., 2008. 2007 Elizabeth Winston Lanier Award Winner. Magnetic resonance imaging of cartilage glycosaminoglycan: Basic principles, imaging technique, and clinical applications. *J Orthop Res*. 26, 281–91.

Gray, M.L., Burstein, D., Lesperance, L.M., Gehrke, L., 1995. Magnetization transfer in cartilage and its constituent macromolecules. *Magn Reson Med*. 34, 319–25.

Gurney, P.T., Hargreaves, B.A., Nishimura, D.G., 2006. Design and analysis of a practical 3D cones trajectory. *Magn Reson Med*. 55, 575–82.

Hagiwara, Y., Ando, A., Chimoto, E., Saijo, Y., Ohmori-Matsuda, K., Itoi, E., 2009. Changes of articular cartilage after immobilization in a rat knee contracture model. *J Orthop Res*. 27, 236–42.

Hanypsiak, B., Spindler, K., Rothrock, C., Calabrese, G., Richmond, B., Herrenbruck, T., Parker, R., 2008. Twelve-year follow-up on anterior cruciate ligament reconstruction: Long-term outcomes of prospectively studied osseous and articular injuries. *Am J Sports Med*. 36, 671–7.

Haralick, R., Shanmugam, K., Dinstein, I., 1973. Textural features for image classification. *IEEE Trans Syst Man Cybern*. 3, 610–18.

Hardy, P.A., Recht, M.P., Piraino, D., Thomasson, D., 1996. Optimization of a dual echo in the steady state (DESS) free-precession sequence for imaging cartilage. *J Magn Reson Imaging*. 6, 329–35.

Hart, D.J., Spector, T.D., 1998. Assessment of changes in joint tissues in patients treated with a disease-modifying osteoarthritis drug (DMOAD): Monitoring outcomes. In: *Osteoarthritis*. K.D. Brandt, M. Doherty, L.S. Lohmander, eds. Oxford: Oxford University Press, pp. 450–8.

Haughom, B., Schairer, W., Souza, R.B., Carpenter, D., Ma, C.B., Li, X., 2011. Abnormal tibiofemoral kinematics following ACL reconstruction are associated with early cartilage matrix degeneration measured by MRI T1rho. *Knee*. 19, 482–7.

Henkelman, R.M., Huang, X., Xiang, Q.S., Stanisz, G.J., Swanson, S.D., Bronskill, M.J., 1993. Quantitative interpretation of magnetization transfer. *Magn Reson Med*. 29, 759–66.

Henkelman, R.M., Stanisz, G.J., Graham, S.J., 2001. Magnetization transfer in MRI: A review. *NMR Biomed*. 14, 57–64.

Hohmann, E., Wortler, K., Imhoff, A.B., 2004. MR imaging of the hip and knee before and after marathon running. *Am J Sports Med*. 32, 55–9.

Holtzman, D.J., Theologis, A.A., Carballido-Gamio, J., Majumdar, S., Li, X., Benjamin, C., 2010. T(1rho) and T(2) quantitative magnetic resonance imaging analysis of cartilage regeneration following microfracture and mosaicplasty cartilage resurfacing procedures. *J Magn Reson Imaging*. 32, 914–23.

Hovis, K.K., Stehling, C., Souza, R.B., Haughom, B.D., Baum, T., Nevitt, M., McCulloch, C., Lynch, J.A., Link, T.M., 2011. Physical activity is associated with magnetic resonance imaging-based knee cartilage T2 measurements in asymptomatic subjects with and those without osteoarthritis risk factors. *Arthritis Rheum*. 63, 2248–56.

Hunter, D.J., Lo, G.H., Gale, D., Grainger, A.J., Guermazi, A., Conaghan, P.G., 2008. The reliability of a new scoring system for knee osteoarthritis MRI and the validity of bone marrow lesion assessment: BLOKS (Boston Leeds Osteoarthritis Knee Score). *Ann Rheum Dis.* 67, 206–11.

Idiyatullin, D., Corum, C., Park, J.Y., Garwood, M., 2006. Fast and quiet MRI using a swept radiofrequency. *J Magn Reson.* 181, 342–9.

Insko, E.K., Kaufman, J.H., Leigh, J.S., Reddy, R., 1999. Sodium NMR evaluation of articular cartilage degradation. *Magn Reson Med.* 41, 30–4.

Jackson, B.D., Teichtahl, A.J., Morris, M.E., Wluka, A.E., Davis, S.R., Cicuttini, F.M., 2004. The effect of the knee adduction moment on tibial cartilage volume and bone size in healthy women. *Rheumatology (Oxford).* 43, 311–14.

Jackson, J.I., Nishimura, D.G., Macovski, A., 1992. Twisting radial lines with application to robust magnetic resonance imaging of irregular flow. *Magn Reson Med.* 25, 128–39.

Jaremko, J.L., Cheng, R.W., Lambert, R.G., Habib, A.F., Ronsky, J.L., 2006. Reliability of an efficient MRI-based method for estimation of knee cartilage volume using surface registration. *Osteoarthritis Cartilage.* 14, 914–22.

Jeffrey, D.R., Watt, I., 2003. Imaging hyaline cartilage. *Br J Radiol.* 76, 777–87.

Jessel, R.H., Zilkens, C., Tiderius, C., Dudda, M., Mamisch, T.C., Kim, Y.J., 2009. Assessment of osteoarthritis in hips with femoroacetabular impingement using delayed gadolinium enhanced MRI of cartilage. *J Magn Reson Imaging.* 30, 1110–15.

Joseph, G.B., Baum, T., Alizai, H., Carballido-Gamio, J., Nardo, L., Virayavanich, W., Lynch, J.A., et al., 2012. Baseline mean and heterogeneity of MR cartilage T2 are associated with morphologic degeneration of cartilage, meniscus, and bone marrow over 3 years—Data from the Osteoarthritis Initiative. *Osteoarthritis Cartilage.* 20, 727–35.

Joseph, G.B., Baum, T., Carballido-Gamio, J., Nardo, L., Virayavanich, W., Alizai, H., Lynch, J.A., McCulloch, C.E., Majumdar, S., Link, T.M., 2011. Texture analysis of cartilage T2 maps: Individuals with risk factors for OA have higher and more heterogeneous knee cartilage MR T2 compared to normal controls—Data from the osteoarthritis initiative. *Arthritis Res Ther.* 13, R153.

Juras, V., Bittsansky, M., Majdisova, Z., Szomolanyi, P., Sulzbacher, I., Gabler, S., Stampfl, J., Schuller, G., Trattnig, S., 2009. In vitro determination of biomechanical properties of human articular cartilage in osteoarthritis using multi-parametric MRI. *J Magn Reson.* 197, 40–7.

Kauffmann, C., Gravel, P., Godbout, B., Gravel, A., Beaudoin, G., Raynauld, J.P., Martel-Pelletier, J., Pelletier, J.P., de Guise, J.A., 2003. Computer-aided method for quantification of cartilage thickness and volume changes using MRI: Validation study using a synthetic model. *IEEE Trans Biomed Eng.* 50, 978–88.

Kellgren, J.H., Lawrence, J.S., 1957. Radiologic assessment of osteoarthritis. *Ann Rheum Dis.* 16, 494–502.

Kessler, M.A., Glaser, C., Tittel, S., Reiser, M., Imhoff, A.B., 2008. Recovery of the menisci and articular cartilage of runners after cessation of exercise: Additional aspects of in vivo investigation based on 3-dimensional magnetic resonance imaging. *Am J Sports Med.* 36, 966–70.

Kettunen, M., Gröhn, O., Silvennoinen, M., Penttonen, M., Kauppinen, R., 2002. Effects of intracellular pH, blood, and tissue oxygen tension on T1rho relaxation in rat brain. *Magn Reson Med.* 48, 470–7.

Kim, D.K., Ceckler, T.L., Hascall, V.C., Calabro, A., Balaban, R.S., 1993. Analysis of water-macromolecule proton magnetization transfer in articular cartilage. *Magn Reson Med*. 29, 211–15.

Kim, Y.J., Jaramillo, D., Millis, M.B., Gray, M.L., Burstein, D., 2003. Assessment of early osteoarthritis in hip dysplasia with delayed gadolinium-enhanced magnetic resonance imaging of cartilage. *J Bone Joint Surg Am*. 85-A, 1987–92.

Kimelman, T., Vu, A., Storey, P., McKenzie, C., Burstein, D., Prasad, P., 2006. Three-dimensional T1 mapping for dGEMRIC at 3.0 T using the Look Locker method. *Invest Radiol*. 41, 198–203.

Koff, M.F., Amrami, K.K., Felmlee, J.P., Kaufman, K.R., 2008. Bias of cartilage T2 values related to method of calculation. *Magn Reson Imaging*. 26, 1236–43.

Koo, S., Andriacchi, T.P., 2007. A comparison of the influence of global functional loads vs. local contact anatomy on articular cartilage thickness at the knee. *J Biomech*. 40, 2961–6.

Kozanek, M., Hosseini, A., de Velde, S.K., Moussa, M.E., Li, J.S., Gill, T.J., Li, G., 2011. Kinematic evaluation of the step-up exercise in anterior cruciate ligament deficiency. *Clin Biomech (Bristol, Avon)*. 26, 950–4.

Krug, R., Larson, P.E., Wang, C., Burghardt, A.J., Kelley, D.A., Link, T.M., Zhang, X., Vigneron, D.B., Majumdar, S., 2011. Ultrashort echo time MRI of cortical bone at 7 tesla field strength: A feasibility study. *J Magn Reson Imaging*. 34, 691–5.

Kurkijarvi, J.E., Nissi, M.J., Kiviranta, I., Jurvelin, J.S., Nieminen, M.T., 2004. Delayed gadolinium-enhanced MRI of cartilage (dGEMRIC) and T2 characteristics of human knee articular cartilage: Topographical variation and relationships to mechanical properties. *Magn Reson Med*. 52, 41–6.

Kwan, M.K., Lai, W.M., Mow, V.C., 1984. Fundamentals of fluid transport through cartilage in compression. *Ann Biomed Eng*. 12, 537–58.

Lammentausta, E., Kiviranta, P., Nissi, M.J., Laasanen, M.S., Kiviranta, I., Nieminen, M.T., Jurvelin, J.S., 2006. T2 relaxation time and delayed gadolinium-enhanced MRI of cartilage (dGEMRIC) of human patellar cartilage at 1.5 T and 9.4 T: Relationships with tissue mechanical properties. *J Orthop Res*. 24, 366–74.

Lammentausta, E., Kiviranta, P., Toyras, J., Hyttinen, M.M., Kiviranta, I., Nieminen, M.T., Jurvelin, J.S., 2007. Quantitative MRI of parallel changes of articular cartilage and underlying trabecular bone in degeneration. *Osteoarthritis Cartilage*. 15, 1149–57.

Larson, P.E., Conolly, S.M., Pauly, J.M., Nishimura, D.G., 2007. Using adiabatic inversion pulses for long-T2 suppression in ultrashort echo time (UTE) imaging. *Magn Reson Med*. 58, 952–61.

Larson, P.E., Gurney, P.T., Nayak, K., Gold, G.E., Pauly, J.M., Nishimura, D.G., 2006. Designing long-T2 suppression pulses for ultrashort echo time imaging. *Magn Reson Med*. 56, 94–103.

Lattanzio, P.J., Marshall, K.W., Damyanovich, A.Z., Peemoeller, H., 2000. Macromolecule and water magnetization exchange modeling in articular cartilage. *Magn Reson Med*. 44, 840–51.

Laurent, D., Wasvary, J., Yin, J., Rudin, M., Pellas, T.C., O'Byrne, E., 2001. Quantitative and qualitative assessment of articular cartilage in the goat knee with magnetization transfer imaging. *Magn Reson Imaging*. 19, 1279–86.

Lesperance, L.M., Gray, M.L., Burstein, D., 1992. Determination of fixed charge density in cartilage using nuclear magnetic resonance. *J Orthop Res*. 10, 1–13.

Li, G., Papannagari, R., DeFrate, L., Yoo, J., Park, S., Gill, T., 2006. Comparison of the ACL and ACL graft forces before and after ACL reconstruction: An in-vitro robotic investigation. *Acta Orthop.* 77, 267–74.

Li, W., Scheidegger, R., Wu, Y., Vu, A., Prasad, P.V., 2008a. Accuracy of T1 measurement with 3-D Look-Locker technique for dGEMRIC. *J Magn Reson Imaging.* 27, 678–82.

Li, X., Cheng, J., Lin, K., Saadat, E., Bolbos, R.I., Jobke, B., Ries, M.D., Horvai, A., Link, T.M., Majumdar, S., 2011a. Quantitative MRI using T1rho and T2 in human osteoarthritic cartilage specimens: Correlation with biochemical measurements and histology. *Magn Reson Imaging.* 29, 324–34.

Li, X., Han, E., Busse, R., Majumdar, S., 2008b. In vivo T1rho mapping in cartilage using 3D magnetization-prepared angle-modulated partitioned k-space spoiled gradient echo snapshots (3D MAPSS). *Magn Reson Med.* 59, 298–307.

Li, X., Han, E., Ma, C., Link, T., Newitt, D., Majumdar, S., 2005. In vivo 3T spiral imaging based multi-slice T(1rho) mapping of knee cartilage in osteoarthritis. *Magn Reson Med.* 54, 929–36.

Li, X., Kuo, D., Theologis, A., Carballido-Gamio, J., Stehling, C., Link, T.M., Ma, C.B., Majumdar, S., 2011b. Cartilage in anterior cruciate ligament-reconstructed knees: MR imaging T1{rho} and T2—Initial experience with 1-year follow-up. *Radiology.* 258, 505–14.

Li, X., Ma, C., Bolbos, R., Stahl, R., Lozano, J., Zuo, J., Lin, K., Link, T., Safran, M., Majumdar, S., 2008c. Quantitative assessment of bone marrow edema pattern and overlying cartilage in knees with osteoarthritis and anterior cruciate ligament tear using MR imaging and spectroscopic imaging. *J Magn Reson Imaging.* 28, 453–61.

Li, X., Ma, C., Link, T., Castillo, D., Blumenkrantz, G., Lozano, J., Carballido-Gamio, J., Ries, M., Majumdar, S., 2007. In vivo T1rho and T2 mapping of articular cartilage in osteoarthritis of the knee using 3 Tesla MRI. *Osteoarthritis Cartilage.* 15, 789–97.

Li, X., Pai, A., Blumenkrantz, G., Carballido-Gamio, J., Link, T., Ma, B., Ries, M., Majumdar, S., 2009. Spatial distribution and relationship of T1rho and T2 relaxation times in knee cartilage with osteoarthritis. *Magn Reson Med.* 61, 1310–18.

Liess, C., Lusse, S., Karger, N., Heller, M., Gluer, C.C., 2002. Detection of changes in cartilage water content using MRI T2-mapping in vivo. *Osteoarthritis Cartilage.* 10, 907–13.

Lin, P.C., Reiter, D.A., Spencer, R.G., 2009. Classification of degraded cartilage through multiparametric MRI analysis. *J Magn Reson.* 201, 61–71.

Ling, W., Regatte, R.R., Navon, G., Jerschow, A., 2008. Assessment of glycosaminogly-can concentration in vivo by chemical exchange-dependent saturation transfer (gagCEST). *Proc Natl Acad Sci USA.* 105, 2266–70.

Ling, W., Regatte, R.R., Schweitzer, M.E., Jerschow, A., 2006. Behavior of ordered sodium in enzymatically depleted cartilage tissue. *Magn Reson Med.* 56, 1151–5.

Logan, M.C., Williams, A., Lavelle, J., Gedroyc, W., Freeman, M., 2004. Ti biofemoral kinematics following successful anterior cruciate ligament reconstruction using dynamic multiple resonance imaging. *Am J Sports Med.* 32, 984–92.

Lohmander, L., Englund, P., Dahl, L., Roos, E., 2007. The long-term consequence of anterior cruciate ligament and meniscus injuries: Osteoarthritis. *Am J Sports Med.* 35, 1756–69.

Lohmander, L.S., Ostenberg, A., Englund, M., Roos, H., 2004. High prevalence of knee osteoarthritis, pain, and functional limitations in female soccer players twelve years after anterior cruciate ligament injury. *Arthritis Rheum.* 50, 3145–52.

Lozano, J., Li, X., Link, T., Safran, M., Majumdar, S., Ma, C.B., 2006. Detection of post-traumatic cartilage injury using quantitative T1rho magnetic resonance imaging. A report of two cases with arthroscopic findings. *J Bone Joint Surg Am.* 88, 1349–52.

Luke, A.C., Stehling, C., Stahl, R., Li, X., Kay, T., Takamoto, S., Ma, B., Majumdar, S., Link, T., 2010. High-field magnetic resonance imaging assessment of articular cartilage before and after marathon running: Does long-distance running lead to cartilage damage? *Am J Sports Med.* 38, 2273–80.

Lusse, S., Claassen, H., Gehrke, T., Hassenpflug, J., Schunke, M., Heller, M., Gluer, C.C., 2000. Evaluation of water content by spatially resolved transverse relaxation times of human articular cartilage. *Magn Reson Imaging.* 18, 423–30.

Lynch, J.A., Zaim, S., Zhao, J., Stork, A., Peterfy, C.G., Genant, H.K., 2000. Cartilage segmentation of 3D MRI scans of the osteoarthritic knee combining user knowl-edge and active contours. In: *Medical Imaging 2000: Image Processing, Proceedings of the SPIE—The International Society for Optical Engineering,* 14–17 February, vol. 3979. San Diego, CA, pp. 925–35.

Madelin, G., Jerschow, A., Regatte, R.R., 2012. Sodium relaxation times in the knee joint in vivo at 7T. *NMR Biomed.* 25, 530–7.

Majumdar, S., Li, X., Blumenkrantz, G., Saldanha, K., Ma, C.B., Kim, H., Lozano, J., Link, T., 2006. MR imaging and early cartilage degeneration and strategies for monitoring regeneration. *J Musculoskelet Neuronal Interact.* 6, 382–4.

Makela, H.I., Grohn, O.H., Kettunen, M.I., Kauppinen, R.A., 2001. Proton exchange as a relaxation mechanism for T1 in the rotating frame in native and immobilized protein solutions. *Biochem Biophys Res Commun.* 289, 813–18.

Mamisch, T.C., Dudda, M., Hughes, T., Burstein, D., Kim, Y.J., 2008a. Comparison of delayed gadolinium enhanced MRI of cartilage (dGEMRIC) using inversion recovery and fast T1 mapping sequences. *Magn Reson Med.* 60, 768–73.

Mamisch, T.C., Kain, M.S., Bittersohl, B., Apprich, S., Werlen, S., Beck, M., Siebenrock, K.A., 2011. Delayed gadolinium-enhanced magnetic resonance imaging of cartilage (dGEMRIC) in Femoacetabular impingement. *J Orthop Res.* 29, 1305–11.

Mamisch, T.C., Menzel, M.I., Welsch, G.H., Bittersohl, B., Salomonowitz, E., Szomolanyi, P., Kordelle, J., Marlovits, S., Trattnig, S., 2008b. Steady-state diffusion imaging for MR in-vivo evaluation of reparative cartilage after matrix-associated autologous chondrocyte transplantation at 3 tesla—Preliminary results. *Eur J Radiol.* 65, 72–9.

Mamisch, T.C., Trattnig, S., Quirbach, S., Marlovits, S., White, L.M., Welsch, G.H., 2010. Quantitative T2 mapping of knee cartilage: Differentiation of healthy control cartilage and cartilage repair tissue in the knee with unloading—Initial results. *Radiology.* 254, 818–26.

Maroudas, A., Muir, H., Wingham, J., 1969. The correlation of fixed negative charge with glycosaminoglycan content of human articular cartilage. *Biochim Biophys Acta.* 177, 492–500.

Maroudas, A., Venn, M., 1977. Chemical composition and swelling of normal and osteoarthrotic femoral head cartilage. II. Swelling. *Ann Rheum Dis.* 36, 399–406.

Mayerhoefer, M.E., Welsch, G.H., Riegler, G., Mamisch, T.C., Materka, A., Weber, M., El-Rabadi, K., Friedrich, K.M., Dirisamer, A., Trattnig, S., 2010. Feasibility of tex-ture analysis for the assessment of biochemical changes in meniscal tissue on T1 maps calculated from delayed gadolinium-enhanced magnetic resonance imag-ing of cartilage data: Comparison with conventional relaxation time measure-ments. *Invest Radiol.* 45, 543–7.

McKenzie, C.A., Williams, A., Prasad, P.V., Burstein, D., 2006. Three-dimensional delayed gadolinium-enhanced MRI of cartilage (dGEMRIC) at 1.5T and 3.0T. *J Magn Reson Imaging*. 24, 928–33.

Meder, R., de Visser, S.K., Bowden, J.C., Bostrom, T., Pope, J.M., 2006. Diffusion tensor imaging of articular cartilage as a measure of tissue microstructure. *Osteoarthritis Cartilage*. 14, 875–81.

Menezes, N.M., Gray, M.L., Hartke, J.R., Burstein, D., 2004. T2 and T1rho MRI in articular cartilage systems. *Magn Reson Med*. 51, 503–9.

Miller, K.L., Hargreaves, B.A., Gold, G.E., Pauly, J.M., 2004. Steady-state diffusion-weighted imaging of in vivo knee cartilage. *Magn Reson Med*. 51, 394–8.

Mithoefer, K., Williams, R.J., III, Warren, R.F., Potter, H.G., Spock, C.R., Jones, E.C., Wickiewicz, T.L., Marx, R.G., 2006. Chondral resurfacing of articular cartilage defects in the knee with the microfracture technique. Surgical technique. *J Bone Joint Surg Am*. 88 Suppl 1 Pt 2, 294–304.

Mlynarik, V., Szomolanyi, P., Toffanin, R., Vittur, F., Trattnig, S., 2004. Transverse relaxation mechanisms in articular cartilage. *J Magn Reson*. 169, 300–7.

Mlynarik, V., Trattnig, S., Huber, M., Zembsch, A., Imhof, H., 1999. The role of relaxation times in monitoring proteoglycan depletion in articular cartilage. *J Magn Reson Imaging*. 10, 497–502.

Morrison, C., Stanisz, G., Henkelman, R.M., 1995. Modeling magnetization transfer for biological-like systems using a semi-solid pool with a super-Lorentzian line-shape and dipolar reservoir. *J Magn Reson B*. 108, 103–13.

Mosher, T., Liu, Y., Yang, Q., Yao, J., Smith, R., Dardzinski, B., Smith, M., 2004a. Age dependency of cartilage magnetic resonance imaging T2 relaxation times in asymptomatic women. *Arthritis Rheum*. 50, 2820–8.

Mosher, T.J., Collins, C.M., Smith, H.E., Moser, L.E., Sivarajah, R.T., Dardzinski, B.J., Smith, M.B., 2004b. Effect of gender on in vivo cartilage magnetic resonance imaging T2 mapping. *J Magn Reson Imaging*. 19, 323–8.

Mosher, T.J., Dardzinski, B.J., Smith, M.B., 2000. Human articular cartilage: Influence of aging and early symptomatic degeneration on the spatial variation of T2— Preliminary findings at 3 T. *Radiology*. 214, 259–66.

Mosher, T.J., Liu, Y., Torok, C.M., 2010. Functional cartilage MRI T2 mapping: Evaluating the effect of age and training on knee cartilage response to running. *Osteoarthritis Cartilage*. 18, 358–64.

Mosher, T.J., Smith, H.E., Collins, C., Liu, Y., Hancy, J., Dardzinski, B.J., Smith, M.B., 2005. Change in knee cartilage T2 at MR imaging after running: A feasibility study. *Radiology*. 234, 245–9.

Mosher, T.J., Zhang, Z., Reddy, R., Boudhar, S., Milestone, B.N., Morrison, W.B., Kwoh, C.K., Eckstein, F., Witschey, W.R., Borthakur, A., 2011. Knee articular cartilage damage in osteoarthritis: Analysis of MR image biomarker reproducibility in ACRIN-PA 4001 multicenter trial. *Radiology*. 258, 832–42.

Nag, D., Liney, G.P., Gillespie, P., Sherman, K.P., 2004. Quantification of T(2) relaxation changes in articular cartilage with in situ mechanical loading of the knee. *J Magn Reson Imaging*. 19, 317–22.

Neidhart, M., Muller-Ladner, U., Frey, W., Bosserhoff, A.K., Colombani, P.C., Frey-Rindova, P., Hummel, K.M., Gay, R.E., Hauselmann, H., Gay, S., 2000. Increased serum levels of non-collagenous matrix proteins (cartilage oligomeric matrix protein and melanoma inhibitory activity) in marathon runners. *Osteoarthritis Cartilage*. 8, 222–9.

Neuman, P., Tjornstrand, J., Svensson, J., Ragnarsson, C., Roos, H., Englund, M., Tiderius, C.J., Dahlberg, L.E., 2011. Longitudinal assessment of femoral knee cartilage quality using contrast enhanced MRI (dGEMRIC) in patients with anterior cruciate ligament injury—Comparison with asymptomatic volunteers. *Osteoarthritis Cartilage.* 19, 977–83.

Nicholson, J.A., Sutherland, A.G., Smith, F.W., Kawasaki, T., 2012. Upright MRI in kinematic assessment of the ACL-deficient knee. *Knee.* 19, 41–8.

Nieminen, M.T., Rieppo, J., Silvennoinen, J., Toyras, J., Hakumaki, J.M., Hyttinen, M.M., Helminen, H.J., Jurvelin, J.S., 2002. Spatial assessment of articular cartilage proteoglycans with Gd-DTPA-enhanced T1 imaging. *Magn Reson Med.* 48, 640–8.

Nieminen, M.T., Rieppo, J., Toyras, J., Hakumaki, J.M., Silvennoinen, J., Hyttinen, M.M., Helminen, H.J., Jurvelin, J.S., 2001. T2 relaxation reveals spatial collagen architecture in articular cartilage: A comparative quantitative MRI and polarized light microscopic study. *Magn Reson Med.* 46, 487–93.

Nieminen, M.T., Toyras, J., Laasanen, M.S., Silvennoinen, J., Helminen, H.J., Jurvelin, J.S., 2004. Prediction of biomechanical properties of articular cartilage with quantitative magnetic resonance imaging. *J Biomech.* 37, 321–8.

Nishii, T., Kuroda, K., Matsuoka, Y., Sahara, T., Yoshikawa, H., 2008. Change in knee cartilage T2 in response to mechanical loading. *J Magn Reson Imaging.* 28, 175–80.

Nissi, M.J., Rieppo, J., Toyras, J., Laasanen, M.S., Kiviranta, I., Nieminen, M.T., Jurvelin, J.S., 2007. Estimation of mechanical properties of articular cartilage with MRI—dGEMRIC, T2 and T1 imaging in different species with variable stages of maturation. *Osteoarthritis Cartilage.* 15, 1141–8.

Noyes, F.R., Stabler, C.L., 1989. A system for grading articular cartilage lesions at arthroscopy. *Am J Sports Med.* 17, 505–13.

Owman, H., Tiderius, C.J., Neuman, P., Nyquist, F., Dahlberg, L.E., 2008. Association between findings on delayed gadolinium-enhanced magnetic resonance imaging of cartilage and future knee osteoarthritis. *Arthritis Rheum.* 58, 1727–30.

Pai, A., Li, X., Majumdar, S., 2008. A comparative study at 3 T of sequence dependence of T2 quantitation in the knee. *Magn Reson Imaging.* 26, 1215–20.

Pakin, S., Schweitzer, M., Regatte, R., 2006. Rapid 3D-T1rho mapping of the knee joint at 3.0T with parallel imaging. *Magn Reson Med.* 56, 563–71.

Pakin, S.K., Tamez-Pena, J.G., Totterman, S., Parker, K.J., 2002. Segmentation, surface extraction, and thickness computation of articular cartilage. In: *SPIE, Medical Imaging*, 23 February, vol. 4684. San Diego, CA, pp. 155–66.

Pan, J., Pialat, J.B., Joseph, T., Kuo, D., Joseph, G.B., Nevitt, M.C., Link, T.M., 2011a. Knee cartilage T2 characteristics and evolution in relation to morphologic abnormalities detected at 3-T MR imaging: A longitudinal study of the normal control cohort from the Osteoarthritis Initiative. *Radiology.* 261, 507–15.

Pan, J., Stehling, C., Muller-Hocker, C., Schwaiger, B.J., Lynch, J., McCulloch, C.E., Nevitt, M.C., Link, T.M., 2011b. Vastus lateralis/vastus medialis cross-sectional area ratio impacts presence and degree of knee joint abnormalities and cartilage T2 determined with 3T MRI—An analysis from the incidence cohort of the Osteoarthritis Initiative. *Osteoarthritis Cartilage.* 19, 65–73.

Pauly, J., Conolly, S., Macovski, A., 1992. Suppression of long T2 components for short T2 imaging. *J Magn Reson Imaging.* 2, 145.

Pelletier, J.P., Raynauld, J.P., Berthiaume, M.J., Abram, F., Choquette, D., Haraoui, B., Beary, J.F., Cline, G.A., Meyer, J.M., Martel-Pelletier, J., 2007. Risk factors associated with the loss of cartilage volume on weight-bearing areas in knee osteoarthritis patients assessed by quantitative magnetic resonance imaging: A longitudinal study. *Arthritis Res Ther.* 9, R74.

Peterfy, C.G., Guermazi, A., Zaim, S., Tirman, P.F., Miaux, Y., White, D., Kothari, M., et al., 2004. Whole-organ magnetic resonance imaging score (WORMS) of the knee in osteoarthritis. *Osteoarthritis Cartilage.* 12, 177–90.

Peterfy, C.G., Schneider, E., Nevitt, M., 2008. The osteoarthritis initiative: Report on the design rationale for the magnetic resonance imaging protocol for the knee. *Osteoarthritis Cartilage.* 16, 1433–41.

Petrou, M., Sevilla, P.G., 2006. *Image Processing: Dealing with Texture.* Chichester: John Wiley & Sons.

Pierce, D.M., Trobin, W., Raya, J.G., Trattnig, S., Bischof, H., Glaser, C., Holzapfel, G.A., 2010. DT-MRI based computation of collagen fiber deformation in human articular cartilage: A feasibility study. *Ann Biomed Eng.* 38, 2447–63.

Pike, G.B., 1996. Pulsed magnetization transfer contrast in gradient echo imaging: A two-pool analytic description of signal response. *Magn Reson Med.* 36, 95–103.

Pike, G.B., Glover, G.H., Hu, B.S., Enzmann, D.R., 1993. Pulsed magnetization transfer spin-echo MR imaging. *J Magn Reson Imaging.* 3, 531–9.

Pollard, T.C., McNally, E.G., Wilson, D.C., Wilson, D.R., Madler, B., Watson, M., Gill, H.S., Carr, A.J., 2010. Localized cartilage assessment with three-dimensional dGEMRIC in asymptomatic hips with normal morphology and cam deformity. *J Bone Joint Surg Am.* 92, 2557–69.

Qian, Y., Williams, A.A., Chu, C.R., Boada, F.E., 2012. High-resolution ultrashort echo time (UTE) imaging on human knee with AWSOS sequence at 3.0 T. *J Magn Reson Imaging.* 35, 204–10.

Rahmer, J., Blume, U., Bornert, P., 2007. Selective 3D ultrashort TE imaging: Comparison of "dual-echo" acquisition and magnetization preparation for improving short-T2 contrast. *MAGMA.* 20, 83–92.

Rahmer, J., Bornert, P., Groen, J., Bos, C., 2006. Three-dimensional radial ultrashort echo-time imaging with T2 adapted sampling. *Magn Reson Med.* 55, 1075–82.

Raya, J.G., Arnoldi, A.P., Weber, D.L., Filidoro, L., Dietrich, O., Adam-Neumair, S., Mutzel, E., et al., 2011a. Ultra-high field diffusion tensor imaging of articular cartilage correlated with histology and scanning electron microscopy. *MAGMA.* 24, 247–58.

Raya, J.G., Melkus, G., Adam-Neumair, S., Dietrich, O., Mutzel, E., Kahr, B., Reiser, M.F., Jakob, P.M., Putz, R., Glaser, C., 2011b. Change of diffusion tensor imaging parameters in articular cartilage with progressive proteoglycan extraction. *Invest Radiol.* 46, 401–9.

Recht, M.P., Piraino, D.W., Paletta, G.A., Schils, J.P., Belhobek, G.H., 1996. Accuracy of fat-suppressed three-dimensional spoiled gradient-echo FLASH MR imaging in the detection of patellofemoral articular cartilage abnormalities. *Radiology.* 198, 209–12.

Reddy, R., Insko, E.K., Noyszewski, E.A., Dandora, R., Kneeland, J.B., Leigh, J.S., 1998. Sodium MRI of human articular cartilage in vivo. *Magn Reson Med.* 39, 697–701.

Reddy, R., Li, S., Noyszewski, E.A., Kneeland, J.B., Leigh, J.S., 1997. In vivo sodium multiple quantum spectroscopy of human articular cartilage. *Magn Reson Med.* 38, 207–14.

Redfield, A.G., 1969. Nuclear spin thermodynamics in the rotating frame. *Science.* 164, 1015–23.

Regatte, R., Akella, S., Lonner, J., Kneeland, J., Reddy, R., 2006. T1rho relaxation mapping in human osteoarthritis (OA) cartilage: Comparison of T1rho with T2. *J Magn Reson Imaging*. 23, 547–53.

Regatte, R.R., Akella, S.V., Borthakur, A., Kneeland, J.B., Reddy, R., 2002. Proteoglycan depletion-induced changes in transverse relaxation maps of cartilage: Comparison of T2 and T1rho. *Acad Radiol*. 9, 1388–94.

Regatte, R.R., Akella, S.V., Wheaton, A.J., Lech, G., Borthakur, A., Kneeland, J.B., Reddy, R., 2004. 3D-T1rho-relaxation mapping of articular cartilage: In vivo assessment of early degenerative changes in symptomatic osteoarthritic subjects. *Acad Radiol*. 11, 741–9.

Reichert, I.L., Robson, M.D., Gatehouse, P.D., He, T., Chappell, K.E., Holmes, J., Girgis, S., Bydder, G.M., 2005. Magnetic resonance imaging of cortical bone with ultrashort TE pulse sequences. *Magn Reson Imaging*. 23, 611–18.

Reiter, D.A., Lin, P.C., Fishbein, K.W., Spencer, R.G., 2009. Multicomponent T2 relaxation analysis in cartilage. *Magn Reson Med*. 61, 803–9.

Reiter, D.A., Roque, R.A., Lin, P.C., Doty, S.B., Pleshko, N., Spencer, R.G., 2011. Improved specificity of cartilage matrix evaluation using multiexponential transverse relaxation analysis applied to pathomimetically degraded cartilage. *NMR Biomed*. 24, 1286–94.

Ristanis, S., Stergiou, N., Patras, K., Tsepis, E., Moraiti, C., Georgoulis, A., 2006. Follow-up evaluation 2 years after ACL reconstruction with bone-patellar tendon-bone graft shows that excessive tibial rotation persists. *Clin J Sport Med*. 16, 111–16.

Rogers, L.Q., Macera, C.A., Hootman, J.M., Ainsworth, B.E., Blairi, S.N., 2002. The association between joint stress from physical activity and self-reported osteoarthritis: An analysis of the Cooper Clinic data. *Osteoarthritis Cartilage*. 10, 617–22.

Roos, E.M., Dahlberg, L., 2005. Positive effects of moderate exercise on glycosaminoglycan content in knee cartilage: A four-month, randomized, controlled trial in patients at risk of osteoarthritis. *Arthritis Rheum*. 52, 3507–14.

Saldanha, K.J., Doan, R.P., Ainslie, K.M., Desai, T.A., Majumdar, S., 2011. Micrometer-sized iron oxide particle labeling of mesenchymal stem cells for magnetic resonance imaging-based monitoring of cartilage tissue engineering. *Magn Reson Imaging*. 29, 40–9.

Samosky, J.T., Burstein, D., Eric Grimson, W., Howe, R., Martin, S., Gray, M.L., 2005. Spatially-localized correlation of dGEMRIC-measured GAG distribution and mechanical stiffness in the human tibial plateau. *J Orthop Res*. 23, 93–101.

Scanlan, S., Chaudhari, A., Dyrby, C., Andriacchi, T., 2010. Differences in tibial rotation during walking in ACL reconstructed and healthy contralateral knees. *J Biomech*. 43, 1817–22.

Scarvell, J., Smith, P., Refshauge, K., Galloway, H., Woods, K., 2005. Association between abnormal kinematics and degenerative change in knees of people with chronic anterior cruciate ligament deficiency: A magnetic resonance imaging study. *Aust J Physiother*. 51, 233–40.

Schmitt, B., Zbyn, S., Stelzeneder, D., Jellus, V., Paul, D., Lauer, L., Bachert, P., Trattnig, S., 2011. Cartilage quality assessment by using glycosaminoglycan chemical exchange saturation transfer and (23)Na MR imaging at 7 T. *Radiology*. 260, 257–64.

Schneider, E., NessAiver, M., White, D., Purdy, D., Martin, L., Fanella, L., Davis, D., Vignone, M., Wu, G., Gullapalli, R., 2008. The osteoarthritis initiative (OAI) magnetic resonance imaging quality assurance methods and results. *Osteoarthritis Cartilage*. 16, 994–1004.

Schueller-Weidekamm, C., Schueller, G., Uffmann, M., Bader, T.R., 2006. Does marathon running cause acute lesions of the knee? Evaluation with magnetic resonance imaging. *Eur Radiol.* 16, 2179–85.

Seon, J.K., Gadikota, H.R., Kozanek, M., Oh, L.S., Gill, T.J., Li, G., 2009. The effect of anterior cruciate ligament reconstruction on kinematics of the knee with combined anterior cruciate ligament injury and subtotal medial meniscectomy: An in vitro robotic investigation. *Arthroscopy.* 25, 123–30.

Sepponen, R., 1992. Rotating frame and magnetization transfer. In: *Magnetic Resonance Imaging.* Vol. 1, D.D. Stark, W.G.J. Bradley, eds. St. Louis, MO: Mosby-Year Book, pp. 204–18.

Shapiro, E.M., Borthakur, A., Dandora, R., Kriss, A., Leigh, J.S., Reddy, R., 2000. Sodium visibility and quantitation in intact bovine articular cartilage using high field (23)Na MRI and MRS. *J Magn Reson.* 142, 24–31.

Shapiro, E.M., Borthakur, A., Gougoutas, A., Reddy, R., 2002. 23Na MRI accurately measures fixed charge density in articular cartilage. *Magn Reson Med.* 47, 284–91.

Shapiro, E.M., Borthakur, A., Kaufman, J.H., Leigh, J.S., Reddy, R., 2001. Water distribution patterns inside bovine articular cartilage as visualized by 1H magnetic resonance imaging. *Osteoarthritis Cartilage.* 9, 533–8.

Shefelbine, S., Ma, C., Lee, K., Schrumpf, M., Patel, P., Safran, M., Slavinsky, J., Majumdar, S., 2006. MRI analysis of in vivo meniscal and tibiofemoral kinematics in ACL-deficient and normal knees. *J Orthop Res.* 24, 208–17.

Siversson, C., Tiderius, C.J., Neuman, P., Dahlberg, L., Svensson, J., 2010. Repeatability of T1-quantification in dGEMRIC for three different acquisition techniques: Two-dimensional inversion recovery, three-dimensional look locker, and three-dimensional variable flip angle. *J Magn Reson Imaging.* 31, 1203–9.

Sled, J.G., Pike, G.B., 2001. Quantitative imaging of magnetization transfer exchange and relaxation properties in vivo using MRI. *Magn Reson Med.* 46, 923–31.

Souza, R.B., Fang, C., Luke, A., Wu, S., Li, X., Majumdar, S., 2012. Relationship between knee kinetics during jumping tasks and knee articular cartilage MRI T1rho and T2 relaxation times. *Clin Biomech (Bristol, Avon).* 27, 403–8.

Souza, R.B., Stehling, C., Wyman, B.T., Hellio Le Graverand, M.P., Li, X., Link, T.M., Majumdar, S., 2010. The effects of acute loading on T1rho and T2 relaxation times of tibiofemoral articular cartilage. *Osteoarthritis Cartilage.* 18, 1557–63.

Spector, T.D., Harris, P.A., Hart, D.J., Cicuttini, F.M., Nandra, D., Etherington, J., Wolman, R.L., Doyle, D.V., 1996. Risk of osteoarthritis associated with long-term weight-bearing sports: A radiologic survey of the hips and knees in female ex-athletes and population controls. *Arthritis Rheum.* 39, 988–95.

Stahl, R., Luke, A., Li, X., Carballido-Gamio, J., Ma, C., Majumdar, S., Link, T., 2009. T1rho, T(2) and focal knee cartilage abnormalities in physically active and sedentary healthy subjects versus early OA patients-a 3.0-Tesla MRI study. *Eur Radiol.* 19, 132–43.

Stammberger, T., Eckstein, F., Englmeier, K.H., Reiser, M., 1999a. Determination of 3D cartilage thickness data from MR imaging: Computational method and reproducibility in the living. *Magn Reson Med.* 41, 529–36.

Stammberger, T., Eckstein, F., Michaelis, M., Englmeier, K.H., Reiser, M., 1999b. Interobserver reproducibility of quantitative cartilage measurements: Comparison of B-spline snakes and manual segmentation. *Magn Reson Imaging.* 17, 1033–42.

Stammberger, T., Hohe, J., Englmeier, K.H., Reiser, M., Eckstein, F., 2000. Elastic registration of 3D cartilage surfaces from MR image data for detecting local changes in cartilage thickness. *Magn Reson Med*. 44, 592–601.

Staroswiecki, E., Bangerter, N.K., Gurney, P.T., Grafendorfer, T., Gold, G.E., Hargreaves, B.A., 2010. In vivo sodium imaging of human patellar cartilage with a 3D cones sequence at 3 T and 7 T. *J Magn Reson Imaging*. 32, 446–51.

Stehling, C., Liebl, H., Krug, R., Lane, N.E., Nevitt, M.C., Lynch, J., McCulloch, C.E., Link, T.M., 2010. Patellar cartilage: T2 values and morphologic abnormalities at 3.0-T MR imaging in relation to physical activity in asymptomatic subjects from the osteoarthritis initiative. *Radiology*. 254, 509–20.

Stehling, C., Luke, A., Stahl, R., Baum, T., Joseph, G., Pan, J., Link, T.M., 2011. Meniscal T1rho and T2 measured with 3.0T MRI increases directly after running a marathon. *Skeletal Radiol*. 40, 725–35.

Stikov, N., Keenan, K.E., Pauly, J.M., Smith, R.L., Dougherty, R.F., Gold, G.E., 2011. Cross-relaxation imaging of human articular cartilage. *Magn Reson Med*. 66, 725–34.

Stobbe, R., Beaulieu, C., 2008. Sodium imaging optimization under specific absorption rate constraint. *Magn Reson Med*. 59, 345–55.

Sur, S., Mamisch, T.C., Hughes, T., Kim, Y.J., 2009. High resolution fast T1 mapping technique for dGEMRIC. *J Magn Reson Imaging*. 30, 896–900.

Sussman, M.S., Pauly, J.M., Wright, G.A., 1998. Design of practical T2-selective RF excitation (TELEX) pulses. *Magn Reson Med*. 40, 890–9.

Tang, S.Y., Souza, R.B., Ries, M., Hansma, P.K., Alliston, T., Li, X., 2011. Local tissue properties of human osteoarthritic cartilage correlate with magnetic resonance T(1) rho relaxation times. *J Orthop Res*. 29, 1312–19.

Tashman, S., Collon, D., Anderson, K., Kolowich, P., Anderst, W., 2004. Abnormal rotational knee motion during running after anterior cruciate ligament reconstruction. *Am J Sports Med*. 32, 975–83.

Tashman, S., Kolowich, P., Collon, D., Anderson, K., Anderst, W., 2007. Dynamic function of the ACL-reconstructed knee during running. *Clin Orthop Relat Res*. 454, 66–73.

Techawiboonwong, A., Song, H.K., Leonard, M.B., Wehrli, F.W., 2008a. Cortical bone water: In vivo quantification with ultrashort echo-time MR imaging. *Radiology*. 248, 824–33.

Techawiboonwong, A., Song, H.K., Wehrli, F.W., 2008b. In vivo MRI of submillisecond T(2) species with two-dimensional and three-dimensional radial sequences and applications to the measurement of cortical bone water. *NMR Biomed*. 21, 59–70.

Theologis, A., Kuo, D., Cheng, J., Bolbos, R., Carballido-Gamio, J., Ma, C., Li, X., 2011. Longitudinal evaluation of biochemical changes in articular cartilage in ACL injured and reconstructed knees—A study using T1ρ MRI at 3 Tesla. *Arthroscopy*. 27, 65–76.

Thorp, L.E., Sumner, D.R., Block, J.A., Moisio, K.C., Shott, S., Wimmer, M.A., 2006. Knee joint loading differs in individuals with mild compared with moderate medial knee osteoarthritis. *Arthritis Rheum*. 54, 3842–9.

Tiderius, C.J., Olsson, L.E., Leander, P., Ekberg, O., Dahlberg, L., 2003. Delayed gadolinium-enhanced MRI of cartilage (dGEMRIC) in early knee osteoarthritis. *Magn Reson Med*. 49, 488–92.

Tiderius, C.J., Olsson, L.E., Nyquist, F., Dahlberg, L., 2005. Cartilage glycosaminoglycan loss in the acute phase after an anterior cruciate ligament injury: Delayed gadolinium-enhanced magnetic resonance imaging of cartilage and synovial fluid analysis. *Arthritis Rheum*. 52, 120–7.

Toffanin, R., Mlynarik, V., Russo, S., Szomolanyi, P., Piras, A., Vittur, F., 2001. Proteoglycan depletion and magnetic resonance parameters of articular cartilage. *Arch Biochem Biophys.* 390, 235–42.

Trattnig, S., Domayer, S., Welsch, G.W., Mosher, T., Eckstein, F., 2009. MR imaging of cartilage and its repair in the knee—A review. *Eur Radiol.* 19, 1582–94.

Trattnig, S., Mlynarik, V., Breitenseher, M., Huber, M., Zembsch, A., Rand, T., Imhof, H., 1999. MRI visualization of proteoglycan depletion in articular cartilage via intravenous administration of Gd-DTPA. *Magn Reson Imaging.* 17, 577–83.

Trattnig, S., Welsch, G.H., Juras, V., Szomolanyi, P., Mayerhoefer, M.E., Stelzeneder, D., Mamisch, T.C., Bieri, O., Scheffler, K., Zbyn, S., 2010. 23Na MR imaging at 7 T after knee matrix-associated autologous chondrocyte transplantation preliminary results. *Radiology.* 257, 175–84.

Trattnig, S., Winalski, C.S., Marlovits, S., Jurvelin, J.S., Welsch, G., Potter, H.G., 2011. Magnetic resonance imaging of cartilage repair: A review. *Cartilage.* 2, 5–26.

van Buul, G.M., Kotek, G., Wielopolski, P.A., Farrell, E., Bos, P.K., Weinans, H., Grohnert, A.U., et al., 2011. Clinically translatable cell tracking and quantification by MRI in cartilage repair using superparamagnetic iron oxides. *PLoS One.* 6, e17001.

Van de Velde, S.K., Bingham, J.T., Hosseini, A., Kozanek, M., DeFrate, L.E., Gill, T.J., Li, G., 2009. Increased tibiofemoral cartilage contact deformation in patients with anterior cruciate ligament deficiency. *Arthritis Rheum.* 60, 3693–702.

van der Hart, C., van den Bekerom, M., Patt, T., 2008. The occurrence of osteoarthritis at a minimum of ten years after reconstruction of the anterior cruciate ligament. *J Orthop Surg Res.* 3, 24.

Vanwanseele, B., Eckstein, F., Knecht, H., Spaepen, A., Stussi, E., 2003. Longitudinal analysis of cartilage atrophy in the knees of patients with spinal cord injury. *Arthritis Rheum.* 48, 3377–81.

Vanwanseele, B., Eckstein, F., Smith, R.M., Lange, A.K., Foroughi, N., Baker, M.K., Shnier, R., Singh, M.A., 2010. The relationship between knee adduction moment and cartilage and meniscus morphology in women with osteoarthritis. *Osteoarthritis Cartilage.* 18, 894–901.

van Zijl, P.C., Yadav, N.N., 2011. Chemical exchange saturation transfer (CEST): What is in a name and what isn't? *Magn Reson Med.* 65, 927–48.

Verstraete, K.L., Almqvist, F., Verdonk, P., Vanderschueren, G., Huysse, W., Verdonk, R., Verbrugge, G., 2004. Magnetic resonance imaging of cartilage and cartilage repair. *Clin Radiol.* 59, 674–89.

von Porat, A., Roos, E., Roos, H., 2004. High prevalence of osteoarthritis 14 years after an anterior cruciate ligament tear in male soccer players: A study of radiographic and patient relevant outcomes. *Ann Rheum Dis.* 63, 269–73.

Wachsmuth, L., Juretschke, H.P., Raiss, R.X., 1997. Can magnetization transfer magnetic resonance imaging follow proteoglycan depletion in articular cartilage? *MAGMA.* 5, 71–8.

Wang, L., Chang, G., Xu, J., Vieira, R.L., Krasnokutsky, S., Abramson, S., Regatte, R.R., 2012. T1rho MRI of menisci and cartilage in patients with osteoarthritis at 3T. *Eur J Radiol.* 81, 2329–36.

Wang, L., Wu, Y., Chang, G., Oesingmann, N., Schweitzer, M.E., Jerschow, A., Regatte, R.R., 2009. Rapid isotropic 3D-sodium MRI of the knee joint in vivo at 7T. *J Magn Reson Imaging.* 30, 606–14.

Wang, N., Xia, Y., 2011. Dependencies of multi-component T2 and T1rho relaxation on the anisotropy of collagen fibrils in bovine nasal cartilage. *J Magn Reson.* 212, 124–32.

Wansapura, J.P., Daniel, B.L., Pauly, J., Butts, K., 2001. Temperature mapping of frozen tissue using eddy current compensated half excitation RF pulses. *Magn Reson Med.* 46, 985–92.

Ward, K.M., Aletras, A.H., Balaban, R.S., 2000. A new class of contrast agents for MRI based on proton chemical exchange dependent saturation transfer (CEST). *J Magn Reson.* 143, 79–87.

Warfield, S.K., Kaus, M., Jolesz, F.A., Kikinis, R., 2000. Adaptive, template moderated, spatially varying statistical classification. *Med Image Anal.* 4, 43–55.

Watrin-Pinzano, A., Ruaud, J.P., Olivier, P., Grossin, L., Gonord, P., Blum, A., Netter, P., Guillot, G., Gillet, P., Loeuille, D., 2005. Effect of proteoglycan depletion on T2 mapping in rat patellar cartilage. *Radiology.* 234, 162–70.

Watts, A., Stobbe, R.W., Beaulieu, C., 2011. Signal-to-noise optimization for sodium MRI of the human knee at 4.7 Tesla using steady state. *Magn Reson Med.* 66, 697–705.

Welsch, G.H., Trattnig, S., Domayer, S., Marlovits, S., White, L.M., Mamisch, T.C., 2009. Multimodal approach in the use of clinical scoring, morphological MRI and biochemical T2-mapping and diffusion-weighted imaging in their ability to assess differences between cartilage repair tissue after microfracture therapy and matrix-associated autologous chondrocyte transplantation: A pilot study. *Osteoarthritis Cartilage.* 17, 1219–27.

Wheaton, A., Borthakur, A., Corbo, M., Charagundla, S., Reddy, R., 2004a. Method for reduced SAR T1rho-weighted MRI. *Magn Reson Med.* 51, 1096–102.

Wheaton, A., Borthakur, A., Reddy, R., 2003. Application of the keyhole technique to T1rho relaxation mapping. *J Magn Reson Imaging.* 18, 745–9.

Wheaton, A., Dodge, G., Elliott, D., Nicoll, S., Reddy, R., 2005. Quantification of cartilage biomechanical and biochemical properties via T1rho magnetic resonance imaging. *Magn Reson Med.* 54, 1087–93.

Wheaton, A.J., Borthakur, A., Kneeland, J.B., Regatte, R.R., Akella, S.V., Reddy, R., 2004b. In vivo quantification of T1rho using a multislice spin-lock pulse sequence. *Magn Reson Med.* 52, 1453–8.

Wheaton, A.J., Borthakur, A., Shapiro, E.M., Regatte, R.R., Akella, S.V., Kneeland, J.B., Reddy, R., 2004c. Proteoglycan loss in human knee cartilage: Quantitation with sodium MR imaging—Feasibility study. *Radiology.* 231, 900–5.

Williams, A., Gillis, A., McKenzie, C., Po, B., Sharma, L., Micheli, L., McKeon, B., Burstein, D., 2004. Glycosaminoglycan distribution in cartilage as determined by delayed gadolinium-enhanced MRI of cartilage (dGEMRIC): Potential clinical applications. *AJR Am J Roentgenol.* 182, 167–72.

Williams, A., Qian, Y., Bear, D., Chu, C.R., 2010. Assessing degeneration of human articular cartilage with ultra-short echo time (UTE) T2* mapping. *Osteoarthritis Cartilage.* 18, 539–46.

Williams, A., Qian, Y., Chu, C.R., 2011. UTE-T2* mapping of human articular cartilage in vivo: A repeatability assessment. *Osteoarthritis Cartilage.* 19, 84–8.

Williams, A., Sharma, L., McKenzie, C.A., Prasad, P.V., Burstein, D., 2005. Delayed gadolinium-enhanced magnetic resonance imaging of cartilage in knee osteoarthritis: Findings at different radiographic stages of disease and relationship to malalignment. *Arthritis Rheum.* 52, 3528–35.

Wirth, W., Benichou, O., Kwoh, C.K., Guermazi, A., Hunter, D., Putz, R., Eckstein, F., 2010a. Spatial patterns of cartilage loss in the medial femoral condyle in osteoarthritic knees: Data from the Osteoarthritis Initiative. *Magn Reson Med*. 63, 574–81.

Wirth, W., Hellio Le Graverand, M.P., Wyman, B.T., Maschek, S., Hudelmaier, M., Hitzl, W., Nevitt, M., Eckstein, F., 2009. Regional analysis of femorotibial cartilage loss in a subsample from the Osteoarthritis Initiative progression subcohort. *Osteoarthritis Cartilage*. 17, 291–7.

Wirth, W., Larroque, S., Davies, R.Y., Nevitt, M., Gimona, A., Baribaud, F., Lee, J.H., et al., 2011. Comparison of 1-year vs 2-year change in regional cartilage thickness in osteoarthritis results from 346 participants from the Osteoarthritis Initiative. *Osteoarthritis Cartilage*. 19, 74–83.

Wirth, W., Nevitt, M., Hellio Le Graverand, M.P., Benichou, O., Dreher, D., Davies, R.Y., Lee, J., et al., 2010b. Sensitivity to change of cartilage morphometry using coronal FLASH, sagittal DESS, and coronal MPR DESS protocols—Comparative data from the Osteoarthritis Initiative (OAI). *Osteoarthritis Cartilage*. 18, 547–54.

Witschey, W., Borthakur, A., Elliott, M., Fenty, M., Sochor, M., Wang, C., Reddy, R., 2008. T1rho-prepared balanced gradient echo for rapid 3D T1rho MRI. *J Magn Reson Imaging*. 28, 744–54.

Witschey, W.R., II, Borthakur, A., Elliott, M., Mellon, E., Niyogi, S., Wallman, D., Wang, C., Reddy, R., 2007. Artifacts in T1 rho-weighted imaging: Compensation for B(1) and B(0) field imperfections. *J Magn Reson*. 186, 75–85.

Witschey, W.R., Borthakur, A., Fenty, M., Kneeland, B.J., Lonner, J.H., McArdle, E.L., Sochor, M., Reddy, R., 2010. T1rho MRI quantification of arthroscopically confirmed cartilage degeneration. *Magn Reson Med*. 63, 1376–82.

Wu, Y., Ackerman, J.L., Chesler, D.A., Graham, L., Wang, Y., Glimcher, M.J., 2003. Density of organic matrix of native mineralized bone measured by water- and fat-suppressed proton projection MRI. *Magn Reson Med*. 50, 59–68.

Xia, Y., 1998. Relaxation anisotropy in cartilage by NMR microscopy (muMRI) at 14-microm resolution. *Magn Reson Med*. 39, 941–9.

Xia, Y., Farquhar, T., Burton-Wuster, N., Ray, E., Jelinski, L., 1994. Difiusion and relaxation mapping of cartilage-bone plugs and excised disks using microscopic magnetic resonance imaging. *Magn Reson Med*. 31, 273–82.

Xia, Y., Farquhar, T., Burton-Wurster, N., Vernier-Singer, M., Lust, G., Jelinski, L.W., 1995. Self-diffusion monitors degraded cartilage. *Arch Biochem Biophys*. 323, 323–8.

Yamazaki, J., Muneta, T., Ju, Y., Sekiya, I., 2010. Differences in kinematics of single leg squatting between anterior cruciate ligament-injured patients and healthy controls. *Knee Surg Sports Traumatol Arthrosc*. 18, 56–63.

Yao, W., Qu, N., Lu, Z., Yang, S., 2009. The application of T1 and T2 relaxation time and magnetization transfer ratios to the early diagnosis of patellar cartilage osteoarthritis. *Skeletal Radiol*. 38, 1055–62.

Zarins, Z.A., Bolbos, R.I., Pialat, J.B., Link, T.M., Li, X., Souza, R.B., Majumdar, S., 2010. Cartilage and meniscus assessment using T1rho and T2 measurements in healthy subjects and patients with osteoarthritis. *Osteoarthritis Cartilage*. 18, 1408–16.

Zhao, J., Li, X., Bolbos, R., Link, T., Majumdar, S., 2010. Longitudinal assessment of bone marrow edema-like lesions and cartilage degeneration in osteoarthritis using 3 T MR T1rho quantification. *Skeletal Radiol*. 36, 523–31.

Zuo, J., Li, X., Banerjee, S., Han, E., Majumdar, S., 2007. Parallel imaging of knee cartilage at 3 Tesla. *J Magn Reson Imaging*. 26, 1001–9.

7

Quantitative Structural and Functional MRI of Skeletal Muscle

**Bruce M. Damon, Theodore Towse, Amanda K.W. Buck,
Ke Li, Nathan Bryant, Zhaohua Ding, and Jane H. Park**
Vanderbilt University

CONTENTS

7.1 Introduction

7.1.1 Background on Skeletal Muscle Structure and Function

Skeletal muscles serve critical mechanical and physiological roles in the body. Their mechanical functions are to generate force and actuate movement by shortening or lengthening underload. These functions allow postural control, locomotion, and many other reflexive and volitional activities. Skeletal muscles also act in glucose uptake and disposal and participate in total body water balance and thermoregulation.

Skeletal muscle cells (fibers) are up to tens of centimeters long and have diameters of 20–80 μm (Polgar et al., 1973). Most of a fiber's volume is occupied by the contractile protein filaments (myofibrils). The smallest functioning unit of contraction is called a sarcomere; myofibrils consist of thousands of sarcomeres in series. At the intermediate structural scale, groups of 100–200 fibers are bundled by connective tissue into structures called fascicles. Muscles connect to the skeleton via tendons, and the muscle–tendon unit runs from a point of origin to a point of insertion. A muscle's origin and insertion define its mechanical line of action.

Muscle contraction is the term used to describe the biological processes that result in muscle activation and force generation. At the molecular level, muscle contraction results from an interaction between the sarcomere proteins actin and myosin. Adenosine triphosphate (ATP) hydrolysis provides the energy for actin–myosin interaction and maintaining ion gradients. ATP is resynthesized by increased flux through the creatine kinase, glycolytic, and oxidative phosphorylation reactions. Oxygen demand and oxygen supply are typically well matched, and so during exercise blood flow increases to enhance the delivery of oxygen, metabolic substrates, and signaling molecules to the muscle and carrying away the metabolic end-products.

7.1.2 Objective of This Chapter

The preceding section has briefly discussed the importance of skeletal muscle to human physiology and the contributions of several body systems to muscle performance. Magnetic resonance imaging (MRI) methods allow investigators to assess many aspects of muscle structure and function. The purpose of this chapter is to describe several MRI methods and show how their application in health and disease can lead to new insights into the structure, function, and health of skeletal muscle. In preparing this chapter, the authors have opted to treat a large number of topics rather superficially and apologize in advance to those whose work is not discussed fully or at all. Where possible, the reader is directed to full review articles.

7.2 MRI Measurements of Muscle Architecture

7.2.1 The Importance of Muscle Architecture to Muscle Function

Muscle architecture is defined as the muscle fibers' geometric properties (including length, curvature, and their orientations with respect to
the muscle's line of action). A recent full review on muscle architecture
is from Lieber and Friden (2001). Here, two basic subtypes are discussed.
A fusiform muscle's fibers are oriented along the muscle's line of action; in
pennate muscles, the fibers' long axes are oblique to the line of action. In
fusiform muscles, more sarcomeres are arrayed in series (causing their displacements to add and optimizing the muscle for high shortening velocities and length excursions). In pennate muscles, more sarcomeres lie in
parallel (causing their forces to add and optimizing the muscle for high
force production). Fiber geometric properties such as curvature influence
the pattern of intramuscular fluid pressure generation during contraction,
thus potentially influencing perfusion patterns (Otten, 1988; Van Leeuwen
and Spoor, 1992, 1993).

7.2.2 Diffusion Tensor Characterization of Water Diffusion

Given the relationship between muscle architecture and mechanics, it is
important to characterize this architecture in three dimensions (3Ds) and
to relate it to function. Diffusion tensor MRI (DT-MRI) is one approach
for measuring muscle architecture *in vivo* (the other main approach
being brightness-mode ultrasound). DT-MRI is an extension of diffusion-
weighted MRI (DW-MRI), a technique used to form spatial maps of the
diffusion coefficient. DW-MRI is typically performed using the pulsed-
gradient, spin-echo sequence (Stejskal and Tanner, 1965), in which a pair
of magnetic field gradients is used to encode diffusion as a reduction in
spin-echo amplitude. The MRI signal, S, is reduced according to the following equation:

$$S = S_0 \cdot e^{-bD}, \tag{7.1}$$

where D is a diffusion coefficient, S_0 is the signal measured in the absence
of diffusion-encoding gradients, and b characterizes the timing, amplitude,
and geometry of the diffusion-encoding gradient pulses.

DT-MRI is used to characterize the anisotropic water diffusion that exists
in some tissues (Basser et al., 1994). The diffusion tensor, $\mathbf{D}$, is a 3×3 symmetric matrix containing diffusion coefficients in the X, Y, Z, XY, XZ, and YZ
directions of the laboratory's coordinate system. To form $\mathbf{D}$, diffusion coefficients are measured in six or more noncollinear directions. These diffusion

coefficients are obtained by analyzing directionally specific signal decays that behave according to the following equation:

$$S(\mathbf{n}) = S_0 \cdot e^{-b\,\mathbf{n}^T \mathbf{D}\mathbf{n}}, \qquad (7.2)$$

where $\mathbf{n}$ describes a diffusion-encoding gradient direction and the superscript T indicates matrix transposition. $\mathbf{D}$ is then formed by using curve-fitting or multiple regression methods.

$\mathbf{D}$ can be diagonalized to produce three eigenvalue/eigenvector pairs. These data describe the profile of water diffusion in the tissue, including the degree of diffusion anisotropy and the magnitude and direction of greatest diffusion. The eigenvalues are symbolized as λ_i, where $i = 1$, 2, or 3 denotes the descending magnitude order of the eigenvalues. The corresponding eigenvectors are denoted as $\mathbf{v}_i$ (the symbol ε is also used to represent the eigenvectors of $\mathbf{D}$, but here this symbol is used to indicate mechanical strain). Further details about DT-MRI can be found in Chapter 2.

7.2.3 Quantitative Diffusion Tensor MRI Fiber Tracking of Skeletal Muscle

DT-MRI fiber tracking is used to map the structural connectivity within certain biological tissues. The most basic fiber tracking approaches propagate fiber tracts from predefined seed points by following each voxel's $\mathbf{v}_1$ and continuing until user-defined termination criteria are met (Basser et al., 2000; Mori et al., 1999). The resulting dataset contains a 3D description of the fiber paths. Additional tractography techniques have been developed to improve functionality of the fiber tracking by mitigating the effects of low signal-to-noise ratio and partial volume artifacts and by accommodating complex fiber structures.

Two properties of water diffusion in skeletal muscle fibers allow DT-MRI to be used to study muscle architecture. The first is that the diffusion coefficient of muscle water is reduced from its value for free water, illustrating the existence of subcellular structures that hinder molecular movement (Chang et al., 1973; Clark et al., 1982; Finch et al., 1971; Rorschach et al., 1973). The second property is that water diffusion in muscle is anisotropic, with larger apparent diffusion coefficients parallel to the fiber's long axis than perpendicular to it (i.e., the $\lambda_1/\mathbf{v}_1$ eigenvalue/vector pair represents diffusion along the long axis of the fibers) (Basser et al., 1994; Cleveland et al., 1976; Napadow et al., 2001; Van Donkelaar et al., 1999).

Diffusion MRI-based fiber tractography has been used to describe the structure of animal (Damon et al., 2002a; Gilbert et al., 2006; Heemskerk et al., 2005) and human (e.g., Budzik et al., 2007; Lansdown et al., 2007;

FIGURE 7.1
(Color version available from crcpress.com; see Preface.) Sample DT-MRI fiber tracking results from the tibialis anterior muscle. The central dark gray structure represents the aponeurosis; the medium gray lines are fiber tracts of the superficial muscle compartment; and the light gray lines are fiber tracts in the deep muscle compartment. (Reproduced from Heemskerk, A.M., et al., *NMR Biomed.* 23, 294–303, 2010. With permission.)

Sinha et al., 2006; Zijta et al., 2011) muscles. By digitizing the internal tendon (aponeurosis) of the muscle and defining seed points along this 3D surface, a single dataset that can be used to quantify fiber tract length, pennation angle, and fiber curvature throughout the muscle is formed (Damon et al., 2012; Damon et al., 2002a; Heemskerk et al., 2009, 2010; Lansdown et al., 2007) (Figure 7.1). In the human tibialis anterior (TA) muscle, there is heterogeneity of pennation angle and fiber tract length throughout the muscle (Heemskerk and Damon, 2009; Heemskerk et al., 2010; Lansdown et al., 2007), with the heterogeneity in pennation angle due to variations in aponeurosis orientation within the axial plane (Lansdown et al., 2007).

Interested readers may wish to read reviews of DT-MRI-based muscle fiber tracking (Damon et al., 2011), general DT-MRI principles (Le Bihan et al., 2001), DT-MRI pitfalls (Jones and Cercignani, 2010), and/or fiber tracking (Lazar, 2010).

7.3 MRI Measurements of Muscle Function

7.3.1 Characterizing Muscle Mechanical Properties

The preceding discussion has shown that a muscle's architecture influences its mechanical performance. The focus of this chapter thus turns now to measurements of mechanical quantities such as strain (symbolized as ε: the relative length change in an object placed under stress) and the shear modulus (G: the ratio of shear stress to shear strain).

One informative method for mapping the mechanical properties of tissues is magnetic resonance elastography (MRE) (Muthupillai et al., 1995). In MRE, oscillating mechanical waves are imparted into a tissue by an external actuator. Shear waves result within the tissue and can be measured using cine-mode phase-contrast (PC) MRI. PC-MRI exploits the fact that spins moving along a magnetic field gradient experience phase shifts that are proportional to gradient strength, the duration of the encoding period, and spin velocity. Because the frequency, amplitude, and damping of the shear waves depend, in part, on tissue mechanical properties, analyzing the cine PC-MRI data using an appropriate mechanical model allows G to be estimated (see Chapter 12 for a detailed discussion of the theory of MRE).

In passive muscle, G has a value of ~30 kPa parallel to the fiber direction and ~5.5 kPa perpendicular to the fiber direction (Papazoglou et al., 2006). The shear modulus, determined under assumptions of isotropic mechanical properties, varies between muscles (Uffmann et al., 2004). This intermuscle variability may relate to variations in fiber orientation with respect to the direction of shear wave generation. G for passive muscle increases from childhood to adulthood (Debernard et al., 2011a, 2011b) and decreases under clinical conditions such as hyperthyroidism (Bensamoun et al., 2007). Several investigators have reported increases in muscle stiffness with contraction load (Debernard et al., 2011b; Dresner et al., 2001; Gennisson et al., 2005; Jenkyn et al., 2003). These data suggest that (1) the stiffness of passive muscle is a physiologically plastic parameter, altered during development and by certain pathological conditions and that (2) muscle stiffness during contraction is the sum of the resting shear stiffness and contraction-associated increases in stiffness (most likely, associated with actin–myosin binding). Although MRE has been explored less than other MRI methods for muscle, it is clear that there is great potential for this method to characterize muscle pathologies, such as high or low muscle tone and fat infiltration. A comprehensive review of MRE in skeletal muscle is that by Ringleb et al. (2007).

During muscle contraction, the relative muscle length changes (i.e., there is strain). Cine PC-MRI (Drace and Pelc, 1994a, 1994b) and spatial tagging methods such as spatial modulation of magnetization (SPAMM; Zerhouni et al., 1988) both have been applied to study strain development during muscle contractions *in vivo*. PC-MRI was discussed earlier. In SPAMM and related

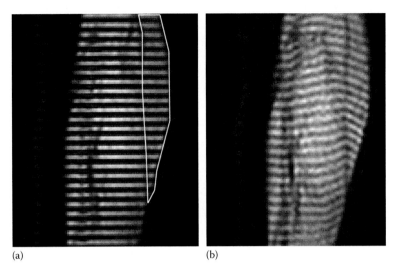

(a)   (b)

FIGURE 7.2
Illustration of spatial tagging in skeletal muscle, sagittal views of the leg. The right side of the images represents the subject's posterior direction. The gastrocnemius muscle is highlighted with the white polygon. Panel (a) shows the image obtained immediately after tag application (i.e., before any tissue motion). The subject then initiated a maximal isometric plantarflexion contraction; images were acquired in a dynamic fashion during the ramp-up in force. The image in panel (b) was obtained after achieving full force. Note that the most superficial margin of the tags does not move, which indicates the relatively fixed position of the muscle fibers' origins during contraction. However, an elevation of the internal margin of the tags occurs, representing the motion of the fibers' insertions during the contraction.

methods, a series of radio frequency (RF) pulses is applied according to a binomial series, creating a first- or higher-order sinusoidal excitation profile in frequency space. Magnetic field gradients applied during the RF train cause a repeated array of bands of tissue to experience saturation effects. A series of images is then acquired; saturated areas show up as lines of low signal (tags). Tissue motion can be inferred from the motion of the tags. SPAMM tagging is illustrated in Figure 7.2.

PC-MRI and spatial tagging methods have provided several important new insights into *in vivo* skeletal muscle mechanics. One such finding is that strain patterns during muscle contraction vary spatially (Finni et al., 2003; Lee et al., 2006; Pappas et al., 2002; Zhong et al., 2008); this may relate to structural heterogeneity (Blemker et al., 2005) and/or a spatially varying distribution of the amount and/or material properties of structures involved in myofascial force transmission (Yucesoy et al., 2002, 2003). MRI has also been used to study multidimensional strains and to relate muscle deformation to muscle architecture. Englund et al. used spatial tagging and DT-MRI to describe multidimensional strains and the association between strain and fiber direction during 50% maximal isometric contractions of the TA muscle (2011). A 3D strain tensor, formed from the tagging data, was diagonalized to describe negative (ε_N) and positive (ε_P) strains; ε_N and ε_P were oriented inferiorly to $\mathbf{v}_1$ and $\mathbf{v}_2$ of the diffusion tensor, respectively. Likewise, DT-MRI and PC-MRI were used to measure lingual muscle architecture and strain rates during swallowing (Felton et al., 2008). These studies provide examples of how multiple MRI approaches can be combined to improve the understanding of muscle structure–function relationships.

7.3.2 Transverse Relaxometry

Transverse relaxometry, referring to the measurement of the permanent and effective transverse relaxation time constants (T_2 and T_2^*, respectively) and/or their respective rates ($1/T_2$, $1/T_2^*$), is the MRI approach that has been most extensively applied to exercising muscle. T_2 and T_2^* can undergo large changes during exercise. To explain why this occurs, the authors begin by describing the organization of water in tissues. Then, the major physiological influences on T_2 and T_2^* in muscle are presented. Finally, some applications of T_2 and T_2^*-based contrast to studies of muscle physiology are noted.

7.3.2.1 Water Compartmentalization in Skeletal Muscle

Biological water is distributed among three spaces: intracellular, interstitial, and vascular. The vascular and interstitial compartments comprise the extracellular space, and the nonvascular portions of the tissue are the parenchyma. Belton and colleagues (1972) reported a multiexponential character to the transverse relaxation of muscle water. They observed three T_2 components with values of <10 ms (which they associated with macromolecular interfacial water), ~40 ms (~82% of the free water signal; intracellular water), and ~170 ms (interstitial free water). After some debate over the correctness of this model, a study by Cole et al. (1993) definitively confirmed its validity, and similar observations have been made in other studies of isolated animal muscles (Damon et al., 2002b; Hazlewood et al., 1974; Louie et al., 2009). However, attempts to replicate the finding of multiexponential relaxation in mammalian muscle *in vivo* have had mixed results (Fan and Does, 2008; Noseworthy et al., 1999; Ploutz-Snyder et al., 1997; Saab et al., 1999).

7.3.2.2 Influences on Transverse Relaxation in Muscle

Three physiologically relevant transverse relaxation mechanisms in skeletal muscle are (1) modulations to intracellular macromolecule-water exchange; (2) alterations to interstitial water content; and (3) blood oxygenation changes. The first important physiological influence on macromolecule-water exchange is the intracellular water content. Because of their slow molecular motion, macromolecules undergo transverse relaxation rapidly; rapid exchange of macromolecular protons with free water lowers the water T_2.

Biophysical properties that affect macromolecule-water proton exchange, such as the relative amounts of protons in the exchanging pools, should therefore influence the T_2. Indeed, increases in intramyocellular water content increase the T_2, and vice versa (Belton and Packer, 1974; Damon et al., 2002b). Another influence on macromolecule-water proton exchange is pH. Amide and hydroxyl groups have chemical shifts separated from water, and so exchange between water and these groups increases the rate of transverse relaxation. Amide-water proton exchange is base-catalyzed for pH > 5 (Englander et al., 1972; Liepinsh and Otting, 1996; Mori et al., 1997), and so intracellular acidification (e.g., due to the metabolic production of weak acids such as lactic acid) would tend to slow this exchange process and lessen its effect on T_2. Several studies have in fact observed linear relationships between $1/T_2$ and pH (Damon et al., 2002b; Fung and Puon, 1981). Also, Louie et al. demonstrated that acidifying a muscle decreases the magnetization transfer rate between macromolecules and water and increases the T_2 (2009).

Secondly, changes in interstitial water content influence a muscle's T_2. Such changes can occur under both physiological and pathological conditions. The underlying idea is that increasing the long T_2 interstitial water fraction increases the whole-muscle T_2 and may even produce a transition from mono- to multiexponential transverse relaxation (Fan and Does, 2008; Ploutz-Snyder et al., 1997). The applicability of this mechanism to whole-muscle T_2 increases will depend on the degree to which the interstitial space expands and the conditions under which T_2 is measured. Context-specific examples for exercise and inflammation are presented below.

The third major source of *in vivo* T_2 and T_2^* contrast is the blood oxygenation level-dependent, or BOLD, contrast effect (Toussaint et al., 1996a). These effects are largest for T_2^*, but there is a T_2 effect also (Donahue et al., 1998; Meyer et al., 1995; Spees et al., 2001; Thulborn et al., 1982; Zhao et al., 2007). In general, BOLD effects result from deoxyhemoglobin's paramagnetism (Pauling and Coryell, 1936), creating a magnetic susceptibility mismatch between deoxyhemoglobin and the diamagnetic water surrounding it. BOLD effects may have intravascular and extravascular components. The intravascular BOLD effect refers to the effect of blood oxyhemoglobin saturation (%HbO$_2$) and hematocrit on the T_2 and T_2^* of water protons in the blood. Free intracellular water protons that exchange with water from deoxyhemoglobin's hydration shell or diffuse through its magnetic susceptibility gradient cause transverse relaxation (Chen and Pike, 2009; Meyer et al., 1995; Stefanovic and Pike, 2004; Thulborn et al., 1982). Because of rapid transmembrane water exchange, the whole-blood's T_2 or T_2^* is affected. The intravascular BOLD effect changes the whole-tissue's T_2 or T_2^* in proportion to the relative blood volume. The extravascular BOLD effect refers to the effect of magnetic susceptibility differences between blood vessels and the tissue parenchyma on the transverse relaxation of extravascular water. As this water diffuses in and out of the magnetic susceptibility gradients formed around the vessels, they experience different Larmor frequencies, and therefore a transverse relaxation effect.

The extravascular BOLD effect introduces a vascular structural dependence to BOLD phenomena (Stables et al., 1998; Yablonskiy and Haacke, 1994). However, several studies have shown that under most reasonably foreseeable experimental conditions, the extravascular BOLD effect is practically unimportant in skeletal muscle at field strengths of 3 T and below (Damon et al., 2007a; Meyer et al., 2004; Sanchez et al., 2010; Utz et al., 2005).

7.3.2.3 Exercise-Induced T_2 Changes

It is well established that the T_2 of muscle water increases due to exercise (Bratton et al., 1965; Fleckenstein et al., 1988); T_2 increases as a function of exercise intensity by as much as 30%, (Adams et al., 1992; Damon et al., 2008; Fisher et al., 1990; Jenner et al., 1994; Price et al., 1998; Yue et al., 1994). During exercise, T_2 or T_2-weighted signal intensity changes evolve in a complex manner as a function of time (Akima et al., 2000; Damon et al., 2003, 2008; Jenner et al., 1994; Kennan et al., 1995). Following exercise, T_2 may initially continue to increase further (Damon et al., 2003; Kennan et al., 1995); full T_2 recovery can require as long as 30–40 min (Reid et al., 2001; Weidman et al., 1991).

Physiological events in the tissue parenchyma generally increase the T_2 during exercise. The largest contribution to these T_2 changes is from the intracellular space (Damon et al., 2002b; Ploutz-Snyder et al., 1997; Prior et al., 2001). Intracellular T_2 increases during exercise are related indirectly to cellular energy metabolism. During exercise, buildup of intracellular metabolites such as lactate and inorganic phosphate creates an osmotic pressure gradient that favors intracellular water accumulation, thereby increasing the intracellular T_2. pH evolves as a function of exercise duration and can eventually produce a significant acidosis. There is an initial alkalosis (because of increased flux through the creatine kinase reaction) that would tend to decrease the T_2; this is followed by an acidosis that would tend to increase the T_2. In addition to these intracellular events, water accumulation in the interstitial space during exercise (Sjogaard et al., 1985) would tend to increase the T_2.

These T_2 changes are modified by the decrease in %HbO$_2$ typically that occurs during exercise. Thus under exercise conditions that maintain a constant blood volume, the BOLD contribution will tend to be a negative effect on whole-muscle T_2 or T_2-weighted signal changes (Damon et al., 2007a). However, under conditions of changing blood volume, the quantitative effect of BOLD contrast on whole-muscle T_2 or T_2^* will depend on the relative relaxation rates in the blood and parenchyma (Towse et al., 2011).

7.3.2.4 Exercise-Induced T_2 Changes: Applications

The preceding discussion has made evident that there is a large number of physiological factors that may influence the T_2 of muscle, each of which relates to some aspect of cellular energy metabolism. This complexity can make it difficult to interpret T_2 changes during exercise, and for some applications more specific MRI methods may be needed. However, T_2 changes have been shown to provide novel physiological information about the muscle activation patterns in complex motor activities. Because of the slow recovery of T_2 following exercise, activities such as cycling can be performed outside of the magnet, and the subject can then be rapidly transferred into the scanner to measure the T_2 changes. This has been used, for instance, to reveal intersubject heterogeneity in muscle use patterns (Hug et al., 2004). Recently, Elder et al. showed that is it possible to infer the temporal nature of muscle activation during cycling, based on T_2 measurements (2010b). Because T_2 changes also have been used to map the spatial pattern of electrically stimulated contractions (Adams et al., 1993), an application of the combined temporal and spatial mapping approach may be helpful in detecting muscle use patterns during electrically stimulated cycling, such as is used in rehabilitation from neurological injuries.

7.3.2.5 Muscle BOLD Contrast: Applications

BOLD contrast has been applied in skeletal muscle to determine %HbO$_2$ (Elder et al., 2010a) and to infer muscle microvascular function following arterial occlusion (Ledermann et al., 2006), the infusion of vasoactive compounds (Utz et al., 2005), and isometric contractions (Meyer et al., 2004). This last protocol is used to provide an example of BOLD contrast in skeletal muscle; in particular, applications are made to the study of obesity and types 1 and 2 diabetes mellitus (T1DM and T2DM, respectively), which have-well described impairments to the skeletal muscle microcirculation [for reviews see Bakker et al. (2009) and Jonk et al. (2007)].

A brief isometric contraction places a relatively small metabolic load on a muscle; however, there is a large postcontraction increase in blood flow. Consequently, Meyer et al. were able to observe a transient rise and fall in small vessel %HbO$_2$ and corresponding changes in BOLD-dependent signals (2004). Because the BOLD-dependent changes are primarily intravascular in nature, resulting largely from a contraction-induced vasodilation, it was suggested that BOLD contrast in skeletal muscle may be an index of microvascular function. Indeed, a subsequent study by Towse et al. (2011) showed that the postcontraction muscle BOLD contrast in chronically active subjects is approximately threefold higher than in sedentary subjects. This group did not see differences in postcontraction BOLD responses among T1DM and T2DM patients and age, body mass index (BMI), and physical activity-matched controls. However, they did see age-dependent variations in BOLD contrast (Slade et al., 2011).

Building on these studies, Damon and colleagues used a dual gradient-recalled echo (GRE) MRI sequence to study blood volume and %HbO$_2$ changes in the muscle microvascular bed following isometric contractions (Damon et al., 2007b; Sanchez et al., 2009). The dual-GRE sequence acquires signals at echo times of 6 ms (which are related to blood volume changes) and 46 ms (which reflect primarily $\%HbO_2$ changes) (Damon et al., 2007b). They found lower postcontraction TE = 6 ms signal intensity changes in the extensor digitorum longus muscle of T2DM and obese non-T2DM subjects than in lean subjects and lower TE = 46 ms signal intensity changes in T2DM patients versus lean nondiabetic subjects (Sanchez et al., 2011). These changes were interpreted as reflecting impaired skeletal muscle microvascular health related to obesity.

Evidence thus exists to support the idea that differences in the time course and magnitude of the postcontraction BOLD contrast may provide powerful insights into muscle microvascular function. This technique has been used to study patients with suspected and diagnosed peripheral vascular complications. Future studies will hopefully continue to address the sensitivity of these measurements to physiological regulators of muscle vascular function, in health and disease.

7.3.3 Muscle Perfusion Measurements Using MRI

As noted, muscles require oxygen, signaling molecules, and nutrient delivery to sustain high rates of metabolism during exercise. These demands are met by increased perfusion, which is defined as the tissue mass-normalized rate of blood delivery to an organ or tumor via its capillary network. There are two broad classes of MRI methods available for measuring perfusion, both of which are based on diffusible tracer methods (Kety and Schmidt, 1948). The first class of methods includes dynamic contrast-enhanced (DCE) MRI and dynamic susceptibility contrast MRI, which employ exogenous molecules (MRI contrast agents) as the tracers. The agent is injected, and the signal intensity is measured in the tissue of interest as a function of time. The tissue signal time course and the rate of contrast delivery by the arterial system are used in concert with a biophysical model to determine tissue parameters related to perfusion. DCE-MRI has been performed in skeletal muscle studies to characterize postocclusive reactive hyperemia in human studies (Thompson et al., 2005) and to characterize ischemia-reperfusion injury in animal models of pressure ulcers (Loerakker et al., 2011). It should be noted that during DCE, the signal enhancement occurs over a period of minutes. Thus it is important that perfusion be constant during this measurement window, and not evolving as a function of time.

The other approach is arterial spin labeling (ASL), in which RF pulses are used to transform the water in blood into an endogenous tracer. In ASL, two types of images are acquired, in alternating fashion. The first is the "control" image, in which no RF labeling of the inflowing arterial blood occurs. There is exchange between blood and tissue water, but the tissue's longitudinal magnetization is essentially unaffected by this exchange. The second type of image is the "tagged" image, in which the magnetization of the spins flowing into the imaging slice plane is inverted. Consequently, the inflowing blood has less than full longitudinal magnetization (this is the "label").

As the labeled blood flows into the tissue and exchanges with the tissue water, the longitudinal magnetization of the tissue parenchyma is reduced. A small signal difference (a few percent) between the labeled and control images results. This signal difference is proportional to the rate of blood perfusion to the tissue and other quantities (such as blood and tissue T_1 and the blood–tissue water partition coefficient) that can be observed or assumed. Mathematical expressions exist that allow quantitative maps of perfusion to be computed from ASL data (e.g., Buxton et al., 1998).

ASL has been successfully applied in a broad variety of skeletal muscle physiology studies. There has been extensive use of postarterial occlusion perfusion measurements, using ASL alone or in concert with other approaches, to characterize the high perfusion rates that occur during this protocol. Toussaint et al. performed the initial ASL studies of muscle, observing perfusion changes during postocclusive hyperemia (1996b). ASL measurements of postocclusive hyperemia were compared quantitatively with venous occlusion plethysmography (Raynaud et al., 2001) and with diffuse correlation spectroscopy signals (Yu et al., 2007). ASL (like other MRI methods) has additional challenges when applied during exercise; these include motion, alterations to shim quality, and changing T_2 values. However, ASL has been successfully applied in both human (Elder et al., 2010a; Frank et al., 1999) and animal (Baligand et al., 2011; Bertoldi et al., 2008; Marro et al., 2005a, 2005b) models of exercise. ASL also has been successfully used to study alterations to muscle perfusion due to knockout of the myostatin gene (Baligand et al., 2010) and peripheral arterial disease (Wu et al., 2009). These studies demonstrate not just the feasibility of ASL in skeletal muscle but also ASL's potential for revealing novel aspects of muscle physiology. There is a recent review article on muscle perfusion measurements by Carlier et al. (2006).

7.4 Assessing Muscle Damage and Disease

7.4.1 The Processes of Muscle Damage and Repair

Skeletal muscle damage can result from lengthening contractions, ischemia-reperfusion injury, disease, or other causes. Following the initial damaging event, a cascade of postinjury repair mechanisms is activated. This response initially includes an autogenic phase, during which proteolytic and lipolytic enzymes are activated and reactive oxygen species are generated, collectively creating cellular debris. Neutrophils then invade the region of damaged muscle tissue, marking the beginning of the phagocytic phase. This is a typical inflammatory response that can last from several hours to up to one week. Activated macrophages migrate to the region, enter the damaged myofibers, secrete digestive enzymes, and engulf digested cellular components. The direction of the final phase depends on the extent of the damage. Recovery of damaged fibers requires the fusion of new myoblasts to increase the number of nuclei and bridge severed fibers. A major source of these nascent myoblasts is satellite cells (Yu et al., 2007). If the damaged sarcolemma cannot be resealed, however, the fiber will be lost to necrosis. If this outcome is frequent in a given region, the fibers will be replaced by fat and scar tissue. The time course and potential success of this repair effort will depend on the severity of the damage and (in the case of chronic disease) the persistence of the underlying cause of damage.

7.4.2 Introduction to MRI of Neuromuscular Disease

Early MRI studies of muscle disease primarily focused on morphological features such as cross-sectional area of preserved muscle, and this remains an extremely important characteristic of muscle to be quantified. In cases of wasting and/or chronic disease, these morphometric measures are primarily aimed at determining cross-sectional area and/or the amount of remaining muscle tissue relative to the amount of pathological fat deposition. Quantitative fat-water imaging can be extremely helpful in this regard. These parameters are common in studies involving the natural history and treatment of atrophy due to cachexia, chronic unloading, denervation, and neuromuscular diseases.

Biopsy and fine-wire electromyography are used in diagnosis and provide information complementary to MRI, but may not be suitable for longitudinal evaluation of chronic muscle diseases. In the clinic, a qualitative assessment is made of images generated with proton density and T_1- and T_2-weighting, with and/or without fat signal saturation. However, recent advancements in the quantitative measures discussed earlier in this chapter, especially T_2 and diffusion, are now being used in clinical and preclinical studies, with the ultimate goal of improving therapeutic interventions. This might be accomplished by gaining greater structural and functional knowledge of the damaged tissue, more accurately or sensitively assessing the efficacy of putative treatments, and/or allowing improved methods for guiding individual treatment.

7.4.3 MRI Biomarker Development for Neuromuscular Diseases

7.4.3.1 T_2-Based Assessments: Animal Models

As noted, T_2 is a commonly used MRI-determined marker of skeletal muscle damage. Although a full, quantitative description of the biophysical mechanisms for muscle damage-induced changes does not exist, it is generally expected that damage processes such as the loss of membrane integrity and the expansion of the interstitial space by edema increase the T_2.

Some examples of T_2 as a biomarker of muscle damage are derived from experimental models of muscle damage. For example, injection of λ-carrageenan into the muscle (Fan and Does, 2008) of a rodent induces a local inflammatory response that includes edema and inflammatory cell infiltration (Radhakrishnan et al., 2003). In muscle, this induces a transition to biexponential transverse relaxation with a significantly enhanced presumptive extracellular T_2 component (Fan and Does, 2008). Significantly elevated T_2 values in response to notexin (snake venom) injection have also been reported (Mattila et al., 1995; Wishnia et al., 2001). T_2 changes, spatially correlated with histological observations and graded to the severity of damage, also have been observed in response to ischemia-reperfusion injury (Heemskerk et al., 2006). T_2 elevations also have been reported in mice with genetic models of muscular dystrophy, including the *mdx* mouse (McIntosh et al., 1998) and γ-sarcoglycan knockout mice (Walter et al., 2005). The T_2 elevations are correlated with the number of Evan's Blue Dye (EBD) positive fibers (which is the classic histological marker of membrane damage) (Frimel et al., 2005). Moreover, the T_2 changes appear to be responsive to clinical interventions. This has been shown, for example, following sildenafil treatment of *mdx* mice to ameliorate defective nitric oxide signaling (Parchen et al., 2008). Also, in a test of proof of principle for T_2 tracking of responsiveness to genetic therapy, restoration toward normal T_2 values was observed following replacement of the γ-sarcoglycan gene in γ-sarcoglycan knockout mice (Walter et al., 2005).

7.4.3.2 Diffusion-Based Assessments: Animal Models

More recent studies have demonstrated the potential value of DT-MRI for characterizing skeletal muscle damage. As with T_2, there is not yet a definitive, quantitative description of the structural and pathological bases for the changes in diffusion indices that are observed following muscle injury. The likely candidates are inflammation, membrane damage, and disruptions to the regular protein lattices that comprise the myofibril.

Diffusion imaging has been used in several muscle damage models to provide a quantitative measurement associated with muscle damage. For example, Heemskerk et al. (2006) observed that during the reperfusion phase of ischemia-reperfusion injury, the mean diffusivity is increased. When more severe damage is created by simultaneous electrical stimulation during the ischemic phase, the diffusion anisotropy, as characterized by the fractional anisotropy (FA), also decreased. The changes in mean diffusivity were due mainly to increases in λ_2 and λ_3, and especially λ_3. A subsequent study from these authors (Heemskerk et al., 2007) used a femoral artery ligation model to reveal different temporal and spatial patterns to the diffusion and T_2 changes, suggesting that these parameters may reflect different aspects of muscle damage and regeneration. Both of these studies provided histological data that supported the MRI observations. The λ-carrageenan injection used by Fan and Does (2008) to induce edema and cause the appearance of the second T_2 component also resulted in the appearance of a second diffusion component. The second diffusion component was associated with the long T_2 component and was characterized by greater and more isotropic diffusion than the diffusion component associated with the short T_2 component. Data from muscle damage models in the *mdx* mouse also support the potential use of diffusion as an early biomarker of muscle damage as well (McMillan et al., 2011).

7.4.3.3 Relaxation and Diffusion Changes in the Autoimmune Inflammatory Myopathies

Dermatomyositis (DM), polymyositis (PM), and inclusion body myositis (IBM) are three typical idiopathic inflammatory myopathies (IIM's). They have autoimmune origins. While the underlying fundamental reasons for these diseases remain unclear, it is known that the antigenic target in PM is the muscle itself, while in DM the target is the vasculature. DM and PM affect mostly proximal muscles, while IBM affects both proximal and distal muscles. On a histological level, these diseases are characterized by muscle inflammation, fat infiltration and replacement, and atrophy; the typical symptoms are weakness, fatigue, and myalgia. These symptoms may be alleviated by treatment with immunosuppressant drugs. In this section, the authors will briefly discuss how relaxation- and diffusion-based MRI measurements are useful for assessing clinical condition and treatments over the course of time.

Figure 7.3 shows examples of MRI images from a control subject and PM and DM patients (Park et al., 2009a). For the control subject, the thigh muscles show a homogeneously dark signal in both T_1- (A) and T_2- (B) weighted images. For the DM patient, the T_1-weighted image (C) appears normal, but the T_2-weighted image (D) shows high signals in many regions. With the PM patient, bright signals are identified in both T_1- and T_2-weighted images (E and F, respectively), indicating both fat infiltration and inflammation. The chronic PM patient (G and H) reveals severe fat replacement in the quadriceps muscles. Altered T_1 and T_2 values in diseased muscles consistent with these qualitative contrast effects have been observed as well (Park et al., 1994).

DW-MRI and DT-MRI also have been performed in DM and PM. Using DW-MRI, Qi et al. (2007) observed ~25% higher mean diffusion coefficients and a pattern of multidimensional diffusion coefficients consistent with lower diffusion anisotropy in inflamed muscles of DM and PM patients than the muscles of healthy controls or of the patients' nonaffected muscles. In chronic PM patients with fat infiltration, the mean diffusivity and T_1 are lowered and the T_2 is elevated. These data illustrate the importance of excluding fat from region-of-interest analyses if the goal is to characterize muscle health using diffusion (Williams et al., 2013) and relaxometry. A recent preliminary report using DT-MRI reported diffusion alterations consistent with those just reported for DW-MRI (Park et al., 2009b). Qi et al. (2007) also

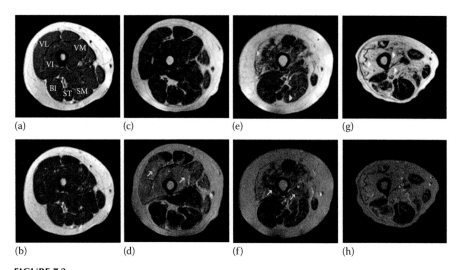

(a) (c) (e) (g)

(b) (d) (f) (h)

FIGURE 7.3
(See color insert.) T_1- and T_2-weighted images of the thigh of a control subject and DM and PM patients. The *top row* (a, c, e, and g) shows T_1-weighted images, and the *bottom row* (b, d, f, and h) shows T_2-weighted images. Separate panels show data for the control subject (a and b) and DM (c and d), PM (e and f), and chronic PM (g and h) patients. Muscles of the thigh are designated in control subject (a): VL, vastus lateralis; VI, vastus intermedius; VM, vastus medialis; BI, biceps femoris; ST, semitendinosus; SM, semimembranosus. T_2-weighted images illustrate fat replacement, and the arrows in the T_2-weighted images illustrate inflammation.

demonstrated the potential for T_2 and mean diffusivity to track changes in clinical condition and responsiveness to therapy. A recent review of imaging in IIM is available (Walker, 2008).

7.5 Summary and Overall Conclusions

MRI methods afford investigators many options for investigating muscle structure and function. Quantitative DT-MRI fiber tracking methods provide insight into muscle structure at high spatial resolution. In addition to being suitable for characterizing muscle architectural properties, these data can be used to help interpret heterogeneous and anisotropic muscle mechanical properties such as strain development and elastic moduli. Other MRI methods for investigating normal and pathological muscle physiology include T_2, BOLD contrast, and perfusion imaging. Applications of these methods to identifying muscle activation patterns during complex motor activities, muscle microvascular function in obesity and diabetes, cardiovascular disease, functional genomic profiling, and assessing muscle damage by injury or disease were noted. These methods will continue to provide new insight into the structure and function of this critical tissue.

Acknowledgments

The authors' work in this field has been supported by grants from the National Institute of Arthritis and Musculoskeletal and Skin Diseases, National Institutes of Health: R01 AR050101 and R01 AR050791.

References

Adams, G.R., Duvoisin, M.R., Dudley, G.A., 1992. Magnetic resonance imaging and electromyography as indexes of muscle function. *J Appl Physiol.* 73, 1578–83.

Adams, G.R., Harris, R.T., Woodard, D., Dudley, G.A., 1993. Mapping of electrical muscle stimulation using MRI. *J Appl Physiol.* 74, 532–7.

Akima, H., Ito, M., Yoshikawa, H., Fukunaga, T., 2000. Recruitment plasticity of neuromuscular compartments in exercised tibialis anterior using echo-planar magnetic resonance imaging in humans. *Neurosci Lett.* 296, 133–6.

Bakker, W., Eringa, E.C., Sipkema, P., van Hinsbergh, V.W., 2009. Endothelial dysfunction and diabetes: Roles of hyperglycemia, impaired insulin signaling and obesity. *Cell Tissue Res.* 335, 165–89.

Baligand, C., Gilson, H., Menard, J.C., Schakman, O., Wary, C., Thissen, J.P., Carlier, P.G., 2010. Functional assessment of skeletal muscle in intact mice lacking myostatin by concurrent NMR imaging and spectroscopy. *Gene Ther.* 17, 328–37.

Baligand, C., Wary, C., Ménard, J.C., Giacomini, E., Hogrel, J.Y., Carlier, P.G., 2011. Measuring perfusion and bioenergetics simultaneously in mouse skeletal muscle: A multiparametric functional-NMR approach. *NMR Biomed.* 24, 281–90.

Basser, P.J., Mattiello, J., Le Bihan, D., 1994. MR diffusion tensor spectroscopy and imaging. *Biophys J.* 66, 259–67.

Basser, P.J., Pajevic, S., Pierpaoli, C., Duda, J., Aldroubi, A., 2000. In vivo fiber tractography using DT-MRI data. *Magn Reson Med.* 44, 625–32.

Belton, P.S., Jackson, R.R., Packer, K.J., 1972. Pulsed NMR studies of water in striated muscle. I. Transverse nuclear spin relaxation times and freezing effects. *Biochim Biophys Acta.* 286, 16–25.

Belton, P.S., Packer, K.J., 1974. Pulsed NMR studies of water in striated muscle. III. The effects of water content. *Biochim Biophys Acta.* 354, 305–14.

Bensamoun, S.F., Ringleb, S.I., Chen, Q., Ehman, R.L., An, K.N., Brennan, M., 2007. Thigh muscle stiffness assessed with magnetic resonance elastography in hyperthyroid patients before and after medical treatment. *J Magn Reson Imaging.* 26, 708–13.

Bertoldi, D., Loureiro de Sousa, P., Fromes, Y., Wary, C., Carlier, P.G., 2008. Quantitative, dynamic and noninvasive determination of skeletal muscle perfusion in mouse leg by NMR arterial spin-labeled imaging. *Magn Reson Imaging.* 26, 1259–65.

Blemker, S.S., Pinsky, P.M., Delp, S.L., 2005. A 3D model of muscle reveals the causes of nonuniform strains in the biceps brachii. *J Biomech.* 38, 657–65.

Bratton, C.B., Hopkins, A.L., Weinberg, J.W., 1965. Nuclear magnetic resonance studies of living muscle. *Science.* 147, 738–9.

Budzik, J.F., Le Thuc, V., Demondion, X., Morel, M., Chechin, D., Cotten, A., 2007. In vivo MR tractography of thigh muscles using diffusion imaging: Initial results. *Eur Radiol.* 17, 3079–85.

Buxton, R.B., Frank, L.R., Wong, E.C., Siewert, B., Warach, S., Edelman, R.R., 1998. A general kinetic model for quantitative perfusion imaging with arterial spin labeling. *Magn Reson Med.* 40, 383–96.

Carlier, P.G., Bertoldi, D., Baligand, C., Wary, C., Fromes, Y., 2006. Muscle blood flow and oxygenation measured by NMR imaging and spectroscopy. *NMR Biomed.* 19, 954–67.

Chang, D.C., Rorschach, H.E., Nichols, B.L., Hazlewood, C.F., 1973. Implications of diffusion coefficient measurements for the structure of cellular water. *Ann N Y Acad Sci.* 204, 434–43.

Chen, J.J., Pike, G.B., 2009. Human whole blood T2 relaxometry at 3 Tesla. *Magn Reson Med.* 61, 249–54.

Clark, M.E., Burnell, E.E., Chapman, N.R., Hinke, J.A., 1982. Water in barnacle muscle. IV. Factors contributing to reduced self-diffusion. *Biophys J.* 39, 289–99.

Cleveland, G.G., Chang, D.C., Hazlewood, C.F., Rorschach, H.E., 1976. Nuclear magnetic resonance measurement of skeletal muscle: Anisotropy of the diffusion coefficient of the intracellular water. *Biophys J.* 16, 1043–53.

Cole, W.C., LeBlanc, A.D., Jhingran, S.G., 1993. The origin of biexponential T2 relaxation in muscle water. *Magn Reson Med.* 29, 19–24.

Damon, B., Wadington, M., Hornberger, J., Lansdown, D., 2007a. Absolute and relative contributions of BOLD effects to the muscle functional MRI signal intensity time course: Effect of exercise intensity. *Magn Reson Med.* 58, 335–45.

Damon, B.M., Buck, A.K.W., Ding, Z., 2011. Diffusion-tensor MRI based skeletal muscle fiber tracking. *Imaging Med.* 3, 675–87.

Damon, B.M., Ding, Z., Anderson, A.W., Freyer, A.S., Gore, J.C., 2002a. Validation of diffusion tensor MRI-based muscle fiber tracking. *Magn Reson Med.* 48, 97–104.

Damon, B.M., Gregory, C.D., Hall, K.L., Stark, H.J., Gulani, V., Dawson, M.J., 2002b. Intracellular acidification and volume increases explain R2 decreases in exercising muscle. *Magn Reson Med.* 47, 14–23.

Damon, B.M., Heemskerk, A.M., Ding, Z., 2012. Polynomial fitting of DT-MRI fiber tracts allows accurate estimation of muscle architectural parameters. *Magn Reson Imaging.* 30, 589–600.

Damon, B.M., Hornberger, J.L., Wadington, M.C., Lansdown, D.A., Kent-Braun, J.A., 2007b. Dual gradient-echo MRI of post-contraction changes in skeletal muscle blood volume and oxygenation. *Magn Reson Med.* 47, 670–9.

Damon, B.M., Wadington, M.C., Lansdown, D.A., Hornberger, J.L., 2008. Spatial heterogeneity in the muscle functional MRI signal intensity time course: Effect of exercise intensity. *Magn Reson Imaging.* 26, 1114–21.

Damon, B.M., Wigmore, D.M., Ding, Z., Gore, J.C., Kent-Braun, J.A., 2003. Cluster analysis of muscle functional MRI data. *J Appl Physiol.* 95, 1287–96.

Debernard, L., Robert, L., Charleux, F., Bensamoun, S.F., 2011a. Characterization of muscle architecture in children and adults using magnetic resonance elastography and ultrasound techniques. *J. Biomech.* 44, 397–401.

Debernard, L., Robert, L., Charleux, F., Bensamoun, S.F., 2011b. Analysis of thigh muscle stiffness from childhood to adulthood using magnetic resonance elastography (MRE) technique. *Clin Biomech.* 26, 836–40.

Donahue, K.M., Van Kylen, J., Guven, S., El-Bershawi, A., Luh, W.M., Bandettini, P.A., Cox, R.W., Hyde, J.S., Kissebah, A.H., 1998. Simultaneous gradient-echo/spin-echo EPI of graded ischemia in human skeletal muscle. *J Magn Reson Imaging*. 8, 1106–13.

Drace, J.E., Pelc, N.J., 1994a. Skeletal muscle contraction: Analysis with use of velocity distributions from phase-contrast MR imaging. *Radiology*. 193, 423–9.

Drace, J.E., Pelc, N.J., 1994b. Measurement of skeletal muscle motion in vivo with phase-contrast MR imaging. *J Magn Reson Imaging*. 4, 157–63.

Dresner, M.A., Rose, G.H., Rossman, P.J., Muthupillai, R., Manduca, A., Ehman, R.L., 2001. Magnetic resonance elastography of skeletal muscle. *J Magn Reson Imaging*. 13, 269–76.

Elder, C.P., Cook, R.N., Chance, M.A., Copenhaver, E.A., Damon, B.M., 2010a. Image-based calculation of perfusion and oxyhemoglobin saturation in skeletal muscle during submaximal isometric contractions. *Magn Reson Med*. 64, 852–61.

Elder, C.P., Cook, R.N., Wilkens, K.L., Chance, M.A., Sanchez, O.A., Damon, B.M., 2010b. A method for detecting the temporal sequence of muscle activation during cycling using MRI. *J Appl Physiol*. 110, 826–33.

Englander, S.W., Downer, N.W., Teitelbaum, H., 1972. Hydrogen exchange. *Ann Rev Biochem*. 41, 903–24.

Englund, E.K., Elder, C.P., Xu, Q., Ding, Z., Damon, B.M., 2011. Combined diffusion and strain tensor MRI reveals a heterogeneous, planar pattern of strain development during isometric muscle contraction. *Am J Physiol Reg Integ Comp Physiol*. 300, R1079–R1090.

Fan, R.H., Does, M.D., 2008. Compartmental relaxation and diffusion tensor imaging measurements in vivo in λ-carrageenan-induced edema in rat skeletal muscle. *NMR Biomed*. 21, 566–73.

Felton, S.M., Gaige, T.A., Benner, T., Wang, R., Reese, T.G., Wedeen, V.J., Gilbert, R.J., 2008. Associating the mesoscale fiber organization of the tongue with local strain rate during swallowing. *J Biomech*. 41, 1782–9.

Finch, E.D., Harmon, J.F., Muller, B.H., 1971. Pulsed NMR measurements of the diffusion constant of water in muscle. *Arch Biochem Biophys*. 147, 299–310.

Finni, T., Hodgson, J.A., Lai, A.M., Edgerton, V.R., Sinha, S., 2003. Nonuniform strain of human soleus aponeurosis-tendon complex during submaximal voluntary contractions in vivo. *J Appl Physiol*. 95, 829–37.

Fisher, M.J., Meyer, R.A., Adams, G.R., Foley, J.M., Potchen, E.J., 1990. Direct relationship between proton T2 and exercise intensity in skeletal muscle MR images. *Invest Radiol*. 25, 480–5.

Fleckenstein, J.L., Canby, R.C., Parkey, R.W., Peshock, R.M., 1988. Acute effects of exercise on MR imaging of skeletal muscle in normal volunteers. *Am J Roentgenol*. 151, 231–7.

Frank, L.R., Wong, E.C., Haseler, L.J., Buxton, R.B., 1999. Dynamic imaging of perfusion in human skeletal muscle during exercise with arterial spin labeling. *Magn Reson Med*. 42, 258–67.

Frimel, T.N., Walter, G.A., Gibbs, J.D., Gaidosh, G.S., Vandenborne, K., 2005. Noninvasive monitoring of muscle damage during reloading following limb disuse. *Muscle Nerve*. 32, 605–12.

Fung, B.M., Puon, P.S., 1981. Nuclear magnetic resonance transverse relaxation in muscle water. *Biophys J*. 33, 27–37.

Gennisson, J.L., Cornu, C., Catheline, S., Fink, M., Portero, P., 2005. Human muscle hardness assessment during incremental isometric contraction using transient elastography. *J Biomech.* 38, 1543–50.

Gilbert, R.J., Wedeen, V.J., Magnusson, L.H., Benner, T., Wang, R., Dai, G., Napadow, V.J., Roche, K.K., 2006. Three-dimensional myoarchitecture of the bovine tongue demonstrated by diffusion spectrum magnetic resonance imaging with tractography. *Anat Rec A Discov Mol Cell Evol Biol.* 288, 1173–82.

Hazlewood, C.F., Chang, D.C., Nichols, B.L., Woessner, D.E., 1974. Nuclear magnetic resonance transverse relaxation times of water protons in skeletal muscle. *Biophys J.* 14, 583–606.

Heemskerk, A., Strijkers, G., Drost, M., van Bochove, G., Nicolay, K., 2007. Skeletal muscle degeneration and regeneration following femoral artery ligation in the mouse: Diffusion tensor imaging monitoring. *Radiology.* 243, 413–21.

Heemskerk, A.M., Damon, B.M., 2009. DTI-based fiber tracking reveals a multifaceted alteration of pennation angle and fiber tract length upon muscle lengthening. *International Society for Magnetic Resonance in Medicine: 17th Annual Meeting,* April 14–24, Honolulu, HI.

Heemskerk, A.M., Drost, M.R., van Bochove, G.S., van Oosterhout, M.F., Nicolay, K., Strijkers, G.J., 2006. DTI-based assessment of ischemia-reperfusion in mouse skeletal muscle. *Magn Reson Med.* 56, 272–81.

Heemskerk, A.M., Sinha, T.K., Wilson, K.J., Ding, Z., Damon, B.M., 2009. Quantitative assessment of DTI-based muscle fiber tracking and optimal tracking parameters. *Magn Reson Med.* 61, 467–72.

Heemskerk, A.M., Sinha, T.K., Wilson, K.J., Ding, Z., Damon, B.M., 2010. Repeatability of DTI-based skeletal muscle fiber tracking. *NMR Biomed.* 23, 294–303.

Heemskerk, A.M., Strijkers, G.J., Vilanova, A., Drost, M.R., Nicolay, K., 2005. Determination of mouse skeletal muscle architecture using three-dimensional diffusion tensor imaging. *Magn Reson Med.* 53, 1333–40.

Hug, F., Bendahan, D., Le Fur, Y., Cozzone, P.J., Grelot, L., 2004. Heterogeneity of muscle recruitment pattern during pedaling in professional road cyclists: A magnetic resonance imaging and electromyography study. *Eur J Appl Physiol.* 92, 334–42.

Jenkyn, T.R., Ehman, R.L., An, K.-N., 2003. Noninvasive muscle tension measurement using the novel technique of magnetic resonance elastography (MRE). *J Biomech.* 36, 1917–21.

Jenner, G., Foley, J.M., Cooper, T.G., Potchen, E.J., Meyer, R.A., 1994. Changes in magnetic resonance images of muscle depend on exercise intensity and duration, not work. *J Appl Physiol.* 76, 2119–24.

Jones, D.K., Cercignani, M., 2010. Twenty-five pitfalls in the analysis of diffusion MRI data. *NMR Biomed.* 23, 803–20.

Jonk, A.M., Houben, A.J.H.M., de Jongh, R.T., Serne, E.H., Schaper, N.C., Stehouwer, C.D.A., 2007. Microvascular dysfunction in obesity: A potential mechanism in the pathogenesis of obesity-associated insulin resistance and hypertension. *Physiology.* 22, 252–60.

Kennan, R.P., Price, T.B., Gore, J.C., 1995. Dynamic echo planar imaging of exercised muscle. *Magn Reson Imaging.* 13, 935–41.

Kety, S.S., Schmidt, C.F., 1948. The nitrous oxide method for the quantitative determination of cerebral blood flow in man: theory, procedure and normal values. *J Clin Invest.* 27, 476–83.

Lansdown, D.A., Ding, Z., Wadington, M., Hornberger, J.L., Damon, B.M., 2007. Quantitative diffusion tensor MRI-based fiber tracking of human skeletal muscle. *J Appl Physiol*. 103, 673–81.

Lazar, M., 2010. Mapping brain anatomical connectivity using white matter tractography. *NMR Biomed*. 23, 821–35.

Le Bihan, D., Mangin, J.F., Poupon, C., Clark, C.A., Pappata, S., Molko, N., Chabriat, H., 2001. Diffusion tensor imaging: Concepts and applications. *J Magn Reson Imaging*. 13, 534–46.

Ledermann, H.P., Schulte, A.-C., Heidecker, H.-G., Aschwanden, M., Jager, K.A., Scheffler, K., Steinbrich, W., Bilecen, D., 2006. Blood oxygenation level-dependent magnetic resonance imaging of the skeletal muscle in patients with peripheral arterial occlusive disease. *Circulation*. 113, 2929–35.

Lee, H.D., Finni, T., Hodgson, J.A., Lai, A.M., Edgerton, V.R., Sinha, S., 2006. Soleus aponeurosis strain distribution following chronic unloading in humans: An in vivo MR phase-contrast study. *J Appl Physiol*. 100, 2004–11.

Lieber, R.L., Friden, J., 2001. Clinical significance of skeletal muscle architecture. *Clin Orthop Relat Res*. 140–51.

Liepinsh, E., Otting, G., 1996. Proton exchange rates from amino acid side chains— implications for image contrast. *Magn Reson Med*. 35, 30–42.

Loerakker, S., Oomens, C.W.J., Manders, E., Schakel, T., Bader, D.L., Baaijens, F.P.T., Nicolay, K., Strijkers, G.J., 2011. Ischemia-reperfusion injury in rat skeletal muscle assessed with T2-weighted and dynamic contrast-enhanced MRI. *Magn Reson Med*. 66, 528–37.

Louie, E.A., Gochberg, D.F., Does, M.D., Damon, B.M., 2009. Magnetization transfer and T2 measurements of isolated muscle: Effect of pH. *Magn Reson Med*. 61, 560–9.

Marro, K.I., Hyyti, O.M., Kushmerick, M.J., 2005a. FAWSETS perfusion measurements in exercising skeletal muscle. *NMR Biomed*. 18, 322–30.

Marro, K.I., Hyyti, O.M., Vincent, M.A., Kushmerick, M.J., 2005b. Validation and advantages of FAWSETS perfusion measurements in skeletal muscle. *NMR Biomed*. 18, 226–34.

Mattila, K.T., Lukka, R., Hurme, T., Komu, M., Alanen, A., Kalimo, H., 1995. Magnetic resonance imaging and magnetization transfer in experimental myonecrosis in the rat. *Magn Reson Med*. 33, 185–92.

McIntosh, L.M., Baker, R.E., Anderson, J.E., 1998. Magnetic resonance imaging of regenerating and dystrophic mouse muscle. *Biochem Cell Biol*. 76, 532–41.

McMillan, A.B., Shi, D., Pratt, S.J., Lovering, R.M., 2011. Diffusion tensor MRI to assess damage in healthy and dystrophic skeletal muscle after lengthening contractions. *J Biomed Biotechnol*. 2011, 970726.

Meyer, M.E., Yu, O., Eclancher, B., Grucker, D., Chambron, J., 1995. NMR relaxation rates and blood oxygenation level. *Magn Reson Med*. 34, 234–41.

Meyer, R.A., Towse, T.F., Reid, R.W., Jayaraman, R.C., Wiseman, R.W., McCully, K.K., 2004. BOLD MRI mapping of transient hyperemia in skeletal muscle after single contractions. *NMR Biomed*. 17, 392–8.

Mori, S., Crain, B.J., Chacko, V.P., van Zijl, P.C., 1999. Three-dimensional tracking of axonal projections in the brain by magnetic resonance imaging. *Ann Neurol*. 45, 265–9.

Mori, S., van Zijl, P.C.M., Shortle, D., 1997. Measurement of water-amide proton exchange rates in the denatured state of staphylococcal nuclease by a magnetization transfer technique. *Proteins*. 28, 325–32.

Muthupillai, R., Lomas, D.J., Rossman, P.J., Greenleaf, J.F., Manduca, A., Ehman, R.L., 1995. Magnetic resonance elastography by direct visualization of propagating acoustic strain waves. *Science.* 269, 1854–7.

Napadow, V.J., Chen, Q., Mai, V., So, P.T., Gilbert, R.J., 2001. Quantitative analysis of three-dimensional-resolved fiber architecture in heterogeneous skeletal muscle tissue using nmr and optical imaging methods. *Biophys J.* 80, 2968–75.

Noseworthy, M.D., Kim, J.K., Stainsby, J.A., Stanisz, G.J., Wright, G.A., 1999. Tracking oxygen effects on MR signal in blood and skeletal muscle during hyperoxia exposure. *J Magn Reson Imaging.* 9, 814–20.

Otten, E., 1988. Concepts and models of functional architecture in skeletal muscle. *Exerc Sport Sci Rev.* 16, 89–137.

Papazoglou, S., Rump, J., Braun, J., Sack, I., 2006. Shear wave group velocity inversion in MR elastography of human skeletal muscle. *Magn Reson Med.* 56, 489–97.

Pappas, G.P., Asakawa, D.S., Delp, S.L., Zajac, F.E., Drace, J.E., 2002. Nonuniform shortening in the biceps brachii during elbow flexion. *J Appl Physiol.* 92, 2381–9.

Parchen, C., Gray, H., Percival, J., Froehner, S., Beavo, J.A., 2008. Effect of sildenafil on mdx skeletal muscle. *Proceedings of the Federation of Societies of Experimental Biology*, April 5–9, San Diego, CA, p. 1137.13.

Park, J.H., Lee, B.C., Olsen, N.J., 2009a. Weakness and fatigue in patients with inflammatory myopathies: Dermatomyositis, polymyositis, and inclusion body myositis. In: *Human Muscle Fatigue.* Williams, C., Ratel, S., eds. Routledge Press, New York, pp. 313–37.

Park, J.H., Lee, B.C., Qi, J., 2009b. Diffusion tensor magnetic resonance imaging (DTI) in the thigh muscles of polymyositis patients. *Proceedings of the 17th International Society for Magnetic Resonance in Medicine*, April 18–24, Honolulu, HI, p. 1907.

Park, J.H., Vital, T.L., Ryder, N.M., Hernanz-Schulman, M., Partain, C.L., Price, R.R., Olsen, N.J., 1994. Magnetic resonance imaging and P-31 magnetic resonance spectroscopy provide unique quantitative data useful in the longitudinal management of patients with dermatomyositis. *Arthritis Rheum.* 37, 736–46.

Pauling, L., Coryell, C., 1936. The magnetic properties and structure of hemoglobin, oxyhemoglobin, and carbonmonoxyhemoglobin. *Proc Natl Acad Sci USA.* 22, 210–6.

Ploutz-Snyder, L.L., Nyren, S., Cooper, T.G., Potchen, E.J., Meyer, R.A., 1997. Different effects of exercise and edema on T2 relaxation in skeletal muscle. *Magn Reson Med.* 37, 676–82.

Polgar, J., Johnson, M.A., Weightman, D., Appleton, D., 1973. Data on fibre size in thirty-six human muscles: An autopsy study. *J Neurol Sci.* 19, 307–18.

Price, T.B., Kennan, R.P., Gore, J.C., 1998. Isometric and dynamic exercise studied with echo planar magnetic resonance imaging (MRI). *Med Sci Sports Exerc.* 30, 1374–80.

Prior, B.M., Ploutz-Snyder, L.L., Cooper, T.G., Meyer, R.A., 2001. Fiber type and metabolic dependence of T2 increases in stimulated rat muscles. *J Appl Physiol.* 90, 615–23.

Qi, J., Olsen, N.J., Price, R.R., Winston, J.A., Park, J.H., 2007. Diffusion-weighted imaging of inflammatory myopathies: Polymyositis and dermatomyositis. *J Magn Reson Imaging.* 27, 212–7.

Radhakrishnan, R., Moore, S.A., Sluka, K.A., 2003. Unilateral carrageenan injection into muscle or joint induces chronic bilateral hyperalgesia in rats. *Pain.* 104, 567–77.

Raynaud, J.S., Duteil, S., Vaughan, J.T., Hennel, F., Wary, C., Leroy-Willig, A., Carlier, P.G., 2001. Determination of skeletal muscle perfusion using arterial spin labeling NMRI: Validation by comparison with venous occlusion plethysmography. *Magn Reson Med.* 46, 305–11.

Reid, R.W., Foley, J.M., Jayaraman, R.C., Prior, B.M., Meyer, R.A., 2001. Effect of aerobic capacity on the T2 increase in exercised skeletal muscle. *J Appl Physiol.* 90, 897–902.

Ringleb, S.I., Bensamoun, S.F., Chen, Q., Manduca, A., An, K.-N., Ehman, R.L., 2007. Applications of magnetic resonance elastography to healthy and pathologic skeletal muscle. *J Magn Reson Imaging.* 25, 301–9.

Rorschach, H.E., Chang, D.C., Hazlewood, C.F., Nichols, B.L., 1973. The diffusion of water in striated muscle. *Ann N Y Acad Sci.* 204, 445–52.

Saab, G., Thompson, R.T., Marsh, G.D., 1999. Multicomponent T2 relaxation of in vivo skeletal muscle. *Magn Reson Med.* 42, 150–7.

Sanchez, O.A., Copenhaver, E.A., Chance, M.A., Fowler, M.J., Towse, T.F., Kent-Braun, J.A., Damon, B.M., 2011. Postmaximal contraction blood volume responses are blunted in obese and type 2 diabetic subjects in a muscle-specific manner. *Am J Physiol Heart Circ Physiol.* 301, H418–H427.

Sanchez, O.A., Copenhaver, E.A., Elder, C.P., Damon, B.M., 2010. Absence of a significant extravascular contribution to the skeletal muscle BOLD effect at 3T. *Magn Reson Med.* 64, 527–35.

Sanchez, O.A., Louie, E.A., Copenhaver, E.A., Damon, B.M., 2009. Repeatability of a dual gradient-recalled echo MRI method for monitoring post-isometric contraction blood volume and oxygenation changes. *NMR Biomed.* 22, 753–61.

Sinha, S., Sinha, U., Edgerton, V.R., 2006. In vivo diffusion tensor imaging of the human calf muscle. *J Magn Reson Imaging.* 24, 182–90.

Sjogaard, G., Adams, R.P., Saltin, B., 1985. Water and ion shifts in skeletal muscle of humans with intense dynamic knee extension. *Am J Physiol Regul Integ Comp Physiol.* 248, R190–R196.

Slade, J.M., Towse, T.F., Gossain, V., Meyer, R.A., 2011. Peripheral microvascular response to muscle contraction is unaltered by early diabetes, but decreases with age. *J Appl Physiol.* 111, 1361–71.

Spees, W.M., Yablonskiy, D.A., Oswood, M.C., Ackerman, J.J., 2001. Water proton MR properties of human blood at 1.5 Tesla: Magnetic susceptibility, T1, T2, T2*, and non-Lorentzian signal behavior. *Magn Reson Med.* 45, 533–42.

Stables, L.A., Kennan, R.P., Gore, J.C., 1998. Asymmetric spin-echo imaging of magnetically inhomogeneous systems: Theory, experiment, and numerical studies. *Magn Reson Med.* 40, 432–42.

Stefanovic, B., Pike, G.B., 2004. Human whole-blood relaxometry at 1.5 T: Assessment of diffusion and exchange models. *Magn Reson Med.* 52, 716–23.

Stejskal, E., Tanner, J., 1965. Spin diffusion measurements: Spin echoes in the presence of a time-dependent field gradient. *J Chem Phys.* 42, 288–92.

Thompson, R.B., Aviles, R.J., Faranesh, A.Z., Raman, V.K., Wright, V., Balaban, R.S., McVeigh, E.R., Lederman, R.J., 2005. Measurement of skeletal muscle perfusion during postischemic reactive hyperemia using contrast-enhanced MRI with a step-input function. *Magn Reson Med.* 54, 289–98.

Thulborn, K.R., Waterton, J.C., Matthews, P.M., Radda, G.K., 1982. Oxygenation dependence of the transverse relaxation time of water protons in whole blood at high field. *Biochim Biophys Acta.* 714, 265–70.

Toussaint, J.F., Kwong, K.K., Mkparu, F., Weisskoff, R.M., LaRaia, P.J., Kantor, H.L., 1996a. Interrelationship of oxidative metabolism and local perfusion demonstrated by NMR in human skeletal muscle. *J Appl Physiol*. 81, 2221–8.

Toussaint, J.F., Kwong, K.K., Mkparu, F.O., Weisskoff, R.M., LaRaia, P.J., Kantor, H.L., 1996b. Perfusion changes in human skeletal muscle during reactive hyperemia measured by echo-planar imaging. *Magn Reson Med*. 35, 62–9.

Towse, T.F., Slade, J.M., Ambrose, J.A., Delano, M.C., Meyer, R.A., 2011. Quantitative analysis of the post-contractile blood-oxygenation-level-dependent (BOLD) effect in skeletal muscle. *J Appl Physiol*. 111, 27–39.

Uffmann, K., Maderwald, S., Ajaj, W., Galban, C.G., Mateiescu, S., Quick, H.H., Ladd, M.E., 2004. In vivo elasticity measurements of extremity skeletal muscle with MR elastography. *NMR Biomed*. 17, 181–90.

Utz, W., Jordan, J., Niendorf, T., Stoffels, M., Luft, F.C., Dietz, R., Friedrich, M.G., 2005. Blood oxygen level-dependent MRI of tissue oxygenation: Relation to endothelium-dependent and endothelium-independent blood flow changes. *Arterioscler Thromb Vasc Biol*. 25, 1408–13.

Van Donkelaar, C.C., Kretzers, L.J., Bovendeerd, P.H., Lataster, L.M., Nicolay, K., Janssen, J.D., Drost, M.R., 1999. Diffusion tensor imaging in biomechanical studies of skeletal muscle function. *J Anat*. 194 (Pt 1), 79–88.

Van Leeuwen, J.L., Spoor, C.W., 1992. Modelling mechanically stable muscle architectures. *Philos Trans R Soc Lond B Biol Sci*. 336, 275–92.

Van Leeuwen, J.L., Spoor, C.W., 1993. Modelling the pressure and force equilibrium in unipennate muscles with in-line tendons. *Philos Trans R Soc Lond B Biol Sci*. 342, 321–33.

Walker, U.A., 2008. Imaging tools for the clinical assessment of idiopathic inflammatory myositis. *Curr Opin Rheumatol*. 20, 656–61.

Walter, G., Cordier, L., Bloy, D., Lee Sweeney, H., 2005. Noninvasive monitoring of gene correction in dystrophic muscle. *Magn Reson Med*. 54, 1369–76.

Weidman, E.R., Charles, H.C., Negro-Vilar, R., Sullivan, M.J., MacFall, J.R., 1991. Muscle activity localization with 31P spectroscopy and calculated T2-weighted 1H images. *Invest Radiol*. 26, 309–16.

Williams, S.E., Heemskerk, A.M., Welch, E.B., Li, K., Damon, B.M., Park, J.H., 2013. Quantitative effects of inclusion of fat on muscle diffusion tensor MRI measurements. *J Magn Reson Imaging*. DOI: 10.1002/jmri.24045.

Wishnia, A., Alameddine, H., Tardif de Géry, S., Leroy-Willig, A., 2001. Use of magnetic resonance imaging for noninvasive characterization and follow-up of an experimental injury to normal mouse muscles. *Neuromuscul Dis*. 11, 50–5.

Wu, W.C., Mohler, E., III, Ratcliffe, S.J., Wehrli, F.W., Detre, J.A., Floyd, T.F., 2009. Skeletal muscle microvascular flow in progressive peripheral artery disease: Assessment with continuous arterial spin-labeling perfusion magnetic resonance imaging. *J Am Coll Cardiol*. 53, 2372–7.

Yablonskiy, D.A., Haacke, E.M., 1994. Theory of NMR signal behavior in magnetically inhomogeneous tissues: The static dephasing regime. *Magn Reson Med*. 32, 749–63.

Yu, G., Floyd, T.F., Durduran, T., Zhou, C., Wang, J., Detre, J.A., Yodh, A.G., 2007. Validation of diffuse correlation spectroscopy for muscle blood flow with concurrent arterial spin labeled perfusion MRI. *Opt Express*. 15, 1064–75.

Yucesoy, C.A., Koopman, B.H., Baan, G.C., Grootenboer, H.J., Huijing, P.A., 2003. Effects of inter- and extramuscular myofascial force transmission on adjacent synergistic muscles: Assessment by experiments and finite-element modeling. *J Biomech.* 36, 1797–811.

Yucesoy, C.A., Koopman, B.H., Huijing, P.A., Grootenboer, H.J., 2002. Three-dimensional finite element modeling of skeletal muscle using a two-domain approach: Linked fiber-matrix mesh model. *J Biomech.* 35, 1253–62.

Yue, G., Alexander, A.L., Laidlaw, D.H., Gmitro, A.F., Unger, E.C., Enoka, R.M., 1994. Sensitivity of muscle proton spin-spin relaxation time as an index of muscle activation. *J Appl Physiol.* 77, 84–92.

Zerhouni, E.A., Parish, D.M., Rogers, W.J., Yang, A., Shapiro, E.P., 1988. Human heart: Tagging with MR imaging—a method for noninvasive assessment of myocardial motion. *Radiology.* 169, 59–63.

Zhao, J., Clingman, C., Närväinen, M., Kauppinen, R., van Zijl, P., 2007. Oxygenation and hematocrit dependence of transverse relaxation rates of blood at 3T. *Magn Reson Med.* 58, 592–7.

Zhong, X., Epstein, F.H., Spottiswoode, B.S., Helm, P.A., Blemker, S.S., 2008. Imaging two-dimensional displacements and strains in skeletal muscle during joint motion by cine DENSE MR. *J Biomech.* 41, 532–40.

Zijta, F., Froeling, M., van der Paardt, M., Lakeman, M., Bipat, S., Montauban van Swijndregt, A., Strijkers, G., Nederveen, A., Stoker, J., 2011. Feasibility of diffusion tensor imaging (DTI) with fibre tractography of the normal female pelvic floor. *Eur Radiol.* 21, 1243–9.

8

Quantitative Techniques in Cardiovascular MRI

**John Biglands, Ananth Kidambi, Peter Swoboda,
Manish Motwani, and Sven Plein**

University of Leeds

CONTENTS

8.1 Introduction

Cardiovascular magnetic resonance is becoming more widely available, and today it has a wide range of clinical indications. While the majority of clinical studies are interpreted by visual inspection, most of the imaging data can also be quantitatively analyzed. Quantitative analysis is mandatory in some instances such as the estimation of iron loading using T_2^* methods. It is commonplace for other instances, such as for the calculation of left ventricular (LV) and right ventricular (RV) volumes and function from cine images, whereas in others quantitation remains largely a research tool, as is the case for quantitative estimates of myocardial blood flow (MBF). This chapter gives an overview of some areas where quantitative analysis of cardiovascular MRI data is often used or particularly promising.

8.2 Quantitative Assessment of Myocardial Function by MRI

Accurate measurement of regional and global myocardial function is a critical part of the management of many cardiovascular diseases. MRI is widely used for the quantitative assessment of cardiac function, and several MRI techniques have been shown to measure regional and global myocardial function both accurately and reproducibly.

8.2.1 Measurement of LV Volumes and Ejection Fraction

Accurate assessment of the volumes and function of the cardiac chambers is the cornerstone of all noninvasive imaging techniques. Cardiac dysfunction can be caused by a wide variety of etiologies and is associated with significant morbidity and mortality. Precise evaluation of cardiac structure and function is therefore needed to provide reliable diagnostic and prognostic information. It is important that noninvasive assessment of cardiac function is reproducible in order to allow monitoring of disease progression and response to therapy.

The most commonly used measurement that describes global cardiac function is ejection fraction (EF). This is calculated by dividing the stroke volume (SV) by the end-diastolic volume (EDV). The SV is calculated by subtracting the end-systolic volume (ESV) from the EDV.

$$EF = (EDV - ESV)/EDV$$

EF can be measured using left ventriculography during cardiac catheterization (Dodge et al., 1960), single-photon emission computed tomography (SPECT) (Underwood et al., 1985), echocardiography (Schiller et al., 1979), computed tomography (CT), and MRI. Because of the rapid cardiac contraction, accurate identification of end-systole, and thus an accurate assessment of EF, requires the imaging modality to have a temporal resolution in excess of 25 frames/s (40 ms/frame). This is best achieved with echocardiography and MRI. In MRI, frame rates of 30 frames/s are routinely achieved, and higher frame rates are possible in short breath-holds with the use of parallel imaging techniques [sensitivity encoding (SENSE), simultaneous acquisition of spatial harmonics (SMASH), generalized autocalibrating partially parallel acquisition (GRAPPA), etc.] (Heidemann et al., 2003).

In MRI, calculations of LV volumes and function are usually derived from two-dimensional (2D) gradient-echo cine images covering the heart in a stack of continuous sections. The most common stack orientation used for measurements of LV function is the LV short-axis plane, aligned parallel to the mitral valve plane and perpendicular to the LV long axis. The use of balanced steady-state free precession (bSSFP) pulse sequences is now standard on all cardiac MRI systems and provides data of high signal-to-noise ratio (SNR) and spatial resolution in short breath-hold durations. With recent data acceleration schemes, it has also become possible to acquire cine images in a single three-dimensional (3D) stack. This reduces acquisition time and potential registration errors. However, small differences have been reported in the LV volumes and EF measured from 3D data sets compared with 2D acquisition (Davarpanah et al., 2010; Greil et al., 2008), and 3D methods have not yet found their way into routine clinical use.

From a stack of cine images, LV volumes are typically calculated by MRI using the "summation of discs" method (Utz et al., 1987). Endocardial contours are manually drawn in end-diastole and end-systole, as shown in Figure 8.1. The volume of each slice within the stack can be calculated (as the slice thickness is known), and by adding them, the LV volumes can be derived while considering any slice gaps between the sections. This method is highly reproducible (Strohm et al., 2001) and is now considered the gold standard of assessing LV volumes and EF with normal ranges

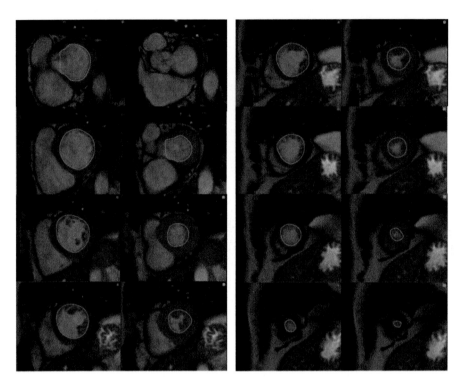

FIGURE 8.1
(See color insert.) SSFP LV stack showing paired images in end-diastole (left columns) and end-systole (right columns) starting from the base of the left ventricle (upper left) to the apex (lower right). Manually drawn endocardial contours, from which EDV, ESV, and EF can be calculated, are shown too.

published for different age groups. The interstudy reproducibility of EF calculated by this method is excellent, and consequently fewer subjects are needed in research studies to detect a significant decrease in EF [87 subjects are needed to detect a 3% change in EF by echocardiography, whereas only 11 subjects are needed to detect the same change by cardiovascular magnetic resonance (CMR)] (Grothues et al., 2002).

SSFP cine sequences have superior contrast between blood pool and myocardium than the formerly used spoiled gradient-echo sequences. Consequently, the values of LV mass and LV volumes tend to be greater in SSFP sequences than in corresponding gradient-echo sequences (Moon et al., 2002), although less effect is seen on the calculated EF.

LV mass can be calculated from the same cine stack used to calculate LV volumes. Both endocardial and epicardial contours are drawn on every slice and volumes are calculated. By subtracting the endocardial volume from the epicardial volume, the LV mass can be calculated. This technique has been demonstrated to be accurate and reproducible (Bottini et al., 1995; Maddahi et al., 1987).

8.2.2 Measurement of RV Volumes and EF

Similar techniques described to calculate LV mass and EF can be used to calculate RV mass and function (Boxt et al., 1992; Katz et al., 1993). This can provide useful diagnostic and prognostic information on conditions including pulmonary hypertension, congenital heart disease, and arrhythmogenic RV cardiomyopathy (ARVC). A contiguous stack of cine images encompassing the entire right ventricle is acquired, and then endocardial and epicardial contours drawn on slices at end-systole and end-diastole, as illustrated in Figure 8.2. Using a summation of discs method (as described for the LV volumes), the RVEDV, ESV, EF, and mass can be calculated both accurately and reproducibly.

The RV stack is often acquired in the transaxial plane as this has been demonstrated to have better interobserver and intraobserver reproducibility than when the stack is acquired in a short-axis orientation (Alfakih et al., 2003). In practice, however, the interstudy reproducibility of RV volumes when calculated from short-axis slices is still good and allows for both LV and RV volumes to be measured from the same series (Grothues et al., 2004).

8.2.3 Quantitative Wall Thickening

From a stack of cine MRI data in the LV short axis, all sectors of the 17-segment American Heart Association (AHA) cardiac model can be evaluated. This allows for accurate qualitative assessment of regional myocardial contractility.

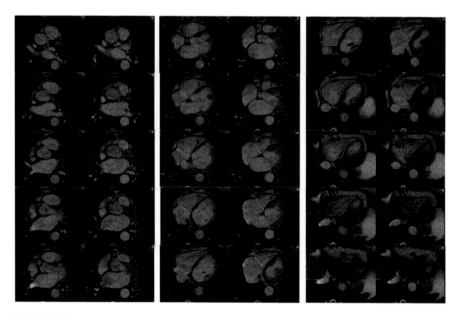

FIGURE 8.2
(See color insert.) SSFP RV stack showing paired images in end-diastole (left columns) and end-systole (right columns). Manually drawn endocardial contours are also shown.

In addition, segmental analysis of LV function during intravenous dobutamine stress is used to detect inducible wall motion abnormalities indicative of underlying coronary heart disease (Pennell et al., 1992; van Rugge et al., 1993). Unlike dobutamine stress echo, MRI offers uniformly high image quality and consequently better sensitivity and specificity for detection of significant coronary stenoses (Nagel et al., 1999).

Cine MRI data also permit quantitative assessment of wall motion in the centerline method, where many radial lines are inserted between endocardial and epicardial contours. The shortening of these lines can then be measured throughout the cardiac cycle, and quantitative measures of regional thickening are calculated (Holman et al., 1997; van Rugge et al., 1994). However, these methods are used infrequently in clinical practice.

8.2.4 Myocardial Tissue Tagging

Myocardial tissue tagging by MRI was first introduced in 1988, when Zerhouni et al. (1988) generated tissue tags with the use of selective radiofrequency saturation prior to the acquisition of cine MR images. The saturated tag lines generate a hypointense signal, and their deformation can be tracked through the subsequent phases of the cine study (Shehata et al., 2009). In 1989, the development of spatial modulation of magnetization (SPAMM) by Axel and Dougherty (1989) allowed the application of tags in two perpendicular planes, thereby creating a grid of tags. As the saturation pulses in myocardial tissue tagging are typically applied on the QRS complex of the electrocardiogram (ECG), tag lines fade from systole to diastole because of T_1 tissue relaxation (Gotte et al., 2006). Complementary SPAMM (CSPAMM), developed by Fischer et al. (1993), improves tag persistence and contrast between tagged and nontagged myocardium by subtracting a positive and a negative tagging grid from the relaxed image.

Several techniques can be used to analyze the images generated by tagging. The most commonly used is harmonic phase analysis (HARP) (Axel et al., 2005; Kuijer et al., 2001), which is a highly automated process that tracks the displacement of tagging lines over time. From these data, information on the deformation of individual segments or the whole myocardium can be generated. It has been demonstrated that HARP generates improved strain maps when CSPAMM is used (Rademakers et al., 1994).

The tagged images can then be tracked throughout the cardiac cycle, and subtle measures of cardiac motion can be measured including radial, longitudinal, and circumferential strain for either the whole heart or a segment of interest. The heart has a complex twisting or torsion motion in systole, where the base rotates clockwise in early systole followed by apical counterclockwise rotation (Burns et al., 2008) as illustrated in Figure 8.3. It is possible to measure and quantify this movement by tagged CMR. Normal myocardial circumferential strain and LV twist patterns derived from CSPAMM from a healthy volunteer are shown in Figure 8.4.

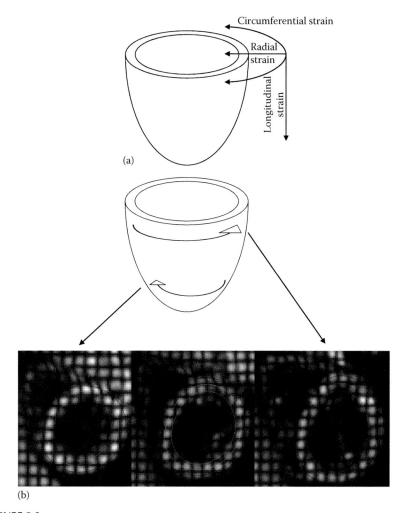

FIGURE 8.3
(a) Schematic representation of circumferential, radial, and longitudinal strain demonstrating their directions of motion. (b) Schematic representation of basal clockwise and apical counter-clockwise rotation during systole. The images below (b) show apical, mid-LV, and basal slices (from left to right) tagged with CSPAMM with endocardial, mid-myocardial, and epicardial contours drawn.

With these technical developments, myocardial tissue tagging has become a reproducible, mature technique for measurements of intramyocardial motion and deformation (Castillo et al., 2005). Tissue tagging has been used in several areas of cardiovascular research including the improved detection of myocardial ischemia during dobutamine stress testing (Kuijpers et al., 2003), quantification of the local (Kramer et al., 1993) and global (Nagel et al., 2000) effects on cardiac function after myocardial infarction, and understanding

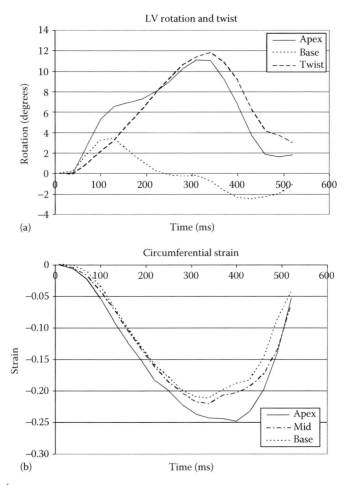

FIGURE 8.4
(a) LV circumferential strain measured by CSPAMM from a healthy volunteer. Strain increases from base to apex. (b) LV rotation measured by CSPAMM from a healthy volunteer. LV twist is calculated by subtracting basal rotation from apical rotation for each time point.

the complex mechanics of hypertrophic (Dong et al., 1994) and dilated (Ennis et al., 2003) cardiomyopathies.

Measurements derived from conventional tagging methods are reliant upon the positioning of the imaging plane. It has recently been proposed that it is possible to overcome this problem by tagging the entire heart in three dimensions (Rutz et al., 2008). An acquisition is performed after line tag preparation in each orthogonal plane. These three acquisitions can then be combined to give a 3D tagged representation of the heart. From this, complex parameters of regional and global deformation can be calculated.

8.2.5 Feature Tracking

Feature tracking is a novel way by which subtle signal variations within myocardial tissue are tracked through the cardiac cycle. This process can be carried out by postprocessing of standard cine images (Maret et al., 2009) rather than requiring acquisition of tagged images (such as those seen in Figure 8.3b). The main advantage of feature tracking over conventional tagging is that as specialized acquisitions (such as SPAMM and CSPAMM) are not required, the scanning time can be significantly reduced. The measures of circumferential strain derived from feature tracking have been shown to be comparable to those derived from HARP tagging (Hor et al., 2010). It has also been shown to provide quantitative information on wall motion abnormalities during dobutamine stress CMR (Schuster et al., 2011).

8.2.6 Strain-Encoded CMR

In conventional tagging methods, the tag surfaces are orthogonal to the imaging plane. Contrast strain-encoded CMR (SENC) utilizes tag surfaces that are parallel to the image plane. This allows for color-coded strain data to be displayed as a cine sequence comparable to SSFP cine sequences (Garot et al., 2004). It has been demonstrated that using SENC sequences, compared to visual analysis of cine sequences, improves the sensitivity of detection of significant coronary stenosis by dobutamine stress CMR (Korosoglou et al., 2009).

8.3 Quantifying Blood Flow: Phase-Contrast MRI

8.3.1 Introduction

The majority of MRI sequences are designed to produce magnitude (modulus) signals, that is, brighter pixels denote more signals. Phase-contrast MRI exploits the vector properties of proton magnetization to generate outputs relating to speed and direction of proton movement. The information gained is analogous to Doppler imaging by ultrasound with the advantages of low interobserver variability (Kondo et al., 1991; Van Rossum et al., 1991) and the ability to measure velocity in any direction. Applications of this method are expanding and include assessment of valve function, vascular stenoses, and cardiac shunting. This section explores the generation of phase-contrast velocity mapping images and their quantitative analysis in common cardiovascular applications.

8.3.2 Principles of Phase-Contrast MRI

Proton spins, which are moving along a magnetic field gradient, acquire a shift in the phase of the transverse spin magnetization in comparison to

G_z

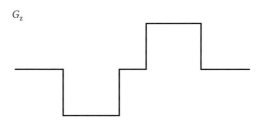

FIGURE 8.5
Simple balanced bipolar flow-sensitizing gradient.

stationary spins. In a linear field gradient, the amount of phase shift (φ) is proportional to the velocity of the moving spin along the gradient. To compensate for dephasing of stationary protons, a gradient of opposite polarity is applied (Figure 8.5). Bipolar gradients are balanced to have a net zero effect on stationary protons but to result in phase shifts for spins moving along the gradient. By acquiring data after both the positive and negative lobes of this bipolar gradient and subtracting these two data sets, the average velocity (*phase difference*) for each voxel can be determined (O'Donnell, 1985).

Because phase is cyclic, the range of phase shifts that can be uniquely identified is limited to 360°, which for most sequences is set as a range of ±180°. If a proton moves faster than the sequence range, a positive shift over 180° will be interpreted as a correspondingly negative phase shift (Figure 8.6). By convention,

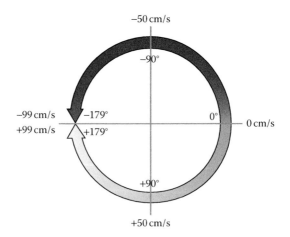

FIGURE 8.6
Phase encoding is conventionally defined with black as −180°, white as +180°, and mid-gray as 0°. In this example, with V_{enc} as 100 cm/s, if a spin was travelling along the gradient at +101 cm/s, this would be resolved as a velocity of −99 cm/s. For faster velocities, the V_{enc} should be increased.

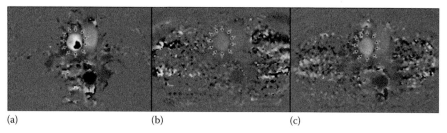

(a)   (b)   (c)

FIGURE 8.7
Optimal velocity encoding. Phase-contrast imaging perpendicular to the ascending aorta (arrowed) in mid-systole. (a) V_{enc} of 150 cm/s produces an aliased image, appearing as an area of dark signal immediately adjacent to the bright signal of aortic blood flow (phase wraparound). (b) V_{enc} of 500 cm/s prevents aliasing but results in poor differentiation of the velocity range, increasing susceptibility to noise. Also note the effect of noise in areas of signal with low magnitude (e.g., lungs). (c) V_{enc} of 250 cm/s is optimal in this case to minimize these effects.

a voxel with zero net phase shift is colored mid-gray, positive phase shifts with increased brightness, and negative phase shifts with increased darkness. If spins are moving faster than the sequence allows, they may be incorrectly resolved as a velocity of the opposite direction, referred to as aliasing (Figure 8.7). To avoid aliasing, an expected threshold velocity (V_{enc}) needs to be estimated prior to the sequence. The amplitudes of the flow-sensitizing gradients need to be set so that the peak velocity corresponds to a phase shift just below 180°. Aliased images can be corrected by software-based "phase unwrapping" of the area of peak velocity (Yang et al., 1996). V_{enc} is inversely proportional to the area of the flow-encoding gradients. For a given imaging time, stronger gradient amplitudes are required to encode smaller velocities. Setting the V_{enc} too high will reduce accuracy, as velocities are compressed to a narrow gray scale, increasing vulnerability to error introduced by noise (Andersen and Kirsch, 1996).

This mechanism can be easily adapted to register flow in any given direction. Note that the flow-encoding direction does not have to be the same as the imaging plane. Both in-plane and through-plane images are commonly used.

8.3.3 Sequence Design

Only the velocity component along the direction of the bipolar gradient contributes to the phase of the MR signal. Therefore, only a single velocity direction can be encoded with an individual measurement, and at least four independent measurements with different arrangements of bipolar gradients need to be performed to acquire data in three orthogonal dimensions. In addition, the bipolar sequence described above is sensitive to a number of potential factors that may cause a relative change in phase. Of these, there are two main effects: (1) phase changes due to motion along a gradient other than the flow-sensitizing gradient and (2) phase changes due to local magnetic field inhomogeneities. These effects can be minimized by performing two similar

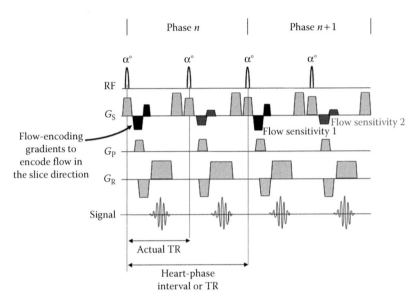

FIGURE 8.8

Sample phase-contrast pulse sequence. Note the bipolar flow-sensitizing gradients. Two flow sensitivity profiles for different threshold velocities are shown. (Courtesy of Dr. John Ridgway, Consultant Clinical Scientist, Leeds Teaching Hospitals NHS Trust, Leeds, UK.)

acquisitions for each phase-encoding step but with differing flow sensitivities in the flow-encoding direction. Phase maps from the two acquisitions can be subtracted to produce a velocity map. This has the effect of filtering out phase changes due to effects other than the flow-sensitizing gradient, leaving a map sensitive to velocity components in the selected direction.

A constant velocity approximation is insufficient for many clinical applications, where acceleration and pulsatile flow are commonplace. Errors introduced by acceleration can be minimized by increasing the temporal resolution of acquisition, (i.e., to reduce TE). Spoiled gradient-echo pulse sequences are commonly used to achieve this (Figure 8.8). To allow for pulsatile or varying flow, data are acquired and gated to the cardiac cycle, producing a cine image sequence. Cine imaging also permits analysis of tissue motion, which can be compensated for during analysis of the velocity maps. Either prospective or retrospective gating, triggered to the R-wave, may be used. Clinical acquisition of data in three spatial dimensions is currently limited by long acquisition times but is a promising development to enable visualization of multidirectional flow (Markl et al., 2011).

8.3.4 Quantitative Measurement of Flow and Velocity

Velocity maps usually require further analysis to provide clinically useful data. A region of interest (ROI) is drawn around the cardiac or vascular structure of interest on the cine sequence; this region is either manually or automatically corrected to account for movement or deformation of the structure during the cardiac cycle. Accurate registration of the ROI to the studied structure is essential to avoid error in flow or volume profiles. Surrounding low signal areas with random pixel speckling may create substantial errors. To aid accurate delineation of the structure, many clinical sequences for phase encoding will also generate a corresponding magnitude image. Software analysis of the ROI can generate a velocity-time profile or flow-time profile (Figure 8.9). Peak velocity can be generated by selecting the pixel within the ROI corresponding to the highest velocity. The limiting temporal and spatial resolution of clinical imaging invariably leads to some averaging of velocities within each

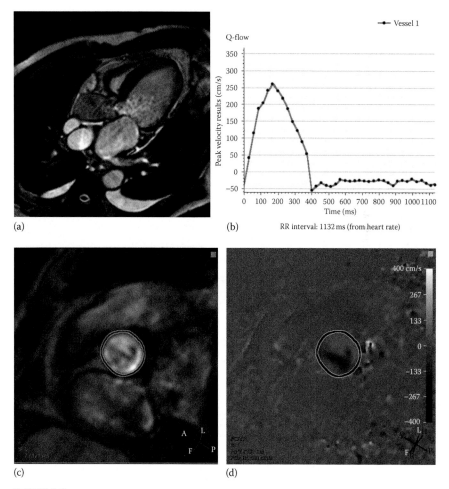

(a)

(b)

(c)

(d)

FIGURE 8.9
Velocity mapping of a stenotic jet. (a) SSFP cine-imaging of severe aortic stenosis. (b) A velocity-time profile is generated by drawing an ROI on (c) the magnitude image of a perpendicular slice (through plane) to the proximal ascending aorta. The ROI is matched to the phase-contrast image (d), from which the pixel corresponding to the highest velocity is charted for each phase.

voxel, and jets of high peak velocity may be underestimated. Mean velocity within the ROI, taken from the mean signal intensity (SI) of all the pixels for each phase, may also be used to quantify valvular stenosis. The instantaneous flow volume for each phase is the product of the area of the ROI and the sum of the velocities within it. When plotted against time, integrating the area under the curve will estimate flow for the cardiac cycle.

8.3.5 Clinical Applications

Measurement of flow and velocity provides useful hemodynamic data in the setting of valvular dysfunction in the heart. For valvular stenosis (Figure 8.10), the peak pressure drop across a stenotic valve can be derived from the peak velocity by using the modified Bernoulli equation (Heys et al., 2010):

$$\text{Peak pressure drop} = 4 \times v_{max}^2,$$

where v_{max} is measured in m/s and peak pressure drop in mm Hg.

Note that this equation assumes that prestenotic velocity is low in comparison to poststenotic velocity and approaches accurate estimation for higher grades of stenosis. Detection of the peak velocity is highly dependent on both temporal and spatial resolution to minimize the averaging of differing velocities within and surrounding the jet.

Valvular regurgitation can be accurately and reproducibly quantified using flow measurements (Honda et al., 1993; Hundley et al., 1995). Quantifying flow in the proximal aorta and main pulmonary artery using phase contrast gives an independent estimate of LV SV and RV SV, respectively. Quantification of regurgitant volume and fraction (a percentage of regurgitant flow divided by forward flow) directly evaluates the severity of valvular regurgitation. The ratio of pulmonary forward flow to aortic forward flow ($Q_p{:}Q_s$; normally around 1:1) is a useful clinical indicator of the severity of pulmonary-to-systemic shunts. For example, the severity of an atrial septal defect with left-to-right shunting could be estimated by measuring the flow through the proximal aorta and main pulmonary artery. For example, if aortic forward flow was 100 mL/s and pulmonary forward flow was 210 mL/s, this would give a $Q_p{:}Q_s$ of 2.1:1, indicating a significant shunting across the atria. These estimations are less accurate in the context of additional valvular regurgitation or cardiac shunting. For example, estimation of the LVSV may be underestimated from the proximal aortic flow in the context of a significant mitral regurgitation or a ventricular septal defect. Other clinical uses include the estimation of the degree of collateral flow around a structure such as aortic coarctation by measuring the flow both proximally and distally. Phase contrast may also be used to provide image contrast in closely apposed structures with differing flow patterns and timings, such as the true and false lumina in aortic dissection (Strotzer et al., 2000).

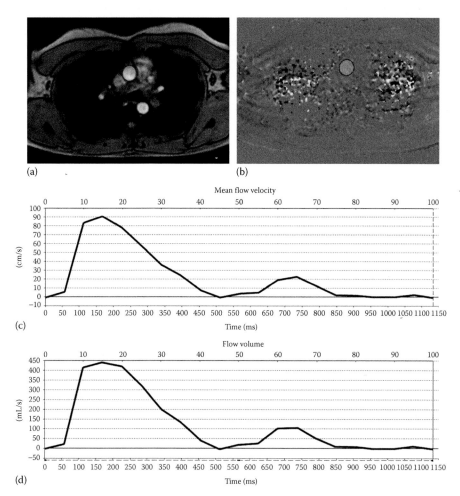

FIGURE 8.10
Generation of velocity and flow profiles from through-plane imaging of the ascending aorta. (a) The magnitude image is helpful to delineate anatomical structure and is used to create an ROI that is transposed to the phase image (b) Velocity-time (c) and flow-time (d) profiles can be calculated from this ROI.

8.3.6 Sources of Error

MRI is advantageous over Doppler echocardiography in its ability to measure velocities in any orientation but is disadvantaged by potentially lower temporal and spatial resolution to pick out the peak of the velocity profile. Partial volume averaging has variable effects on velocity measurements, notably for smaller structures, and is dependent on not just the adjoining velocities but also T_1 and T_2 characteristics of the structures involved (Firmin et al., 1990). These effects can be reduced by optimizing echo and repetition times of the sequence, as well as improving spatial resolution. Intravoxel dephasing and turbulent flow may further increase error in measurement (O'Brien et al., 2008), although the effects of these can be minimized with optimal gradient profile design (Firmin et al., 1990). Use of non-Cartesian *k*-space acquisition, such as interleaved spiral acquisition (Meyer et al., 1992) or parallel imaging techniques (Thunberg et al., 2003), also helps optimize the balance between acquisition time and resolution. Both in-plane resolution and slice thickness are also important to minimize partial volume effects.

8.4 Quantitative Analysis of Myocardial Perfusion MRI

8.4.1 Introduction

Myocardial perfusion MRI is increasingly used for the noninvasive detection of ischemic heart disease (Al-Saadi et al., 2000; Costa et al., 2007; Lelieveldt et al., 2004; Nagel et al., 2003; Plein et al., 2002). The quantitative analysis of MBF values from myocardial perfusion MRI data improves objectivity over qualitative analysis. However, the acquisition of myocardial perfusion MRI data that are adequate for quantitative analysis imposes additional imaging requirements on this already challenging imaging area. This section describes the key requirements and trade-offs for imaging myocardial perfusion before discussing the additional requirements for quantitative analysis, including the additional constraints in the contrast agent (CA) injection regime. Finally, some common methods for generating semiquantitative and quantitative MBF estimates are described.

8.4.2 Dynamic Contrast-Enhanced MRI

In order to assess myocardial perfusion, the blood passing into the myocardium needs to alter image SI so that it can be visualized. This is typically achieved using a signal-enhancing CA, which is injected intravenously, while multiple images of the heart in the same anatomical position and the same point in the cardiac cycle are acquired in successive heartbeats. The acquisition of a dynamic series of MR images during the passage of CA through the body is known as dynamic contrast-enhanced MRI (DCE-MRI) (Alfakih et al., 2003; Tofts, 1997). Ideally, the resulting image series shows a motion-free cross-section of cardiac tissue whose SI rises and falls over time as the bolus of CA passes through the myocardial tissue (see Figure 8.11).

Myocardial ischemia is characterized by reduced MBF, usually due to coronary artery disease (CAD). Unless coronary stenosis is severe, MBF is maintained at rest due to autoregulation. In the clinical application of DCE-MRI, it is therefore necessary to study MBF reserve under stress conditions

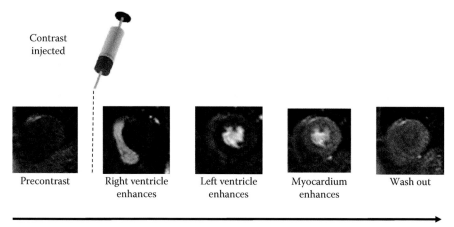

FIGURE 8.11
CA is injected while the dynamic scan is in progress. CA can be seen as signal enhancement in the right ventricle followed by the left ventricle and more gradually in the myocardium, before finally washing out.

in order to detect myocardial perfusion defects resulting from myocardial ischemia. Typically, this is induced pharmacologically using a vasodilator such as adenosine or dipyridamole.

8.4.3 Requirements for DCE-MRI Perfusion Imaging

The essential requirements of a DCE-MRI cardiac perfusion imaging sequence are as follows:

- Each image must be acquired with a single shot in one heartbeat due to the temporally variant nature of the image sequence.
- Images must be T_1-weighted to maximize the effect of the CA on image SI.
- Cardiac and respiratory motion must be dealt with.

8.4.3.1 Acquisition

In DCE-MRI, the image appearance changes significantly between contiguous frames due to the passage of the CA through the heart, so segmented imaging strategies that build up the image over multiple cardiac cycles are not possible. Therefore, in order to acquire images quickly, DCE-MRI perfusion imaging is generally performed as a single-shot technique with a fast read-out sequence such as fast-spoiled gradient echo (FGE), SSFP, or echo planar imaging (EPI). There is no consensus as to which of these is the optimal read-out sequence for myocardial perfusion MRI, but it is generally accepted that all provide diagnostic results.

8.4.3.2 T₁-Weighting

At the lower concentrations of CA used in DCE-MRI, CAs predominantly enhance SI by reducing T_1 relaxation times (TRs). Therefore, myocardial perfusion MRI sequence images should be T_1-weighted in order to maximize the effect of the CA on SI. To reduce acquisition time, the perfusion sequences described above employ small flip angles and very short TRs resulting in poor T_1-contrast. For this reason, a preparation pulse with a sufficiently long saturation time (TS) to establish a high T_1-contrast is applied prior to the read-out pulse sequence. For instance, an FGE read-out sequence utilizes a rapid succession of small flip angle RF pulses to allow longitudinal magnetization to recover quickly enabling short TR but also limiting the changes in contrast between the tissue types (see Figure 8.12a). If a 90° preparation pulse is applied with a single long delay before the FGE read-out, then the image contrast is increased (see Figure 8.12b) before the read-out sequence is applied. The time between the preparation pulses and the central, contrast defining line of k-space is known as the saturation time (TS).

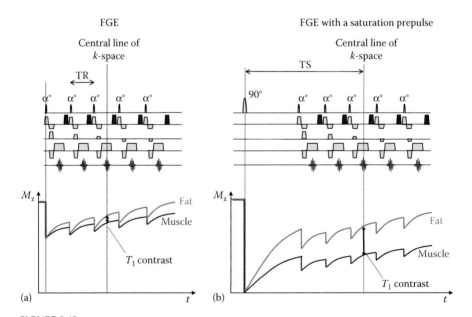

FIGURE 8.12
(Color version available from crcpress.com; see Preface.) (a) FGE uses small flip angles, which generate small differences in M_z magnetization during recovery, corresponding to poorer image contrast. (b) by applying a 90° preparation pulse and a significant TS before applying the FGE read-out a higher level of image contrast is established.

8.4.3.3 Motion, Cardiac Phase, Trigger Delay

Typically, breathing motion is dealt with in perfusion MRI by patient breath-holding and cardiac motion by triggering acquisition to the patient's ECG. The cardiac phase of the image is set by the *trigger delay* (TD), which is the time from the ECG R-wave to the time of the acquisition of the central line of *k*-space (k_o). This is the dominant contributor to the image contrast (see Figure 8.13). Typically, several slices are acquired at different TDs distributed through the cardiac cycle.

8.4.4 Optimizing the Sequence (Choices and Trade-Offs)

The ideal myocardial perfusion MRI pulse sequence would maximize T_1-weighting, spatial resolution, temporal resolution, and coverage of the heart. As an increase to any of these requirements increases the acquisition time, trade-offs must be made between these factors to obtain the desired image properties. Figure 8.14 compares some illustrative imaging choices. If only one slice through the heart were required, then a single image with a constant TS and TD per RR interval could be acquired with the advantages of image contrast and cardiac phase maintained throughout the imaging series. Acquiring two slices per RR interval requires two TD values so that each image slice corresponds to a different cardiac phase that may not be ideal (Radjenovic et al., 2010). In order to image three slices, it is possible to use a shared prepulse for all slices. As the TS is the longest time delay in the acquisition, this can yield a significant increase in coverage; however, each slice will now have a different TS and thus different image contrast. For this

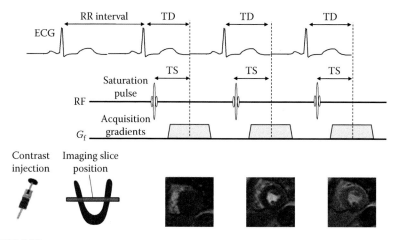

FIGURE 8.13
(Color version available from crcpress.com; see Preface.) Within each RR interval, the TD sets the point within the cardiac cycle that the image is acquired, and the TS sets the time between the saturation pulse and the acquisition, thereby controlling the T_1-weighting of the image.

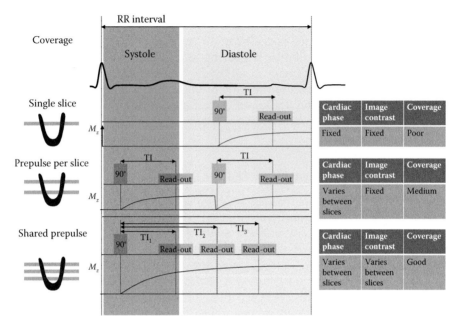

FIGURE 8.14
(Color version available from crcpress.com; see Preface.) With a single acquisition per RR interval, there is no variation in cardiac phase and image contrast but poor coverage. Two prepulse acquisitions will give a fixed image contrast, but the two images will be in different cardiac phases. Using a shared prepulse will give different image contrast for each image but can increase coverage.

reason, the majority of myocardial perfusion pulse sequences use individual preparation pulses for each acquired slice.

8.4.5 Quantitative Myocardial Perfusion Analysis

In order to obtain a quantitative perfusion estimate from DCE-MRI, perfusion data sets, SI versus time curves must be generated from ROIs representing the myocardium and the arterial input function (AIF) of the myocardium. The AIF represents how the CA entering the myocardium varies over time and thus would ideally be taken from the coronary arteries themselves. In practice, an ROI is placed in the LV blood pool because the coronary arteries are too small to obtain an uncorrupted signal and an extra image per RR interval is impractical (Figure 8.15).

8.4.6 Further Requirements for Quantitation

8.4.6.1 Temporal Resolution

For visual analysis, it is possible to image more slices of the heart over two RR intervals, thereby increasing coverage. For quantitative perfusion, the heart must be imaged every RR interval (Jerosch-Herold, 2010; Kroll et al., 1996).

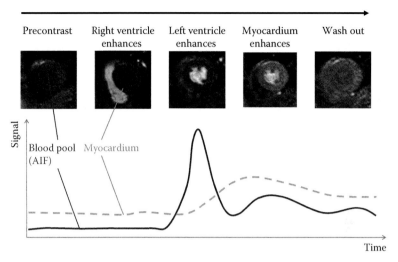

FIGURE 8.15
(Color version available from crcpress.com; see Preface.) For every image in the dynamic sequence, contours describing the myocardium and a region in the blood pool are drawn. The signal from each region is plotted over time to generate myocardial and blood pool uptake curves, which can be analyzed to give an estimate of MBF.

8.4.6.2 Concentration/Signal Linearity

For visual assessment of cardiac perfusion data sets, the aim is to provide the highest contrast between the healthy myocardium (bright) and the hypoperfused region (see Figure 8.16). Quantitative analysis on the other hand seeks to interpret changes in SI as a measurable change in CA concentration and hence blood flow. In other words, it requires that the relationship between the observed SI and the CA concentration be linear. Figure 8.17 illustrates how non-linearities in the SI versus concentration curve cause blunting of the AIF peak,

FIGURE 8.16
Under stress conditions, an area of reduced SI is observed in the septal and anterior walls of the left ventricle (the white arrows mark the extremities of this region). This is consistent with disease in the coronary arteries supplying these regions. (Image courtesy of John Greenwood.)

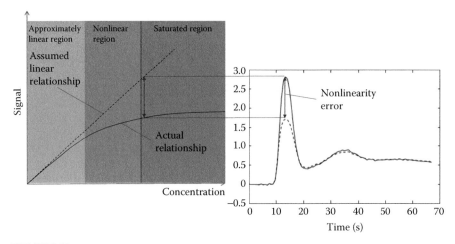

FIGURE 8.17
SI vs. CA concentration is shown on the left-hand side. Three regions are highlighted: the approximately linear region, where the linearity assumption holds; the nonlinear region, where CA concentration may still be calculated from a given SI value; and the saturated region, where CA concentration cannot be calculated. The right-hand graph shows how this propagates into a peak height error in the measured blood pool curve that causes an overestimate in MBF.

yielding underestimates in MBF. At low CA concentrations, the relationship with SI is approximately linear, and the data in this region should not be significantly affected by nonlinearity effects. At higher doses, the relationship becomes nonlinear, but if the relationship between SI and CA concentration is known, then it should be possible to calculate the concentration from a given SI value. At even higher concentrations, the relationship plateaus, and it is no longer possible to calculate the CA concentration from a given SI reliably.

The relationship between CA concentration and SI is dependent on the particular imaging sequence used. In general, it is more linear at lower CA doses and with reduced T_1-weighting of the scan sequence. The degree of nonlinearity depends on the following:

- Dose of administered CA
- MR pulse sequence type (EPI, FGE, and SSFP)
- TS.

Acquisition protocols for quantitative perfusion imaging attempt to optimize these factors to ameliorate the effect of this nonlinearity on the MBF estimate.

8.4.7 Avoiding Nonlinearity Errors

A number of strategies for avoiding errors due to the nonlinearity between SI and CA concentrations have been presented, each associated with their own set of further problems.

8.4.7.1 Low-Dose Strategies

The simplest method is to use a low dose of CA so that the SI and concentration relationship is in the approximately linear region (see Figure 8.17). CA doses need to be around 0.01 mmol/kg to ensure linearity in the blood pool (Utz et al., 2007). These low doses reduce the CNR and SNR of the images, rendering visual analysis (still the mainstay of clinical reporting) difficult. The myocardial curve enhances much less dramatically than the AIF (see Figure 8.15), and the low doses can reduce the SNR in the myocardium to an extent that MBF estimates are highly susceptible to noise, compromising the precision of the MBF estimate.

8.4.7.2 Dual-Bolus Strategies

To tackle these issues, dual-bolus strategies have been proposed that employ two contrast injections. First, a low-dose bolus is injected from which the AIF will be acquired, but the poor SNR myocardial data will be discarded. This is followed by a higher dose bolus from which only the myocardial curves will be used. The method is practically challenging as it requires the patient to undergo a total of four contrast injections (if they are to be imaged under rest and stress conditions) but has shown good agreement with positron emission tomography MBF (PET MBF) values (Ritter et al., 2006). However, the increased accuracy of the dual-bolus technique has not been shown to increase its diagnostic value over single bolus. This might be explained by the introduction of extra noise from the separate prebolus analysis (Groothuis et al., 2010).

8.4.7.3 Dual-Sequence Strategies

An alternative approach is to reduce nonlinearities by altering sequence parameters. The dual-sequence strategy uses a sequence with very short TS (~10 ms) and low image resolution followed by a more typical TS (~100 ms) acquisition (Gatehouse et al., 2004). The short TS images will exhibit a more linear SI and concentration relationship and are used to generate the AIF curve, whereas the longer TS images are used for the myocardial enhancement curve. Performing two acquisitions for each heart position requires a corresponding reduction in coverage, and as with the dual-bolus technique, the AIF images will suffer from lower CNR than the myocardial images.

8.4.7.4 Nonlinearity Correction

Nonlinearity correction attempts to convert the SI versus time curve to a concentration versus time curve *post hoc* (Cernicanu and Axel, 2006; Hsu et al., 2008). In this approach, the MR imaging sequence equation is used to convert each SI value into a T_1 value (using prior knowledge of the imaging parameters) (Fritz-Hansen et al., 1996, 2008; Larsson et al., 1996). If the native tissue T_1 has been measured prior to the perfusion scan, then the contrast-enhanced T_1 values can be converted to concentrations. If the SI value lies in the approximately linear or nonlinear region (Figure 8.17), then it should be possible to convert it to CA concentration. However, if the curve has passed into the plateau region, then the solution to the pulse sequence equation will become error prone, and the correction becomes useless. Such conversions are also susceptible to errors in the native T_1 measurement and errors in the pulse sequence parameters.

8.4.8 Quantification of MBF

Quantitative and semiquantitative measurements of MBF have been shown to be useful in the diagnosis of ischemic heart disease (Costa et al., 2007; Nagel et al., 2003; Schwitter et al., 2001), and the methods for quantitative analysis of perfusion data have been reviewed in detail (Jerosch-Herold, 2010; Jerosch-Herold et al., 2004). After a brief overview of semiquantitative methods, the following sections explain the basic concepts of mathematical methods for obtaining a quantitative estimate for MBF.

8.4.9 Semiquantitative Methods

Semiquantitative perfusion metrics measure a defining property of the enhancement curve, which is thought to correlate with blood flow. These parameters include the area under the first-pass curve, the peak SI value, and the time from the onset of contrast in the region to the peak of the curve. By far, the most widely used semiquantitative parameter is the up-slope of the first-pass enhancement curve, which measures signal enhancement and is related to blood flow. However, the shape of the myocardial enhancement curve depends on the shape of the AIF, so up-slope measurements must be normalized to the input function, usually by dividing the AIF up-slope to generate an MBF index (MBFI):

$$\text{MBFI} = \text{Myocardial up-slope} / \text{AIF up-slope.}$$

In order to determine the presence of ischemia and the myocardial perfusion reserve index (MPRI), the stress MBFI can be divided by the rest MBFI:

$$\text{MPRI} = \text{Stress MBFI} / \text{Rest MBFI.}$$

In practice, dividing by the rest measure is also useful for compensating for SI differences in the images due to MR acquisition factors such as coil inhomogeneities (Nagel et al., 2003).

Measurements using up-slope characteristics have been shown to yield high specificity and sensitivity in the diagnosis of CAD in humans

(Nagel et al., 2003; Schwitter et al., 2001). However, up-slope measurements do not give absolute quantitative flow values and have been shown to give a nonlinear response to increasing MBF at hyperemic flow rates (Christian et al., 2004). They cannot be directly compared in magnitude to quantitative measures, such as coronary flow rates or quantitative MBF measurements from other imaging modalities.

8.4.10 Quantitative Methods

Absolute quantitation of myocardial perfusion involves building a mathematical model, describing how CA passes through the myocardium. Any model makes certain assumptions and generalizations in order to simplify the complexity of real-world observations into a comprehensible mathematical equation. Some models used in quantitative myocardial perfusion take into account factors, such as the leakage of CA through capillary walls into the surrounding tissue (Fritz-Hansen et al., 2008; Pack and DiBella, 2010). However, for measuring MBF, the most widely accepted and utilized methods treat the myocardium as a single compartment with a single inflow channel and a single outflow channel within which the CA is instantaneously well mixed and does not leak.

Figure 8.18 represents such a model. In the figure, $i(t)$ represents the concentration of CA entering the myocardium at time t, which is the AIF taken from the blood pool in the MR images. The contrast leaving the myocardium (the contrast in the veins) is denoted as $o(t)$; however, this is not measured from MR perfusion images. Instead, the quantity of CA resident in the myocardial tissue over time $q(t)$ is measured. These three terms are related by the fact that the rate at which a substance accumulates within the compartment is given by the difference of concentrations flowing into and out of the compartment multiplied by the flow rate. Mathematically, this is expressed in the mass balance equation:

$$q(t) = F \int_0^t [i(\tau) - o(\tau)] d\tau. \tag{8.1}$$

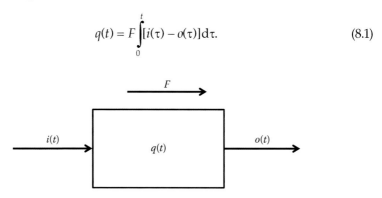

FIGURE 8.18
Diagram representing the myocardium as a single, well-mixed compartment across which tracer flows with a flow F, having a single input function $i(t)$ and single output function $o(t)$.

The quantity of tracer in the compartment at time t is the difference between what has flowed into the compartment and what has flowed out, that is, tracer that has entered the compartment and not yet left it stays in the compartment. If we have measured $i(t)$ and $q(t)$, it is not possible to obtain F from Equation 8.1 without knowledge of $o(t)$, which has not been measured. Thus, Equation 8.1 is rearranged to relate the AIF and myocardial curves directly by means of convolution with the myocardial tissue response function $R(t)$:

$$q(t) = FR(t) * i(t),\qquad (8.2)$$

where the symbol "*" denotes the convolution operation (see Section 8.4.10.1) and $R(t)$ is the tissue response function of the myocardium.

The tissue response function $R(t)$ represents the probability that a tracer molecule remains in the ROI at time t. Therefore, at time 0 (when the tracer first reaches the myocardium), $R(t)$ must be 1, meaning that all of the tracer remains in the tissue at this point. For any tracer, to leave at this point would require an infinitely small myocardium or an infinite myocardial flow. Over time, as tracer passes through the myocardium, the residue function decreases exponentially until no tracer remains ($R(t) = 0$). Therefore, at time 0,

$$FR(t = 0) = F \times 1 = F.$$

So, if Equation 8.2 can be solved for $FR(t)$ for a given $i(t)$ and $q(t)$, then F can be found by simply taking the value of the curve at $FR(t = 0)$.

8.4.10.1 Convolution

The symbol "*" in Equation 8.2 denotes the convolution operation. Convolution is a mathematical operation that takes two input functions (f and g) and produces a third output function ($f * g$). Mathematically, it is expressed as follows:

$$f(t) * g(t) = \int_0^t f(\tau)g(t-\tau)d\tau.\qquad (8.3)$$

It is perhaps easier to understand convolution if Equation 8.3 is expressed in a stepwise fashion. To generate $f * g$,

- take the mirror image of g in the line $\tau = 0$;
- offset the function g by t (i.e., slide it to the left extremity of f), $[g(t-\tau)]$;
- multiply the two functions together, $[f(\tau)g(t-\tau)]$;

- take the area under the resulting curve as the *t*th value of the output curve $f(t)*g(t)$, $\left[f(t)^* g(t) = \int f(\tau)g(t-\tau)d\tau \right]$; and

- Slide *g* one time step to the right and repeat over the full range of *t* values, $\left[f(t)^* g(t) = \int_0^t f(\tau)g(t-\tau)d\tau \right]$.

Figure 8.19 illustrates this process diagrammatically. The result of this process is that the time-varying function *f* is transformed by the function *g*. The process is analogous to the way an electrical component might transform a time-varying electrical signal or in two dimensions how a filter might affect an image. In quantitative myocardial perfusion, the convolution operation models how the myocardial tissue affects the shape of the AIF as it passes through the myocardium.

8.4.10.2 Deconvolution

To find $FR(t)$ for a given $i(t)$ and $q(t)$, the inverse of the convolution operation, known as deconvolution, is required. Deconvolution is a mathematically ill-posed problem. This means that for a given pair of functions in Equation 8.2, there are multiple solutions for $FR(t)$. Practically, this makes algorithms for performing deconvolution unstable, meaning that very small differences in $i(t)$ and $q(t)$ can generate large differences in $FR(t)$, which is problematic for quantitation. To minimize this problem, the deconvolution process must be constrained in some way. As we expect $R(t)$ to be a monotonically decaying function, one solution is to represent $R(t)$ as a monotonically

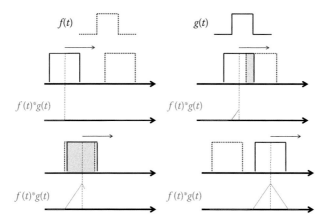

FIGURE 8.19
(Color version available from crcpress.com; see Preface.) Convolution of two square functions $g(t)$ (solid line) and $f(t)$ (dotted line). The $g(t)$ function is flipped in the *y*-axis. The new $g(t)$ slides along the time axis and the area under the product of the two functions at each time point is taken as the convolution result $f(t)*g(t)$ for that time point.

decaying function with a small number of parameters. A popular parametrization for MBF quantitation is the Fermi function that has only three parameters. Its use in perfusion quantitation is based on the similarity it bears to response functions measured with intravascular tracers (tracers that do not leak from the vascular space), and it reduces the deconvolution problem to a simple three-parameter fitting procedure (Jerosch-Herold et al., 1998). Because the Fermi function is based on observations from intravascular tracers, it can only be used in the first pass of contrast through the myocardium (before leakage from the vascular space has become significant). Attempting to include post first-pass data in Fermi-constrained analysis has been shown to produce errors (Pack and DiBella, 2010). Other approaches to deconvolution have imposed less rigorous constraints on the tissue response function, requiring only that $R(t)$ be smooth (Jerosch-Herold et al., 2002; Pack et al., 2008), and are able to deal with post first-pass data.

8.5 Summary

MRI is established in clinical and research studies as the gold standard in the assessment of LV and RV mass, volumes, and EF. Several techniques are also available to quantitatively assess regional and global myocardial contractility, including quantitative wall thickening, tissue tagging, SENC, and more recently feature tracking. Tissue tagging is able to provide previously unmeasured information on the complex dynamics of myocardial contraction and relaxation in a wide range of disease processes. Ongoing technical developments to lengthen tag persistence, decrease scanning time, and improve analysis will serve to increase the usefulness of this technique in understanding the mechanics of the normal and pathological heart in the future.

MRI is inherently sensitive to both phase and magnitude changes. Phase-contrast MRI enables accurate and reproducible measurement of blood flow and velocity within cardiovascular structures. These sequences can be combined with concurrent magnitude images to delineate anatomy and with cine imaging to account for movement of the ROI and pulsatile flow. Both in-plane and through-plane 2D acquisitions are commonly used for accurate quantification of valvular heart disease, shunting, and vascular patency.

MRI can be used to quantify MBF at rest and during stress. The necessary requirements for acquiring myocardial perfusion MRI data series have been described. The trade-offs to be considered when designing pulse sequences for perfusion MRI have been discussed, with a particular emphasis on the special requirements for quantitative perfusion. The quantitative analysis of MBF has a wide range of potential clinical applications but remains underused in practice because of time-consuming postprocessing.

References

Alfakih, K., Plein, S., Bloomer, T., Jones, T., Ridgway, J., Sivananthan, M., 2003. Comparison of right ventricular volume measurements between axial and short axis orientation using steady-state free precession magnetic resonance imaging. *J Magn Reson Imaging*. 18, 25–32.

Al-Saadi, N., Nagel, E., Gross, M., Bornstedt, A., Schnackenburg, B., Klein, C., Klimek, W., Oswald, H., Fleck, E., 2000. Noninvasive detection of myocardial ischemia from perfusion reserve based on cardiovascular magnetic resonance. *Circulation*. 101, 1379–83.

Andersen, A.H., Kirsch, J.E., 1996. Analysis of noise in phase contrast MR imaging. *Med Phys*. 23, 857–69.

Axel, L., Dougherty, L., 1989. MR imaging of motion with spatial modulation of magnetization. *Radiology*. 171, 841–5.

Axel, L., Montillo, A., Kim, D., 2005. Tagged magnetic resonance imaging of the heart: A survey. *Med Image Anal*. 9, 376–93.

Bottini, P.B., Carr, A.A., Prisant, L.M., Flickinger, F.W., Allison, J.D., Gottdiener, J.S., 1995. Magnetic resonance imaging compared to echocardiography to assess left ventricular mass in the hypertensive patient. *Am J Hypertens*. 8, 221–8.

Boxt, L.M., Katz, J., Kolb, T., Czegledy, F.P., Barst, R.J., 1992. Direct quantitation of right and left ventricular volumes with nuclear magnetic resonance imaging in patients with primary pulmonary hypertension. *J Am Coll Cardiol*. 19, 1508–15.

Burns, A.T., McDonald, I.G., Thomas, J.D., Macisaac, A., Prior, D., 2008. Doin' the twist: New tools for an old concept of myocardial function. *Heart*. 94, 978–83.

Castillo, E., Osman, N.F., Rosen, B.D., El-Shehaby, I., Pan, L., Jerosch-Herold, M., Lai, S., Bluemke, D.A., Lima, J.A., 2005. Quantitative assessment of regional myocardial function with MR-tagging in a multi-center study: Interobserver and intraobserver agreement of fast strain analysis with Harmonic Phase (HARP) MRI. *J Cardiovasc Magn Reson*. 7, 783–91.

Cernicanu, A., Axel, L., 2006. Theory-based signal calibration with single-point T1 measurements for first-pass quantitative perfusion MRI studies. *Acad Radiol*. 13, 686–93.

Christian, T.F., Rettmann, D.W., Aletras, A.H., Liao, S.L., Taylor, J.L., Balaban, R.S., Arai, A.E., 2004. Absolute myocardial perfusion in canines measured by using dual-bolus first-pass MR imaging. *Radiology*. 232, 677–84.

Costa, M.A., Shoemaker, S., Futamatsu, H., Klassen, C., Anglolillo, D.J., Nguyen, M., Siuciak, A., et al., 2007. Quantitative magnetic resonance perfusion imaging detects anatomic and physiologic coronary artery disease as measured by coronary angiography and fractional flow reserve. *J Am Coll Cardiol*. 50, 514–22.

Davarpanah, A.H., Chen, Y.P., Kino, A., Farrelly, C.T., Keeling, A.N., Sheehan, J.J., Ragin, A.B., Weale, P.J., Zuehlsdorff, S., Carr, J.C., 2010. Accelerated two- and three-dimensional cine MR imaging of the heart by using a 32-channel coil. *Radiology*. 254, 98–108.

Dodge, H.T., Sandler, H., Ballew, D.W., Lord, J.D., Jr., 1960. The use of biplane angiocardigraphy for the measurement of left ventricular volume in man. *Am Heart J*. 60, 762–76.

Dong, S.J., MacGregor, J.H., Crawley, A.P., McVeigh, E., Belenkie, I., Smith, E.R., Tyberg, J.V., Beyar, R., 1994. Left ventricular wall thickness and regional systolic function in patients with hypertrophic cardiomyopathy. A three-dimensional tagged magnetic resonance imaging study. *Circulation.* 90, 1200–9.

Ennis, D.B., Epstein, F.H., Kellman, P., Fananapazir, L., McVeigh, E.R., Arai, A.E., 2003. Assessment of regional systolic and diastolic dysfunction in familial hypertrophic cardiomyopathy using MR tagging. *Magn Reson Med.* 50, 638–42.

Firmin, D.N., Nayler, G.L., Kilner, P.J., Longmore, D.B., 1990. The application of phase shifts in NMR for flow measurement. *Magn Reson Med.* 14, 230–41.

Fischer, S.E., McKinnon, G.C., Maier, S.E., Boesiger, P., 1993. Improved myocardial tagging contrast. *Magn Reson Med.* 30, 191–200.

Fritz-Hansen, T., Hove, J.D., Kofoed, K.F., Kelbaek, H., Larsson, H.B.W., 2008. Quantification of MRI measured myocardial perfusion reserve in healthy humans: A comparison with positron emission tomography. *J Magn Reson Imaging.* 27, 818–24.

Fritz-Hansen, T., Rostrup, E., Larsson, H.B., Søndergaard, L., Ring, P., Henriksen, O., 1996. Measurement of the arterial concentration of Gd-DTPA using MRI: A step toward quantitative perfusion imaging. *Magn Reson Med.* 36, 225–31.

Garot, J., Lima, J.A., Gerber, B.L., Sampath, S., Wu, K.C., Bluemke, D.A., Prince, J.L., Osman, N.F., 2004. Spatially resolved imaging of myocardial function with strain-encoded MR: Comparison with delayed contrast-enhanced MR imaging after myocardial infarction. *Radiology.* 233, 596–602.

Gatehouse, P.D., Elkington, A.G., Ablitt, N.A., Yang, G.-Z., Pennell, D.J., Firmin, D.N., 2004. Accurate assessment of the arterial input function during high-dose myocardial perfusion cardiovascular magnetic resonance. *J Magn Reson Imaging.* 20, 39–45.

Gotte, M.J., Germans, T., Russel, I.K., Zwanenburg, J.J., Marcus, J.T., van Rossum, A.C., van Veldhuisen, D.J., 2006. Myocardial strain and torsion quantified by cardiovascular magnetic resonance tissue tagging: Studies in normal and impaired left ventricular function. *J Am Coll Cardiol.* 48, 2002–11.

Greil, G.F., Germann, S., Kozerke, S., Baltes, C., Tsao, J., Urschitz, M.S., Seeger, A., et al., 2008. Assessment of left ventricular volumes and mass with fast 3D cine steady-state free precession k-t space broad-use linear acquisition speed-up technique (k-t BLAST). *J Magn Reson Imaging.* 27, 510–15.

Groothuis, J.G.J., Kremers, F.P.P.J., Beek, A.M., Brinckman, S.L., Tuinenburg, A.C., Jerosch-Herold, M., van Rossum, A.C., Hofman, M.B.M., 2010. Comparison of dual to single contrast bolus magnetic resonance myocardial perfusion imaging for detection of significant coronary artery disease. *J Magn Reson Imaging.* 32, 88–93.

Grothues, F., Moon, J.C., Bellenger, N.G., Smith, G.S., Klein, H.U., Pennell, D.J., 2004. Interstudy reproducibility of right ventricular volumes, function, and mass with cardiovascular magnetic resonance. *Am Heart J.* 147, 218–23.

Grothues, F., Smith, G.C., Moon, J.C., Bellenger, N.G., Collins, P., Klein, H.U., Pennell, D.J., 2002. Comparison of interstudy reproducibility of cardiovascular magnetic resonance with two-dimensional echocardiography in normal subjects and in patients with heart failure or left ventricular hypertrophy. *Am J Cardiol.* 90, 29–34.

Heidemann, R.M., Ozsarlak, O., Parizel, P.M., Michiels, J., Kiefer, B., Jellus, V., Muller, M., et al., 2003. A brief review of parallel magnetic resonance imaging. *Eur Radiol.* 13, 2323–37.

Heys, J.J., Holyoak, N., Calleja, A.M., Belohlavek, M., Chaliki, H.P., 2010. Revisiting the simplified bernoulli equation. *Open Biomed Eng J.* 4, 123–8.

Holman, E.R., Buller, V.G., de Roos, A., van der Geest, R.J., Baur, L.H., van der Laarse, A., Bruschke, A.V., Reiber, J.H., van der Wall, E.E., 1997. Detection and quantification of dysfunctional myocardium by magnetic resonance imaging. A new three-dimensional method for quantitative wall-thickening analysis. *Circulation.* 95, 924–31.

Honda, N., Machida, K., Hashimoto, M., Mamiya, T., Takahashi, T., Kamano, T., Kashimada, A., Inoue, Y., Tanaka, S., Yoshimoto, N., 1993. Aortic regurgitation: Quantitation with MR imaging velocity mapping. *Radiology.* 186, 189–94.

Hor, K.N., Gottliebson, W.M., Carson, C., Wash, E., Cnota, J., Fleck, R., Wansapura, J., et al., 2010. Comparison of magnetic resonance feature tracking for strain calculation with harmonic phase imaging analysis. *JACC Cardiovasc Imaging.* 3, 144–51.

Hsu, L.-Y., Kellman, P., Arai, A.E., 2008. Nonlinear myocardial signal intensity correction improves quantification of contrast-enhanced first-pass MR perfusion in humans. *J Magn Reson Imaging.* 27, 793–801.

Hundley, W.G., Li, H.F., Willard, J.E., Landau, C., Lange, R.A., Meshack, B.M., Hillis, L.D., Peshock, R.M., 1995. Magnetic resonance imaging assessment of the severity of mitral regurgitation: Comparison with invasive techniques. *Circulation.* 92, 1151–8.

Jerosch-Herold, M., 2010. Quantification of myocardial perfusion by cardiovascular magnetic resonance. *J Cardiovasc Magn Reson.* 12, 57.

Jerosch-Herold, M., Seethamraju, R.T., Swingen, C.M., Wilke, N.M., Stillman, A.E., 2004. Analysis of myocardial perfusion MRI. *J Magn Reson Imaging.* 19, 758–70.

Jerosch-Herold, M., Swingen, C., Seethamraju, R.T., 2002. Myocardial blood flow quantification with MRI by model-independent deconvolution. *Med Phys.* 29, 886–97.

Jerosch-Herold, M., Wilke, N., Stillman, A.E., 1998. Magnetic resonance quantification of the myocardial perfusion reserve with a Fermi function model for constrained deconvolution. *Med Phys.* 25, 73–84.

Katz, J., Whang, J., Boxt, L.M., Barst, R.J., 1993. Estimation of right ventricular mass in normal subjects and in patients with primary pulmonary hypertension by nuclear magnetic resonance imaging. *J Am Coll Cardiol.* 21, 1475–81.

Kondo, C., Caputo, G.R., Semelka, R., Foster, E., Shimakawa, A., Higgins, C.B., 1991. Right and left ventricular stroke volume measurements with velocity-encoded cine MR imaging: In vitro and in vivo validation. *AJR Am J Roentgenol.* 157, 9–16.

Korosoglou, G., Lossnitzer, D., Schellberg, D., Lewien, A., Wochele, A., Schaeufele, T., Neizel, M., et al., 2009. Strain-encoded cardiac MRI as an adjunct for dobutamine stress testing: Incremental value to conventional wall motion analysis. *Circ Cardiovasc Imaging.* 2, 132–40.

Kramer, C.M., Lima, J.A., Reichek, N., Ferrari, V.A., Llaneras, M.R., Palmon, L.C., Yeh, I.T., Tallant, B., Axel, L., 1993. Regional differences in function within noninfarcted myocardium during left ventricular remodeling. *Circulation.* 88, 1279–88.

Kroll, K., Wilke, N., Jerosch-Herold, M., Wang, Y., Zhang, Y., Bache, R.J., Bassingthwaighte, J.B., 1996. Modeling regional myocardial flows from residue functions of an intravascular indicator. *Am J Physiol.* 271, H1643–H1655.

Kuijer, J.P., Jansen, E., Marcus, J.T., van Rossum, A.C., Heethaar, R.M., 2001. Improved harmonic phase myocardial strain maps. *Magn Reson Med.* 46, 993–9.

Kuijpers, D., Ho, K.Y., van Dijkman, P.R., Vliegenthart, R., Oudkerk, M., 2003. Dobutamine cardiovascular magnetic resonance for the detection of myocardial ischemia with the use of myocardial tagging. *Circulation*. 107, 1592–7.

Larsson, H.B.W., FritzHansen, T., Rostrup, E., Sondergaard, L., Ring, P., Henriksen, O., 1996. Myocardial perfusion modeling using MRI. *Magn Reson Med*. 35, 716–26.

Lelieveldt, B.P.F., Geest, R.J., Reiber, J.H.C., 2004. Towards "one-stop" cardiac MR image analysis. *Imaging Decis (Berl)*. 8, 2–12.

Maddahi, J., Crues, J., Berman, D.S., Mericle, J., Becerra, A., Garcia, E.V., Henderson, R., Bradley, W., 1987. Noninvasive quantification of left ventricular myocardial mass by gated proton nuclear magnetic resonance imaging. *J Am Coll Cardiol*. 10, 682–92.

Maret, E., Todt, T., Brudin, L., Nylander, E., Swahn, E., Ohlsson, J.L., Engvall, J.E., 2009. Functional measurements based on feature tracking of cine magnetic resonance images identify left ventricular segments with myocardial scar. *Cardiovasc Ultrasound*. 7, 53.

Markl, M., Kilner, P.J., Ebbers, T., 2011. Comprehensive 4D velocity mapping of the heart and great vessels by cardiovascular magnetic resonance. *J Cardiovasc Magn Reson*. 13, 1–22.

Meyer, C.H., Hu, B.S., Nishimura, D.G., Macovski, A., 1992. Fast spiral coronary artery imaging. *Magn Reson Med*. 28, 202–13.

Moon, J.C., Lorenz, C.H., Francis, J.M., Smith, G.C., Pennell, D.J., 2002. Breath-hold FLASH and FISP cardiovascular MR imaging: Left ventricular volume differences and reproducibility. *Radiology*. 223, 789–97.

Nagel, E., Klein, C., Paetsch, I., Hettwer, S., Schnackenburg, B., Wegscheider, K., Fleck, E., 2003. Magnetic resonance perfusion measurements for the noninvasive detection of coronary artery disease. *Circulation*. 108, 432–7.

Nagel, E., Lehmkuhl, H.B., Bocksch, W., Klein, C., Vogel, U., Frantz, E., Ellmer, A., Dreysse, S., Fleck, E., 1999. Noninvasive diagnosis of ischemia-induced wall motion abnormalities with the use of high-dose dobutamine stress MRI: Comparison with dobutamine stress echocardiography. *Circulation*. 99, 763–70.

Nagel, E., Stuber, M., Lakatos, M., Scheidegger, M.B., Boesiger, P., Hess, O.M., 2000. Cardiac rotation and relaxation after anterolateral myocardial infarction. *Coron Artery Dis*. 11, 261–7.

O'Brien, K.R., Cowan, B.R., Jain, M., Stewart, R.A.H., Kerr, A.J., Young, A.A., 2008. MRI phase contrast velocity and flow errors in turbulent stenotic jets. *J Magn Reson Imaging*. 28, 210–18.

O'Donnell, M., 1985. NMR blood flow imaging using multiecho, phase contrast sequences. *Med Phys*. 12, 59.

Pack, N.A., DiBella, E.V.R., Rust, T.C., Kadrmas, D.J., McGann, C.J., Butterfield, R., Christian, P.E., Hoffman, J.M., 2008. Estimating myocardial perfusion from dynamic contrast-enhanced CMR with a model-independent deconvolution method. *J Cardiovasc Magn Reson*. 10, 52.

Pack, N.A., DiBella, V.R., 2010. Comparison of myocardial perfusion estimates from dynamic contrast-enhanced magnetic resonance imaging with four quantitative analysis methods. *Magn Reson Med*. 64, 125–37.

Pennell, D.J., Underwood, S.R., Manzara, C.C., Swanton, R.H., Walker, J.M., Ell, P.J., Longmore, D.B., 1992. Magnetic resonance imaging during dobutamine stress in coronary artery disease. *Am J Cardiol*. 70, 34–40.

Plein, S., Ridgway, J.P., Jones, T.R., Bloomer, T.N., Sivananthan, M.U., 2002. Coronary artery disease: Assessment with a comprehensive MR imaging protocol initial results. *Radiology*. 225, 300–7.

Rademakers, F.E., Rogers, W.J., Guier, W.H., Hutchins, G.M., Siu, C.O., Weisfeldt, M.L., Weiss, J.L., Shapiro, E.P., 1994. Relation of regional cross-fiber shortening to wall thickening in the intact heart. Three-dimensional strain analysis by NMR tagging. *Circulation*. 89, 1174–82.

Radjenovic, A., Biglands, J.D., Larghat, A., Ridgway, J.P., Ball, S.G., Greenwood, J.P., Jerosch-Herold, M., Plein, S., 2010. Estimates of systolic and diastolic myocardial blood flow by dynamic contrast-enhanced MRI. *Magn Reson Med*. 64, 1696–703.

Ritter, C., Brackertz, A., Sandstede, J., Beer, M., Hahn, D., Kostler, H., 2006. Absolute quantification of myocardial perfusion under adenosine stress. *Magn Reson Med*. 56, 844–9.

Rutz, A.K., Ryf, S., Plein, S., Boesiger, P., Kozerke, S., 2008. Accelerated whole-heart 3D CSPAMM for myocardial motion quantification. *Magn Reson Med*. 59, 755–63.

Schiller, N.B., Acquatella, H., Ports, T.A., Drew, D., Goerke, J., Ringertz, H., Silverman, N.H., et al., 1979. Left ventricular volume from paired biplane two-dimensional echocardiography. *Circulation*. 60, 547–55.

Schuster, A., Kutty, S., Padiyath, A., Parish, V., Gribben, P., Danford, D.A., Makowski, M.R., Bigalke, B., Beerbaum, P., Nagel, E., 2011. Cardiovascular magnetic resonance myocardial feature tracking detects quantitative wall motion during dobutamine stress. *J Cardiovasc Magn Reson*. 13, 58.

Schwitter, J., Nanz, D., Kneifel, S., Bertschinger, K., Büchi, M., Knüsel, P.R., Marincek, B., Lüscher, T.F., von Schulthess, G.K., 2001. Assessment of myocardial perfusion in coronary artery disease by magnetic resonance: A comparison with positron emission tomography and coronary angiography. *Circulation*. 103, 2230–5.

Shehata, M.L., Cheng, S., Osman, N.F., Bluemke, D.A., Lima, J.A., 2009. Myocardial tissue tagging with cardiovascular magnetic resonance. *J Cardiovasc Magn Reson*. 11, 55.

Strohm, O., Schulz-Menger, J., Pilz, B., Osterziel, K.J., Dietz, R., Friedrich, M.G., 2001. Measurement of left ventricular dimensions and function in patients with dilated cardiomyopathy. *J Magn Reson Imaging*. 13, 367–71.

Strotzer, M., Aebert, H., Lenhart, M., Nitz, W., Wild, T., Manke, C., Volk, M., Feuerbach, S., 2000. Morphology and hemodynamics in dissection of the descending aorta. Assessment with MR imaging. *Acta Radiol*. 41, 594–600.

Thunberg, P., Karlsson, M., Wigstrom, L., 2003. Accuracy and reproducibility in phase contrast imaging using SENSE. *Magn Reson Med*. 50, 1061–8.

Tofts, P.S., 1997. Modeling tracer kinetics in dynamic Gd-DTPA MR imaging. *J Magn Reson Imaging*. 7, 91–101.

Underwood, S.R., Walton, S., Laming, P.J., Jarritt, P.H., Ell, P.J., Emanuel, R.W., Swanton, R.H., 1985. Left ventricular volume and ejection fraction determined by gated blood pool emission tomography. *Br Heart J*. 53, 216–22.

Utz, J.A., Herfkens, R.J., Heinsimer, J.A., Bashore, T., Califf, R., Glover, G., Pelc, N., Shimakawa, A., 1987. Cine MR determination of left ventricular ejection fraction. *AJR Am J Roentgenol*. 148, 839–43.

Utz, W., Niendorf, T., Wassmuth, R., Messroghli, D., Dietz, R., Schulz-Menger, J., 2007. Contrast-dose relation in first-pass myocardial MR perfusion imaging. *J Magn Reson Imaging*. 25, 1131–5.

Van Rossum, A.C., Sprenger, M., Visser, F.C., Peels, K.H., Valk, J., Roos, J.P., 1991. An in vivo validation of quantitative blood flow imaging in arteries and veins using magnetic resonance phase-shift techniques. *Eur Heart J*. 12, 117–26.

van Rugge, F.P., van der Wall, E.E., de Roos, A., Bruschke, A.V., 1993. Dobutamine stress magnetic resonance imaging for detection of coronary artery disease. *J Am Coll Cardiol*. 22, 431–9.

van Rugge, F.P., van der Wall, E.E., Spanjersberg, S.J., de Roos, A., Matheijssen, N.A., Zwinderman, A.H., van Dijkman, P.R., Reiber, J.H., Bruschke, A.V., 1994. Magnetic resonance imaging during dobutamine stress for detection and localization of coronary artery disease. Quantitative wall motion analysis using a modification of the centerline method. *Circulation*. 90, 127–38.

Yang, G.Z., Burger, P., Kilner, P.J., Karwatowski, S.P., Firmin, D.N., 1996. Dynamic range extension of cine velocity measurements using motion registered spatio-temporal phase unwrapping. *J Magn Reson Imaging*. 6, 495–502.

Zerhouni, E.A., Parish, D.M., Rogers, W.J., Yang, A., Shapiro, E.P., 1988. Human heart: Tagging with MR imaging—A method for noninvasive assessment of myocardial motion. *Radiology*. 169, 59–63.

9

Quantitative MRI of Tumors

James P.B. O'Connor

University of Manchester

Richard A.D. Carano

Genentech Inc.

CONTENTS

9.1 Introduction

9.1.1 Cancer, Society, and Role of Imaging

Nearly one in four deaths in the United States and Europe are due to cancer (American Cancer Society, 2009; Jemal et al., 2011). Despite the continued high death rate due to cancer, great advances made in the treatment of cancer have occurred over the last several decades. The average 5-year relative survival rate for all cancers has increased from 50% between 1975 and 1977 to 66% between 1996 and 2004, and marked improvements have been seen in specific cancers including breast and colorectal cancers (American Cancer Society, 2009). These advances have been due, in large part, to early detection of disease and the therapeutic advances made over that period.

Noninvasive imaging (x-ray, ultrasound, CT, MRI, PET) has played a substantial role in the improvement of treatments by assisting in early detection and providing a noninvasive means of monitoring therapies. The ability to monitor a therapeutic response noninvasively has greatly aided patient management and staging. The timely identification of effective therapies has allowed rapid dose optimization and discontinuation of ineffective therapies, minimizing potentially harmful side effects in favor of alternative therapeutic avenues. MRI has proven to be a highly sensitive technique for detecting tumors because MRI provides unparalleled soft tissue contrast compared to other imaging modalities. Although MRI is primarily used to identify tumor and quantify tumor size, the wide array of contrast mechanisms available to MRI has enabled development of many techniques that quantify both structural and physiological feature of tumors.

9.1.2 Biology of Tumors

Tumors arise from normal tissue due to the transformation of a normal cell to a cancerous cell. This transformation arises from a genetic or somatic mutation resulting in a cell undergoing uncontrolled growth. Nearly all solid tumors exhibit degrees of tissue spatial heterogeneity (Heppner, 1984). In general, solid tumors are comprised of several macroscopic tissue components consisting of regions of viable tumor, stroma, and necrosis. In addition, regions of microhemorrhage, cystic change, and calcification are present in certain tumor types and contribute to the spatial heterogeneity.

Regions of viable tumor are generally areas of densely packed neoplastic cells, although the cell packing varies between different tumor types. The stroma is comprised of tissue structures that exist between the malignant cells and the adjacent normal tissue of the host. Stromal tissue development is generally a product of the neighboring normal host tissue and is produced from interactions between the tumor cells and the host (Liotta and Kohn, 2001; Mueller and Fusenig, 2004). Stroma is comprised of nonmalignant supporting tissue that includes connective tissue, blood vessels, and often the presence of infiltrating inflammatory cells. Tumors have been described as wounds that do not heal because the tumor hijacks many of the host processes for wound healing (inflammation, new tissue formation, tissue remodeling, angiogenesis) to form the stroma and supply the required growth factors and nutrients necessary for its continued and rapid growth (Dvorak, 1986). The heterogeneous nature of tumor tissue morphology is due in large part to this rapid growth rate of tumor cells, which leads to spatial heterogeneity of perfusion, oxygenation, and metabolism. Tumors quickly outgrow their blood supply leading to hypoxia, regions of necrosis due to the lack of sufficient blood flow and areas of new blood vessel growth to provide nutrients and oxygen to the viable tumor. Several studies have provided data showing that this rapid growth often leaves tumors with a central necrotic core and an expanding, invading viable rim (Checkley et al., 2003; Choi et al., 2011; Rajendran et al., 2004), although in many clinical tumors the spatial heterogeneity is less clearly structured. As tumors grow and invade neighboring tissues and vessels, hemorrhage can occur leading to the pooling of blood and edema within and adjacent to the tumor mass.

In 1971, Folkman proposed that a tumor's growth is limited to around 2–3 mm in diameter without the growth of new blood vessels and that the inhibition of new blood vessel growth is a potential therapeutic approach (1971). Tumors would remain a small, avascular nodule, without this transition to active angiogenesis that has been termed the angiogenic switch (Folkman and Hanahan, 1991). A number of activators and inhibitors of angiogenesis have been identified (Bergers and Benjamin, 2003). These modulators of vascular growth remain in balance, keeping the vasculature in a quiescent state. The rapid growth of a tumor can create hypoxic conditions that push this equation toward new vessel formation. Vascular endothelial growth factor A (VEGF-A or VEGF) has been identified as a central mediator of normal and pathological angiogenesis by the promotion of endothelial cell proliferation and migration as well as increased vascular permeability (Bates and Curry, 1996; Ferrara and Henzel, 1989; Keck et al., 1989; Leung et al., 1989). VEGF levels have been found to be elevated in a variety of cancer types including colorectal, hematological malignancies, and liver (Lee et al., 2000; List et al., 2001; Poon et al., 2001a, 2001b). VEGF promotes greater blood flow and nutrient delivery by promoting vessel growth, increasing vasodilation, and vascular permeability.

9.2 MR ^{1}H Relaxometry Characteristics of Tumors

The triggering of the angiogenic switch contributes greatly to the visibility of tumors by MRI. Angiogenesis allows for tumors to increase in size beyond 2–3 mm in diameter (Folkman, 1971), resulting in a tumor size that is amenable to the clinical spatial resolution limits of MRI. But, in addition to an increase in tumor size, up-regulation of VEGF and other proangiogenic growth factors makes tumor vessels more leaky and contributes to intratumoral edema (Ferrara et al., 2003). This biology makes identification of tumors possible using multiple contrast mechanisms available to MRI. The increase in permeability contributes to a rise in edema within the tumor and this leads to increases in proton density, the longitudinal relaxation time (T_1), and the transverse relaxation time (T_2) relative to surrounding tissues

TABLE 9.1

The Range of Reported Mean NMR Relaxation Parameters for Normal Tissues and Tumors Estimated at 1.5 T

Organ	Tissue/lesion	T_1 (ms)	T_2 (ms)	Description
Brain	Gray matter	950–1136	80–102	
	White matter	600–889	60–80	
	CSF (Ibrahim et al., 1998; Wong et al., 2001)	2390–4591	230–1721	
	Edema	1032–1649	140–271	Peritumoral edema
	Tumor	–	337–396	Adenoma
	Tumor	1159–1500	103–174	Anaplastic Astrozytoma
	Tumor	–	155–363	Astrocytoma I–II
	Tumor		105–172	Astrocytoma III–IV
	Tumor	1718–3036	205–317	Glioblastoma
	Tumor	1135	133	Oligodendroglioma
	Tumor	1168	191	–
Liver	Normal	197–675	44–59	
	Tumor	–	86–142	Hemangioma
	Tumor	1191	61–92	Metastases
Kidney	Normal	559	84	
Muscle	Normal	797–1206	31–43	
Spleen	Normal	1183	87	

Source: Summarized from reviews of published studies for normal (Adapted from Akber, S.F., *Physiol Chem Phys Med NMR.* 28, 205–38, 1996) and abnormal (Adapted from Akber, S.F., *Physiol Chem Phys Med NMR.* 40, 1–42, 2008) tissues unless otherwise noted.

Notes: NMR, nuclear magnetic resonance; CSF, cerebrospinal fluid.

(Damadian, 1971; Kiricuta and Simplaceanu, 1975; Weisman et al., 1972). Table 9.1 lists literature values of T_1 and T_2 for both normal and pathological tissues. Although, differences exist between mean relaxation parameters for tumor and normal tissues, the overlap in range has resulted in the failure of attempts to employ individual parameters (proton density, T_1 or T_2) as a quantitative diagnostic marker or to define individual tissues by their native relaxation properties or proton density (Houdek et al., 1988; Just and Thelen, 1988; Just et al., 1988; Kjaer et al., 1991; Komiyama et al., 1987; Mills et al., 1984).

Differences in the proton density, T_1 and T_2, between tumor and the surrounding normal tissues do, however, allow tumor visualization within weighted MR images; for example, many tumors appear hypointense in T_1-weighted images and hyperintense in T_2-weighted images (Figure 9.1), and these differences allow for reliable clinical identification of tumors in solid organs. These endogenous MRI contrast mechanisms were employed in the earliest MRI attempts to visualize tumors by exploiting differences with normal tissues surrounding the tumor (Alfidi et al., 1982; Kerlan et al., 1983). The choice of contrast is dependent on tumor location and the properties of the neighboring normal tissues. Examples of

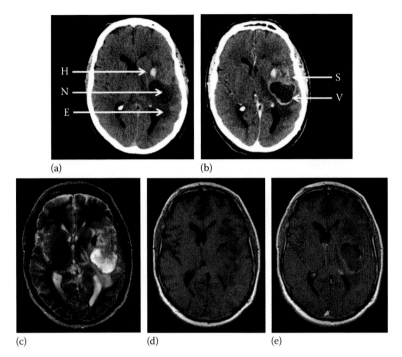

FIGURE 9.1
CT and MRI images of a 47-year-old man with a left temporal lobe GBM. (a and b) Pre- and postcontrast CT images show a heterogeneous tumor exhibiting cystic change and necrosis (*N*, focal low density), hemorrhage (*H*, high density on precontrast images), soft tissue components (*S*), and vascular ring enhancement (*V*, increasing density). Surrounding edema is present (*E*, ill-defined low density). (c) T_2-weighted image shows corresponding necrosis (focal high signal), hemorrhage (very low signal), and edema (ill-defined high signal). (d and e) Tissue discrimination is less pronounced on pre- and postcontrast T_1-weighted images, but vascular contrast enhancement is clearly demonstrated (peripheral high signal). Other key radiological features are tumor size, position and relationship to other structures, and degree of mass effect.

these contrast differences include the contrast between an edematous liver mass with normal liver (shorter T_2), a breast tumor and neighboring fat (shorter T_1), a lung tumor and neighboring air (lower proton density), and a brain tumor and neighboring gray and white matter (shorter T_2). The elevated T_2 has made tumors and the surrounding peritumoral edema visible on T_2 weighted images (Figure 9.1), although the presence of peritumoral edema makes it difficult to precisely identify tumor boundaries (Els et al., 1995). More recently, the native T_1 of tumors has been investigated as a biomarker of response to therapy. Though nonspecific, studies have shown that reductions in tumor T_1 occur in response to various cytotoxic and anti-VEGF therapies (bevacizumab and cediranib) and the vascular disrupting agent ZD6126 (Jamin et al., 2013; McSheehy et al., 2010; O'Connor et al., 2009).

Elevated tumor vascular permeability and increased vascularity are the primary mechanisms that make tumors highly visible with contrast-enhanced volumetric imaging. Both MRI and CT imaging exploit the leakage of imaging contrast agents to identify solid tumors and measure tumor dimensions. This increased contrast leakage by tumor vasculature has made contrast-enhanced MRI and CT valuable imaging modalities for tumor identification (Figure 9.1). For MRI, paramagnetic gadolinium chelates (Magnevist, Berlex Laboratories: Gadopentetate dimeglumine, Gd-DTPA; Omniscan, Nycomed: Gadodiamide, Gd-DTPA-BMA; ProHance, Bracco Diagnostics: Gadoteriodol, Gd-HP-DO3A) have been employed as extracellular contrast agents due to their ability to shorten T_1, T_2, and T_2^* (Carr et al., 1984c; Weinmann et al., 1984b). From their initial introduction, gadolinium chelate contrast agents have been used to enhance the appearance of tumors (Carr et al., 1984a, 1984b). These agents are given via an intravenous injection several minutes prior to the contrast-enhanced sequence, usually toward the end of the MRI examination. After injection, gadolinium-based agents leak more rapidly out of the tumor vasculature than vessels in the neighboring normal tissue due to the elevated permeability of the tumor vasculature. This results in the agent accumulating in the interstitial space yielding increased image contrast for the tumors in T_1-weighted images.

9.3 Novel ^{1}H Contrast Mechanisms for Imaging Tumors

9.3.1 Diffusion

One of the greatest strengths of MRI is the multiple means by which image contrast can be generated. One of the more elegant contrast mechanisms is the exploitation of the mobility of water molecules by diffusion-weighted imaging (DWI). Molecular diffusion is a process where thermal energy produces random translational movement of molecules, termed Brownian motion (Brown, 1828). The random molecular translation is characterized by the Einstein (1905, 1956) relation that defines the mean-squared displacement in three dimensions as follows:

$$\langle r \rangle = 6Dt_{\mathrm{d}}, \tag{9.1}$$

where r is the displacement from the origin, t_{d} is the diffusion time, and D is the self-diffusion coefficient (usually given in cm^2/s or mm^2/s). Carr and Purcell (1954) showed that molecular self-diffusion influence the nuclear magnetic resonance (NMR) signal by shortening the T_2 of liquids and that the diffusion coefficient can be measured by making NMR spin echoes sensitized to diffusion by applying a constant gradient throughout the Hahn spin echo experiment. This work was followed by a modification of the Bloch equations by H.C. Torrey to include the contribution of diffusion and the application of a spatially varying magnetic field gradient (1956). Stejskal and Tanner (1965) introduced pulsed field gradient NMR (PFG-NMR), where pulsed magnetic field gradients were employed to measure diffusion by imparting a spatially dependent phase shift to diffusing molecules as a means to quantifying the displacement that occurs between the gradients pulses. PFG-NMR is the means by which diffusion weighting was generated in the first DWI experiments by Le Bihan et al. (1986). PFG-NMR yields the following signal attenuation relationship when the gradient pulses are applied in one direction (Le Bihan, 1991):

$$A(b) = A(0) \times \exp(-bD), \tag{9.2}$$

where $A(b)$ is the signal amplitude, $A(0)$ represents the theoretical signal intensity (SI) for $b = 0$. Diffusion is represented by the scalar diffusion coefficient, D. The b value is defined as $\gamma^2 \mathbf{g}^2 \delta^2 \alpha^2 t_d$. Here, γ is the gyromagnetic ratio, $\mathbf{g}$ and δ are commonly used to represent the strength and duration, respectively, of the applied diffusion gradients. t_d is the diffusion time, which is the time that molecules are allowed to diffuse between the diffusion gradients. t_d equals $\Delta - \delta/4$ for half-sine-shaped gradients and $\Delta - \delta/3$ for rectangular-shaped gradients, where Δ is the separation between the two diffusion gradient pulses. α $(= 2/\pi)$ and $\delta/4$ are correction factors for using half-sine-shaped gradient pulses (instead of $\delta/3$ for rectangular gradient pulses). Equation 9.2 is appropriate for isotropic diffusion, where diffusion measurements are independent of measurement direction. For anisotropic media, the measured diffusion values differ when the diffusion gradients are applied in different directions. The diffusion properties of a tissue can only be completely characterized by measuring the full 3×3 effective diffusion tensor (**D**) (Basser et al., 1994). Readers can refer to Chapter 2 for a comprehensive discussion of the theory of diffusion imaging.

The PFG-NMR diffusion imaging experiment can be influenced by water molecules in multiple compartments found within the imaging voxel. The relative contribution of intra- and extravascular water molecules to the diffusion signal depends on the choice of b-values. A PFG-NMR diffusion approach that focused on the intravascular spins was proposed as a means to measure microvascular flow by quantifying the signal loss due to blood moving through capillaries and employing a model of the microvasculature as a randomly oriented network of vessels, leading to signal attenuation similar to the diffusion losses from Brownian motion (Ahn et al., 1987; Le Bihan et al., 1986; Nalcioglu and Cho, 1987). Le Bihan et al. introduced the use of a biexponential model to describe the signal losses due to "intravoxel incoherent motions" (IVIM) to quantify the random losses produced by microvascular flow and molecular diffusion (1986). This approach modeled tissue water movement within a voxel as consisting of two compartments: an intravascular compartment (f, intravascular volume fraction) that corresponds to microvascular flow (D^*, the pseudodiffusion coefficient) and an extravascular compartment $(1 - f)$ that corresponds to extravascular water diffusion (D). The IVIM method employs a biexponential model to extract parametric estimates of f, D^*, and D. Ahn et al. introduced an alternative approach to extract microvascular flow information that exploited the phase coherence of the vascular signal to separate the microvascular flow signal from the diffusion signal where phase coherence is lost (1987). Since the motion (flow) in each capillary segment is coherent, phase cancellation occurs at even echoes due to spin rephasing. For the diffusing molecules, the diffusion phenomenon is truly a random process where incoherent summation takes place, preventing even-echo rephasing. After these initial efforts toward quantifying tissue perfusion noninvasively with PFG-NMR, research continued; but a robust practical application of these imaging techniques to quantify perfusion across a wide range of tissues has not been realized due to a number of practical considerations that include the need for very high signal-to-noise ratio (SNR) to extract the multiexponential components, very strong gradients, and signal corruption due to motion artifacts; and indeed may not be possible in many tissues such as the brain where the blood volume (~5%) may be too low to make accurate measurements (King et al., 1992; Pekar et al., 1992; Rosen et al., 1989). Despite these limitations, *in vivo* IVIM-based perfusion estimates have been successful for several applications, outside the brain, where the blood volume is high and, thus, provides a greater intravascular signal. As an example, successful kidney, liver, and placenta IVIM studies have been performed clinically where f values were found to be 33%, 29%, and 26%, respectively (Moore et al., 2000; Yamada et al., 1999).

Diffusion within tissues: The major focus and success of diffusion imaging has been in quantifying the extravascular signal because of its dependence on the microstructural characteristics of tissue. The choice of b-values will dictate the relative contributions of the intra- and extravascular signals. The contributions of the rapidly moving intravascular spins are eliminated when the lowest b-value is greater than 100–150 s/mm^2 (Padhani et al., 2009). This results in the diffusion estimate being dominated by the extravascular signal (both extra- and intracellular spins). When water molecules are diffusing within systems with internal structures, Equation 9.2 will no longer strictly apply because of the interaction of the water molecules with these internal barriers to diffusion (Hurlimann et al., 1995; Woessner, 1963). This case is referred to restricted diffusion (Figure 9.2), where interactions with barriers increase as t_d is increased, further restricting the movement of water molecules and, thus, making the measured diffusion coefficient dependent on t_d (Woessner, 1963). Given that the measured diffusion coefficient is influenced by restricting structures and other processes (microvascular flow, bulk motion errors, etc.) and, thus, does not represent free diffusion, the NMR estimated value for diffusion has been referred to as the apparent diffusion coefficient (ADC). It has been shown that when the signal attenuation due to diffusion is small, Equation 9.2 can still be employed (with D replaced by ADC) (Woessner, 1963). Figure 9.3 shows the diffusion-weighted images of a tumor for a range of b-values and the ADC map that was generated by fitting the data to Equation 9.2 on a voxel-by-voxel basis.

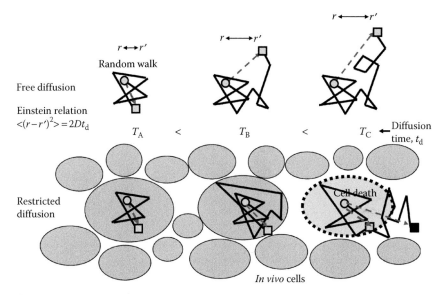

FIGURE 9.2
A drawing describing MRI measurements of diffusion in tissues. MRI measures diffusion by employing spatially dependent magnetic field (diffusion) gradients to encode the position of water molecules. This provides an estimate of the average translational movement of molecules along the axis of the applied diffusion gradient. For free diffusion (top row), the mean-squared displacement $(r - r')$ of a population of molecules is proportional to the product of the diffusion coefficient and the diffusion time (t_d) (Einstein relation) and increases with increasing t_d. For restricted diffusion (bottom row), barriers to diffusion (hydrophobic structures) restrict the movement of water molecules. Due to the random nature of the movement of water molecules, the interactions with barriers to diffusion increase with increasing diffusion time and this limits the measured displacement and, thus, the estimate of the "apparent" diffusion coefficient. As tissues break down in response to injury or cytotoxic therapies, barriers to diffusion are reduced and this leads to an increase in the measured diffusion coefficient.

Restricted diffusion yields a signal attenuation curve that is nonmonoexponential in nature and cannot be completely defined by a single ADC value (Hurlimann et al., 1995; Woessner, 1963). In addition, deviations from monoexponential behavior can also arise from the fact that the signal is the weighted average of multiple compartments (e.g., extra- and intracellular compartments) (Niendorf et al., 1996). Thus, restricted diffusion and multiple compartments increase the complexity of the diffusion attenuation signal, and this has led to varying approaches to deconvolve the measured signal into multiple components (Clark and Le Bihan, 2000; Cory and Garroway, 1990; Mulkern et al., 1999; Niendorf et al., 1996; Pfeuffer et al., 1998, 1999). Due to restricted diffusion, ADC estimates are dependent on t_d employed in the experiment, where $D(t)$ represents the diffusion time-dependent ADC (Figure 9.2). At short t_d, $D(t)$ is reduced from the bulk diffusion coefficient (D_0) in direct proportion to the relative volume of a surface layer of water molecules since only the water molecules near the restrictive boundaries experience the

Diffusion-weighted images (*b*-values in s/mm²)

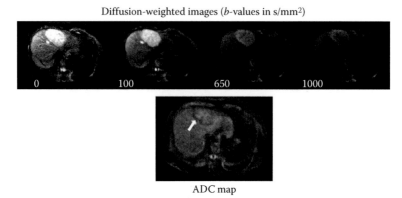

ADC map

FIGURE 9.3

Diffusion-weighted MRI of a colorectal cancer patient with a metastatic tumor in the liver. Diffusion-weighted images at different *b*-values (top row) and the calculated ADC spatial map (bottom row). Images were acquired at four different diffusion weightings. The tumor consists of viable rim (solid white arrow) that is characterized by low ADC surrounded by a likely necrotic core that exhibits higher ADC values. (Courtesy of Dr. Alexandre Coimbra, Clinical Imaging Group, Early Clinical Development, Genentech, Inc., South San Francisco, CA.)

restrictions and reflect local properties of the environment (surface-to-volume ratio) (Mitra et al., 1992, 1993). As t_d increases, $D(t)$ approaches a diffusion time-independent constant value (D_{eff}), where all diffusing water molecules within the restrictive environment have sufficient time to effectively experience the same restrictions that comprised the confining space (Figure 9.2). At long t_d, $D(t)$ estimates are made within the effective media regime, where $D(t)$ has become constant and the deviation from D_0 is proportional to the tortuosity of the restrictive space (Helmer et al., 1995).

Barriers to diffusion can also lead to preferred directions for diffusion. As an example, white matter tracts exhibit a high degree of diffusion anisotropy due to the hydrophobic nature of myelin and the organization of the tracts. Alternatively, diffusion within tumors has been found to be isotropic and this is consistent with tumor biology since tumors are produced from undifferentiated growth (Taouli et al., 2003). Often, outside the brain, tumor DWI is performed in only one direction to reduce scan times (Berry et al., 2008; Carano et al., 2004b; Henning et al., 2007). Due to the high anisotropic nature of brain tissue, diffusion imaging of brain tumors is usually performed along three orthogonal directions or by employing diffusion tensor imaging (DTI) to measure the diffusion tensor on a voxel-by-voxel basis. Generally, the choice of *b*-values (number, magnitude, range, direction) is a trade-off between the imaging time and the diffusion properties of tissue that are of interest for the particular application.

Diffusion within tumors: The rapid growth of tumor's cells lead to a highly heterogeneous mass that includes regions of densely packed viable tumor cells and regions of necrosis (produced due to a lack of an adequate blood supply). DWI has been applied to address the high degree of intratumor heterogeneity observed in solid tumors by employing b-values that focus on the extravascular spins and eliminate the vascular signal [minimum b-value $\geq$100 s/mm² (Padhani et al., 2009)]. Generally, viable tumor exhibits a relatively low ADC when compared to other tissues. Table 9.2 provides examples of measured ADC values obtained from both preclinical and clinical studies. ADC values show no dependence on field strength. The range of reported mean ADC values of viable tumor is likely to have been widened by the fact that a number of studies employ low b-values that are too low to suppress flow contamination to the estimates, even on occasions including b-values of 0 s/mm² (Table 9.2).

With the exception of brain (gray and white matter) and spinal cord, ADC values for solid tumors are generally lower for viable tumor than ADC estimates obtained from neighboring normal tissues. The low diffusion values within tumors are very apparent in whole-body applications and contribute to the success of whole-body DWI tumor visualization studies (Figure 9.4). Whole-body DWI has been proposed as a noninvasive and radiation-free alternative to FDG-PET as a means for tumor detection (Takahara et al., 2004) and initial comparisons have demonstrated equivalent accuracy (Komori et al., 2007; Ohno et al., 2008). The low ADC values for viable tumor are generally attributed to the high cellular density of viable tumor, but other factors may contribute, as an example the presence of extracellular fibrosis will restrict water mobility. ADC values have been shown to be correlated with cellular density for a number of tumors (Guo et al., 2002; Lyng et al., 2000; Sugahara et al., 1999). Another example of tissue that becomes very dense is the enlarged spleen (splenomegaly) due to extramedullary hematopoiesis that can occur due to cancer. Mouse models of cancer can induce splenic changes, where the spleen can increase 15 times in size and lead to a high cellular density within the enlarged spleen. This has been shown to result in a 60% reduction in splenic ADC (Barck et al., 2009). Splenic hyperintensity has also been observed in whole-body diffusion images of cancer patients, which is consistent with low diffusion rates (Komori et al., 2007).

ADC estimates have shown promise in differentiating malignant lesions from benign lesions when two fairly homogeneous groups (such as benign and malignant) with relatively narrow ADC ranges are compared. Researchers have found differences between mean ADC values for malignant and benign lesions in patients. Malignant breast tumors were found to have lower ADC values than were found for benign lesions (Guo et al., 2002; Woodhams et al., 2005). An ADC threshold (160 × 10⁻⁵ mm²/s) was found to have a sensitivity of 95% and specificity of 46% in breast cancer differentiation (Woodhams et al., 2005). DWI has also shown promise for liver lesions, where malignant lesions were found to have lower ADC values than benign lesions (Kilickesmez et al., 2009; Parikh et al., 2008; Taouli et al., 2003). For liver lesion characterization, malignancy identification was found to have a sensitivity of 74.2%, specificity of 77.3%, and an accuracy of 75.3% for a threshold ADC of 160 × 10⁻⁵ mm²/s (Parikh et al., 2008). In addition,

TABLE 9.2

Reported Values for the ADC for Both Normal Tissues and Tumors from Preclinical and Clinical Studies

Location	Tissue	Host	Mean ADC ($\times 10^{-5}$ mm²/s)	b-Values (s/mm²)	B_0 field (T)	Description	References
SC models	Viable tumor	Mouse	47 ± 8	85, 201, 368, 584, 849, 1165	4.7	SC flank, Colo 205 colorectal	Carano et al. (2004b)
	Viable tumor	Mouse	55 ± 9	84, 198, 362, 575, 837, 1147	4.7	SC leg, HM7 colorectal	Berry et al. (2008)
	Viable tumor, hypoxic	Mouse	112 ± 1	15, 60, 140, 390, 560, 760	2.0	SC leg, RIF-1, fibrosarcoma	Henning et al. (2007)
	Viable tumor, oxygenated	Mouse	76 ± 1	15, 60, 140, 390, 560, 760	2.0	SC leg, RIF-1, Fibrosarcoma	Henning et al. (2007)
	Necrosis, cystic	Mouse	161 ± 41	85, 201, 368, 584, 849, 1165	4.7	SC Flank, Colo 205 colorectal	Carano et al. (2004b)
	Necrosis, cystic	Mouse	156 ± 24	84, 198, 362, 575, 837, 1147	4.7	SC leg, HM7 colorectal	Berry et al. (2008)
	Necrosis, cystic	Mouse	212 ± 5	15, 60, 140, 390, 560, 760	2.0	SC leg, RIF-1, fibrosarcoma	Henning et al. (2007)
	Necrosis w/hemorrhage	Mouse	126 ± 13	85, 201, 368, 584, 849, 1165	4.7	SC flank, Colo 205 colorectal	Carano et al. (2004b)
	Necrosis w/hemorrhage	Mouse	151 ± 12	84, 198, 362, 575, 837, 1147	4.7	SC leg, HM7 colorectal	Berry et al. (2008)
	Necrosis w/hemorrhage	Mouse	179 ± 1	15, 60, 140, 390, 560, 760	2.0	SC leg, RIF-1, fibrosarcoma	Henning et al. (2007)
Brain	Viable tumor	Rat	91 ± 5	87–1669 (42 b-values)	2.0	Orthotopic rat 9L glioma	Chenevert et al. (1997)
	Solid tumor enhancing	Human	131 ± 23	79.4, 401.4		Glioma	Brunberg et al. (1995)
	Necrosis, cystic	Human	235 ± 35	79.4, 401.4		Glioma	Brunberg et al. (1995)
	Gray matter	Human	90 ± 4	79.4, 401.4		Glioma	Brunberg et al. (1995)
	White matter	Human	83 ± 6	79.4, 401.4		Glioma	Brunberg et al. (1995)
Breast	Normal	Human	209 ± 27	0, 750, 1000	1.5	Normal breast tissue	Woodhams et al. (2005)
	Viable tumor	Mouse	68 ± 23	100, 1000	9.4	Orthotopic, 4T1 mouse breast cancer	Barck et al. (2009)
	Solid tumor	Human	97 ± 20	0, 1000	1.5	Malignant lesion	Guo et al. (2002)
	Solid tumor	Human	122 ± 31	0, 750, 1000	1.5	Malignant lesion	Woodhams et al. (2005)
	Benign lesion	Human	157 ± 23	0, 1000	1.5	Benign lesions	Guo et al. (2002)
	Benign lesion	Human	167 ± 54	0, 750, 1000	1.5	Benign lesions	Woodhams et al. (2005)
	Cyst	Human	235 ± 8	0, 1000	1.5	Cyst	Guo et al. (2002)

Organ	Tissue	Species	ADC value	b-values	Field	Lesion	Reference
Liver	Normal	Human	151 ± 49	0, 134, 267 400	1.5	Normal liver	Taouli et al. (2003)
	Normal	Human	156 ± 14	0, 500, 1000	1.5	Normal liver	Kilickesmez et al. (2009)
	Tumor	Human	85 ± 50	0, 134, 267, 400	1.5	Metastatic Lesion	Taouli et al. (2003)
	Tumor	Human	139 ± 24	0, 134, 267, 400	1.5	HCC	Taouli et al. (2003)
	Tumor	Human	139 ± 38	0, 50, 500	1.5	Malignant lesion	Parikh et al. (2008)
	Tumor	Human	115 ± 36	0, 500, 1000	1.5	HCC	Kilickesmez et al. (2009)
	Benign lesion	Human	219 ± 67	0, 50, 500	1.5	Benign lesion	Parikh et al. (2008)
	Benign lesion	Human	164 ± 33	0, 134, 267, 400	1.5	Benign lesion	Taouli et al. (2003)
	Hemangiomas	Human	204 ± 42	0, 50, 500	1.5	Hemangiomas	Parikh et al. (2008)
	Hemangiomas	Human	284 ± 71	0, 134, 267, 400	1.5	Hemangiomas	Taouli et al. (2003)
	Hemangiomas	Human	197 ± 49	0, 500, 1000	1.5	Hemangiomas	Kilickesmez et al. (2009)
	Cysts	Human	254 ± 67	0, 50, 500	1.5	Liver cyst	Parikh et al. (2008)
	Cysts	Human	289 ± 83	0, 134, 267, 400	1.5	Liver cyst	Taouli et al. (2003)
Lung	Tumor	Human	101.9 ± 36	0, 1000	1.5	Malignant tumor	Uto et al. (2009)
	Benign lesion	Human	115.1 ± 31	0, 1000	1.5	Benign lesion	Uto et al. (2009)
Spleen	Normal	Mouse	115 ± 6	100, 1000	9.4	Mouse spleen	Barck et al. (2009)
	Splenomegaly	Mouse	47 ± 4	100, 1000	9.4	Orthotopic, 4T1 mouse breast cancer	Barck et al. (2009)

Notes: ADC, apparent diffusion coefficient; SC, subcutaneous.

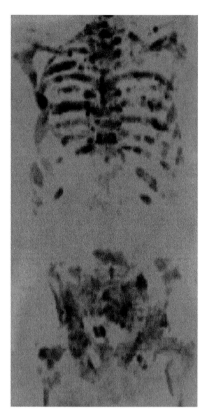

FIGURE 9.4
Whole-body diffusion-weighted image of a 63-year-old man with widespread metastatic bone disease from prostate cancer. The image is an inverted grayscale maximum intensity projection of the diffusion-weighted image acquired with a b-value of 900 s/mm^2. (Courtesy of Dr. Matthew Blackledge and Dr. Dow-Mu Koh, Department of Radiology, Royal Marsden Hospital and Institute of Cancer Research, Surrey, UK.)

overall DWI liver lesion detection rate was significantly higher for DWI over T_2-weighted imaging (87.7% vs. 70.1%, respectively) (Parikh et al., 2008). DWI has been found to match, or, in some cases, exceed the performance of T_2-weighted imaging for the characterization of prostate lesions (Chen et al., 2008; Miao et al., 2007). Despite these successes, several studies did not detect ADC differences between malignant and benign liver lesions (Vandecaveye et al., 2009; Wang et al., 2010); although one of these studies did find that the ratio of tumor to normal tissue SI obtained for an intermediate b-value (600 s/mm^2) image provided high sensitivity and specificity for malignant/benign lesion differentiation (95.2% and 82.7%, respectively) (Vandecaveye et al., 2009). DWI has demonstrated strong promise in prostate cancer (Figure 9.5). The combination of ADC data with T_2-weighted images has been shown by a number of groups to improve prostate cancer

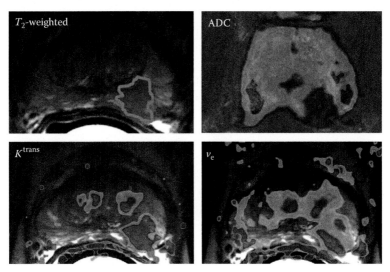

FIGURE 9.5

(See color insert.) DWI and DCE-MRI images in a 62-year-old man with T3a Gleason 4 + 3 prostatic carcinoma. The T_2-weighted image shows an ill-defined low signal nodule in the left peripheral zone (outlined in red). ADC mapping demonstrates the tumor extent (low ADC indicated by maroon-blue color). DCE-MRI enhancement–time curve in this region (not shown) shows rapid enhancement followed by washout, indicating malignancy. K^{trans} and v_e maps show the tumor extent (high K^{trans} and high v_e indicated by reddish color), with early extra-capsular infiltration into surrounding periprostatic fat. (Courtesy of Dr. Thomas Hambrock, Department of Radiology, University Medical Centre, Nijmegen, the Netherlands.)

detection and characterization over T_2-weighted images alone (Haider et al., 2007; Morgan et al., 2007; Yoshimitsu et al., 2008) and has clinical use in targeting repeat biopsy in those patients known to have cancer but with persistent negative biopsy. DWI also has use in detecting early increase in malignant potential (based on the Gleason scoring system) (Figure 9.5). DWI has also shown promise as a means to characterize lung tumors based on image intensity ratios (Uto et al., 2009), whereas ADC-based discrimination has had limited success for lung tumors (Koyama et al., 2010; Uto et al., 2009). Absolute ADC values were found to poorly discriminate between malignant and benign lesions, but the ratio of the high b-value image SI for the lung lesion to the spine (lung-to-spine ratio, LSR) produces a strong endpoint, with a sensitivity and specificity of 83.3% and 90%, respectively (Uto et al., 2009). The inability of quantitative ADC values to provide similar performance as the ratio metric (LSR) may be due to the inherent difficulties associated with lung imaging (low SNR, motion, low proton density, and magnetic susceptibility effects). The ADC values were calculated from two images, while the LSR is a ratio formed from voxel values within a single image and, thus, may benefit from the cancellation of noise effects that are present within the single image.

The application of DWI to oncology has achieved its most success as a means to differentiate regions of cell death, from regions of viable tumor (Brunberg et al., 1995; Helmer et al., 1995; Le Bihan et al., 1986; Maeda et al., 1992). Necrotic regions have ADC values that are two- to fourfold higher (Table 9.2) than ADC values for viable tumor. Both necrosis and apoptosis lead to cell death, and these processes are followed by the eventual removal of cellular debris by macrophage activity. These processes produce an expansion of the extracellular space causing a reduction in the barriers that restrict movement of water molecules. As these barriers (cell membranes, organelles, etc.) break down, a rise in ADC occurs (Figure 9.2). This relationship between ADC and necrosis has been confirmed in both experimental (Helmer et al., 1995; Maier et al., 1997) and clinical tumors (Brunberg et al., 1995; Krabbe et al., 1997). The contrast between viable tumor and necrotic regions improves as t_d is increased due to an increase likelihood that diffusing water molecules will interact with the restrictions found in their local environment, and, thus, allow greater time for the water molecules to sample their environment (Helmer et al., 1998). Despite DWI's ability to differentiate between necrotic and viable tumor, DWI's ability to distinguish viable tumor and other neighboring viable tissues can be hindered by edema and is limited in regions where ADC values are relatively similar such as the brain (Els et al., 1995). DWI has been applied to cytotoxic therapies and vascular disrupting agents in attempt to detect an early therapeutic response prior to changes in tumor volume. DWI has been shown to be able to detect an antitumor therapeutic response in animal models of cancer (Chenevert et al., 1997; Galons et al., 1999; Poptani et al., 1998) and clinically (Chenevert et al., 2000; Mardor et al., 2003; Pickles et al., 2006; Theilmann et al., 2004), with mean tumor ADC values increasing within days after treatment (Chenevert et al., 1997, 2000; Galons et al., 1999; Mardor et al., 2003; Pickles et al., 2006; Poptani et al., 1998; Theilmann et al., 2004). Mean ADC values have been found to be more sensitive to the early effects of anticancer therapy than conventional MRI parameters (T_1 and T_2) (Chenevert et al., 1997).

An early transient decrease in ADC has also been observed within hours after treatment with vascular disruption agent (VDA), combretastatin A4 phosphate, in a rat tumor model (Thoeny et al., 2005). Researchers employed a wide range of b-values to calculate an ADC_{low}, ADC_{high}, and ADC_{total}, from the low range, high range, and full range of b-values employed, respectively. ADC_{low} was used as a surrogate for microvascular perfusion and ADC_{high} provided a measure of extravascular diffusion as a means to investigate microstructural changes of the tissue. They saw a decrease in ADC_{low} at 1 and 6 hours post treatment, but it became elevated at 48 hours. ADC_{high} was reduced at 6 hours and elevated at 48 hours post treatment (Thoeny et al., 2005). The decrease in ADC_{low} was attributed to a reduction in perfusion. The reduction in ADC_{high} was attributed to cell swelling that occurred from the reduction in perfusion, possible hypoxia and energy failure. The elevation in ADC_{low} and ADC_{high} at 48 hours was due to a large increase in percentage of necrotic tissue and that the mean values were averaged over the entire tumor (necrotic core and viable rim). Combretastatin A4 phosphate has been evaluated clinically by DWI where a similar increase in ADC_{high} was seen (Koh et al., 2009). ADC_{low} values were highly variable and did not detect a treatment response. Antiangiogenic agents that target the VEGF pathway have been linked to a reduction in ADC. The reduction in ADC observed in glioma patients has been found to correspond to an area of reduced perfusion and hypoxia (Rieger et al., 2010). The reduction in ADC may be due to a VDA-like response. Or, it is likely that a prolonged reduction in ADC is due to a reduction in edema and a contraction of the extracellular space (Padhani et al., 2009).

DWI has also shown potential to predict a successful treatment response prior to initiation of therapy in a number of tumor types (De Vries et al., 2003; Dzik-Jurasz et al., 2002; Koh et al., 2007; Mardor et al., 2004; Park et al., 2010a, 2010c; Pope et al., 2009, 2011; Yuan et al., 2009). Low mean tumor ADC values have been shown to be predictive of a positive therapeutic response in colorectal (De Vries et al., 2003; Dzik-Jurasz et al., 2002; Koh et al., 2007), brain (Mardor et al., 2004), liver (Yuan et al., 2009), prostate (Park et al., 2010c), and breast (Park et al., 2010a) cancers for both chemotherapy and chemoradiation therapy. Dzik-Jurasz et al. found a strong negative correlation between mean pretreatment ADC and the percent regression in tumor size for both chemotherapy and combined chemoradiation therapy in rectal cancer patients (2002). Mardor et al. found a similar relationship for radiotherapy of brain lesions and concluded that low pretreatment diffusion values indicate that highly viable tumors respond better to radiation therapy than highly necrotic (high ADC) tumors (2004). This is consistent with the well-established phenomena that highly necrotic tumors respond poorly to radiation because of low oxygen levels in these highly hypoxic tumors (Brown and Wilson, 2004).

The ability for ADC to predict the response of antiangiogenic therapies that target VEGF appears to be more complex (Pope et al., 2009, 2011). For newly diagnosed glioblastoma multiforme (GBM) patients, lower pretreatment ADC values were found to predict longer progression-free survival (PFS) (Pope et al., 2011). Alternatively, higher pretreatment ADC values have been shown by the same research group to be predictive of an improved therapeutic response to anti-VEGF in recurrent GBM patients (Pope et al., 2009). The approach employed in these two studies was to fit the whole enhancing tumor ADC histograms with a two normal distribution mixture model. The pretreatment ADC histogram analysis determined that the mean of the lower ADC histogram population (ADC_L) was able to stratify bevacizumab-treated patients based on ADC threshold value.

It should be emphasized, however, that most of these studies are small and retrospective analyses, often on relatively old clinical data. These are not insignificant limitations, and well-designed, larger prospective studies are required to confirm these data and to determine the clinical use, if any, for DWI-based techniques in monitoring response to therapy and predicting clinical outcome.

9.3.2 Dynamic Contrast-Enhanced MRI

Dynamic contrast-enhanced (DCE)-MRI enables noninvasive quantification of microvascular structure and function and has been increasingly used in cancer radiology since the early 1990s. In this technique, an intravenous bolus of gadolinium contrast agent is injected and rapid serial measurements of image SI are made as the contrast agent enters tumor arterioles, passes through capillary beds, and washes out of the tumor. Since gadolinium ions are paramagnetic, they interact with nearby hydrogen nuclei to shorten the T_1 times of local tissue water and thus increase SI on T_1-weighted images. The degree of signal enhancement reflects the physiological and physical factors including tissue perfusion, the concentration–time course of contrast agent in the artery that supplies the vascular bed [the arterial input function (AIF)], capillary surface area, capillary permeability, and the volume of the extracellular extravascular space (EES) (Figure 9.6) (O'Connor et al., 2007).

In most clinical applications, DCE-MRI is performed in nonspecialist centers and analysis is tailored toward describing simple qualitative metrics such as curve shape or rapidity of enhancement and washout. In most scenarios, the technique is used to characterize lesions as benign or malignant and is used as part of a wider MRI examination, including anatomical and other advanced techniques such as DWI. Here, the derived MRI parameters do not have a clear defined relationship to anatomical or physiological

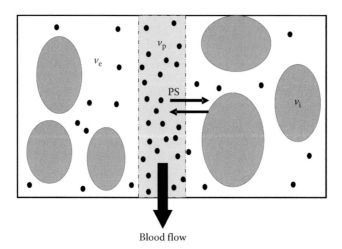

Blood flow

FIGURE 9.6
Blood flows through a tumor enabling contrast media molecules (represented as black dots) to distribute in two compartments—the blood plasma volume (v_p) and the volume of the extracellular extravascular space (v_e). Most clinically available MRI contrast agents do not leak into the intracellular space (v_i). Contrast agent leakage is governed by the concentration difference between the plasma and the extracellular extravascular space and by the permeability and surface area of the capillary endothelia, expressed as PS. The relationship between these compartments is considered differently in various pharmacokinetic models.

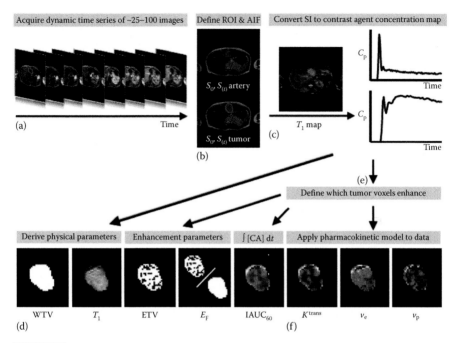

FIGURE 9.7
(See color insert.) Overview of DCE-MRI data acquisition and analysis. (a) Multiple images are acquired as a bolus of contrast agent passes through tissue capillaries. (b) ROI for a tumor and the feeding vessel AIF are defined. (c) SI values or each voxel are converted into contrast agent concentration, using a map of T_1 values. (d) These steps allow calculation of tumor volume and T_1 values. (e) Next, voxels are classified as enhancing or not after which parameters based on the amount and proportional enhancement can be defined along with the $IAUC_{60}$. (f) Finally, a pharmacokinetic model may be applied to derive parameters such as K^{trans}.

correlates and interpretation is largely subjective and reliant on an expert opinion by a highly trained radiologist on a per-patient basis. In distinction, many research applications rely on the derivation of quantitative metrics that estimate values of flow, blood volume, or vessel permeability. These methods require highly refined acquisition and analysis protocols (Figure 9.7), and the interpretation is largely quantitative and based on statistical analysis of data across a group of patients.

One of the major clinical uses for DCE-MRI is in breast cancer, where various analysis techniques are used in routine practice (Turnbull, 2009). At one end of the spectrum, simple visual inspection of SI curves can be performed with data acquired every 30–60 s or so to define the rate of enhancement and shape of the SI curve (Kuhl et al., 1999). This approach is simple to perform with a region of interest (ROI) drawn on a series of images (with or without correction for motion), can help distinguish benign from malignant lesions (Figure 9.8), and is still used daily by experienced radiologists in the evaluation of breast lesions. Several more complex techniques that can be performed on conventional clinical workstations offer further applications.

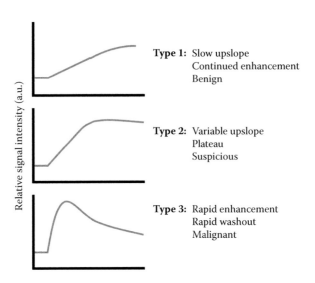

FIGURE 9.8

Schematic cartoon of SI curves used to characterize breast lesions in clinical practice: Most (though not all) malignant tumors are characterized by rapid arterial enhancement and rapid washout. In clinical practice, these features are important but are always used in conjunction with other morphological MRI parameters to determine the likelihood that a lesion is malignant.

These range from deriving parameters based on the SI curve (such as the time to peak, TTP) through to pharmacokinetic analysis using one of various available models to describe the passage of contrast agent through tissue, all of which are variants of the kinetic model described by Kety (1951) (detailed below). These more advanced analyses typically utilize data registration to reduce any motion, and require calculation or estimation of tissue T_1 and an AIF. A formal ROI must be defined either by the radiologist or by using an automated software package (Szabo et al., 2004).

DCE-MRI has been used since the early to mid-1990s to characterize malignant breast lesions. In general terms, malignant lesions were found to have washout of contrast agent after initial rapid enhancement, the latter feature relating in part to greater microvessel density (Buckley et al., 1997) and greater microvascular permeability (Knopp et al., 1999b). Over time, the relative value of various features such as rim enhancement and lesion speculation have been realized and incorporated into scoring systems such as the American College of Radiology's BI-RADS-MRI lexicon that provides a simple method of reporting likelihood of malignancy in a breast lesion (Erguvan-Dogan et al., 2006).

While there has been considerable variation in the spatial coverage, spatial resolution, and temporal resolution of analysis used in studies, DCE-MRI has a sensitivity of over 95% in characterizing breast lesions (Benndorf et al., 2010; Shimauchi et al., 2010). Cited specificity is highly variable (between 40% and 95%) dependent on particular methods for acquisition and analysis (Benndorf et al., 2010). Multiple studies suggest that the area under the curve for receiver-operator characteristic (ROC) curve analysis is typically

0.8–0.9 depending on each study analysis (El Khouli et al., 2009). In practice, most patients still undergo ultrasound or x-ray mammographic assessment and DCE-MRI is reserved for problem solving of cases that remain equivocal. Therefore, the true positive predictive value (PPV) for the technique as used in clinical practice may be less than the above figures imply.

While x-ray mammography is an effective screening method in the normal population, particularly in those over 50, it is less effective in younger high-risk women with BRACA-1, 2, and other mutations, where the higher proportion of breast parenchyma to fat often results in dense mammograms that are hard to interpret. Multiple large multicenter studies have evaluated the role of DCE-MRI in screening for cancers in women with a high genetic risk of breast cancer (Kriege et al., 2004; Leach et al., 2005a). While there are differences in the age range of patients and in the percentage risk for developing breast cancer relative to the normal population required for entry into each study, results are very consistent with DCE-MRI, showing clearly that DCE-MRI is the most sensitive method of detecting cancer in women with strong familial risk of breast cancer. For example, in the magnetic resonance imaging in breast screening (MARIBS) study (Leach et al., 2005a), a standardized acquisition and analysis approach was used in 22 centers, following training to ensure compliance. Prescriptive analyses were followed, using both morphology (well or poorly defined, spiculation), enhancement pattern (homogeneous, heterogeneous, rim like), simple quantification of enhancement upslope, and washout pattern [based on the grading system of Kuhl et al. (1999)]. This system allowed lesions to be graded as malignant, suspicious, or benign. Across these eight studies, the sensitivity of DCE-MRI was approximately twice that of x-ray mammography (75–90% vs. 40%). Combined with mammography, DCE-MRI has a sensitivity of over 90%, with a recall rate of around 10%. The additional benefit was largely from detection of small, node-negative tumors. The strength of these results has enabled guidelines to be issued regarding the high-risk groups who may benefit from annual breast screening using DCE-MRI (NICE, 2006).

Around 5% of patients with breast cancer receive cytotoxic neoadjuvant chemotherapy (NAC). Since approximately one-third respond poorly to this therapy, there is interest in using advanced imaging techniques to identify those patients who are responding to NAC as soon as possible, rather than wait to the end of a course of chemotherapy after many months. Multiple independent studies have reported statistically significant reduction in either the area under the contrast agent concentration–time curve at 60 s ($IAUC_{60}$), or the volume transfer constant (K^{trans}) in breast cancer at or around two completed cycles of therapy in patients treated with a variety of different cytotoxic regimens. These small studies suggest that larger reductions in K^{trans} appear to indicate subsequent favorable response (Baar et al., 2009; Hayes et al., 2002; Martincich et al., 2004). This finding that is consistent with preliminary evidence from 120 patients suggests that those with persistent higher K^{trans} values after around two cycles of treatment with NAC have a significantly worse overall survival compared to those with a lower/reduced K^{trans} value, using multivariate analysis that included other imaging and nonimaging indicators of prognosis (Li et al., 2011a). Further studies are required to test if these findings are repeatable, robust, and can be translated into altering treatment selection and improving outcome for those patients who show early indicators of poor response to NAC. Applications for DCE-MRI and T_2^*-weighted dynamic MRI are well described in other tumor types, notably in prostate cancer (Figure 9.5), liver metastases, and in glioma, but are beyond the scope of this chapter.

Calculation of quantitative microvascular parameters from DCE-MRI data is complex, varies between studies, and requires conversion of signal enhancement into information concerning contrast agent concentration [CA]. Unlike CT, the relationship between SI and [CA] is not linear except over a narrow range of [CA] (Roberts, 1997). In MRI, SI data must be converted into tissue T_1, using the relationship between T_1 and [CA]:

$$\frac{1}{T_1} = \frac{1}{T_{1(0)}} + r_1[CA], \tag{9.3}$$

where $1/T_1$ is the measured R_1 after contrast, $1/T_{1(0)}$ is the baseline R_1, and r_1 is the relaxivity constant for CA in a pure solvent. From this, it is apparent that it is possible to calculate [CA] by measuring change in T_1 and having prior knowledge of r_1. This enables calculation of [CA] and from there the initial area under the gadolinium concentration–time curve (IAUC) can be derived:

$$\text{IAUC}_t = \int_0^t [CA](t')\,dt', \tag{9.4}$$

where [CA] (t') represents the concentration of contrast agent measured in the tissue at time t'. IAUC can be measured at various points after injection of contrast agent, but is typically measured at 60 s. Although IAUC_{60} is biologically nonspecific, reflecting a combination of flow, blood volume, vessel permeability, vessel surface area, and EES volume, it is a reasonably reliable and reproducible quantity and is widely used as a surrogate biomarker in clinical studies (Evelhoch, 1999; O'Connor et al., 2012).

Next, voxel enhancement (following administration of contrast agent) must be defined and a decision made as to whether subsequent data analysis is made on all voxels or simply on those that enhance. Inclusion of voxels that do not enhance (zero values) artificially lowers the mean and/or median parameter values. Exclusion of zero values can obscure important heterogeneity information and can increase motion errors by effectively increasing the surface area of the normal tissue boundary around the sampled tumor (Jackson et al., 2007). Voxel enhancement can be used to generate useful vascular biomarkers, based on the proportion of voxels that enhance within a tumor. Few oncology studies have reported the enhancing fraction (E_F), but it has shown promise as a simple biomarker of the tumor vasculature. In a preclinical study of the tubulin-binding agent ZD6126, high enhancing pixels were defined as those that had greater IAUC values than the median IAUC value in muscle (Robinson et al., 2003). Over 90% of tumors showed a reduction in the proportion of highly enhancing pixels demonstrated following therapy. Similarly, this parameter is sensitive to antivascular therapy as seen in colorectal cancer liver metastases treated with bevacizumab (O'Connor et al., 2009) and cediranib (Mitchell et al., 2011).

Pharmacokinetic modeling techniques of varying complexity enable *estimates* of physiological characteristics such as blood flow and capillary endothelial permeability to be calculated. Measured data (the tumor concentration–time course curve and an AIF) are "fitted" to mathematical equations, from which parameters that reflect microvascular function are derived (Parker and Buckley, 2005). In theory, these parameters are independent of acquisition protocol and allow use in multicenter studies employing varying image acquisition protocols and equipment. Choice of analysis technique is crucial since it will determine not only the range of parameters available but also their *precise meaning*. In all cases, there is an inherent compromise between extracting relatively simple parameters with poor specificity versus physiologically congruent but less statistically stable parameters. Commonly used standard terms are listed in Table 9.3 and explained in Figures 9.6 and 9.7.

The simple Tofts and Kermode variant (1991) of the Kety model (1951) describes the concentration of contrast agent within each voxel, thus allowing calculation of the size of the EES (v_e) as well as K^{trans}:

$$C_t(t) = K^{trans} \int_0^t C_p(t') \exp\left(\frac{-K^{trans}}{v_e}\right) t' dt', \tag{9.5}$$

where $C_t(t)$ is the concentration of agent in the voxel at time t, C_p is the concentration of agent in the plasma volume, and K^{trans} is the volume transfer constant in one direction. Other commonly used models vary in their terminology and in some details (Brix et al., 1991; Larsson et al., 1990) but are based on the same tracer kinetic model. This model has been extended by Tofts (1997) to enable calculation of plasma volume in addition to v_e and K^{trans}, given by

$$C_t(t) = v_p C_p(t) + K^{trans} \int_0^t C_p(t') \exp\left(\frac{-K^{trans}}{v_e}\right) t' dt', \tag{9.6}$$

where v_p is the fractional plasma volume, and has been used in several clinical studies. As well as producing an estimate of v_p, this model alters the scalar magnitude of K^{trans} values, typically reducing them several fold (O'Connor et al., 2012). Alternatively, more complex models based on the tissue homogeneity model introduced by Johnson and Wilson allow direct estimation of permeability surface area product (PS), F, and mean capillary transit time (τ) and derivation of K^{trans} (Johnson and Wilson, 1966; St Lawrence and Lee, 1998), where

$$C_t(t) = F_p \int_0^\tau C_p(t-t') dt' + EF_p \int_\tau^t C_p(t-t') \exp\left(\frac{-EF_p(t'-\tau)}{v_e}\right) dt'. \tag{9.7}$$

TABLE 9.3

Commonly Used Parameters in DCE-MRI Studies

Symbol	Short name	Unit	Notes
Recommended primary endpoints (Leach et al., 2005b)			
K^{trans}	Volume transfer constant between plasma and the EES	min^{-1}	Composite measure of permeability, capillary surface area, and flow
$IAUC_{60}$	Initial area under Gd concentration agent–time curve at 60 s	mM min	Similar measure to K^{trans}, but also influenced by v_e
Alternative functional biomarkers derived from tracer kinetic modeling			
k_{ep}	Rate constant between EES and plasma	min^{-1}	
v_e	Volume of EES per unit volume of tissue	None	
v_p	Blood plasma volume	None	Relatively poor reproducibility
F	Blood flow	$ml/g\ min^{-1}$	Requires faster temporal resolution than achieved in most studies
PS	Permeability surface area product per unit mass of tissue	$ml/g\ min^{-1}$	Requires faster temporal resolution than achieved in most studies
Alternative biophysical measurements that are not derived from tracer kinetic modeling			
WTV	Whole tumor volume	mm^3	Simple to measure
ETV	Enhancing tumor volume	mm^3	Simple to measure
E_F	Enhancing fraction (ETV:WTV ratio)	None	Simple to measure
T_1	Longitudinal relaxation time	ms	Simple to measure

Notes: DCE-MRI, dynamic contrast-enhanced magnetic resonance imaging; EES, extracellular extravascular leakage space.

The models described in Equations 9.5 and 9.6 have an important limitation: change in blood flow (F), endothelial permeability, and endothelial surface area can all alter K^{trans} measurement. When contrast agent delivery is ample, K^{trans} represents the PS per unit volume of tissue for transendothelial transport between plasma and EES (i.e., $K^{trans} \sim$ PS). This assumption is not necessarily true and in these cases, K^{trans} represents the blood plasma flow per unit volume of tissue (Tofts et al., 1999). In distinction, the model described in Equation 9.7 can separate the perfusion and leakage phases. However, curve fitting is more complex and higher temporal resolution data are required (Parker and Buckley, 2005). The summary presented here is necessarily simplistic; a full discussion of the various kinetic models available, their uses, and their assumptions is found in an excellent review by Sourbron and Buckley (2012).

In practice, nearly all clinical trial DCE-MRI protocols to date (of around 100 studies) have employed the simple version of the Kety model outlined in Equation 9.5 and restricted tracer kinetic data analysis to calculating $IAUC_{60}$ or K^{trans} (O'Connor et al., 2012). Since blood flow and capillary endothelial permeability cannot be separated in this model, measured changes in K^{trans} cannot be assigned to either F or permeability (Leach et al., 2005b), but represent a composite of the two processes. Application of the extended Kety model enables v_p to be calculated, which also appears to be a useful pharmacodynamic biomarker of response to anti-VEGF antibodies and tyrosine kinase inhibitors (Hahn et al., 2008; O'Connor et al., 2009).

The models described above require measurement of an AIF along with the tumor contrast agent concentration–time course curve. These two functions are then used to quantify the passage of contrast agent through the tumor. Ideally, the AIF should be measured for each examination, since it varies between individuals and different examinations, reflecting physiological variation in cardiac output, vascular tone, renal function, and injection timing. However, AIF measurement is technically demanding and at best produces an indirect measurement from a nearby large artery that may differ from the vessel supplying the tumor (Parker and Buckley, 2005). Consequently, many groups do not measure the AIF and instead use an idealized mathematical function (Weinmann et al., 1984a). Although idealized functions do not reflect the true blood supply to the tumor at each examination, this approach does relax requirements on temporal resolution, slice positioning, and sequence choice in the imaging protocol.

Over 100 studies have used MRI-based measurement of tumor perfusion to evaluate drugs with proposed antivascular mechanisms of action (O'Connor et al., 2012). These compounds include anti-VEGF antibodies, tyrosine kinase receptor inhibitors (of VEGF and other growth factors that mediate tumor angiogenesis), and vascular disrupting agents. Despite this (now common) use of advanced MRI in many trials, it should be noted that the overwhelming majority of such studies do not use functional imaging of any kind, where imaging, if included, is used to provide tumor volume estimates.

Most DCE-MRI studies have calculated either $IAUC_{60}$ or K^{trans}, and there is compelling evidence that these and related parameters (influx constant K_i, k_{ep}, and PS) are consistently reduced following administration of several antivascular compounds with proven antivascular mechanism of action (Table 9.4). For example, statistically significant reductions in K^{trans} have been reported in nine published DCE-MRI studies of bevacizumab monotherapy with pooled data from 162 patients (Barboriak et al., 2007; Gururangan et al., 2010; Gutin et al., 2009; Kreisl et al., 2011; Li et al., 2011b; O'Connor et al., 2009; Siegel et al., 2008; Wedam et al., 2006; Zhang et al., 2009) (example parameter maps shown in Figure 9.9). This finding is consistent across multiple patient populations, at various time points between 4 hour and 28 days and is congruent with data from CT perfusion studies of bevacizumab-treated patients (Willett et al., 2009). These findings support the notion that VEGF blockade causes rapid and sustained antivascular effects that evolve during a single cycle of anti-VEGF therapy (O'Connor et al., 2009). Comparable results have been reported in studies of patients with solid tumors with clinically effective tyrosine kinase inhibitors (Drevs et al., 2007; Hahn et al., 2008; Morgan et al., 2003).

TABLE 9.4

Summary of Key Data from DCE-MRI Studies Investigating Novel Antiangiogenic and Vascular Disrupting Agents

Study (author)	Agent	N	Cancer type	Biomarkers derived	Imaging time points	Evidence for drug efficacy (statistically significant unless stated otherwise)	Dose-related biomarker change
Anti-VEGF antibody							
Wedam (Wedam et al., 2006)	Bevacizumab	19	Breast	K^{trans}, k_{ep}, v_e	0, d21	↓ K^{trans} 34%, k_{ep} 15%, v_e 14%	—
Li (Li et al., 2011b)	Bevacizumab	17	Breast	$IAUC_{60}$, K^{trans}, v_e, R_2^*	0, 14	↓ K^{trans} 30%, ↓ $IAUC_{60}$ 29%	—
O'Connor (O'Connor et al., 2009)	Bevacizumab	10	mCRC	$IAUC_{60}$, K^{trans}, v_e, v_p, E_F, T_1	Base x2, 4 h, 48 h, d8, d12	↓ K^{trans} 19–25% 4 h–d8, ↓ v_p 33–41% 4 h–d12, ↓ E_F 18–22% 48 h–d12, ↓ T_1 18% d12	—
Barboriak (Barboriak et al., 2007)	Bevacizumab	7	GBM	K^{trans}, v_e, v_p	0, 24 h	↓ K^{trans} 43%, ↓ v_p 33%, ↓ v_e 60%	—
Gutin (Gutin et al., 2009)	Bevacizumab	10	GBM	K^{trans}, v_p	0, d27	↓ K^{trans} 30% and ↓ v_p 27%	—
Zhang (Zhang et al., 2009)	Bevacizumab	45	GBM	K^{trans}, k_{ep}, v_p, ETV	0, ≤d4	↓ K^{trans} 46%, ↓ ETV 39% and significant ↓ v_p	—
Gururangan (Gururangan et al., 2010)	Bevacizumab	15	GBM/AA	K^{trans}	0, d15	↓ K^{trans} in 6/15 patients only	
Kreisl (Kreisl et al., 2011)	Bevacizumab	31	AA	K^{trans}, k_{ep}, v_p, ETV	0, ≤d4, d28	↓ K^{trans} 31–52% d4 and d28, ↓ ETV 44–57% d4 and d28 and ↓ v_p 21–46% d4 and d28	—
Siegel (Siegel et al., 2008)	Bevacizumab	8	Liver	SI enhancement	0, d56	↓ enhancement 87%	—
VEGFR inhibitor							
O'Donnell (O'Donnell et al., 2005)	Semaxanib	24	Mixed	K^{trans}, v_e	0, 4 h, <d14, d28	None	NS
Drevs (Drevs et al., 2007)	Cediranib	31	Mixed	$IAUC_{60}$	0, d2, d28, d56	↓ $IAUC_{60}$ of ≥40% in majority of pts	↓ $IAUC_{60}$ correlated to AUC_{ss} and C_{max}

Study	Drug	n	Tumor	Parameters	Time points	Results	Correlation
Mitchell (Mitchell et al., 2011)	Cediranib	7	Mixed	IAUC$_{60}$, K^{trans}, v_e, v_p	0, d28, d56, d112	↓K^{trans} 33–44%, d28–d112, ↓v_p 64% d56	—
VEGFR/PDGFR inhibitor							
Morgan (Morgan et al., 2003)	Vatalanib	26	mCRC	K$_i$	0, d2, d28, d56	↓K$_i$ 58% d2 and 53% at d28 in ≥1000 mg	↓K$_i$ correlated to AUC$_{ss}$ and C$_{max}$
Mross (Mross et al., 2005)	Vatalanib	27	mCRC or breast	K$_i$	0, d2, d28, d56	↓K$_i$ 54% d2 and 51% at d28 in ≥1000 mg	↓K$_i$ correlated to AUC$_{ss}$ and C$_{max}$
Thomas (Thomas et al., 2005)	Vatalanib	35	Mixed	K$_i$	0, d2, d28, d56	↓K$_i$ 46% d2 and 40% at d28 in ≥1000 mg	↓K$_i$ correlated to AUC$_{ss}$ and C$_{max}$
Hahn (Hahn et al., 2008)	Sorafenib	48	mRCC	IAUC$_{90}$, K^{trans}, v_e, v_p	0, d28	↓K^{trans} 24%, ↓v_p 34%, ↓IAUC$_{90}$ 30% at 400 mg dose	Dose dependent ↓ in IAUC$_{90}$, K^{trans} and v_p
Flaherty (Flaherty et al., 2008)	Sorafenib	17	mRCC	K^{trans}, v_e	0, variable d19–77	↓K^{trans} 60%	—
VEGFR/FGFR inhibitor							
Jonker (Jonker et al., 2011)	Brivanib	52	Mixed	IAUC$_{60}$, K^{trans}	0, d2, d8, d26	↓K^{trans} 31–46% d2 at 800 mg dose, persists at d8 and d26; scheduling effect seen	↓K^{trans} correlated to AUC$_{ss}$
Antitubulin							
Galbraith (Galbraith et al., 2003)	Fosbretabulin	18	Mixed	IAUC$_{60}$, K^{trans}, k_{ep}, v_e, E$_F$	Base x2, 4–6 h or 24 h, d28	↓K^{trans} 29–37% at 4–24 h; 4/8 pts at doses ≥88 mg/m^2 had ↓v_e and ↓E$_F$	Yes—defined biologically active dose ≥52 mg/m^2
Stevenson (Stevenson et al., 2003)	Fosbretabulin	10	Mixed	K^{trans}, v_e, T$_1$	0, d5	↓K^{trans} >30% in 6/10 patients	↓K^{trans} at 52–75 mg/m^2
Akerley (Akerley et al., 2007)	Fosbretabulin	13	Mixed	K^{trans}	0, 24 h	↓K^{trans} 19–46%	↓K^{trans} at 45–63 mg/m^2
Nathan (Nathan et al., 2012)	Fosbretabulin	9	Mixed	K^{trans}	0, 4 h	↓K^{trans} ~25%	↓K^{trans} at 45–63 mg/m^2

Notes: DCE-MRI, dynamic contrast-enhanced magnetic resonance imaging; EES, extracellular extravascular leakage space; AA, anaplastic astrocytoma; FGFR, fibroblast growth factor receptor thyroid cancer; GBM, glioblastoma multiforme; mCRC, metastatic colorectal cancer; mRCC, metastatic renal cell cancer; PDGFR, platelet derived growth factor receptor; VEGFR, vascular endothelial growth factor receptor.

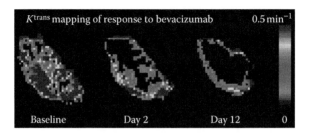

FIGURE 9.9

(See color insert.) DCE-MRI parameter maps of K^{trans} in a patient with a colorectal liver metastasis. At baseline, the distribution of K^{trans} is spatially heterogeneous with high values in the periphery and low values in the core. Following treatment with 10 mg/kg of bevacizumab, tumor K^{trans} was reduced at 2 days and further reduced after 12 days of treatment compared with baseline.

The relationship between an image biomarker and antitumor efficacy must also be accompanied by change in plasma or tumor PK if the biomarker is to be regarded as a true pharmacological indicator of drug efficacy. Vascular response (reduction in $IAUC_{60}$ or K^{trans} from baseline) shows clear dose-dependent relationships in trials of cediranib (Drevs et al., 2007), vatalanib (Morgan et al., 2003), sorafenib (Hahn et al., 2008), and pazopanib (Murphy et al., 2010), where relationships between drug dose and vascular response were assumed to be linear over the dose range examined. In many of these and other phase I studies, pharmacokinetic data demonstrate that active drug concentrations are reached within individual patients' plasma. The area under the plasma concentration–time curve at a steady state (AUC_{ss}) or the maximum drug concentration in the plasma (C_{max}) have correlated significantly with change in $IAUC_{60}$ or K^{trans} following numerous tyrosine kinase inhibitors, including cediranib (Drevs et al., 2007), pazopanib (Murphy et al., 2010), and vatalanib (Morgan et al., 2003; Mross et al., 2005; Thomas et al., 2005).

Two further examples of how DCE-MRI measurement of perfusion can benefit drug development are in selecting optimum drug dose and schedule for future studies. In four unrelated studies of the antitubulin agent fosbretabulin, vascular response was observed at 45 mg/m^2 or greater, defining a therapeutic window between the biologically active dose and the clinical maximum tolerated dose of 68 mg/m^2 (Akerley et al., 2007; Galbraith et al., 2003; Nathan et al., 2012; Stevenson et al., 2003). A study of brivanib (Jonker et al., 2011) evaluated different dose schedules in multiple cohorts with 12–15 patients per cohort, to determine optimum schedule for phase II studies. Statistically significant reductions in $IAUC_{60}$ and K^{trans} were seen with brivanib either 400 mg twice daily or 800 mg once daily in continuous daily dosing, but not with 800 mg intermittent dosing or at lower dose levels.

Importantly, many drugs with little demonstrable clinical benefits show little or no significant reduction in $IAUC_{60}$ or K^{trans}, for examples in studies of semaxanib (O'Donnell et al., 2005) and vandetanib in tumors that show little clinical response to the drug (Miller et al., 2005). Unfortunately, this clarity—significant reduction in biomarkers by active compounds and lack of change in biomarkers by ineffectual compounds—is not seen in all studies. Reductions in K^{trans} seen in studies of vatalanib (Morgan et al., 2003; Mross et al., 2005; Thomas et al., 2005) were comparable to those seen with bevacizumab, cediranib, sorafenib, and fosbretabulin, yet vatalanib has failed to demonstrate a survival benefit at phase III testing in CRC when used in combination therapy with cytotoxic agents (Hecht et al., 2011). Thus antivascular activity does not guarantee phase III success (Ellis, 2003), but it appears that statistically significant change in MRI-based perfusion biomarkers is necessary but not sufficient evidence of clinically significant efficacy of therapies secondary to antivascular mechanisms.

There are still many unresolved issues surrounding use of "perfusion" MRI in early phase clinical trials of antivascular agents. These include choice of biomarker (e.g., established primary end points such as $IAUC_{60}$ or K^{trans}, versus or alternative parameters such as enhancing tumor volume or fractional plasma volume); choice of time for image measurements; how T_1 is calculated or estimated; how an AIF is calculated or estimated; and whether the ROI is limited to a few slices or includes the whole tumor (O'Connor et al., 2007). These discrepancies are present in the existing literature and limit direct comparison between study data. Despite these factors, there is now considerable evidence for using DCE-MRI in phase I/II studies to help indicate biological efficacy, elucidate mechanism of action, and determine biologically active dose and schedule.

One rapidly developing research area is how imaging biomarkers may be used to inform clinical outcome irrespective of the treatment prescribed (a prognostic indicator) or directly relating to the therapy given (a predictive indicator). Both baseline parameters and early pharmacokinetic changes in a parameter have been investigated for this end, for example, as reviewed in the section on breast cancer above. Elsewhere, there is emerging evidence that image biomarkers relate to survival in several phase I and phase II clinical trials of novel therapies. An illustrative example is found in studies of anti-VEGF monotherapy in patients with metastatic renal cell cancer (RCC), with multiple studies evaluating image biomarkers of tumor vascularity measured both before treatment and between 3 and 12 weeks following therapy. Two independent studies of patients with metastatic RCC treated with sorafenib have reported that high baseline values of K^{trans} were associated with longer PFS (Flaherty et al., 2008; Hahn et al., 2008). In one of these studies, higher baseline plasma volume was also related to a longer PFS (Hahn et al., 2008). An equivalent relationship between K^{trans} and PFS has been shown in a study of sunitinib in metastatic RCC (Bjarnason et al., 2011). Since K^{trans} is a composite parameter reflecting flow and permeability, it is not possible to determine exactly which biological correlate relates to outcome. Further data will be required to determine if these findings are predictive or, more likely, simply prognostic of outcome.

9.3.3 Other Vascular Applications

Tissue perfusion can be quantified using arterial spin labeling (ASL). This technique adopts the same principle as bolus tracking studies, where a "tracer" is introduced into the blood and the concentration of this tracer is tracked over time as it passes through the tissue of interest. In ASL, protons within flowing blood act as an endogenous tracer once they are "labeled" magnetically, using an RF inversion pulse. ASL image acquisition technique involves collection of a pair of images—the "label" and the "control" image. The labeled image includes additional signal from labeled blood. On subtraction of the two images a perfusion-weighted image can be obtained. The difference in signal between the two images is on the order of only 1%, so a number of averages are required to attain adequate signal. ASL is attractive compared with other MRI measurements of perfusion, since it is completely noninvasive, permits serial measurements to be made during one examination, and enables absolute quantification of perfusion. However, ASL has an inherently low SNR requiring a number of averages to produce a good quality image, so that acquisitions are typically 10 min for whole brain coverage and often are substantially longer for other organs, although this improves at higher field strengths (Petersen et al., 2006).

There are two main methods used in ASL. In continuous ASL (CASL), a spatially localized RF field is positioned through the feeding arteries and continuously inverts the longitudinal magnetization of the blood protons as they flow through the plane. The labeling pulse can cause large power deposition that is a particular problem at higher field strengths. Due to good coverage and high signal, CASL is particularly suited to resting state perfusion measurements (Detre and Alsop, 1999; Parkes and Tofts, 2002). Alternatively, in pulsed ASL (PASL) a large volume of spins are labeled using a brief RF pulse, resulting in faster acquisition of image pairs, making PASL particularly suited to functional imaging studies (Tourdias et al., 2008; Wong, 2005).

In order to produce accurate perfusion maps, a number of factors need to be measured or assumed, including the degree of arterial spin inversion, the transit time from label to slice, the T_1 of blood and tissue, and the equilibrium magnetization of arterial blood. Most applications use a single-tissue compartment model (Detre et al., 1992), where the signal in the difference image $M(t)$, following labeling at time $t = 0$, can be described quite simply:

$$\frac{dM(t)}{dt} = -\frac{M(t)}{T_1} + f(m_a - m_v), \tag{9.8}$$

where T_1 is the T_1 of tissue, f is perfusion; m_a is the arterial magnetization, and m_v the venous magnetization. There is evidence that the labeled water remains largely intravascular at the time of imaging, and hence replacing T_1 with T_1 of blood has been shown to be more accurate (Parkes and Tofts, 2002).

Most applications of ASL have been in the brain. Preclinical studies of glioma have shown the feasibility of ASL techniques in measuring differential cerebral blood flow (CBF) in the periphery and center of brain tumors, demonstrating significant intratumoral heterogeneity (Silva et al., 2000; Sun et al., 2004). Relatively few studies have investigated the clinical application of ASL in cancer management compared with dynamic susceptibility contrast MRI (DSC-MRI) and DCE-MRI. However, there are indications that the technique may be promising and in some cases outperform gadolinium-based methods. Clinical evaluation in 36 patients with various grades of glioma has shown that ASL and DSC-MRI provide comparable estimates of perfusion on a per-tumor basis ($R^2 = 0.83$; $p < .005$) (Warmuth et al., 2003). Other studies have also shown a correlation between values within a study but difference in the overall magnitude of estimates of perfusion between ASL and other MRI-based methods (Ludemann et al., 2009).

Like DSC-MRI (Aronen et al., 1994) and DCE-MRI (Patankar et al., 2005), ASL-based methods of quantifying tumor perfusion can noninvasively grade glioma (Wolf et al., 2005). Reduction in relative CBF at 6 weeks measured by both techniques has been shown to relate to clinical outcome (change in tumor volume) in a study of 25 patients with cerebral metastases from various solid malignancies undergoing stereotactic radiosurgery (Weber et al., 2004). At present, therefore, there are indications that ASL is feasible as a technique to evaluate primary and secondary tumors within the brain. However, it is not clear if this technique, which requires considerable local expertise in implementation and has relatively poor SNR, provides additional benefit to more widely available MRI methods of quantifying tumor perfusion.

The requirement for a relatively high arterial blood flow and the sensitivity of ASL to organ motion has limited applications outside the brain in studies of cancer. Although imaging the kidney presents some challenges due to presence of cardiac, respiratory, and bowel motion, ASL has been demonstrated feasible in clinical studies of normal kidneys and in patients with renal tumors (De Bazelaire et al., 2005). Subsequent studies by the same investigators have shown that ASL detected changes in perfusion in patients with metastases from renal cancer in the chest and abdomen treated with the anti-VEGF tyrosine kinase inhibitor vatalanib; where patients who had stable disease or better at 4 months demonstrated statistically significant reductions in perfusion 1 month after therapy ($42 \pm 22\%$; $p = .02$) (de Bazelaire et al., 2008); and that two xenograft models responsive to sorafenib-induced VEGF inhibition showed reduction in ASL (Schor-Bardach et al., 2009). Collectively, these three studies provide promising data to support further investigation of ASL derived measurement of perfusion as a biomarker of response to anti-angiogenic therapy and as a possible prognostic indicator. As with studies in cerebral tumors, at present there is not enough data to allow meaningful comparison of the relative merits of ASL with other MRI and non-MRI alternative imaging techniques of measuring tumor perfusion.

9.4 Non-¹H Contrast Mechanisms for Imaging Tumors

9.4.1 ³¹P Spectroscopy, ²³Na Imaging, ¹³C and Dynamic Nuclear Polarization

Nuclei that contain an odd mass number have a net angular momentum and as such are MRI visible. These MRI active nuclei can align their axis of rotation to an applied magnetic field and can, in theory, be used to derive spectra or images. These techniques can be performed on conventional clinical and preclinical systems but require tuning a dedicated RF coil. Despite their potential benefits, most of these techniques remain in the research arena and have not impacted significantly on clinical practice to date.

Phosphorus (^{31}P) spectroscopy is a long-established technique and has been used to investigate tumor biology for several decades (Tozer and Griffiths, 1992). ^{31}P has a lower Larmor frequency than hydrogen and, thus, requires a dedicated RF coil. Phosphorus is much less abundant than hydrogen in the body (by a factor of over 1000-fold). The low abundance results in SNR being relatively poor and, generally, higher field strengths are required to achieve acceptable spatial resolution. This has resulted in mainly limiting investigation of phosphorous to spectroscopy and often at higher field strength (4 T or higher) in preclinical systems. One application of ^{31}P in oncology is to measure tumor adenosine triphosphate (ATP) metabolism, where cellular energetics can be followed in animal models of tumors following chemotherapy (Winter et al., 2001).

Sodium (^{23}Na) is MRI visible within the human body. A handful of studies have evaluated the role of ^{23}Na in measuring the EES of tumors, based on the fact that the concentration of Na$^+$ ions in the EES is regulated to be around 135–145 mM, compared to ~5 mM in cells. Although the Na$^+$ ion concentration in the vasculature is also between 135 and 145 mM (as it is in equilibrium with the EES), the microvasculature is a relatively small compartment in most tumor voxels (around 2–10%) and so in simple terms, the concentration of Na$^+$ ions can be used to monitor the change in the size of the EES. This method has been used to monitor the temporal evolution of necrosis and other cell death processes, following various chemotherapy regimens in animal models (Babsky et al., 2005; Schepkin et al., 2005, 2006; Sharma et al., 2005).

There is considerable current interest in imaging carbon with NMR techniques, since carbon is an abundant atom that is a vital component of organic substrates and metabolites. Nearly all carbon exists in the ^{12}C isotope with only a fraction in an MRI-visible form, ^{13}C. While only 1.1% of carbon exists as ^{13}C, a technique called dynamic nuclear polarization (DNP) has been described that enables molecules containing ^{13}C to become hyperpolarized. This process results in an increase in sensitivity of more than 10 000-fold, allowing the spatial distribution of a labeled and injected molecule to be imaged, along with its metabolites, which can be distinguished since they resonate at different frequencies. DNP has demonstrated a great deal of promise as a tool for investigating tumor metabolism (Kurhanewicz et al., 2011). Hyperpolarized [1–^{13}C]pyruvate has been one of the most widely explored hyperpolarized imaging agents as a means to investigate tumor metabolism (Albers et al., 2008; Chen et al., 2007a; Day et al., 2007; Golman et al., 2006). Hyperpolarized [1–^{13}C]pyruvate imaging provides a means to quantify pyruvate metabolism by monitoring the formation of hyperpolarized metabolites. Figure 9.10 provides an example of hyperpolarized [1–^{13}C]pyruvate imaging in mouse prostate tumor model (TRAMP), where hyperpolarized [1–^{13}C]pyruvate and [1–^{13}C]lactate images were acquired after the injection of 350 µl of hyperpolarized [1–^{13}C]pyruvate. Lactate dehydrogenase conversion of [1–^{13}C]pyruvate to [1–^{13}C]lactate has been found by DNP hyperpolarization MRI to decrease after chemotherapy in an animal model providing support that DNP can

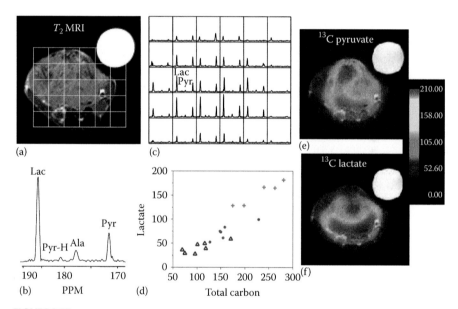

FIGURE 9.10

(See color insert.) Hyperpolarized [1–^{13}C]pyruvate imaging in mouse prostate tumor model. (a) Axial T_2-weighted ^{1}H image depicting the primary late stage tumor in TRAMP mouse and the overlay of hyperpolarized [1–^{13}C]lactate image after the injection of 350 µl of hyperpolarized [1–^{13}C]pyruvate. Hyperpolarized [1–^{13}C]lactate increased in going from normal to prostate cancer and with disease progression. (b and c) Corresponding array of 0.034 cc ^{13}C MRSI spectra acquired 35 s after injection of hyperpolarized [1–^{13}C]pyruvate. (d) Plot of hyperpolarized lactate vs. total hyperpolarized carbon (Lac + Pyr + Ala + Pyr-H) demonstrating that both hyperpolarized lactate and total hyperpolarized carbon increase with prostate cancer evolution and progression. (e and f) Overlaid hyperpolarized [1–^{13}C] pyruvate and ^{13}C lactate images. (Courtesy of Prof. John Kurhanewicz, Departments of Radiology and Biomedical Imaging, Urology and Pharmaceutical Chemistry, University of California, San Francisco, CA.)

be used as an early biomarker for a therapeutic response (Day et al., 2007). A proof of concept hyperpolarized [1–^{13}C]pyruvate study has been performed in prostate cancer patients demonstrating clinical feasibility of the technique (Nelson et al., 2012). DNP has also been applied to endogenous molecules such as CO_2 and bicarbonate ions and from this the pH of a tumor can be mapped (from the Henderson–Hasselbalch equation) (Gallagher et al., 2008a). Other applications include measuring amino acid metabolism as a biomarker of tumor proliferation (Gallagher et al., 2008b). The major limitation of this technique is the short half-life of the hyperpolarized state, which is often less than a minute and therefore requires rapid image acquisition. The widespread translation of this technique into the clinic faces many challenges. At the moment, hyperpolarized ^{13}C imaging is far from routine clinical use, but because the technique has great potential to aid diagnosis, to target tumor biopsy, to determine response to therapy and to predict outcome there is considerable current interest in developing these methods.

9.5 Morphometric Analysis

9.5.1 Tumor Size Estimates

The most commonly derived quantitative measurement in cancer imaging is tumor size, typically measured in either one or two dimensions (O'Connor et al., 2008). These measurements have a long history, initially in plain film radiology, where standardized and objective outcome criteria to measure response to anticancer treatment were first formulated over fifty years ago. In 1960, the Eastern Cooperative Oncology Group proposed that objective measurement of tumor size should be used as a clinical endpoint in the evaluation of patients treated with cytotoxic chemotherapeutic drugs, based on the notion that killing tumor cells would reduce tumor dimensions. In 1979, these principles were first codified in a World Health Organization (WHO, 1979) document, with definition of four categories of response: complete response (CR), partial response (PR), stable disease (SD), and progressive disease (PD), relying on plain radiographs and computed tomography.

Despite attempts at standardization, the calculation methods and reporting of the WHO criteria exhibited considerable variation between centers and this led to formation of new agreed standard criteria for measuring tumor response, the response evaluation criteria in solid tumors (RECIST), published in 2000 (Therasse et al., 2000). The criteria were proposed to unify existing radiological evaluation of tumor response, while at the same time acknowledging the need for subsequent development. The RECIST criteria are founded on the same principles as those proposed by WHO. Like its predecessor, RECIST criteria were developed to evaluate response to chemotherapy agents that typically produced substantial changes in tumor size.

One key difference between the RECIST and WHO criteria was the precise definition of some of the response categories; most crucially, the definition of what constitutes PD was changed, as well as adopting measurements based on one dimension. Emphasis was placed on cross-sectional imaging with CT and MRI (by then in widespread use throughout the world), although use of chest radiographs was permitted, along with some clinical and ultrasound measurements. The magnitude of change required to demonstrate partial response or disease progression was largely derived from retrospective statistical evaluation of measurements (not image data) from drugs trials. The quality control for these measurements was not known. Details of the WHO and RECIST response criteria are discussed thoroughly in the original report of the RECIST criteria by Therasse and colleagues (2000). The RECIST criteria have been further updated in 2009 to modify the guidelines for lesion selection and measurement of lymph node disease (Eisenhauer et al., 2009).

The main strengths of codified radiological response criteria lie in their objective standardized method for reporting response to therapy. No extra equipment or analysis steps are required, and this has allowed widespread adoption in both trials reporting and clinical practice. Since radiotherapy and cytotoxic chemotherapies typically produce a change in tumor size, the criteria have been adopted as an acceptable surrogate endpoint for many clinical trials and form the basis to defining tumor progression (Michaelis and Ratain, 2006). However, criticisms of current radiological response criteria have been widely cited in recent years, particularly the limitations in measuring small lesions (it is difficult to accurately measure a 1.0 cm tumor which, although small, may contain ~10^9 cells). This can translate into scenarios such as categorizing some patients as cured despite later relapse from micrometastases. Criteria such as RECIST are also difficult to apply to some tumor types, including pleural disease, meningeal deposits, and cystic and bone lesions (Padhani and Ollivier, 2001). Finally in most tumor types, response rate and PFS are poor surrogates of overall survival, questioning the value of measuring tumor size and using this as a lynchpin upon which clinical management decisions are made (Booth and Eisenhauer, 2012). This is particularly relevant in the era of targeted molecular therapies which work by cytostatic mechanisms that halt tumor growth, rather than reduce lesion size. Several studies of antivascular drugs have demonstrated survival benefit with minimal response rates as defined by RECIST or WHO criteria and have been approved by the FDA despite these data (Escudier et al., 2007; Hurwitz et al., 2004). These drugs would have been considered ineffective if assessed by the RECIST criteria alone.

There is, therefore, widespread consensus that while traditional radiological response criteria such as the WHO and RECIST approaches provide a useful objective codification of response to therapy, significant limitations exist (Michaelis and Ratain, 2006; Padhani and Ollivier, 2001; Ratain and Eckhardt, 2004). This has prompted discussion of how improvements may be made in lesion evaluation, particularly in the context of clinical trials of novel anticancer agents, posing the question "Is size worth measuring?" The answer to this question appears fivefold. First, in some scenarios traditional size-based assessment of tumor response and progression appears to be a good surrogate of overall survival. Examples include colorectal and ovarian cancers treated with cytotoxic chemotherapy (Bast et al., 2007; Tang et al., 2007). Second, there is evidence that size-based measurements are useful in some circumstances if less stringent response criteria are applied. For example, a decrease of tumor size of just 10% has been shown to relate to progression in various tumor types treated with targeted therapies, particularly in metastatic renal cancer (Krajewski et al., 2011). Third, as imaging capability develops, measurement techniques employed to assess changes in size require modification. Three-dimensional image acquisition now enables accurate assessment of tumor volumes. For example, in a study of patients with recurrent high-grade glioma, volumetric measurements were predictive of overall survival, but WHO- and RECIST-based measurements were not (Dempsey et al., 2005). At present, the RECIST working group has concluded that there is insufficient evidence for widespread adoption of volumetric measurement in clinical practice (Eisenhauer et al., 2009), but it remains likely that such analyses will find a place in radiology in the future in limited scenarios. Fourth, there is evidence that size measurements should be used as a continuous variable within drug development where individual patient response is less important than detecting a group effect. Several studies of novel targeted therapies have shown small but statistically significant reduction in size measured by MRI in one, two, or three dimensions (Mitchell et al., 2011; Mross et al., 2009; O'Connor et al., 2009). Fifth, there is evolving evidence that it may be important to quantify the amount of tumor with certain physiological characteristics such as the volume of enhancing or necrotic tissue (discussed below).

9.5.2 Tumor Shape, Fractals, and Texture

There has been interest over the last few decades in measuring tumor shape and morphology using a variety of techniques based around texture and fractal analyses. These approaches are complex and the details of the precise analyses are outside the scope of this chapter. In general, texture- and fractal-based analyses work on the assumption that lesions may have a pattern of spatial arrangement of a parameter (such as K^{trans}, ADC, etc.). The most frequently used texture analysis technique using DCE-MRI data is the co-occurrence matrix method that was initially described by Haralick and colleagues (1973). Here, a co-occurrence matrix element denoted Pd, θ (i, j) measures the probability of starting from any voxel in an image with a designated value i, moving d voxels along the image in a direction θ, and then arriving at another voxel with value j. This co-occurrence matrix is a two-dimensional histogram describing a joint distribution of all the possible moves with step size d and direction θ on the image. From this, various features such as lesion contrast and correlation can be exacted and used as parameters of shape and/or spatial complexity. Most applications of MRI-based texture analysis have been in breast cancer to improve distinction between benign and malignant lesions (Chen et al., 2007b; Sinha et al., 1997).

Fractal analysis differs in its methodology but aims to achieve a similar descriptor of a tumor. Fractal dimensions are estimates of object complexity that were originally developed to characterize geometrical patterns result-ing from abstract recursive procedures (Peitgen et al., 2004). Fractal dimen-sion analyses can be applied to objects such as image parameter maps that do not arise from fractal processes. The simplest fractal dimension is the box-counting dimension, computed by imposing regular grids of a range of scales on the object in question and then counting the number of grid ele-ments (boxes) that are occupied by the object at each scale. By plotting the number of occupied boxes against the reciprocal of the scale on log–log axes, the box-counting dimension is computed as the slope of the line of best fit. More complex techniques involve assigning color-scale values to the fractal parameter. In all cases, the number of scales will be limited by tumor size and imaging resolution. Preliminary data suggest that fractal analyses can distin-guish tumor from normal tissue in the prostate (Lv et al., 2009), low and high grade tumors from one another (Rose et al., 2009) and are sensitive to treat-ment effect (Alic et al., 2011). However, these utilities are available from far simpler image analysis and routine application to clinical data is therefore of uncertain benefit. One area that has shown greater clinical utility is in using baseline imaging data to determine patient outcome. At present, few quanti-tative imaging biomarkers have a clear role here, but data from colorectal can-cer liver metastases highlight that measurements of tumor shape (from fractal analysis) may relate to prognosis in patients where simple measurements of tumor size and perfusion fail to do so (O'Connor et al., 2011) (Figure 9.11).

9.6 Multiparametric Analysis of Tumors

MRI provides a powerful tool for investigating solid tumors because of the many methods by which image contrast can be generated. As stated above, conventional MRI parameters, T_1, T_2, and proton density (M_0), can all be used to generate contrast in MR images. In addition to conventional NMR para-meters, MRI can be used to measure diffusion, vascular characteristics (DCE-MRI, DSC-MRI, ASL, VSI, etc.), and employ non-^{1}H contrast mechanisms to characterized the tumor and its complex microenvironment. For these MR techniques, often parametric-weighted images or quantitative spatial maps of the specific parameters can be generated. But, spatial heterogeneity of per-fusion, oxygenation, and metabolism contributes to the heterogeneous tis-sue morphology that is a common characteristic of solid tumors, both at the

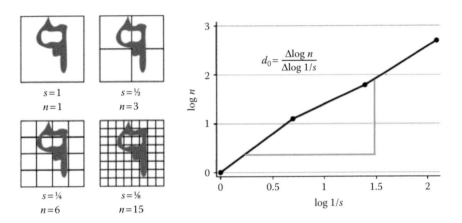

FIGURE 9.11
Example of fractal analysis: an object is successively divided using boxes of decreasing scale, *s*. The number of boxes, *n*, occupied by the object at each scale is recorded. This count is plotted against the reciprocal of the scale on log–log axes. Box-counting dimension (d_0) is computed as the slope of the line.

genotypic and phenotypic levels (Eberhard et al., 2000; Gerlinger et al., 2012; Gonzalez-Garcia et al., 2002; Heppner, 1984). As individual parameters are averaged across a ROI, spatial heterogeneity will often minimize the diagnostic potential of that parameter. As an example, individual MRI parameters, including T_2, have shown promise in differentiating tumor from neighboring tissue but have been limited by presence of peritumor edema (Els et al., 1995). Although DWI is able to differentiate between necrotic and viable tissue, it is unable to consistently differentiate between viable tumor tissue and other neighboring viable tissues and is also hindered by edema (Els et al., 1995). Contrast-enhanced T_1-weighted MRI has been the most successful method for identifying brain tumor boundaries based on the vascular permeability differences that exist between tumor and neighboring tissues, but some necrotic regions show contrast leakage rates that are equivalent to viable tumor and can corrupt viable tumor estimates (Berry et al., 2008; Els et al., 1995). Based on the limitations associated with individual parametric methods, multiple imaging parameters are often acquired to provide supporting or complementary information about the tumor. Often, these parameters are analyzed individually, providing a range for normal and abnormal values. The individual summary statistics for each parameter are then combined to provide improved diagnostic performance over the individual parameters employed alone. These multiparametric analyses have shown great promise in improving diagnostic estimates in a number of tumor types, including brain malignancies and prostate cancer.

The independent analysis of multiple parameters does not exploit the relationship between these parameters at the voxel level. The field of multispectral (MS) analysis has been advanced as a means to analyze multiple

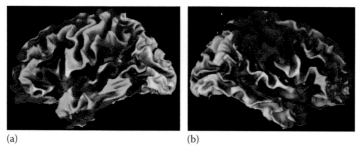

(a)   (b)

FIGURE 1.7
Results from a study in healthy brain aging using AAL: region-specific significant loss of GM with age, in the left (a) and right (b) brain hemispheres; z-scores are color-coded from −3 (magenta) to −10 (red).

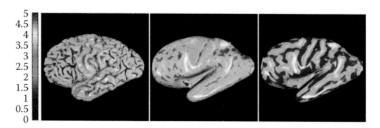

FIGURE 1.8
Lateral view of pial surfaces from a high-resolution dataset: neocortical thickness (left), inflated thickness maps (middle), and inflated convexity maps (right). (From Osechinskiy, S. and Kruggel, F., *Int J Biomed Imaging*, doi:10.1155/2012/870196, 2012.)

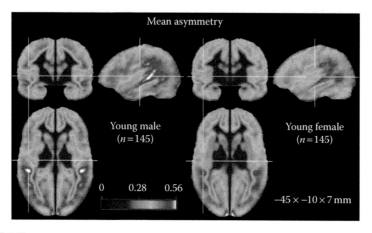

FIGURE 1.10
Orthogonal sections of mean asymmetry maps for young males (left panel, n = 145, 18–32 years) and females (right panel, n = 145, 18–32 years). Note the prominent asymmetry in males apparent in the vicinity of Heschl's gyrus, the superior temporal gyrus, postcentral gyrus, and the optic radiation. In females (as in males), the only prominent asymmetry is found at the occipital pole, and much less at the postcentral gyrus. (From Kovalev, V.A., et al., *Neuroimage* 19, 895–905, 2003. With permission.)

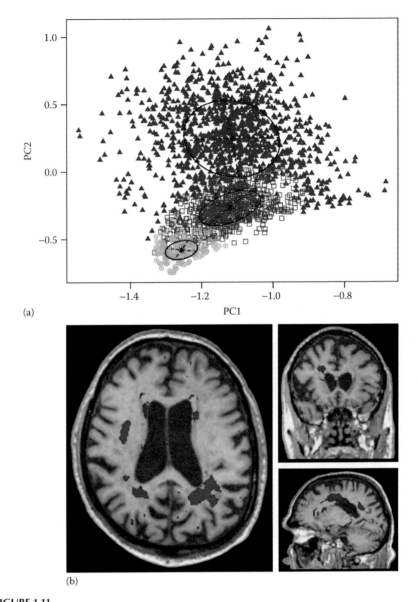

(a)

(b)

FIGURE 1.11
(a) Discrimination of feature vectors corresponding to WM (blue), diffuse white matter hyper-intensities (DWMH, red), and periventricular lesions (PVL, green). (b) Overlay of classification result onto the T_1-weighted image with DWMH (red) and PVL (green). (From Kruggel, F., et al., *Neuroimage* 39, 987–996, 2007. With permission.)

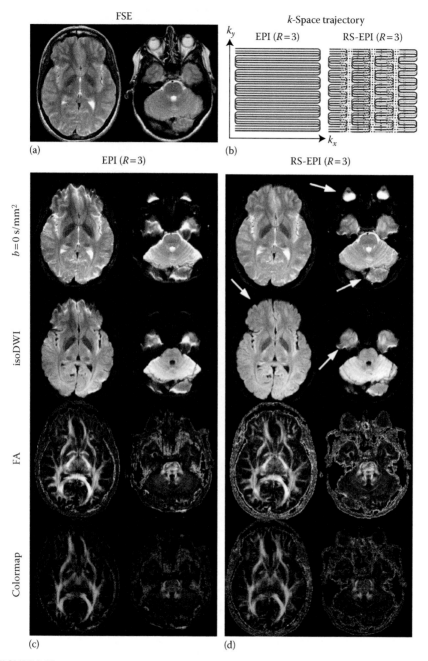

FIGURE 2.10
High-resolution DTI using multishot EPI and RS-EPI with a resolution of 288 × 288. (b) The RS-EPI readout had seven "blinds," each of which had an acquisition matrix size of 64 × 288. For both EPI and RS-EPI, three shots in the k_y direction were used and each shot was reconstructed independently using parallel imaging. (c, d) The use of parallel imaging reduced the distortions for both single-shot EPI (SS-EPI) and RS-EPI. However, RS-EPI has considerably lower geometric distortions compared to EPI due to the multishot readout used in RS-EPI acquisition (white arrows). The non-diffusion-weighted FSE acquisition is free of any geometric distortions and is shown in (a) for comparison.

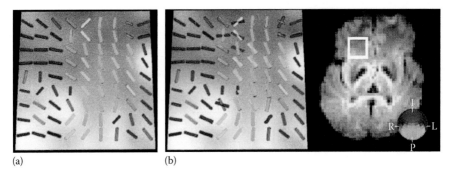

(a)   (b)

FIGURE 2.11
Comparison of the principal eigenvectors (i.e., fiber orientation) estimated with single-tensor model (a) and two-tensor model in the ROI specified in the reference image (b). The single-tensor model fails to represent fiber crossings that are better represented by a two-tensor model. (From Tuch, D.S., Reese, T.G., Wiegell, M.R., Makris, N., Belliveau, J.W., Wedeen, V.J: High angular resolution diffusion imaging reveals intravoxel white matter fiber heterogeneity. *Magn Reson Med.* 2002. 48. 577–82. Copyright Wiley-VCH Verlag GmbH & Co. KGaA. Reproduced with permission.)

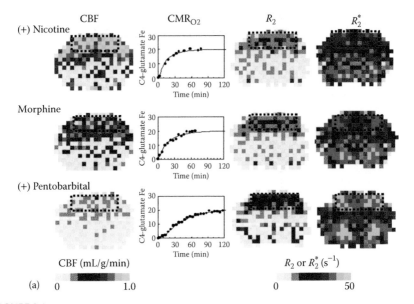

FIGURE 3.1
Summary of multimodal investigations of calibrated fMRI in rat brain at 7.0 T: (a) Examples of CBF, CMR_{O2}, R_2, and R_2^* data from baseline condition of morphine. Cortical values of CBF, CMR_{O2}, and CBV increased with nicotine, whereas these parameters decreased with pentobarbital. The reverse was observed for by R_2 or R_2^*.

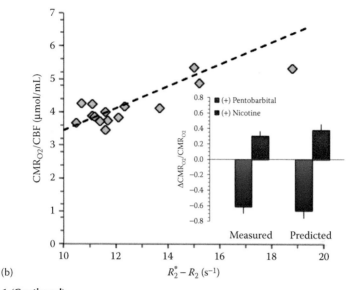

(b)

FIGURE 3.1 (Continued)
(b) Relationship between CMR_{O_2}/CBF and $R_2^*-R_2$ for all conditions. Inset: comparison between $\Delta CMR_{O_2}/CMR_{O_2}$ measured by ^{13}C MRS and predicted by calibrated fMRI. (Modified from Kida, I., et al., *J Cereb Blood Flow Metab.* 20, 847–60, 2000.)

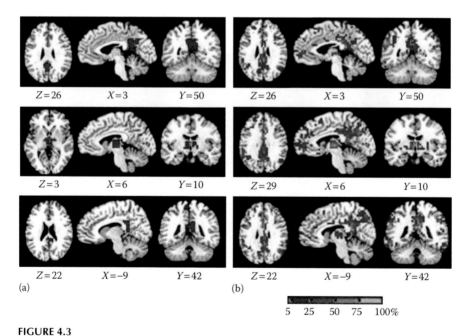

(a)   (b)

5   25   50   75   100%

FIGURE 4.3
An example result using the x-guided clustering method. (a) The clusters identified using group difference information between amnestic mild cognitively impaired and age-matched cognitively normal subjects. (b) The decreased connectivity pattern to the clusters shown on the left. Color bar indicates the percentage of disconnection. For example, light blue indicates that a voxel is disconnected to 75–100% of the voxels in the corresponding cluster.

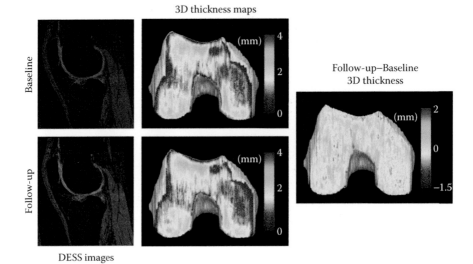

3D thickness maps

Follow-up–Baseline
3D thickness

DESS images

FIGURE 6.4
An example of intrasubject longitudinal cartilage thickness matching using the technique
described by Carballido-Gamio et al. (Modified from Carballido-Gamio, J., et al., *Med Image
Anal.* 12, 120–35, 2008a.)

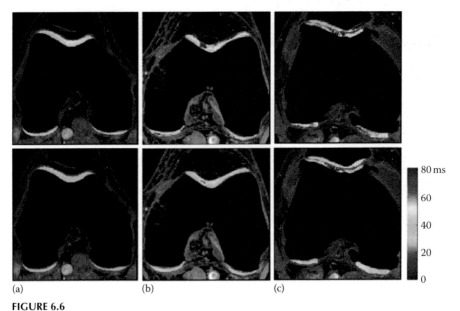

(a)   (b)   (c)

FIGURE 6.6
$T_{1\rho}$ (top row) and T_2 maps (bottom row) of a healthy control (a), a subject with mild OA (b), and
a subject with severe OA (c). Significant elevation of $T_{1\rho}$ and T_2 values was observed in subjects
with OA. $T_{1\rho}$ and T_2 elevation had different spatial distribution and may provide complementary information associated with the cartilage degeneration.

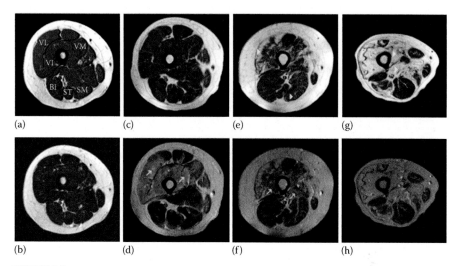

FIGURE 7.3

T_1- and T_2-weighted images of the thigh of a control subject and DM and PM patients. The *top row* (a, c, e, and g) shows T_1-weighted images, and the *bottom row* (b, d, f, and h) shows T_2-weighted images. Separate panels show data for the control subject (a and b) and DM (c and d), PM (e and f), and chronic PM (g and h) patients. Muscles of the thigh are designated in control subject (a): VL, vastus lateralis; VI, vastus intermedius; VM, vastus medialis; BI, biceps femoris; ST, semitendinosus; SM, semimembranosus. T_2-weighted images illustrate fat replacement, and the arrows in the T_2-weighted images illustrate inflammation.

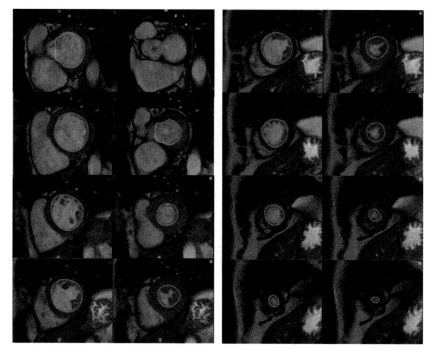

FIGURE 8.1

SSFP LV stack showing paired images in end-diastole (left columns) and end-systole (right columns) starting from the base of the left ventricle (upper left) to the apex (lower right). Manually drawn endocardial contours, from which EDV, ESV, and EF can be calculated, are shown too.

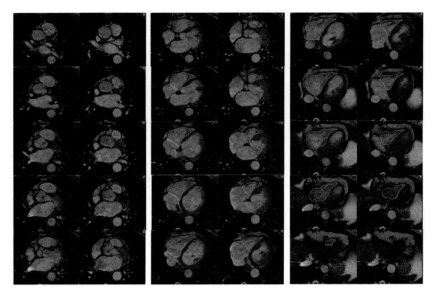

FIGURE 8.2
SSFP RV stack showing paired images in end-diastole (left columns) and end-systole (right columns). Manually drawn endocardial contours are also shown.

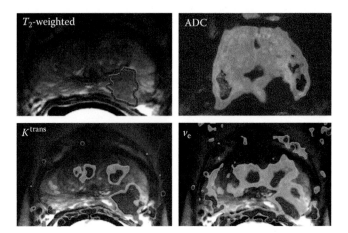

FIGURE 9.5
DWI and DCE-MRI images in a 62-year-old man with T3a Gleason 4 + 3 prostatic carcinoma. The T_2-weighted image shows an ill-defined low signal nodule in the left peripheral zone (outlined in red). ADC mapping demonstrates the tumor extent (low ADC indicated by maroon-blue color). DCE-MRI enhancement–time curve in this region (not shown) shows rapid enhancement followed by washout, indicating malignancy. K^{trans} and v_e maps show the tumor extent (high K^{trans} and high v_e indicated by reddish color), with early extracapsular infiltration into surrounding periprostatic fat. (Courtesy of Dr. Thomas Hambrock, Department of Radiology, University Medical Centre, Nijmegen, the Netherlands.)

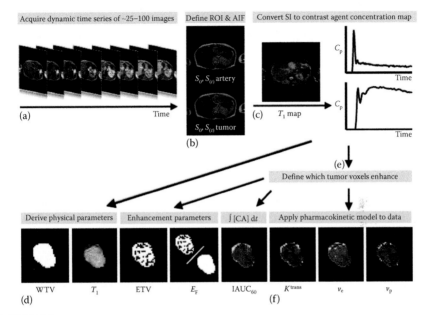

FIGURE 9.7

Overview of DCE-MRI data acquisition and analysis. (a) Multiple images are acquired as a bolus of contrast agent passes through tissue capillaries. (b) ROI for a tumor and the feeding vessel AIF are defined. (c) SI values or each voxel are converted into contrast agent concentration, using a map of T_1 values. (d) These steps allow calculation of tumor volume and T_1 values. (e) Next, voxels are classified as enhancing or not after which parameters based on the amount and proportional enhancement can be defined along with the $IAUC_{60}$. (f) Finally, a pharmacokinetic model may be applied to derive parameters such as K^{trans}.

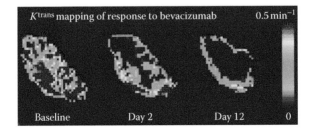

FIGURE 9.9

DCE-MRI parameter maps of K^{trans} in a patient with a colorectal liver metastasis. At baseline, the distribution of K^{trans} is spatially heterogeneous with high values in the periphery and low values in the core. Following treatment with 10 mg/kg of bevacizumab, tumor K^{trans} was reduced at 2 days and further reduced after 12 days of treatment compared with baseline.

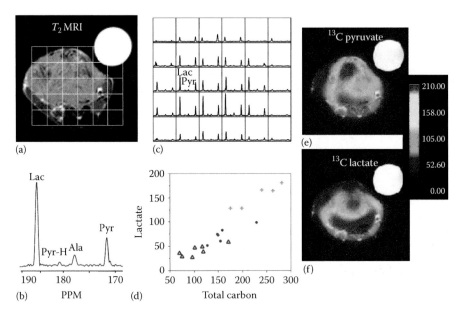

FIGURE 9.10
Hyperpolarized [1–¹³C]pyruvate imaging in mouse prostate tumor model. (a) Axial T_2-weighted ¹H image depicting the primary late stage tumor in TRAMP mouse and the overlay of hyperpolarized [1–¹³C]lactate image after the injection of 350 μl of hyperpolarized [1–¹³C]pyruvate. Hyperpolarized [1–¹³C]lactate increased in going from normal to prostate cancer and with disease progression. (b and c) Corresponding array of 0.034 cc ¹³C MRSI spectra acquired 35 s after injection of hyperpolarized [1–¹³C]pyruvate. (d) Plot of hyperpolarized lactate vs. total hyperpolarized carbon (Lac + Pyr + Ala + Pyr-H) demonstrating that both hyperpolarized lactate and total hyperpolarized carbon increase with prostate cancer evolution and progression. (e and f) Overlaid hyperpolarized [1–¹³C]pyruvate and ¹³C lactate images. (Courtesy of Prof. John Kurhanewicz, Departments of Radiology and Biomedical Imaging, Urology and Pharmaceutical Chemistry, University of California, San Francisco, CA.)

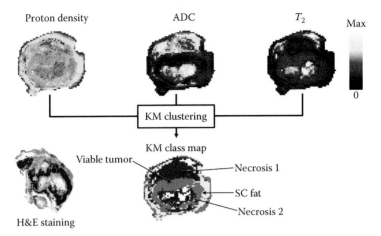

FIGURE 9.13
MS segmentation of a mouse xenograft tumor model employing a human colorectal cancer cell line (Colo 205). The MS feature space was constructed from three parameter maps: ADC, T_2, and proton density (M_0). The tissue class map derived by k-means (KM) multiscale segmentation depicts the segmentation of the tumor into viable tumor (orange), subcutaneous (SC) adipose tissue (yellow), and two necrotic classes (necrosis 1: red and necrosis 2: white). The corresponding hematoxylin and eosin (H&E)-stained histological section is also shown.

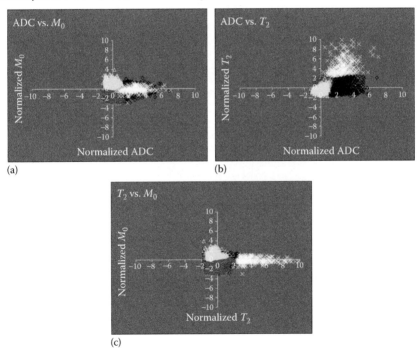

FIGURE 9.14
Cluster plots of the parameters employed in the MS classification example provided in Figure 9.13. Cluster points are color-coded based on their respective KM classification: viable tumor (orange), subcutaneous adipose tissue (yellow), necrosis 1 (red), and necrosis 2 (white). (a) ADC vs. M_0, (b) ADC vs. T_2, and (c) T_2 vs. M_0. Feature values represent the normalized feature space, where normalization consisted of a linear transformation of the parameter space (each feature was normalized to a zero mean and unit variance).

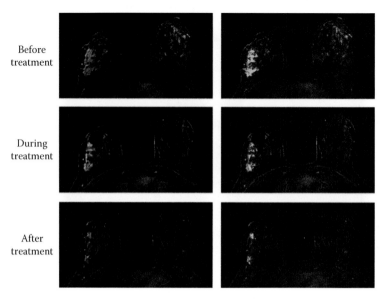

FIGURE 10.6

An example of a non-mass-type breast cancer undergoing NAC treatment. The postcontrast fat sat images are shown. The tumor breaks up into scattered pieces. The pixel-by-pixel pharmacokinetic analysis is applied to obtain the parameters K^{trans} and k_{ep}. The right column shows color-coded K^{trans} values overlaid on the original images. It can be seen that although the tumor is responding well, some hot spots still exist in the tumor. The histogram of each parameter can be analyzed from the pixel-by-pixel fitting results and compared in these studies acquired before, during, and after chemotherapy.

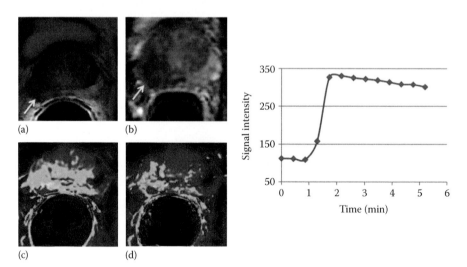

FIGURE 10.7

An example of prostate cancer. A lesion is visible on T_2-weighted image (a) showing a hypointense area in the periphery zone compared to the contralateral side. It shows decreased ADC (b) due to increased cell density. The color-coded enhancement maps clearly show that the tumor has a higher IAUC (c) and also a higher wash-in slope (d) compared to the contralateral normal side. The DCE kinetics measured from the tumor shows the wash-out pattern.

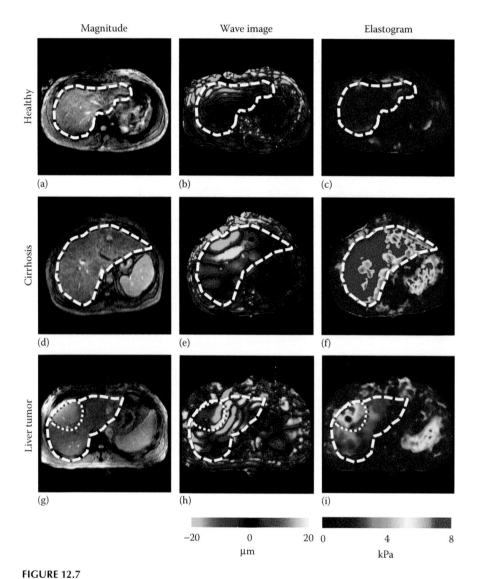

Magnitude　　　　Wave image　　　　Elastogram

Healthy

(a)　　　　　　　　(b)　　　　　　　　(c)

Cirrhosis

(d)　　　　　　　　(e)　　　　　　　　(f)

Liver tumor

(g)　　　　　　　　(h)　　　　　　　　(i)

−20　　　0　　　20 0　　　4　　　8
μm　　　　　　　kPa

FIGURE 12.7

Hepatic MRE. (a, b, c) The left, middle, and right columns show the magnitude image, a single wave image, and the shear stiffness map of a patient with a healthy liver, respectively. The boundaries of the liver are indicated with the dotted region and the stiffness was found to be 1.7 kPa. (d, e, f) The middle row shows the corresponding data from a patient with a cirrhotic liver. The longer shear wavelength and the higher shear stiffness (17.8 kPa) compared to the healthy liver are visible. (g, h, i) The bottom row shows data obtained from a patient with a malignant tumor in the liver, whose borders are indicated with the inner dotted region of interest. The high stiffness of the tumor is well evident from the elastogram (i). The shear stiffness of the tumor was found to be 9.23 kPa, well above the malignancy cutoff value of 5 kPa. (From Mariappan, Y.K., et al., *Clinical Anatomy*. 23, 497–511, 2010.)

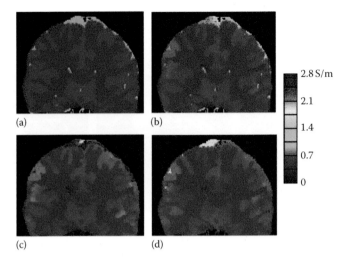

(a)   (b)

2.8 S/m

2.1

1.4

0.7

0

(c)   (d)

FIGURE 13.8
Conductivity reconstruction of FDTD simulation of a realistic head model. (a) Conductivity distribution used in FDTD simulations. (b) Exact reconstruction using Equation 13.13, that is, based on complex H_p. (c) Approximated reconstruction using Equation 13.15, that is, based only on the transmit phase φ_p. (d) Approximated reconstruction based on the transceive phase assumption (Equation 13.40). A corresponding, quantitative analysis can be found in Table 13.1. (From Voigt, T., et al., Quantitative conductivity and permittivity imaging of the human brain using electric properties tomography. 2011a, 456–66, Copyright Wiley-VCH Verlag GmbH & Co. KGaA. Reproduced with permission.)

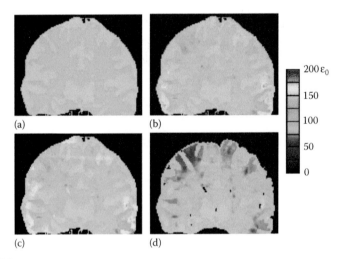

(a)   (b)

200 ε_0

150

100

50

0

(c)   (d)

FIGURE 13.9
Permittivity reconstruction of FDTD simulation data of a realistic head model. (a) Permittivity distribution used in FDTD simulations. (b) Exact reconstruction using Equation 13.14, that is, based on complex H_p. (c) Same as (b), however, based on the transceive phase assumption Equation 13.40. (d) Approximated reconstruction using Equation 13.16, that is, based only on B_1 magnitude. CSF is not shown in (d) since magnitude-based reconstruction of CSF permittivity violates $\omega\varepsilon \gg \sigma$, and thus does not yield a meaningful value (cf. Figure 13.3). A corresponding, quantitative analysis can be found in Table 13.1. (From Voigt, T., et al., Quantitative conductivity and permittivity imaging of the human brain using electric properties tomography. 2011a, 456–66, Copyright Wiley-VCH Verlag GmbH & Co. KGaA. Reproduced with permission.)

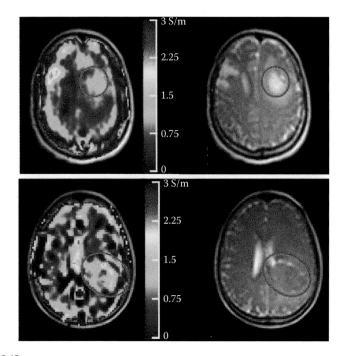

FIGURE 13.13
Tumor pilot study. Tumor conductivity appears to be pathologically increased by a factor of 2–3 compared with the surrounding WM. Quantitative values are given in Table 13.6. (From Voigt, T., et al., *Proceedings of the 19th Annual Meeting of ISMRM*, 7–13 May, Montreal, QC, Canada, 2011b, p. 127.)

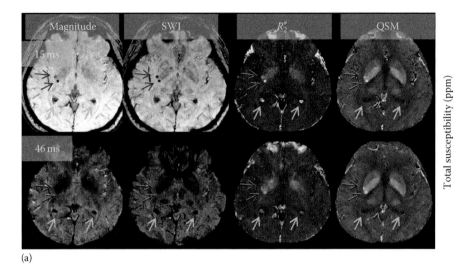

(a)

Total susceptibility (ppm)

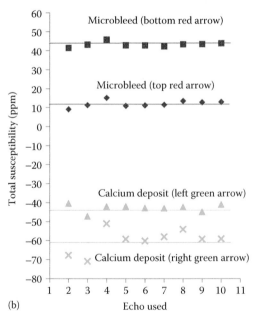

(b)

FIGURE 14.4
(a) CMB appearances (red arrows) change with TE drastically in magnitude, SWI, and moderately R_2^* map but little in QSM. (b) QSMs reconstructed using various echoes confirm its invariance with TE.

images or parametric maps at the voxel level. MS analysis generally refers to the application of multidimensional pattern recognition techniques to image analysis, where multiple images, that provide complementary information of the same scene, are analyzed simultaneously (Duda and Hart, 1973; Vannier et al., 1985). The focus of this section will be to review the areas where tumor MRI has benefited from a multiparametric imaging (parameters analyzed individually) approach to improve diagnostic accuracy and where multidimensional pattern recognition methodologies have been employed to further exploit the spatial information contained within the multiparametric feature space to address image segmentation problems associated with tumor heterogeneity and improve diagnostic accuracy.

9.6.1 Multiparametric Tumor Imaging

Multiparametric approaches to tumor MRI have shown great promise in a number of tumor imaging applications. Studies have shown that the acquisition and analysis of additional MR endpoints have improved the accuracy of MR assessments over monoparametric estimates for multiple cancers, including brain, prostate, and breast tumors. Many of these studies have focused on combining conventional imaging techniques with physiological and metabolic endpoints to improve performance. Generally, a conventional brain MRI exam consists of noncontrast T_2-weighted and contrast-enhanced T_1-weighted imaging. Despite providing strong imaging contrast for tumor visualization, conventional MRI has been found to perform poorly in tumor characterization (Knopp et al., 1999a). Conventional MRI accuracy for grading gliomas has been reported to range from 55% to 83.3% (Dean et al., 1990; Knopp et al., 1999a; Kondziolka et al., 1993; Moller-Hartmann et al., 2002; Watanabe et al., 1992). The addition of DSC-MRI estimates of relative cerebral blood volume (rCBV) has been found to improve the accuracy of MRI characterization of tumor grade (Knopp et al., 1999a; Law et al., 2003). rCBV estimates have been combined with proton MRS estimates of metabolite ratios to increase the accuracy of imaging-based tumor grade assessments (Di Costanzo et al., 2006; Law et al., 2003; Zonari et al., 2007). The combination of rCBV and metabolite ratios (Cho/C and Cho/NAA) were found to provide a sensitivity, specificity, PPV and negative-predictive value (NPV) of 93.3%, 60.0%, 87.5%, and 75%, respectively, when maximizing the average of the sensitivity and specificity values (Law et al., 2003). In this same study, conventional imaging provided a sensitivity, specificity, PPV, and NPV of 72.5%, 65.0%, 86.1%, and 44.1%, respectively (Law et al., 2003). Another study found that the combination of rCBV and NAA/Cr ratio resulted in a sensitivity, specificity, PPV, and NPV of 87.7%, 72.5%, 83.8%, and 78.4%, respectively, for differentiating low grade (grade 2) from high grade gliomas (grade 3 or 4) (Zonari et al., 2007). Diffusion estimates have also been combined with rCBV and proton MRS metabolite ratios, but did not improve the classification results (Di Costanzo et al., 2006; Zonari et al., 2007). Although, diffusion estimates have been shown to improve performance when combined with other endpoints (Kim and Kim, 2007; Park et al., 2010b). A semiquantitative analysis of ADC maps combined with semiquantitative scoring of ASL perfusion maps provided strong classification results of tumor grade (sensitivity = 90.9%, specificity = 90.9%, PPV = 95.2%, NPV = 83.3%) (Kim and Kim, 2007). In addition, a noncontrast-based protocol (diffusion- and susceptibility-weighted imaging) provided similar classification results (sensitivity = 90.9%, specificity = 90.9%, PPV = 95.2%, NPV = 83.3%) (Park et al., 2010b).

Another area where multiparametric imaging has demonstrated utility is in prostate cancer imaging (Figure 9.5). Multiple MR techniques have been employed in combination to address prostate cancer characterization, including T_2-weighted imaging, diffusion, DCE-MRI, and MRS. The addition of diffusion and DCE-MRI has been found to improve the accuracy of the MRI assessment over T_2-weighted imaging alone (Delongchamps et al., 2011a, 2011b; Tamada et al., 2011). T_2-weighted MRI provides modest sensitivity and specificity in prostate tumor detection (sensitivity = 27–71%, specificity = 68–99%) (Delongchamps et al., 2011a, 2011b; Tamada et al., 2011). The combination of T_2-weighted, diffusion, and DCE-MRI has been found to significantly improve the performance of MRI prostate tumor detection over T_2-weighted imaging (sensitivity = 53–85%, specificity = 83–98%) (Delongchamps et al., 2011a, 2011b; Tamada et al., 2011).

9.6.2 Application of Multidimensional Pattern Recognition Techniques

MS analysis generally refers to the application of multidimensional pattern recognition techniques to image analysis, where multiple images that provide complementary information of the same scene are analyzed simultaneously (Duda and Hart, 1973). The term "MS" analysis is derived from the initial imaging applications of multidimensional pattern recognition techniques, where satellite images of multiple frequencies (e.g., visible, near-infrared, far-infrared, etc.) were analyzed simultaneously. For MS analysis of MRI data, multiple images (of differing contrast mechanisms) are acquired of the same subject and are viewed as a multidimensional feature space (Vannier et al., 1985), where the discipline of multidimensional pattern recognition is well suited as a method of analysis (Duda and Hart, 1973; Tou and Gonzalez, 1974). MS analysis is usually employed to perform classification and resultant image segmentation. MS analysis employs a view of the image processing chain as a pattern recognition problem (Figure 9.12). The function of a pattern recognition system is to perform the assignment of a physical object or event to one of several predefined categories. A pattern recognition system can be viewed as consisting of a transducer, feature extractor, and classifier (Figure 9.12) (Duda and Hart, 1973). For imaging applications, the transducer may be a camera, x-ray system, or for MRI applications, the transducer is an MRI system. The function of the transducer is

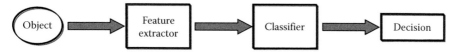

FIGURE 9.12
A schematic model of a pattern recognition system.

to convert the measured signal to an observable form, such as an image. Within an image or set of images, there exists useful information (features) along with extraneous or useless information (noise). The function of the feature extractor is to reduce the data by measuring certain properties or features found within the data. The features are then passed to a classifier that evaluates the information and makes the final decision about the class or classes (categories) that comprise the object. Although a single feature may provide some information about a scene or object, additional features may provide supporting or unique information about the scene. MRI provides a large number of possible features for use in a classification algorithm. Once the features are chosen, each voxel location is represented by a multidimensional feature vector:

$$\mathbf{f} = \begin{Bmatrix} f_1 \\ f_2 \\ \cdot \\ \cdot \\ f_n \end{Bmatrix}, \tag{9.9}$$

where n represents the number of features and f_i is the value for feature i. An MS image is a two-dimensional array, where each voxel, $\mathbf{I}(x, y)$, is represented by vector in the form of $\mathbf{f}$:

$$\mathbf{I}(x, y) = \begin{Bmatrix} I(x, y)_1 \\ I(x, y)_2 \\ \cdot \\ \cdot \\ I(x, y)_n \end{Bmatrix}. \tag{9.10}$$

Each element of the vector is the voxel value, $I_i(x, y)$, for the x, y spatial location in the feature image i. Ideally, features should be orthogonal (uncorrelated), where each feature provides unique information about the object. Usually, orthogonal features are difficult to obtain since many parameters possess a degree of correlation (negative or positive) with each other since they are obtained from the same object. Several techniques can be applied to decorrelate the data. The Karhunen–Loeve transform (principal components analysis) can be employed to rotate the feature space to derive a set of uncorrelated feature vectors (Duda and Hart, 1973).

The function of the classifier is to partition the feature space into regions that correspond to predefined categories. The ideal partitions would divide the feature space where the data are classified without error. Unfortunately, this is usually not possible, and the partitions are chosen to minimize the probability of error or the average cost of errors based on a particular cost function. Classification algorithms can be divided into supervised and unsupervised classifiers. Supervised classifiers require a training step to identify sample data points that represent each class in the test data set. Supervised classifiers include parametric and nonparametric classifiers (Duda and Hart, 1973). An example of a parametric classifier is the multivariate Gaussian (MVG) or Bayesian classifier that assumes a MVG distribution for each class and employs training data to generate parametric estimates. Nonparametric classifiers make no *a priori* assumptions about the statistical distribution of the data. The k-nearest neighbor (k-NN) classifier is an example of a nonparametric classifier, where voxels are assigned to a class based on their distance from classified voxels in the training data set. Unsupervised approaches do not require prior statistics or labeled training data. Clustering algorithms, such as k-means (KM) and fuzzy C-means (FCM), are examples of unsupervised segmentation, where they perform image segmentation by iteratively separating the data into groups or clusters of similar voxels based on the distance between the voxels within the chosen feature space (Bezdek et al., 1993; Tou and Gonzalez, 1974). Both supervised and unsupervised approaches can achieve robust results. The choice of classifier will depend on a number of factors, including data quality, availability of robust training data, computations resources, and, ultimately, the desired application (Bezdek et al., 1993; Clarke et al., 1995; Duda and Hart, 1973).

Figure 9.13 provides an example of an MS classification application that employs a multiparametric feature space and an unsupervised classifier. The MS application is the segmentation of an experimental mouse xenograft tumor into a viable tumor class, two necrotic classes, and subcutaneous fat class (Carano et al., 2004a). The feature space consists of ADC, T_2, and proton density (M_0) parameter estimates. Diffusion data were included in the feature space to increase the power of segmenting viable tumor from necrotic regions. The MS data were classified by the application of the KM clustering algorithm (Carano et al., 2004b). The resultant KM classification and the corresponding histological slice for a xenograft tumor are also included in Figure 9.13. Prior to clustering, each feature is normalized to have a zero mean and unit variance. This provides equalization of the dynamic range of each feature to ensure that one feature cannot dominate the clustering by biasing the distance measured employed in the KM clustering algorithm. The KM clustering algorithm is applied to the normalized feature space and iteratively sorts the data into regions of similar voxels by assigning each voxel to the nearest cluster center based on the distance measure. The cluster means are updated after the completion of every iterative pass through the data and this continues until the algorithm converges (no change in cluster means). Figure 9.14

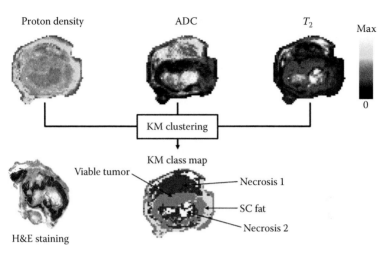

FIGURE 9.13
(See color insert.) MS segmentation of a mouse xenograft tumor model employing a human colorectal cancer cell line (Colo 205). The MS feature space was constructed from three parameter maps: ADC, T_2, and proton density (M_0). The tissue class map derived by k-means (KM) multiscale segmentation depicts the segmentation of the tumor into viable tumor (orange), subcutaneous (SC) adipose tissue (yellow), and two necrotic classes (necrosis 1: red and necrosis 2: white). The corresponding hematoxylin and eosin (H&E)-stained histological section is also shown.

provides clusters plots of the normalized features and the corresponding KM classification (color-coded) for each voxel. The KM clustering algorithm extracted a viable tumor class that generally exhibits low ADC, intermediate M_0, and low to intermediate T_2. Subcutaneous adipose tissue is characterized by low ADC, high M_0, and low-to-intermediate T_2. The two necrosis classes are differentiated from the viable tissues (tumor and subcutaneous adipose tissue) primarily by their high ADC values, but differ from each other based on their T_2 values. Figure 9.14 demonstrates some of the advantages of a multidimensional pattern recognition approach to image segmentation. The classes are color-coded and the decision boundaries at the color transitions. The KM clustering algorithm in this example employed a Mahalanobis distance measure that classifies the data by forming hyperplane decision boundaries that need not be orthogonal to the parameter axes (Carano et al., 2004b). As can be seen Figure 9.14, the optimal decision boundaries from this MS analysis generally form oblique angles to the parameter axes. This is a major advantage over analysis methods where each parameter in a multiparameter dataset is analyzed individually. Individual parametric analysis (e.g., normal vs. abnormal values), as described in the previous section, restricts the decision boundaries to being orthogonal to the feature axis and will likely result in suboptimal classifications (Duda and Hart, 1973).

MRI MS analysis techniques have been employed for tissue characterization of a wide range of normal and abnormal tissues, where MR parametric maps or MR-weighted images have been used to form the multidimensional

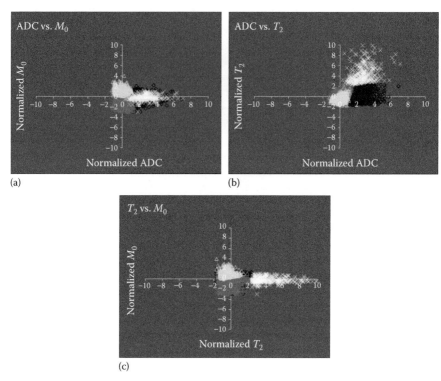

FIGURE 9.14
(See color insert.) Cluster plots of the parameters employed in the MS classification example provided in Figure 9.13. Cluster points are color-coded based on their respective KM classification: viable tumor (orange), subcutaneous adipose tissue (yellow), necrosis 1 (red), and necrosis 2 (white). (a) ADC vs. M_0, (b) ADC vs. T_2, and (c) T_2 vs. M_0. Feature values represent the normalized feature space, where normalization consisted of a linear transformation of the parameter space (each feature was normalized to a zero mean and unit variance).

feature space (Berry et al., 2008; Carano et al., 1998, 2004a, 2004b; Cline et al., 1990; Taxt et al., 1992; Vannier et al., 1985, 1987). In these studies, MS techniques have been employed to successfully segment brain abnormalities, such as brain tumors (Cline et al., 1990; Gordon et al., 1999) and ischemic lesions (Carano et al., 1998), abdominal tumors (Taxt et al., 1992), and joint abnormalities due to rheumatoid arthritis (Carano et al., 2004a). MS techniques, based on conventional MRI sequences, have been employed to segment tumors in experimental cancer models (Gordon et al., 1999) and clinical disease (Jacobs et al., 2003; McClain et al., 1995; Taxt et al., 1992; Vaidyanathan et al., 1997). MS analysis of T_1-, T_2-, and PD-weighted images was able to provide acceptable segmentation of uterine tumors from normal tissues with a supervised MVG-based classifier (Taxt et al., 1992). Contrast-enhanced T_1-weighted images combined with T_2- and PD-weighted images formed a feature space to successfully segment brain tumors from other brain tissues where automated segmentations compared well with manual segmentations

(McClain et al., 1995; Vaidyanathan et al., 1997). MS segmentation of brain tumors, employing conventional MRI feature images, has also been validated histologically in animal models and by analysis of human biopsy samples (Gordon et al., 1999; Vinitski et al., 1998). An MS analysis approach, applied to a T_2- and T_1-weighted (pre- and postcontrast) data set, showed promise in differentiating malignant from benign lesions (sensitivity = 89%, specificity = 78%) in breast cancer patients (Jacobs et al., 2003).

MS applications have expanded to include nonconventional sequences (diffusion, DSC-MRI, DCE-MRI) to better characterize the tumor and increase diagnostic potential. As described above (Figure 9.14), diffusion parameters have been combined with conventional parameters (T_2 and proton density) to identify viable tumor and differentiate it from necrotic and normal tissue populations (Carano et al., 2004b; Henning et al., 2007). Unsupervised KM clustering, employed in a hierarchical manner, was able to extract regions of viable tumor, where volume estimates were highly correlated with histological estimates (Carano et al., 2004b). An MS KM clustering algorithm was able to differentiate viable tumor from two necrotic tissue classes and subcutaneous fat (Figure 9.13). The two necrotic populations were differentiated based on their T_2 values (Figure 9.13) (Carano et al., 2004b). Histologically, the one necrotic class was found to represent an acellular, cyst-like region (high T_2, high ADC) and the other necrotic class (low T_2, high ADC) was found to contain high levels of deoxyhemoglobin that is likely due to recent or active hemorrhage (Carano et al., 2004b). This diffusion-based MS approach has been found to benefit tumor vascular imaging (DCE-MRI and VSI) by limiting the vascular analysis to the viable tumor and excluding necrotic regions (Berry et al., 2008; Ungersma et al., 2010). The necrotic regions are insensitive to a therapeutic response and cannot be differentiated (low T_2, high ADC necrotic class) from viable tumor based on its DCE-MRI parameters alone (Berry et al., 2008). Further hierarchical segmentation of the viable tumor based on its T_2 and ADC values found that the viable tumor can be divided into two regions: a well-oxygenated viable region and hypoxic viable, perinecrotic region that was confirmed by HIF-1 alpha immunohistochemistry staining (Henning et al., 2007).

Perfusion data (DCE-MRI, DSC-MRI, etc.) has also been included as MS features to aid in tumor segmentation and characterization (Kale et al., 2008; Lee and Nelson, 2008; Mouthuy et al., 2012; Ozer et al., 2010; Witjes et al., 2003). A k-NN classifier was applied to a feature space that included MRS, DWI, and DSC-MRI parameters for the purpose of predicting regions of new contrast enhancement in brain tumor patients (Lee and Nelson, 2008). A large number of features were evaluated to determine their value to the classification question. This approach achieved a sensitivity of 78% and specificity of 79% for new contrast-enhancement prediction employing seven of the 38 possible features (Lee and Nelson, 2008). An MS GBM characterization study employed a feature space generated from T_2, DSC-MRI, postcontrast T_1-weighted images to differentiate a single brain metastasis from GBM tumors (Mouthuy et al., 2012). Physiological, morphological, and textural parameters were generated from ROIs within the images to form the feature space. The application of KM clustering to a feature space comprised of rCBV and texture parameters yielded a sensitivity of 92% and a specificity of 71% in differentiating GBMs from a single brain metastasis (Mouthuy et al., 2012). MS approaches have also been applied to prostate cancer for the purpose of tumor segmentation (Ozer et al., 2010). Automated MS classification algorithms (support vector machines, relevance vector machines, and a fuzzy Markov random field method) were evaluated in their abilities to segment prostate tumors by employing a feature space constructed from DCE-MRI, T_2 and diffusion parametric spatial maps. The study found that an MS approach improved segmentation results over single parameter analysis (Ozer et al., 2010). A result that was similar to studies that employed human readings, but in this case, the analysis was fully automated.

In summary, multiparametric methods have been shown to provide superior segmentation and characterization results over single parametric analysis methods for a range of cancers. The field of multidimensional pattern recognition provides a means to formally address the analysis of multiple parameters simultaneously and has been found to be well suited for use in image analysis (MS analysis). The application of MS analysis techniques to MRI provides an eloquent and robust analysis methodology to address the ever-growing demand to better characterized the tumor by including structural, physiological, and molecular MRI sequences as part of the imaging exam.

9.7 Conclusions

Solid tumors arise from genetic or somatic mutations resulting in a cell undergoing uncontrolled growth. This leads to heightened angiogenesis, metabolism, and heterogeneity in structure and physiology. Given the now well-known difficulties in treating such a varied and adaptable target, oncology therapies are moving beyond conventional cytotoxic radiation and chemotherapy-based regimens to more targeted approaches that limit tumor growth by altering such processes as angiogenesis, tumor metabolism, and other physiology mechanisms. Many of these new therapies will likely not produce dramatic reductions in tumor size and, thus, will require alternative tumor biomarkers to assess their activity during their clinical development and, if successful, during patient management.

Given the wide array of contrast mechanisms available, MRI is well positioned to provide many of the biomarkers that will be required for oncology patient management and drug development by quantifying both the structural and physiological features of solid tumors. The focus of this chapter has been to introduce the reader to many of these techniques, their current utility, and their future potential to provide valuable oncology endpoints. The ultimate utility of these endpoints will depend on their contribution to the development and clinical use of effective therapies against cancer.

References

Ahn, C.B., et al., 1987. The effects of random directional distributed flow in nuclear magnetic resonance imaging. *Med Phys.* 14, 43–8.

Akber, S.F., 1996. NMR relaxation data of water proton in normal tissues. *Physiol Chem Phys Med NMR.* 28, 205–38.

Akber, S.F., 2008. Water proton relaxation times of pathological tissues. *Physiol Chem Phys Med NMR.* 40, 1–42.

Akerley, W.L., et al., 2007. A randomized phase 2 trial of combretatstatin A4 phosphate (CA4P) in combination with paclitaxel and carboplatin to evaluate safety and efficacy in subjects with advanced imageable malignancies. *J Clin Oncol (Meetings Abstracts).* 25 (18S), 14060.

Albers, M.J., et al., 2008. Hyperpolarized 13C lactate, pyruvate, and alanine: Noninvasive biomarkers for prostate cancer detection and grading. *Cancer Res.* 68, 8607–15.

Alfidi, R.J., et al., 1982. Preliminary experimental results in humans and animals with a superconducting, whole-body, nuclear magnetic resonance scanner. *Radiology.* 143, 175–81.

Alic, L., et al., 2011. Heterogeneity in DCE-MRI parametric maps: A biomarker for treatment response? *Phys Med Biol.* 56, 1601–16.

American Cancer Society, 2009. *Cancer Facts & Figures 2009.* American Cancer Society, Atlanta.

Aronen, H.J., et al., 1994. Cerebral blood volume maps of gliomas: Comparison with tumor grade and histologic findings. *Radiology.* 191, 41–51.

Baar, J., et al., 2009. A vasculature-targeting regimen of preoperative docetaxel with or without bevacizumab for locally advanced breast cancer: Impact on angiogenic biomarkers. *Clin Cancer Res.* 15, 3583–90.

Babsky, A.M., et al., 2005. Application of 23Na MRI to monitor chemotherapeutic response in RIF-1 tumors. *Neoplasia.* 7, 658–66.

Barboriak, D.P., et al., 2007. Treatment of recurrent glioblastoma multiforme with bevacizumab and irinotecan leads to rapid decreases in tumor plasma volume and Ktrans. *RSNA.* 93, SST09–SST05.

Barck, K.H., et al., 2009. Viable tumor tissue detection in murine metastatic breast cancer by whole-body MRI and multispectral analysis. *Magn Reson Med.* 62, 1423–30.

Basser, P.J., Mattiello, J., LeBihan, D., 1994. Estimation of the effective self-diffusion tensor from the NMR spin echo. *J Magn Reson B.* 103, 247–54.

Bast, R.C., et al., 2007. Clinical trial endpoints in ovarian cancer: Report of an FDA/ASCO/AACR Public Workshop. *Gynecol Oncol.* 107, 173–6.

Bates, D.O., Curry, F.E., 1996. Vascular endothelial growth factor increases hydraulic conductivity of isolated perfused microvessels. *Am J Physiol.* 271, H2520–H2528.

Benndorf, M., et al., 2010. Breast MRI as an adjunct to mammography: Does it really suffer from low specificity? A retrospective analysis stratified by mammographic BI-RADS classes. *Acta Radiol.* 51, 715–21.

Bergers, G., Benjamin, L.E., 2003. Tumorigenesis and the angiogenic switch. *Nat Rev Cancer.* 3, 401–10.

Berry, L.R., et al., 2008. Quantification of viable tumor microvascular characteristics by multispectral analysis. *Magn Reson Med.* 60, 64–72.

Bezdek, J.C., Hall, L.O., Clarke, L.P., 1993. Review of MR image segmentation techniques using pattern recognition. *Med Phys.* 20, 1033–48.

Bjarnason, G.A., et al., 2011. Microbubble ultrasound (DCE-US) compared to DCE-MRI and DCE-CT for the assessment of vascular response to sunitinib in renal cell carcinoma (RCC). *J Clin Oncol (Meetings Abstracts).* 29 (S), 4627.

Booth, C.M., Eisenhauer, E.A., 2012. Progression-free survival: Meaningful or simply measurable? *J Clin Oncol.* 30, 1030–3.

Brix, G., et al., 1991. Pharmacokinetic parameters in CNS Gd-DTPA enhanced MR imaging. *J Comput Assist Tomogr.* 15, 621–8.

Brown, J.M., Wilson, W.R., 2004. Exploiting tumour hypoxia in cancer treatment. *Nat Rev Cancer.* 4, 437–47.

Brown, R., 1828. A brief account of microscopical observations made in the months of June, July and August, 1827, on the particles contained in the pollen of plants; and on the general existence of active molecules in organic and inorganic bodies. *Philos. Mag.* 4 (New Series), 161–73.

Brunberg, J.A., et al., 1995. In vivo MR determination of water diffusion coefficients and diffusion anisotropy: Correlation with structural alteration in gliomas of the cerebral hemispheres. *AJNR Am J Neuroradiol.* 16, 361–71.

Buckley, D.L., et al., 1997. Microvessel density of invasive breast cancer assessed by dynamic Gd-DTPA enhanced MRI. *J Magn Reson Imaging.* 7, 461–4.

Carano, R.A., et al., 1998. Determination of focal ischemic lesion volume in the rat brain using multispectral analysis. *J Magn Reson Imaging.* 8, 1266–78.

Carano, R.A., et al., 2004a. Multispectral analysis of bone lesions in the hands of patients with rheumatoid arthritis. *Magn Reson Imaging.* 22, 505–14.

Carano, R.A., et al., 2004b. Quantification of tumor tissue populations by multispectral analysis. *Magn Reson Med.* 51, 542–51.

Carr, D.H., et al., 1984a. Gadolinium-DTPA as a contrast agent in MRI: Initial clinical experience in 20 patients. *Am J Roentgenol.* 143, 215–24.

Carr, D.H., et al., 1984b. Intravenous chelated gadolinium as a contrast agent in NMR imaging of cerebral tumours. *Lancet.* 1, 484–6.

Carr, D.H., et al., 1984c. Iron and gadolinium chelates as contrast agents in NMR imaging: Preliminary studies. *J Comput Assist Tomogr.* 8, 385–9.

Carr, H.Y., Purcell, E.M., 1954. Effects of diffusion on free precession in nuclear magnetic resonance experiments. *Phys Rev.* 94, 630–8.

Checkley, D., et al., 2003. Use of dynamic contrast-enhanced MRI to evaluate acute treatment with ZD6474, a VEGF signalling inhibitor, in PC-3 prostate tumours. *Br J Cancer.* 89, 1889–95.

Chen, A.P., et al., 2007a. Hyperpolarized C-13 spectroscopic imaging of the TRAMP mouse at 3T-initial experience. *Magn Reson Med.* 58, 1099–106.

Chen, M., et al., 2008. Prostate cancer detection: Comparison of T2-weighted imaging, diffusion-weighted imaging, proton magnetic resonance spectroscopic imaging, and the three techniques combined. *Acta Radiol.* 49, 602–10.

Chen, W., et al., 2007b. Volumetric texture analysis of breast lesions on contrast-enhanced magnetic resonance images. *Magn Reson Med.* 58, 562–71.

Chenevert, T.L., McKeever, P.E., Ross, B.D., 1997. Monitoring early response of experimental brain tumors to therapy using diffusion magnetic resonance imaging. *Clin Cancer Res.* 3, 1457–66.

Chenevert, T.L., et al., 2000. Diffusion magnetic resonance imaging: An early surrogate marker of therapeutic efficacy in brain tumors. *J Natl Cancer Inst.* 92, 2029–36.

Choi, Y.P., et al., 2011. Molecular portraits of intratumoral heterogeneity in human ovarian cancer. *Cancer Lett.* 307, 62–71.

Clark, C.A., Le Bihan, D., 2000. Water diffusion compartmentation and anisotropy at high b values in the human brain. *Magn Reson Med.* 44, 852–9.

Clarke, L.P., et al., 1995. MRI segmentation: Methods and applications. *Magn Reson Imaging.* 13, 343–68.

Cline, H.E., et al., 1990. Three-dimensional segmentation of MR images of the head using probability and connectivity. *J Comput Assist Tomogr.* 14, 1037–45.

Cory, D.G., Garroway, A.N., 1990. Measurement of translational displacement probabilities by NMR: An indicator of compartmentation. *Magn Reson Med.* 14, 435–44.

Damadian, R., 1971. Tumor detection by nuclear magnetic resonance. *Science.* 171, 1151–3.

Day, S.E., et al., 2007. Detecting tumor response to treatment using hyperpolarized 13C magnetic resonance imaging and spectroscopy. *Nat Med.* 13, 1382–7.

Dean, B.L., et al., 1990. Gliomas: classification with MR imaging. *Radiology.* 174, 411–5.

de Bazelaire, C., et al., 2005. Arterial spin labeling blood flow magnetic resonance imaging for the characterization of metastatic renal cell carcinoma. *Acad Radiol.* 12, 347–57.

de Bazelaire, C., et al., 2008. Magnetic resonance imaging-measured blood flow change after antiangiogenic therapy with PTK787/ZK 222584 correlates with clinical outcome in metastatic renal cell carcinoma. *Clin Cancer Res.* 14, 5548–54.

Delongchamps, N.B., et al., 2011a. Multiparametric MRI is helpful to predict tumor focality, stage, and size in patients diagnosed with unilateral low-risk prostate cancer. *Prostate Cancer Prostatic Dis.* 14, 232–7.

Delongchamps, N.B., et al., 2011b. Multiparametric magnetic resonance imaging for the detection and localization of prostate cancer: Combination of T2-weighted, dynamic contrast-enhanced and diffusion-weighted imaging. *BJU Int.* 107, 1411–8.

Dempsey, M.F., Condon, B.R., Hadley, D.M., 2005. Measurement of tumor "size" in recurrent malignant glioma: 1D, 2D, or 3D? *AJNR Am J Neuroradiol.* 26, 770–6.

Detre, J.A., Alsop, D.C., 1999. Perfusion magnetic resonance imaging with continuous arterial spin labeling: Methods and clinical applications in the central nervous system. *Eur J Radiol.* 30, 115–24.

Detre, J.A., et al., 1992. Perfusion imaging. *Magn Reson Med.* 23, 37–45.

De Vries, A.F., et al., 2003. Tumor microcirculation and diffusion predict therapy outcome for primary rectal carcinoma. *Int J Radiat Oncol Biol Phys.* 56, 958–65.

Di Costanzo, A., et al., 2006. Multiparametric 3T MR approach to the assessment of cerebral gliomas: Tumor extent and malignancy. *Neuroradiology.* 48, 622–31.

Drevs, J., et al., 2007. Phase I clinical study of AZD2171, an oral vascular endothelial growth factor signaling inhibitor, in patients with advanced solid tumors. *J Clin Oncol.* 25, 3045–54.

Duda, R.O., Hart, P.E., 1973. *Pattern Classification and Scene Analysis.* John Wiley & Sons, New York.

Dvorak, H.F., 1986. Tumors: Wounds that do not heal. Similarities between tumor stroma generation and wound healing. *N Engl J Med.* 315, 1650–9.

Dzik-Jurasz, A., et al., 2002. Diffusion MRI for prediction of response of rectal cancer to chemoradiation. *Lancet.* 360, 307–8.

Eberhard, A., et al., 2000. Heterogeneity of angiogenesis and blood vessel maturation in human tumors: Implications for antiangiogenic tumor therapies. *Cancer Res.* 60, 1388–93.

Einstein, A., 1905. Über die von der molekularkinetischen Theorie der Wärme geforderte Bewegung von in ruhenden Flüssigkeiten suspendierten Teilchen. *Ann Phys.* 17, 549–60.

Einstein, A., 1956. *Investigations on the Theory of the Brownian Movement.* Dover, New York.

Eisenhauer, E.A., et al., 2009. New response evaluation criteria in solid tumours: Revised RECIST guideline (version 1.1). *Eur J Cancer.* 45, 228–47.

El Khouli, R.H., et al., 2009. Dynamic contrast-enhanced MRI of the breast: Quantitative method for kinetic curve type assessment. *AJR Am J Roentgenol.* 193, W295–W300.

Ellis, L.M., 2003. Antiangiogenic therapy: More promise and, yet again, more questions. *J Clin Oncol.* 21, 3897–9.

Els, T., et al., 1995. Diffusion-weighted MR imaging of experimental brain tumors in rats. *MAGMA.* 3, 13–20.

Erguvan-Dogan, B., et al., 2006. BI-RADS-MRI: A primer. *AJR Am J Roentgenol.* 187, W152–W160.

Escudier, B., et al., 2007. Sorafenib in advanced clear-cell renal-cell carcinoma. *N Engl J Med.* 356, 125–34.

Evelhoch, J.L., 1999. Key factors in the acquisition of contrast kinetic data for oncology. *J Magn Reson Imaging.* 10, 254–9.

Ferrara, N., Gerber, H.P., LeCouter, J., 2003. The biology of VEGF and its receptors. *Nat Med.* 9, 669–76.

Ferrara, N., Henzel, W.J., 1989. Pituitary follicular cells secrete a novel heparin-binding growth factor specific for vascular endothelial cells. *Biochem Biophys Res Commun.* 161, 851–8.

Flaherty, K.T., et al., 2008. Pilot study of DCE-MRI to predict progression-free survival with sorafenib therapy in renal cell carcinoma. *Cancer Biol Ther.* 7, 496–501.

Folkman, J., 1971. Tumor angiogenesis: Therapeutic implications. *N Engl J Med.* 285, 1182–6.

Folkman, J., Hanahan, D., 1991. Switch to the angiogenic phenotype during tumorigenesis. *Princess Takamatsu Symp.* 22, 339–47.

Galbraith, S.M., et al., 2003. Combretastatin A4 phosphate has tumor antivascular activity in rat and man as demonstrated by dynamic magnetic resonance imaging. *J Clin Oncol.* 21, 2831–42.

Gallagher, F.A., et al., 2008a. Magnetic resonance imaging of pH in vivo using hyperpolarized 13C-labelled bicarbonate. *Nature.* 453, 940–3.

Gallagher, F.A., et al., 2008b. 13C MR spectroscopy measurements of glutaminase activity in human hepatocellular carcinoma cells using hyperpolarized 13C-labeled glutamine. *Magn Reson Med.* 60, 253–7.

Galons, J.P., et al., 1999. Early increases in breast tumor xenograft water mobility in response to paclitaxel therapy detected by non-invasive diffusion magnetic resonance imaging. *Neoplasia.* 1, 113–7.

Gerlinger, M., et al., 2012. Intratumor heterogeneity and branched evolution revealed by multiregion sequencing. *N Engl J Med*. 366, 883–92.

Golman, K., et al., 2006. Metabolic imaging by hyperpolarized 13C magnetic resonance imaging for in vivo tumor diagnosis. *Cancer Res*. 66, 10855–60.

Gonzalez-Garcia, I., Sole, R.V., Costa, J., 2002. Metapopulation dynamics and spatial heterogeneity in cancer. *Proc Natl Acad Sci USA*. 99, 13085–9.

Gordon, J., et al., 1999. Utililization of experimental animal model for correlative multispectral MRI and pathological analysis of brain tumors. *Magn Reson Imaging*. 17, 1495–502.

Guo, Y., et al., 2002. Differentiation of clinically benign and malignant breast lesions using diffusion-weighted imaging. *J Magn Reson Imaging*. 16, 172–8.

Gururangan, S., et al., 2010. Lack of efficacy of bevacizumab plus irinotecan in children with recurrent malignant glioma and diffuse brainstem glioma: A Pediatric Brain Tumor Consortium study. *J Clin Oncol*. 28, 3069–75.

Gutin, P.H., et al., 2009. Safety and efficacy of bevacizumab with hypofractionated stereotactic irradiation for recurrent malignant gliomas. *Int J Radiat Oncol Biol Phys*. 75, 156–63.

Hahn, O.M., et al., 2008. Dynamic contrast-enhanced magnetic resonance imaging pharmacodynamic biomarker study of sorafenib in metastatic renal carcinoma. *J Clin Oncol*. 26, 4572–8.

Haider, M.A., et al., 2007. Combined T2-weighted and diffusion-weighted MRI for localization of prostate cancer. *AJR Am J Roentgenol*. 189, 323–8.

Haralick, R.M., Shanmuga, K., Dinstein, I., 1973. Textural features for image classification. *IEEE Trans Syst Man Cybern*. 6, 610–21.

Hayes, C., Padhani, A.R., Leach, M.O., 2002. Assessing changes in tumour vascular function using dynamic contrast-enhanced magnetic resonance imaging. *NMR Biomed*. 15, 154–63.

Hecht, J.R., et al., 2011. Randomized, placebo-controlled, phase III study of first-line oxaliplatin-based chemotherapy plus PTK787/ZK 222584, an oral vascular endothelial growth factor receptor inhibitor, in patients with metastatic colorectal adenocarcinoma. *J Clin Oncol*. 29, 1997–2003.

Helmer, K.G., Dardzinski, B.J., Sotak, C.H., 1995. The application of porous-media theory to the investigation of time-dependent diffusion in in vivo systems. *NMR Biomed*. 8, 297–306.

Helmer, K.G., Han, S., Sotak, C.H., 1998. On the correlation between the water diffusion coefficient and oxygen tension in RIF-1 tumors. *NMR Biomed*. 11, 120–30.

Henning, E.C., et al., 2007. Multispectral quantification of tissue types in a RIF-1 tumor model with histological validation. Part I. *Magn Reson Med*. 57, 501–12.

Heppner, G.H., 1984. Tumor heterogeneity. *Cancer Res*. 44, 2259–65.

Houdek, P.V., et al., 1988. MR characterization of brain and brain tumor response to radiotherapy. *Int J Radiat Oncol Biol Phys*. 15, 213–8.

Hurlimann, M.D., et al., 1995. Spin echoes in a constant gradient and in the presence of simple restriction. *J Magn Reson A*. 113, 260–4.

Hurwitz, H., et al., 2004. Bevacizumab plus irinotecan, fluorouracil, and leucovorin for metastatic colorectal cancer. *N Engl J Med*. 350, 2335–42.

Ibrahim, M.A., Emerson, J.F., Cotman, C.W., 1998. Magnetic resonance imaging relaxation times and gadolinium-DTPA relaxivity values in human cerebrospinal fluid. *Invest Radiol*. 33, 153–62.

Jackson, A., et al., 2007. Imaging tumor vascular heterogeneity and angiogenesis using dynamic contrast-enhanced magnetic resonance imaging. *Clin Cancer Res.* 13, 3449–59.

Jacobs, M.A., et al., 2003. Benign and malignant breast lesions: Diagnosis with multiparametric MR imaging. *Radiology.* 229, 225–32.

Jamin, Y., et al., 2013. Evaluation of clinically translatable MR imaging biomarkers of therapeutic response in the TH-MYCN transgenic mouse model of neuroblastoma. *Radiology.* 266, 130–40.

Jemal, A., et al., 2011. Global cancer statistics. *CA Cancer J Clin.* 61, 69–90.

Johnson, J.A., Wilson, T.A., 1966. A model for capillary exchange. *Am J Physiol.* 210, 1299–303.

Jonker, D.J., et al., 2011. A phase I study to determine the safety, pharmacokinetics and pharmacodynamics of a dual VEGFR and FGFR inhibitor, brivanib, in patients with advanced or metastatic solid tumors. *Ann Oncol.* 22, 1413–9.

Just, M., Thelen, M., 1988. Tissue characterization with T1, T2, and proton density values: results in 160 patients with brain tumors. *Radiology.* 169, 779–85.

Just, M., et al., 1988. Tissue characterization of benign brain tumors: Use of NMR-tissue parameters. *Magn Reson Imaging.* 6, 463–72.

Kale, M.C., et al., 2008. Multispectral co-occurrence with three random variables in dynamic contrast enhanced magnetic resonance imaging of breast cancer. *IEEE Trans Med Imaging.* 27, 1425–31.

Keck, P.J., et al., 1989. Vascular permeability factor, an endothelial cell mitogen related to PDGF. *Science.* 246, 1309–12.

Kerlan, R.K., Jr., et al., 1983. Nuclear magnetic resonance and computed tomographic angiography in the detection of hepatic metastases. *J Clin Gastroenterol.* 5, 461–4.

Kety, S.S., 1951. The theory and applications of the exchange of inert gas at the lungs and tissues. *Pharmacol Rev.* 3, 1–41.

Kilickesmez, O., et al., 2009. Value of apparent diffusion coefficient measurement for discrimination of focal benign and malignant hepatic masses. *J Med Imaging Radiat Oncol.* 53, 50–5.

Kim, H.S., Kim, S.Y., 2007. A prospective study on the added value of pulsed arterial spin-labeling and apparent diffusion coefficients in the grading of gliomas. *AJNR Am J Neuroradiol.* 28, 1693–9.

King, M.D., et al., 1992. Perfusion and diffusion MR imaging. *Magn Reson Med.* 24, 288–301.

Kiricuta, I.C., Jr., Simplaceanu, V., 1975. Tissue water content and nuclear magnetic resonance in normal and tumor tissues. *Cancer Res.* 35, 1164–7.

Kjaer, L., et al., 1991. Tissue characterization of intracranial tumors by MR imaging. In vivo evaluation of T1- and T2-relaxation behavior at 1.5 T. *Acta Radiol.* 32, 498–504.

Knopp, E.A., et al., 1999a. Glial neoplasms: Dynamic contrast-enhanced T2*-weighted MR imaging. *Radiology.* 211, 791–8.

Knopp, M.V., et al., 1999b. Pathophysiologic basis of contrast enhancement in breast tumors. *J Magn Reson Imaging.* 10, 260–6.

Koh, D.M., et al., 2007. Predicting response of colorectal hepatic metastasis: Value of pretreatment apparent diffusion coefficients. *AJR Am J Roentgenol.* 188, 1001–8.

Koh, D.M., et al., 2009. Reproducibility and changes in the apparent diffusion coefficients of solid tumours treated with combretastatin A4 phosphate and bevacizumab in a two-centre phase I clinical trial. *Eur Radiol.* 19, 2728–38.

Komiyama, M., et al., 1987. MR imaging: Possibility of tissue characterization of brain tumors using T1 and T2 values. *AJNR Am J Neuroradiol.* 8, 65–70.

Komori, T., et al., 2007. 2-[Fluorine-18]-fluoro-2-deoxy-D-glucose positron emission tomography/computed tomography versus whole-body diffusion-weighted MRI for detection of malignant lesions: Initial experience. *Ann Nucl Med.* 21, 209–15.

Kondziolka, D., Lunsford, L.D., Martinez, A.J., 1993. Unreliability of contemporary neurodiagnostic imaging in evaluating suspected adult supratentorial (low-grade) astrocytoma. *J Neurosurg.* 79, 533–6.

Koyama, H., et al., 2010. Comparison of STIR turbo SE imaging and diffusion-weighted imaging of the lung: Capability for detection and subtype classification of pulmonary adenocarcinomas. *Eur Radiol.* 20, 790–800.

Krabbe, K., et al., 1997. MR diffusion imaging of human intracranial tumours. *Neuroradiology.* 39, 483–9.

Krajewski, K.M., et al., 2011. Comparison of four early posttherapy imaging changes (EPTIC; RECIST 1.0, tumor shrinkage, computed tomography tumor density, Choi criteria) in assessing outcome to vascular endothelial growth factor-targeted therapy in patients with advanced renal cell carcinoma. *Eur Urol.* 59, 856–62.

Kreisl, T.N., et al., 2011. A phase II trial of single-agent bevacizumab in patients with recurrent anaplastic glioma. *Neuro Oncol.* 13, 1143–50.

Kriege, M., et al., 2004. Efficacy of MRI and mammography for breast-cancer screening in women with a familial or genetic predisposition. *N Engl J Med.* 351, 427–37.

Kuhl, C.K., et al., 1999. Dynamic breast MR imaging: Are signal intensity time course data useful for differential diagnosis of enhancing lesions? *Radiology.* 211, 101–10.

Kurhanewicz, J., et al., 2011. Analysis of cancer metabolism by imaging hyperpolarized nuclei: Prospects for translation to clinical research. *Neoplasia.* 13, 81–97.

Larsson, H.B., et al., 1990. Quantitation of blood-brain barrier defect by magnetic resonance imaging and gadolinium-DTPA in patients with multiple sclerosis and brain tumors. *Magn Reson Med.* 16, 117–31.

Law, M., et al., 2003. Glioma grading: Sensitivity, specificity, and predictive values of perfusion MR imaging and proton MR spectroscopic imaging compared with conventional MR imaging. *AJNR Am J Neuroradiol.* 24, 1989–98.

Le Bihan, D., 1991. Molecular diffusion nuclear magnetic resonance imaging. *Magn Reson Q.* 7, 1–30.

Le Bihan, D., et al., 1986. MR imaging of intravoxel incoherent motions: Application to diffusion and perfusion in neurologic disorders. *Radiology.* 161, 401–7.

Leach, M.O., et al., 2005a. Screening with magnetic resonance imaging and mammography of a UK population at high familial risk of breast cancer: A prospective multicentre cohort study (MARIBS). *Lancet.* 365, 1769–78.

Leach, M.O., et al., 2005b. The assessment of antiangiogenic and antivascular therapies in early-stage clinical trials using magnetic resonance imaging: Issues and recommendations. *Br J Cancer.* 92, 1599–610.

Lee, J.C., et al., 2000. Prognostic value of vascular endothelial growth factor expression in colorectal cancer patients. *Eur J Cancer.* 36, 748–53.

Lee, M.C., Nelson, S.J., 2008. Supervised pattern recognition for the prediction of contrast-enhancement appearance in brain tumors from multivariate magnetic resonance imaging and spectroscopy. *Artif Intell Med.* 43, 61–74.

Leung, D.W., et al., 1989. Vascular endothelial growth factor is a secreted angiogenic mitogen. *Science.* 246, 1306–9.

Li, S.P., et al., 2011a. Use of dynamic contrast-enhanced MR imaging to predict survival in patients with primary breast cancer undergoing neoadjuvant chemotherapy. *Radiology.* 260, 68–78.

Li, S.P., et al., 2011b. Evaluating the early effects of anti-angiogenic treatment in human breast cancer with Intrinsic susceptibility-weighted and diffusion-weighted MRI: Initial observations. *Proc Int Soc Magn Reson Med.* 19, 342.

Liotta, L.A., Kohn, E.C., 2001. The microenvironment of the tumour-host interface. *Nature.* 411, 375–9.

List, A.F., 2001. Vascular endothelial growth factor signaling pathway as an emerging target in hematologic malignancies. *Oncologist.* 6, 24–31.

Ludemann, L., et al., 2009. Brain tumor perfusion: Comparison of dynamic contrast enhanced magnetic resonance imaging using T1, T2, and T2* contrast, pulsed arterial spin labeling, and H2(15)O positron emission tomography. *Eur J Radiol.* 70, 465–74.

Lv, D., et al., 2009. Computerized characterization of prostate cancer by fractal analysis in MR images. *J Magn Reson Imaging.* 30, 161–8.

Lyng, H., Haraldseth, O., Rofstad, E.K., 2000. Measurement of cell density and necrotic fraction in human melanoma xenografts by diffusion weighted magnetic resonance imaging. *Magn Reson Med.* 43, 828–36.

Maeda, M., et al., 1992. Intravoxel incoherent motion (IVIM) MRI in intracranial, extraaxial tumors and cysts. *J Comput Assist Tomogr.* 16, 514–8.

Maier, C.F., et al., 1997. Quantitative diffusion imaging in implanted human breast tumors. *Magn Reson Med.* 37, 576–81.

Mardor, Y., et al., 2003. Early detection of response to radiation therapy in patients with brain malignancies using conventional and high b-value diffusion-weighted magnetic resonance imaging. *J Clin Oncol.* 21, 1094–100.

Mardor, Y., et al., 2004. Pretreatment prediction of brain tumors' response to radiation therapy using high b-value diffusion-weighted MRI. *Neoplasia.* 6, 136–42.

Martincich, L., et al., 2004. Monitoring response to primary chemotherapy in breast cancer using dynamic contrast-enhanced magnetic resonance imaging. *Breast Cancer Res Treat.* 83, 67–76.

McClain, K.J., Zhu, Y., Hazle, J.D., 1995. Selection of MR images for automated segmentation. *J Magn Reson Imaging.* 5, 485–92.

McSheehy, P.M., et al., 2010. Quantified tumor T1 is a generic early-response imaging biomarker for chemotherapy reflecting cell viability. *Clin Cancer Res.* 16, 212–25.

Miao, H., Fukatsu, H., Ishigaki, T., 2007. Prostate cancer detection with 3-T MRI: Comparison of diffusion-weighted and T2-weighted imaging. *Eur J Radiol.* 61, 297–302.

Michaelis, L.C., Ratain, M.J., 2006. Measuring response in a post-RECIST world: From black and white to shades of grey. *Nat Rev Cancer.* 6, 409–14.

Miller, K.D., et al., 2005. A multicenter phase II trial of ZD6474, a vascular endothelial growth factor receptor-2 and epidermal growth factor receptor tyrosine kinase inhibitor, in patients with previously treated metastatic breast cancer. *Clin Cancer Res.* 11, 3369–76.

Mills, C.M., et al., 1984. Cerebral abnormalities: Use of calculated T1 and T2 magnetic resonance images for diagnosis. *Radiology.* 150, 87–94.

Mitchell, C.L., et al., 2011. A two-part Phase II study of cediranib in patients with advanced solid tumours: The effect of food on single-dose pharmacokinetics and an evaluation of safety, efficacy and imaging pharmacodynamics. *Cancer Chemother Pharmacol.* 68, 631–41.

Mitra, P.P., Sen, P.N., Schwartz, L.M., 1993. Short-time behavior of the diffusion coefficient as a geometrical probe of porous media. *Phys Rev B Condens Matter.* 47, 8565–74.

Mitra, P.P., et al., 1992. Diffusion propagator as a probe of the structure of porous media. *Phys Rev Lett.* 68, 3555–8.

Moller-Hartmann, W., et al., 2002. Clinical application of proton magnetic resonance spectroscopy in the diagnosis of intracranial mass lesions. *Neuroradiology.* 44, 371–81.

Moore, R.J., et al., 2000. In vivo intravoxel incoherent motion measurements in the human placenta using echo-planar imaging at 0.5 T. *Magn Reson Med.* 43, 295–302.

Morgan, B., et al., 2003. Dynamic contrast-enhanced magnetic resonance imaging as a biomarker for the pharmacological response of PTK787/ZK 222584, an inhibitor of the vascular endothelial growth factor receptor tyrosine kinases, in patients with advanced colorectal cancer and liver metastases: Results from two phase I studies. *J Clin Oncol.* 21, 3955–64.

Morgan, V.A., et al., 2007. Evaluation of the potential of diffusion-weighted imaging in prostate cancer detection. *Acta Radiol.* 48, 695–703.

Mouthuy, N., et al., 2012. Multiparametric magnetic resonance imaging to differentiate high-grade gliomas and brain metastases. *J Neuroradiol.* 39, 301–7.

Mross, K., et al., 2005. Phase I clinical and pharmacokinetic study of PTK/ZK, a multiple VEGF receptor inhibitor, in patients with liver metastases from solid tumours. *Eur J Cancer.* 41, 1291–9.

Mross, K., et al., 2009. DCE-MRI assessment of the effect of vandetanib on tumor vasculature in patients with advanced colorectal cancer and liver metastases: A randomized phase I study. *J Angiogenes Res.* 1, 5.

Mueller, M.M., Fusenig, N.E., 2004. Friends or foes—Bipolar effects of the tumour stroma in cancer. *Nat Rev Cancer.* 4, 839–49.

Mulkern, R.V., et al., 1999. Multi-component apparent diffusion coefficients in human brain. *NMR Biomed.* 12, 51–62.

Murphy, P.S., et al., 2010. Vascular response of hepatocellular carcinoma to pazopanib measured by dynamic contrast-enhanced MRI: Pharmacokinetic and clinical activity correlations. *Proc Int Soc Magn Reson Med.* 18, 2720.

Nalcioglu, O., Cho, Z.H., 1987. Measurement of bulk and random directional velocity fields by NMR imaging. *IEEE Trans Med Imaging.* 6, 356–9.

Nathan, P., et al., 2012. Phase I trial of combretastatin A4 phosphate (CA4P) in combination with bevacizumab in patients with advanced cancer. *Clin Cancer Res.* 18, 3428–39.

Nelson, S., et al., 2012. Proof of concept clinical trial of hyperpolarized C-13 pyruvate in patients with prostate cancer. *Proceedings of International Society Magnetic Resonance in Medicine.* Melbourne, Australia, p. 274.

NICE, 2006. *Familial Breast Cancer: The Classification and Care of Women at Risk of Familial Breast Cancer in Primary, Secondary and Tertiary Care—Update.* National Institute for Clinical Excellence. London. http://www.nice.org.uk/guidance/cg41 (accessed 3 January 2012).

Niendorf, T., et al., 1996. Biexponential diffusion attenuation in various states of brain tissue: Implications for diffusion-weighted imaging. *Magn Reson Med.* 36, 847–57.

O'Connor, J.P., et al., 2007. DCE-MRI biomarkers in the clinical evaluation of antiangiogenic and vascular disrupting agents. *Br J Cancer*. 96, 189–95.

O'Connor, J.P., et al., 2008. Quantitative imaging biomarkers in the clinical development of targeted therapeutics: Current and future perspectives. *Lancet Oncol*. 9, 766–76.

O'Connor, J.P., et al., 2009. Quantifying antivascular effects of monoclonal antibodies to vascular endothelial growth factor: Insights from imaging. *Clin Cancer Res*. 15, 6674–82.

O'Connor, J.P., et al., 2011. DCE-MRI biomarkers of tumour heterogeneity predict CRC liver metastasis shrinkage following bevacizumab and FOLFOX-6. *Br J Cancer*. 105, 139–45.

O'Connor, J.P., et al., 2012. Dynamic contrast-enhanced MRI in clinical trials of antivascular therapies. *Nat Rev Clin Oncol*. 9, 167–77.

O'Donnell, A., et al., 2005. A phase I study of the angiogenesis inhibitor SU5416 (semaxanib) in solid tumours, incorporating dynamic contrast MR pharmacodynamic end points. *Br J Cancer*. 93, 876–83.

Ohno, Y., et al., 2008. Non-small cell lung cancer: Whole-body MR examination for M-stage assessment—utility for whole-body diffusion-weighted imaging compared with integrated FDG PET/CT. *Radiology*. 248, 643–54.

Ozer, S., et al., 2010. Supervised and unsupervised methods for prostate cancer segmentation with multispectral MRI. *Med Phys*. 37, 1873–83.

Padhani, A.R., Ollivier, L., 2001. The RECIST (response evaluation criteria in solid tumors) criteria: Implications for diagnostic radiologists. *Br J Radiol*. 74, 983–6.

Padhani, A.R., et al., 2009. Diffusion-weighted magnetic resonance imaging as a cancer biomarker: Consensus and recommendations. *Neoplasia*. 11, 102–25.

Parikh, T., et al., 2008. Focal liver lesion detection and characterization with diffusion-weighted MR imaging: Comparison with standard breath-hold T2-weighted imaging. *Radiology*. 246, 812–22.

Park, S.H., et al., 2010a. Diffusion-weighted MR imaging: Pretreatment prediction of response to neoadjuvant chemotherapy in patients with breast cancer. *Radiology*. 257, 56–63.

Park, S.M., et al., 2010b. Combination of high-resolution susceptibility-weighted imaging and the apparent diffusion coefficient: Added value to brain tumour imaging and clinical feasibility of non-contrast MRI at 3 T. *Br J Radiol*. 83, 466–75.

Park, S.Y., et al., 2010c. Prediction of biochemical recurrence following radical prostatectomy in men with prostate cancer by diffusion-weighted magnetic resonance imaging: Initial results. *Eur Radiol*. 21, 1111–8.

Parker, G.J.M., Buckley, D.L., 2005. Tracer kinetic modelling for T1-weighted DCE-MRI. In: *Dynamic Contrast-Enhanced Magnetic Resonance Imaging in Oncology. Medical Radiology: Diagnostic Radiology*, A. Jackson, D.L. Buckley, G.J.M. Parker, eds. Springer, Berlin, pp. 81–92.

Parkes, L.M., Tofts, P.S., 2002. Improved accuracy of human cerebral blood perfusion measurements using arterial spin labeling: Accounting for capillary water permeability. *Magn Reson Med*. 48, 27–41.

Patankar, T.F., et al., 2005. Is volume transfer coefficient (K(trans)) related to histologic grade in human gliomas? *AJNR Am J Neuroradiol*. 26, 2455–65.

Peitgen, H.-O., Jurgens, H., Saupe, D., 2004. *Chaos and Fractals*. Springer, Berlin.

Pekar, J., Moonen, C.T., van Zijl, P.C., 1992. On the precision of diffusion/perfusion imaging by gradient sensitization. *Magn Reson Med*. 23, 122–9.

Petersen, E.T., et al., 2006. Non-invasive measurement of perfusion: A critical review of arterial spin labelling techniques. *Br J Radiol.* 79, 688–701.

Pfeuffer, J., Provencher, S.W., Gruetter, R., 1999. Water diffusion in rat brain in vivo as detected at very large b values is multicompartmental. *MAGMA.* 8, 98–108.

Pfeuffer, J., et al., 1998. Water signal attenuation in diffusion-weighted 1H NMR experiments during cerebral ischemia: Influence of intracellular restrictions, extracellular tortuosity, and exchange. *Magn Reson Imaging.* 16, 1023–32.

Pickles, M.D., et al., 2006. Diffusion changes precede size reduction in neoadjuvant treatment of breast cancer. *Magn Reson Imaging.* 24, 843–7.

Poon, R.T., Fan, S.T., Wong, J., 2001a. Clinical implications of circulating angiogenic factors in cancer patients. *J Clin Oncol.* 19, 1207–25.

Poon, R.T., Ng, I.O., et al., 2001b. Serum vascular endothelial growth factor predicts venous invasion in hepatocellular carcinoma: A prospective study. *Ann Surg.* 233, 227–35.

Pope, W.B., et al., 2009. Recurrent glioblastoma multiforme: ADC histogram analysis predicts response to bevacizumab treatment. *Radiology.* 252, 182–9.

Pope, W.B., et al., 2011. Apparent diffusion coefficient histogram analysis stratifies progression-free survival in newly diagnosed bevacizumab-treated glioblastoma. *AJNR Am J Neuroradiol.* 32, 882–9.

Poptani, H., et al., 1998. Monitoring thymidine kinase and ganciclovir-induced changes in rat malignant glioma in vivo by nuclear magnetic resonance imaging. *Cancer Gene Ther.* 5, 101–9.

Rajendran, J.G., et al., 2004. Hypoxia and glucose metabolism in malignant tumors: Evaluation by [18F]fluoromisonidazole and [18F]fluorodeoxyglucose positron emission tomography imaging. *Clin Cancer Res.* 10, 2245–52.

Ratain, M.J., Eckhardt, S.G., 2004. Phase II studies of modern drugs directed against new targets: if you are fazed, too, then resist RECIST. *J Clin Oncol.* 22, 4442–5.

Rieger, J., et al., 2010. Bevacizumab-induced diffusion-restricted lesions in malignant glioma patients. *J Neurooncol.* 99, 49–56.

Roberts, T.P., 1997. Physiologic measurements by contrast-enhanced MR imaging: Expectations and limitations. *J Magn Reson Imaging.* 7, 82–90.

Robinson, S.P., et al., 2003. Tumour dose response to the antivascular agent ZD6126 assessed by magnetic resonance imaging. *Br J Cancer.* 88, 1592–7.

Rose, C.J., et al., 2009. Quantifying spatial heterogeneity in dynamic contrast-enhanced MRI parameter maps. *Magn Reson Med.* 62, 488–99.

Rosen, B.R., Belliveau, J.W., Chien, D., 1989. Perfusion imaging by nuclear magnetic resonance. *Magn Reson Q.* 5, 263–81.

Schepkin, V.D., et al., 2005. Sodium magnetic resonance imaging of chemotherapeutic response in a rat glioma. *Magn Reson Med.* 53, 85–92.

Schepkin, V.D., et al., 2006. Proton and sodium MRI assessment of emerging tumor chemotherapeutic resistance. *NMR Biomed.* 19, 1035–42.

Schor-Bardach, R., et al., 2009. Does arterial spin-labeling MR imaging-measured tumor perfusion correlate with renal cell cancer response to antiangiogenic therapy in a mouse model? *Radiology.* 251, 731–42.

Sharma, R., et al., 2005. Rapid in vivo Taxotere quantitative chemosensitivity response by 4.23 Tesla sodium MRI and histo-immunostaining features in N-Methyl-N-Nitrosourea induced breast tumors in rats. *Cancer Cell Int.* 5, 26.

Shimauchi, A., et al., 2010. Breast cancers not detected at MRI: Review of false-negative lesions. *Am J Roentgenol.* 194, 1674–9.

Siegel, A.B., et al., 2008. Phase II trial evaluating the clinical and biologic effects of bevacizumab in unresectable hepatocellular carcinoma. *J Clin Oncol*. 26, 2992–8.

Silva, A.C., Kim, S.G., Garwood, M., 2000. Imaging blood flow in brain tumors using arterial spin labeling. *Magn Reson Med*. 44, 169–73.

Sinha, S., et al., 1997. Multifeature analysis of Gd-enhanced MR images of breast lesions. *J Magn Reson Imaging*. 7, 1016–26.

Sourbron, S.P., Buckley, D.L., 2012. Tracer kinetic modelling in MRI: Estimating perfusion and capillary permeability. *Phys Med Biol*. 57, R1–R33.

St Lawrence, K.S., Lee, T.Y., 1998. An adiabatic approximation to the tissue homogeneity model for water exchange in the brain: I. Theoretical derivation. *J Cereb Blood Flow Metab*. 18, 1365–77.

Stejskal, E.O., Tanner, J.E., 1965. Spin diffusion measurements: Spin echoes in the presence of time-dependent field gradient. *J Chem Phys*. 42, 288–92.

Stevenson, J.P., et al., 2003. Phase I trial of the antivascular agent combretastatin A4 phosphate on a 5-day schedule to patients with cancer: Magnetic resonance imaging evidence for altered tumor blood flow. *J Clin Oncol*. 21, 4428–38.

Sugahara, T., et al., 1999. Usefulness of diffusion-weighted MRI with echo-planar technique in the evaluation of cellularity in gliomas. *J Magn Reson Imaging*. 9, 53–60.

Sun, Y., et al., 2004. Perfusion MRI of U87 brain tumors in a mouse model. *Magn Reson Med*. 51, 893–9.

Szabo, B.K., Aspelin, P., Wiberg, M.K., 2004. Neural network approach to the segmentation and classification of dynamic magnetic resonance images of the breast: Comparison with empiric and quantitative kinetic parameters. *Acad Radiol*. 11, 1344–54.

Takahara, T., et al., 2004. Diffusion weighted whole body imaging with background body signal suppression (DWIBS): Technical improvement using free breathing, STIR and high resolution 3D display. *Radiat Med*. 22, 275–82.

Tamada, T., et al., 2011. Locally recurrent prostate cancer after high-dose-rate brachytherapy: The value of diffusion-weighted imaging, dynamic contrast-enhanced MRI, and T2-weighted imaging in localizing tumors. *Am J Roentgenol*. 197, 408–14.

Tang, P.A., et al., 2007. Surrogate end points for median overall survival in metastatic colorectal cancer: Literature-based analysis from 39 randomized controlled trials of first-line chemotherapy. *J Clin Oncol*. 25, 4562–8.

Taouli, B., et al., 2003. Evaluation of liver diffusion isotropy and characterization of focal hepatic lesions with two single-shot echo-planar MR imaging sequences: Prospective study in 66 patients. *Radiology*. 226, 71–8.

Taxt, T., et al., 1992. Multispectral analysis of uterine corpus tumors in magnetic resonance imaging. *Magn Reson Med*. 23, 55–76.

Theilmann, R.J., et al., 2004. Changes in water mobility measured by diffusion MRI predict response of metastatic breast cancer to chemotherapy. *Neoplasia*. 6, 831–7.

Therasse, P., et al., 2000. New guidelines to evaluate the response to treatment in solid tumors. European Organization for Research and Treatment of Cancer, National Cancer Institute of the United States, National Cancer Institute of Canada. *J Natl Cancer Inst*. 92, 205–16.

Thoeny, H.C., et al., 2005. Diffusion-weighted MR imaging in monitoring the effect of a vascular targeting agent on rhabdomyosarcoma in rats. *Radiology*. 234, 756–64.

Thomas, A.L., et al., 2005. Phase I study of the safety, tolerability, pharmacokinetics, and pharmacodynamics of PTK787/ZK 222584 administered twice daily in patients with advanced cancer. *J Clin Oncol*. 23, 4162–71.

Tofts, P.S., 1997. Modeling tracer kinetics in dynamic Gd-DTPA MR imaging. *J Magn Reson Imaging*. 7, 91–101.

Tofts, P.S., Kermode, A.G., 1991. Measurement of the blood-brain barrier permeability and leakage space using dynamic MR imaging. 1. Fundamental concepts. *Magn Reson Med*. 17, 357–67.

Tofts, P.S., et al., 1999. Estimating kinetic parameters from dynamic contrast-enhanced T(1)-weighted MRI of a diffusable tracer: Standardized quantities and symbols. *J Magn Reson Imaging*. 10, 223–32.

Torrey, H.C., 1956. Bloch equations with diffusion terms. *Phys Rev*. 104, 563–5.

Tou, T., Gonzalez, R.C., 1974. *Pattern Recognition Principles*. Addison-Wesley, Reading.

Tourdias, T., et al., 2008. Pulsed arterial spin labeling applications in brain tumors: Practical review. *J Neuroradiol*. 35, 79–89.

Tozer, G.M., Griffiths, J.R., 1992. The contribution made by cell death and oxygenation to 31P MRS observations of tumour energy metabolism. *NMR Biomed*. 5, 279–89.

Turnbull, L.W., 2009. Dynamic contrast-enhanced MRI in the diagnosis and management of breast cancer. *NMR Biomed*. 22, 28–39.

Ungersma, S.E., et al., 2010. Vessel imaging with viable tumor analysis for quantification of tumor angiogenesis. *Magn Reson Med*. 63, 1637–47.

Uto, T., et al., 2009. Higher sensitivity and specificity for diffusion-weighted imaging of malignant lung lesions without apparent diffusion coefficient quantification. *Radiology*. 252, 247–54.

Vaidyanathan, M., et al., 1997. Monitoring brain tumor response to therapy using MRI segmentation. *Magn Reson Imaging*. 15, 323–34.

Vandecaveye, V., et al., 2009. Diffusion-weighted MRI provides additional value to conventional dynamic contrast-enhanced MRI for detection of hepatocellular carcinoma. *Eur Radiol*. 19, 2456–66.

Vannier, M.W., et al., 1985. Multispectral analysis of magnetic resonance images. *Radiology*. 154, 221–4.

Vannier, M.W., et al., 1987. Multispectral magnetic resonance image analysis. *Crit Rev Biomed Eng*. 15, 117–44.

Vinitski, S., et al., 1998. In vivo validation of tissue segmentation based on a 3D feature map using both a hamster brain tumor model and stereotactically guided biopsy of brain tumors in man. *J Magn Reson Imaging*. 8, 814–9.

Wang, H., et al., 2010. Comparison of diffusion-weighted with T2-weighted imaging for detection of small hepatocellular carcinoma in cirrhosis: Preliminary quantitative study at 3-T. *Acad Radiol*. 17, 239–43.

Warmuth, C., Gunther, M., Zimmer, C., 2003. Quantification of blood flow in brain tumors: Comparison of arterial spin labeling and dynamic susceptibility-weighted contrast-enhanced MR imaging. *Radiology*. 228, 523–32.

Watanabe, M., Tanaka, R., Takeda, N., 1992. Magnetic resonance imaging and histopathology of cerebral gliomas. *Neuroradiology*. 34, 463–9.

Weber, M.A., et al., 2004. Assessment of irradiated brain metastases by means of arterial spin-labeling and dynamic susceptibility-weighted contrast-enhanced perfusion MRI: Initial results. *Invest Radiol*. 39, 277–87.

Wedam, S.B., et al., 2006. Antiangiogenic and antitumor effects of bevacizumab in patients with inflammatory and locally advanced breast cancer. *J Clin Oncol*. 24, 769–77.

Weinmann, H.J., Laniado, M., Mutzel, W., 1984a. Pharmacokinetics of GdDTPA/dimeglumine after intravenous injection into healthy volunteers. *Physiol Chem Phys Med NMR*. 16, 167–72.

Weinmann, H.J., et al., 1984b. Characteristics of gadolinium-DTPA complex: A potential NMR contrast agent. *Am J Roentgenol.* 142, 619–24.

Weisman, I.D., et al., 1972. Recognition of cancer in vivo by nuclear magnetic resonance. *Science.* 178, 1288–90.

Willett, C.G., et al., 2009. Efficacy, safety, and biomarkers of neoadjuvant bevacizumab, radiation therapy, and fluorouracil in rectal cancer: A multidisciplinary phase II study. *J Clin Oncol.* 27, 3020–6.

Winter, P.M., Poptani, H., Bansal, N., 2001. Effects of chemotherapy by 1,3-bis(2-chloroethyl)-1-nitrosourea on single-quantum- and triple-quantum-filtered 23Na and 31P nuclear magnetic resonance of the subcutaneously implanted 9L glioma. *Cancer Res.* 61, 2002–7.

Witjes, H., et al., 2003. Multispectral magnetic resonance image analysis using principal component and linear discriminant analysis. *J Magn Reson Imaging.* 17, 261–9.

Woessner, D.E., 1963. NMR spin-echo self-diffusion measurements on fluids undergoing restricted diffusion. *J. Phys. Chem.* 67, 1365–1367.

Wolf, R.L., et al., 2005. Grading of CNS neoplasms using continuous arterial spin labeled perfusion MR imaging at 3 Tesla. *J Magn Reson Imaging.* 22, 475–82.

Wong, E.C., 2005. Quantifying CBF with pulsed ASL: Technical and pulse sequence factors. *J Magn Reson Imaging.* 22, 727–31.

Wong, E.C., et al., 2001. T(1) and T(2) selective method for improved SNR in CSF-attenuated imaging: T(2)-FLAIR. *Magn Reson Med.* 45, 529–32.

Woodhams, R., et al., 2005. ADC mapping of benign and malignant breast tumors. *Magn Reson Med Sci.* 4, 35–42.

World Health Organization, 1979. *WHO Handbook for Reporting Results of Cancer Treatment.* World Health Organization, Geneva.

Yamada, I., et al., 1999. Diffusion coefficients in abdominal organs and hepatic lesions: Evaluation with intravoxel incoherent motion echo-planar MR imaging. *Radiology.* 210, 617–23.

Yoshimitsu, K., et al., 2008. Usefulness of apparent diffusion coefficient map in diagnosing prostate carcinoma: Correlation with stepwise histopathology. *J Magn Reson Imaging.* 27, 132–9.

Yuan, Z., et al., 2009. Role of magnetic resonance diffusion-weighted imaging in evaluating response after chemoembolization of hepatocellular carcinoma. *Eur J Radiol.* 75, e9–e14.

Zhang, W., et al., 2009. Acute effects of bevacizumab on glioblastoma vascularity assessed with DCE-MRI and relation to patient survival. *Proc Int Soc Magn Reson Med.* 17, 282.

Zonari, P., Baraldi, P., Crisi, G., 2007. Multimodal MRI in the characterization of glial neoplasms: The combined role of single-voxel MR spectroscopy, diffusion imaging and echo-planar perfusion imaging. *Neuroradiology.* 49, 795–803.

10

Quantitative Measurements of Tracer Transport Parameters Using Dynamic Contrast-Enhanced MRI as Vascular Perfusion and Permeability Indices in Cancer Imaging

Min-Ying Su

University of California, Irvine

Jeon-Hor Chen

University of California, Irvine
China Medical University Hospital
China Medical University

CONTENTS

10.1 Introduction

Dynamic contrast-enhanced MRI (DCE-MRI) is a rapidly evolving imaging technique for measuring vascular volume and permeability by using injection of exogenous gadolinium (Gd)-based contrast agents. The main clinical application is for diagnosis and therapy response monitoring of tumors based on the enhancements as well as the transport properties of contrast agents. It is the current standard for breast MRI many research studies have also been conducted to evaluate its applications in different organs of the body. The development of Gd-based contrast agents in the early 1980s opened a new era for imaging of tumors and vascular systems. Tumors require a higher supply of nutrients to support the rapid growth; therefore, it is necessary to induce the formation of new blood vessels, termed "angiogenesis." One of the earliest applications of Gd-based contrast agents is for the enhancement of viable tumors on postcontrast T_1-weighted images. The typical scan protocol includes pre- and postcontrast images, and the enhanced tumors can be detected by visual inspection of postcontrast images referencing to precontrast images. If the contrast injection is done while keeping the patient still inside the scanner, the subtraction images can be generated to reveal the contrast-enhanced lesions more clearly. However, while this technique was proven very helpful for imaging of lesions in the brain and musculoskeletal system, the application in the breast encountered a great difficulty. The tumors (malignant and benign), as well as normal breast tissues, all showed contrast enhancements, and it was difficult to differentiate them based on one set of postcontrast image acquired several minutes after the injection (Figure 10.1). Further research has found that if a series of postcontrast images were acquired after injection the enhancement time course (enhancement kinetics) could be measured. This provided additional information to aid in distinguishing malignant lesions from benign lesions and normal breast tissues (El Khouli et al., 2009; Kuhl et al., 1999). This technique was termed "dynamic contrast-enhanced MRI" or "DCE-MRI."

This chapter is focused on the quantitative analysis of DCE-MRI and its clinical applications. In Section 10.2, the measurement of enhancement kinetics by DCE-MRI and the qualitative and quantitative analysis to obtain parameters, such as the exchange rate constants K^{trans} and k_{ep}, are described. The debate about whether it is necessary to measure the arterial input function (AIF) from each individual patient will be discussed. Then, in Section 10.3, the quantitative DCE-MRI analysis tools offered in several widely used commercial computer-aided analysis platforms are described. Since DCE-MRI is the current standard for breast MRI it is used as an example to show how the quantitative analysis can be used for diagnosis (Section 10.4) and therapy response monitoring (Section 10.5). In general, for diagnosis of breast cancer, the hotspot approach to characterize tissues with the most aggressive pathology should be taken (Liney et al., 1999); but for therapy response monitoring, the whole tumor should be

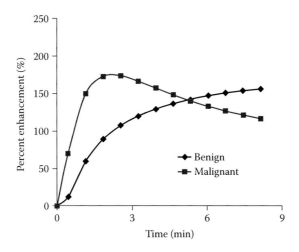

FIGURE 10.1

(Color version available from crcpress.com; see Preface.) The example of DCE enhancement kinetics measured from malignant and benign breast tumors. The malignant tumor shows a rapid wash-in, reaches to the maximum quickly, and then starts to show wash-out. The benign tumor shows a persistent enhancing pattern. If only a postcontrast imaging set is acquired at 5 min after injection, these two tumors may show exactly the same level of enhancements and cannot be differentiated. However, when the DCE kinetic is measured using multiple postcontrast frames, these two curves can be easily differentiated based on the pattern.

analyzed (Martincich et al., 2004). Lastly, in Section 10.6, the application of quantitative DCE-MRI in other organs of the body is described.

10.2 Measurement and Analysis of Dynamic Contrast-Enhanced Kinetics

10.2.1 Measurement of Enhancement Kinetics

The T_1-weighted three-dimensional gradient echo sequence is the most commonly used sequence for acquiring DCE-MRI. It generates high-quality images, has a good temporal resolution, and can cover a large imaging slab with good in-plane spatial resolution and slice thickness without gaps. All major scanner manufactures offer DCE-MRI sequences that can be easily prescribed with desired spatial and temporal resolution and the total number of pre- and postcontrast imaging frames. However, there is no standard about how to perform the contrast injection. There are some variations from site to site, and it is necessary for every study to describe the injection protocol.

The DCE kinetics can be analyzed using a straightforward approach based on the increased signal intensity or using a more sophisticated approach to convert the measured signal enhancement to the concentration of the contrast agents. If the purpose is to obtain physiological parameters based on transport kinetics using MR contrast agents as the tracer, the concentration of the contrast agents needs to be measured to allow for pharmacokinetic modeling analysis. All Gd-based contrast agents used in routine clinical practice are extracellular agents, and they can be distributed in the vascular and interstitial spaces. Under the assumption of the fast exchange regime, the concentration is proportional to the increased T_1 relaxivity ($R_1 = 1/T_1$); thus, the T_1 relaxation time before injection (T_{10}) and at postinjection time points has to be measured. The gradient echo sequences using different flip angles are commonly used. The T_{10} can be estimated using three flip angles (5°, 10°, and 15°). Given the needed temporal resolution during the DCE acquisition, most studies only use one flip angle after injection (e.g., 15°), and the $T_1(t)$ is estimated by referencing to the proton density-weighted images acquired before injection using 5° flip angle. The next step is to convert the increased R_1 to the concentration of the contrast agents based on the linear relationship in Equation 10.1:

$$\frac{1}{T_1(t)} = \frac{1}{T_1(0)} + \alpha[\text{Gd}].\tag{10.1}$$

The proportional constant α is the molar relaxivity, which is different in different tissues and cannot be accurately measured. The common approach is to use the constant measured in water or saline at 1.5 T and 3.0 T as a reasonable estimate. After the [Gd] concentration time course is obtained, it can be analyzed using the pharmacokinetic model to obtain the parameters that are associated with vascular properties (vascular perfusion, volume, and permeability). The low-molecular-weight Gd-based contrast agents cannot yield precise measurements that are respectively associated with perfusion (or volume) and permeability; the combined effects are rather seen.

10.2.2 Qualitative and Semiquantitative Analysis of Enhancement Kinetics

The signal enhancement is calculated by subtracting the precontrast signal intensity $S(0)$ from the postcontrast signal intensity $S(t)$, and then the enhancement can be normalized to the precontrast signal intensity $S(0)$ to calculate the percent enhancement, as expressed in Equation 10.2:

$$\text{Percent enhancement}(\%) = \frac{\left[S(t) - S(0)\right]}{S(0)} \times 100\%.\tag{10.2}$$

Normalization to $S(0)$ is necessary to handle the problem of varying coil sensitivity, allowing for comparison of tissue enhancements across the entire imaging field of view. This approach is easy and does not require multiple calibration scans and thus is commonly used in clinical examinations. The enhancement kinetics can be evaluated using three distinct features: the wash-in phase, the maximum enhancement, and the wash-out phase. Several heuristic parameters can be analyzed from the curve, such as the wash-in slope (maximum slope or the slope within a time period), the percent maximum enhancement, the time to reach the maximum enhancement, and the wash-out slope (within a time period). Since these parameters may be affected by the noise level at different data points, a more robust and commonly used parameter is the initial area under the curve (IAUC), which integrates the area under the kinetic curve, usually during the early time period such as the first 90 s. This parameter IAUC reflects how fast and how much the contrast material is delivered into the lesion.

10.2.3 Quantitative Pharmacokinetic Analysis Based on Modeling

A more sophisticated analysis method is to perform pharmacokinetic analysis based on two compartmental models, as shown in Figure 10.2. The most widely used method is the unified Tofts model (Tofts, 1997; Tofts and Kermode, 1991). The two compartments are the vascular space and the interstitial space, with the transfer constant K^{trans} to leak from the vascular to the interstitial space and the rate constant k_{ep} from the interstitial space back to the vascular space. Another parameter considered in the model is the distribution volume in the extravascular–extracellular space v_e (within the interstitial space). The change of concentration in the interstitial space (C_e) is expressed in Equation 10.3:

$$\frac{dC_e}{dt} = K^{trans}[C_b] - k_{ep}[C_e]. \tag{10.3}$$

The total concentration in the tissue can be written as the contribution from both vascular and interstitial compartments as $C_t = v_b \times C_b + v_e \times C_e$.

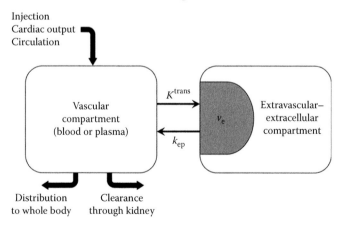

FIGURE 10.2
The two-compartmental model. The blood (or plasma) kinetics is dependent on the injection of contrast agents, the cardiac output, and the circulation for the contrast agent to arrive at the feeding artery. The blood kinetics is usually modeled by a bi-exponential decay function. The exchange between the vascular and the extravascular–extracellular compartment is characterized by the in-flux transfer constant (K^{trans}) and the out-flux rate constant (k_{ep}). The distribution volume in the extravascular–extracellular compartment is noted as v_e.

Many parameters can be included in the fitting model, but the problem of over-fitting needs to be considered. Although the vascular compartment is present, whether the fitted parameter, v_b, is truly reflecting the vascular space is of great concern. Since all clinical Gd-based contrast agents are small agents that can quickly diffuse from the vascular space to the interstitial space, the fitted parameter v_b is very likely to contain the early leakage into the interstitial space, which is not an accurate measurement of the true vascular volume. Due to this concern, the most commonly used model (Tofts model) assumes a relatively small vascular space compared to the interstitial space and ignores the term v_b. The tissue concentration is expressed as $C_t = v_e \times C_e$, where C_e is dependent on the blood concentration C_b. However, when the analysis is performed for highly vascularized lesion, the term v_b may not be ignored.

The blood kinetics C_b is required for the pharmacokinetic fitting. Usually, it is modeled as a bi-exponential decay function as expressed in Equation 10.4:

$$C_b(t) = D \left[a1 \times \exp(-m1 \times t) + a2 \times \exp(-m2 \times t) \right], \tag{10.4}$$

where D is the injection dose of the contrast agent in [mmol/L] (the standard dose used for clinical imaging is 0.1 mmol/L), $a1$ is the fraction of the fast decay component with the decay rate $m1$, and $a2$ is the fraction of the slow decay component with the decay rate $m2$. The fast decay component is related to the quick distribution of the injected contrast agents to the whole body, and the slow delay component is related to the back diffusion of contrast agents from the whole body to the blood to be cleared from the kidney. Since the contrast agents can only be mixed in the plasma, not into the blood cells, the more precise model uses the plasma concentration C_p, which is related to C_b with the correction term $(1 - \text{Hct})$, where Hct is the hematocrit, as shown in Equation 10.5:

$$[C_b] = [C_p](1 - \text{Hct}). \tag{10.5}$$

When the absolute concentration of the contrast agents (such as mmol/L) in both the tissue and the blood is measured and used in the fitting, the unit for K^{trans} and k_{ep} is 1/time [commonly used as (1/min)]. If diffusion is the only process involved for the transport of contrast agents between vascular and interstitial compartments, the exchange rate between the blood compartment and the extravascular–extracellular distribution space v_e should be equal, that is, $k_{ep} = K^{trans}/v_e$ ($0 < v_e < 1$). If the absolute concentration is not obtained for either the tissue or the blood, the fitted parameters K^{trans} will carry an arbitrary unit (depending on which parameter is used in fitting, such as the percent enhancement), but the unit for k_{ep} is always (1/min). When the fitted pharmacokinetic parameters are analyzed for therapy monitoring studies, a percentage change in the follow-up study compared to the baseline study is usually calculated. As long as the acquisition and the analysis are performed consistently, it is not necessary to convert the signal enhancement and the blood kinetics to the [Gd] concentration.

10.2.4 Different Transport Regimes Limited by Perfusion and Permeability

Depending on how fast the contrast agents can leak out from the vascular space into the interstitial space, and how fast the contrast agents in the vascular space can be replenished, the transport kinetics will be governed by different parameters. They can be classified into three regimes. The first regime is "perfusion-dependent." This happens when the permeability is very high and the contrast agents can leak out very quickly into the interstitial space; therefore, the delivery of contrast agents into the tissue is dependent on how fast the blood flow can bring in more contrast agents to replenish the depleted contrast agents in the vascular space. The transport equation is described by Equation 10.6, where F is the blood perfusion [the volume of blood per gram of tissue per minute (cc/g/min)] and ρ is the tissue density (g/cc):

$$\frac{dC_e}{dt} = F\rho(1 - Hct)\left[C_p - \frac{C_e}{v_e}\right]. \tag{10.6}$$

The opposite regimen is "permeability-dependent." This happens when the blood perfusion is very high and can quickly replenish the contrast agents to maintain a high blood concentration; therefore, the delivery of contrast agents into the tissue is dependent on the vascular permeability (P) and the exchange surface area (S) on the vessel wall for the contrast agents to leak into the interstitial space, as shown in Equation 10.7:

$$\frac{dC_e}{dt} = PS\rho\left[C_p - \frac{C_e}{v_e}\right]. \tag{10.7}$$

The transport between these two extreme regimes is referred as "extraction-limited." As there is no strong dominating factor in either perfusion or permeability, the amount of contrast agents that can be delivered into the tissue is described by the extraction parameter E, as shown in Equation 10.8. E is dependent on the vascular permeability and the exchange surface area, and together with the perfusion parameter F, all three parameters will affect the transport kinetics:

$$\frac{dC_e}{dt} = EF\rho(1 - Hct)\left[C_p - \frac{C_e}{v_e}\right]. \tag{10.8}$$

Depending on the expected regimes, the appropriate transport equation can be chosen to obtain the governing parameters by pharmacokinetic model fitting.

10.2.5 The Individually Measured versus the General Population AIF

Since the blood concentration is required for fitting, one area of debate is whether the AIF should be measured from each individual patient. The advantage is to provide the most accurate blood kinetics by considering the different hemodynamics of individual patients. But the problem is that it is difficult to measure the AIF accurately. If a wrong measurement is used, it may lead to large errors in the fitted parameters. The preferred artery for measuring the AIF is the one directly feeding the lesion. Consequently, a high spatial resolution is needed to delineate the vessel and to avoid the partial volume effect (the voxel has to be completely contained within the vessel); a very high temporal resolution is also needed to catch the maximum enhancement. Further, the image quality has to be good to obtain smooth enhancement kinetics without too much noise or fluctuation. The motion artifact, including respiratory motion and pulsation of vessels, may lead to degradation of the vessel image. All these problems lead to difficulty in measuring AIF accurately from the small feeding artery. The alternative approach is to measure the AIF from the large artery such as the aorta; but the high velocity of blood flow needs to be handled properly. The AIF measured from the aorta may not truly reflect the AIF of the artery directly feeding the lesion.

Due to these difficulties, the blood kinetics measured from the healthy general population provides a reasonable and acceptable approach. As long as the patient does not have cardiovascular problems or kidney diseases, the general population blood kinetics can be used in the pharmacokinetic model fitting to obtain K^{trans} and k_{ep}. Since the results are obtained using the unified model, the fitted parameters may be compared between different studies. However, while this is acceptable for studies dealing with lesion diagnosis or characterization, it may not be appropriate for therapy monitoring studies, particularly for those designed to measure the vascular changes in clinical studies of antiangiogenic or antivascular therapeutic agents. It is recommended that for such studies the individual AIF be measured and used as the reference (Leach et al., 2005). Given the requirement of both high spatial and temporal resolution, the reliability of the measured AIF needs to be investigated when designing the imaging protocol (Ashton et al., 2008), so at least the possible errors introduced by the variation in the measured AIF can be evaluated in subsequent data analysis and interpretation (Cheng, 2008).

10.2.6 Image Registration between Pre- and Postcontrast DCE Frames

One major problem leading to poor quality of the DCE kinetics is the motion during the DCE acquisition, including both patient movement and physiological motion such as breathing and heartbeat pulsation. A rigid coregistration process is usually applied to spatially align all DCE frames with respect to the selected reference frame. It should be noted that the coregistration is applied to the organ of interest, not the entire image. For example, in breast imaging, the coregistration should not be done to align the thoracic body region while compromising the coregistration of the breast region. Regional coregistration within a small field of view is the commonly used approach.

10.2.7 Lesion Region of Interest-Based Analysis and Pixel-by-Pixel Analysis

The choice of the region of interest (ROI)-based analysis or pixel-by-pixel analysis is usually dependent on the clinical application. In general, for diagnostic purpose, the hotspot ROI approach should be used, and for therapy monitoring study, the pixel-by-pixel analysis should be applied. One of the advantages of ROI-based analysis is that it is less susceptible to noise and signal fluctuation through averaging over many pixels, and the fitting to obtain K^{trans} and k_{ep} is unlikely to fail so that the obtained results can be directly used in the analysis. However, it cannot reveal the degree of heterogeneity within the tumor. The advantage of the pixel-by-pixel analysis is the rich data obtained from the entire lesion that allows for histogram analysis within the lesion. The major disadvantage is that the kinetics measured from some pixels may be very noisy, and the fitting quality needs to be carefully inspected. Usually, the pixels with unsatisfactory fitting quality need to be discarded in the analysis.

10.3 Computer-Aided Quantitative Analysis Systems

There are several commercially available computer-aided analysis software dedicated to the analysis of DCE-MRI, such as Merge CADstream, DynaCAD, and iCAD. All these have specially designed product for the breast, as well as for other organs. The features provided in these different systems are similar. The software detects the DCE sequence and imports all images for analysis. The typical display includes rendering maximum intensity projection (MIP) and the corresponding view from three planes (axial, sagittal, and coronal)—one acquired and two reformatted. The color coding to label the suspicious enhancements can be turned on based on the choice of the threshold enhancement to show color coding. Two postcontrast frames are selected for determining the wash-out pattern. Typically, the red color is used to label the voxel showing wash-out, indicating a high suspicion of malignancy. Within the selected tumor ROI, the percentage of voxels that show wash-out, plateau, and persistent enhancing pattern can be calculated and displayed. The program also provides the lesion segmentation tool based on connectivity of enhanced tissues.

The software for performing quantitative pharmacokinetic modeling based on DCE-MRI kinetics to extract the fitting parameters (K^{trans}, k_{ep}, v_p, and v_e) is also commercially available. One example is "Tissue 4D." The signal enhancement is converted to the concentration using an estimated proportional constant provided by the software. Three blood kinetic curves (fast, medium, and slow flow) are built-in and can be selected based on the organ of interest. The software also supports the measurement of individual AIF from the analyzed case and is used in the fitting. Two fitting models are provided: the Tofts model (ignoring the vascular space) and Tofts + v_p model (considering the vascular space). The obtained fitting parameters (K^{trans} and k_{ep} or in addition v_e and v_p) from the pixel-by-pixel analysis within the selected ROI can be displayed as overlaying color maps and histograms, and the data can be exported for further analysis. It is anticipated that the availability of such a standard analysis tool will facilitate comparison of quantitative DCE studies, which could not be done when using different in-house programs developed by different research groups.

10.4 DCE-MRI for Diagnosis of Breast Cancer

Tumors need angiogenesis to support the rapid tumor growth. In general, malignant lesions are more aggressive and require a higher angiogenic activity. For example, they have a higher expression of vascular endothelial growth factor (VEGF) that stimulates the formation of new vessels. These new vessels are immature and leakier (a wider endothelial junction) and allow the contrast agents to quickly leak from the vascular space into the interstitial space and back diffuse to the vascular space to be cleared (Daldrup-Link et al., 2004; Raatschen et al., 2008). DCE-MRI can be used to measure the transport kinetics of contrast agents in the tissue, allowing for analysis of transfer rates associated with vascular perfusion, volume, and permeability. The fibroglandular tissues in the breast have a relatively high vascular supply and interstitial space, which may allow leakage of contrast agents to show a strong background tissue enhancement that increases with time after the contrast injection (King et al., 2011). Therefore, the lesions can be better differentiated at early times before the background tissue shows strong enhancements (Kuhl, 2007). The earlier DCE-MRI studies were performed using a relatively high temporal resolution (Schorn et al., 1999), and different postcontrast frames acquired at different times can be used to find the optimal frame that shows the best signal contrast between lesions and normal tissues. Usually, the maximum enhancement is seen around 2 min after injection of contrast agents.

Many studies have investigated the diagnostic capability of quantitative DCE-MRI to differentiate between benign and malignant breast lesions (Bassett et al., 2008; Kuhl et al., 2005b, 2007; Saslow et al., 2007; Schelfout et al., 2004; Tillman et al., 2002; Zhang et al., 2002). When the contrast agents are injected, the higher vascular space and the higher vascular permeability allow more contrast agents to be quickly delivered into the interstitial space of the lesion, and the agents can quickly back diffuse to the vascular space to be cleared. As such, the enhancement kinetics shows a rapid wash-in, reaches to the maximum enhancement quickly, and then starts to show wash-out. In pharmacokinetic analysis, this kinetic pattern will yield a high K^{trans} and a high k_{ep}. In contrast, benign lesions may not have a large number of angiogenic vessels, but still have a high interstitial space to uptake contrast agents and show enhancement kinetics with persistent enhancing pattern (Agrawal et al., 2009). Such curve will yield a low K^{trans} and a low k_{ep}. Although these two patterns have been proven as reliable diagnostic features, many lesions show the enhancement kinetics in-between, reaching a plateau during the imaging period without showing a clear wash-out or a clear persistent enhancement pattern (Agrawal et al., 2009). Given the heterogeneous nature of breast lesions, the enhancement kinetics also varies with the tissue location or the placement of the ROI within the lesion (Liney et al., 1999; Turnbull, 2009). For diagnostic purpose, the rule of the most aggressive pathology is applied; therefore, the hotspot approach should be used (Liney et al., 1999). The ROI should be placed on the tissue regions with the strongest enhancement, and when the wash-out pattern is seen in some part of the lesion, this lesion is considered suspicious of malignancy. In contrast, the benign lesions are often more homogeneous, and if the enhancement kinetics measured from the entire lesion shows the persistent enhancing pattern, this lesion is more likely benign (Agrawal et al., 2009).

The research results obtained over a decade for diagnosis of breast cancer suggest that a higher spatial resolution that reveals the morphology of the lesion in a greater detail has a better diagnostic value (Kuhl et al., 2005a; Pinker et al., 2009; Tozaki and Fukuda, 2006). The general rule is that one set of high-quality postcontrast images should be acquired within 3 min after injection of contrast agents for lesion detection and morphological characterization, and the kinetics over a period of 5–10 min should be measured for the determination of the DCE pattern. Both lesion morphology and kinetic pattern need to be considered for giving a final diagnostic impression (Newell et al., 2010; Nie et al., 2008; Shimauchi et al., 2011). The breast lesions can be separated into mass- and non-mass-like enhancement (Newell et al., 2010; Tozaki and Fukuda, 2006). For mass lesions, the enhancement kinetics may provide very helpful diagnostic information (Figure 10.3); but for non-mass lesions, the pattern of DCE kinetics cannot be used to differentiate between malignant and benign lesions (Figure 10.4). The diagnosis will need to focus on analyzing the morphological distributions of the enhanced tissues (Newell et al., 2010). Developing intellectual Computer-Aided Diagnosis (CAD) systems that can consider all morphological and DCE kinetic information and make a diagnostic impression is an active research area (Newell et al., 2010; Nie et al., 2008; Shimauchi et al., 2011). Such a system will be very helpful to provide a second reader and will particularly help inexperienced radiologists in interpreting breast MRI.

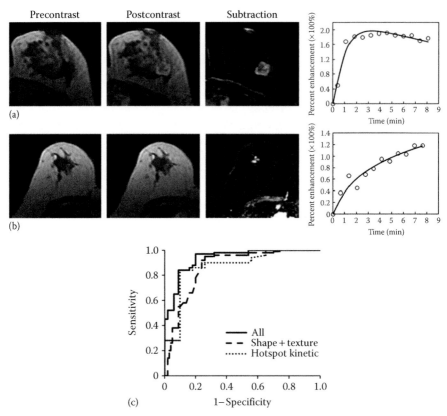

FIGURE 10.3

Two examples of mass-type malignant and benign breast tumors are shown. The images are acquired using a no-fat-saturation sequence: (a) a mass-type invasive ductal carcinoma showing the typical malignant wash-out DCE curve; (b) a mass-type benign lesion (mixed fibroadenoma and adenosis) showing the typical benign persistent DCE curve; and (c) the ROC curve based on the morphological parameters, the DCE kinetic parameters, and the combined morphological and DCE parameters. The hotspot DCE kinetic parameters can achieve the AUC of 0.88, and it can be added to the morphological parameter to improve the AUC from 0.87 to 0.93.

10.5 DCE-MRI for Neoadjuvant Chemotherapy Response Monitoring and Prediction

In addition to diagnosis, another major application of DCE-MRI is for monitoring the response of breast cancer undergoing neoadjuvant chemotherapy (NAC; preoperative chemotherapy) (Bahri et al., 2009; Chen et al., 2008, 2009, 2011; De Los Santos et al., 2011; Loo et al., 2011; Yu et al., 2007). NAC is conventionally used to downstage inoperable, locally advanced breast cancer. The advantages of NAC also include that (1) it may facilitate breast-conserving surgery; (2) when a patient achieves pathologic complete response

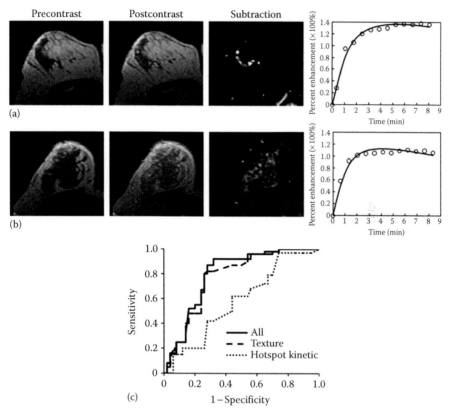

FIGURE 10.4
Two examples of non-mass-type malignant and benign breast tumors that present as regional enhancements without clear boundary. The images are acquired using a no-fat-saturation sequence: (a) a non-mass-type ductal carcinoma *in situ* (DCIS) showing the plateau DCE pattern; (b) a non-mass-type benign fibrocystic changes also showing the plateau DCE pattern; and (c) the ROC curve based on the morphological parameters, the DCE kinetic parameters, and the combined morphological and DCE parameters. The hotspot DCE kinetic parameters only achieve the AUC of 0.59, which is slightly better than random guess (AUC = 0.5), and when it is added to the morphological parameter, AUC does not increase much (from 0.76 to 0.78).

(pCR) or near pCR, the prognosis is favorable; and (3) the sensitivity of each individual patient's cancer to different drug regimens can be tested *in vivo*. Due to these advantages, patients with operable cancer may also choose to receive NAC, and it has become a very important treatment modality for patients diagnosed with stage 2 or higher breast cancer. Of all breast imaging modalities, MRI is considered the most accurate for assessing the treatment response (Balu-Maestro et al., 2002; Choyke et al., 2003; Esserman et al., 2001; Rieber et al., 2002). MRI can also help in the selection of patients suitable for receiving breast-conserving surgery after NAC (Chen et al., 2009; Straver et al., 2010). The size or the extent of disease lesions before starting therapy should be established as the baseline reference. For diagnosis of residual disease after therapy, the patterns of the DCE kinetics should no longer be used as diagnostic criteria. Rather, any enhanced tissues within the tumor bed are considered as the residual disease (Bahri et al., 2009; Chen et al., 2008, 2011).

Changes in lesion size on DCE-MRI are usually not detected until several weeks following NAC (Rieber et al., 2002). If early surrogate response indicators could be established to predict final treatment outcome, it would help achieve the goal of pCR. There are other attempts to investigate whether functional information provided by MRI may serve as earlier response indicators than the size change. They included pharmacokinetic parameters [area under the curve (AUC), K^{trans}, k_{ep}, etc.], the apparent diffusion coefficient (ADC) measured by diffusion-weighted imaging (DWI), and the total choline peak measured by proton MR spectroscopy.

It is well known that the cancer therapy also causes vascular damage and that the enhancement kinetic pattern will change from the wash-out pattern to a less aggressive pattern of plateau or persistent enhancement (Ah-See et al., 2008; Padhani et al., 2006; Yu et al., 2007). Using DCE pharmacokinetic

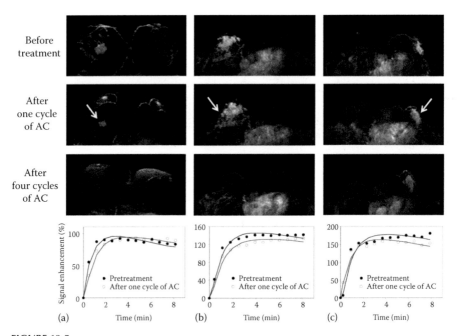

FIGURE 10.5
(Color version available from crcpress.com; see Preface.) The DCE kinetics measured before and after one cycle of chemotherapy from (a) a responder after one cycle of AC regimen, which shows a slower wash-in and a slower wash-out after chemotherapy; (b) a confirmed responder after four cycles, which has not yet shown a good response after one cycle, but the change of DCE kinetic is similar to the responder in (a) showing slower wash-in and slower wash-out, thus indicating that it may be a responder; and (c) a nonresponder that does not show any tumor shrinkage during the treatment, and the DCE shows a faster wash-in and a faster wash-out after one cycle of chemotherapy.

parameters to serve as an earlier response indicator has been extensively investigated, as illustrated in Figures 10.5 and 10.6, but the results were controversial (He et al., 2012; Loo et al., 2008; Manton et al., 2006; Padhani et al., 2006; Pickles et al., 2005; Yu et al., 2007). Some researchers found that the early change in pharmacokinetic parameters before size shrinkage could predict the final treatment outcome (He et al., 2012), but others found that pharmacokinetic parameters were not reliable that the size change was the best response indicator (He et al., 2012; Loo et al., 2008; Manton et al., 2006; Padhani et al., 2006; Pickles et al., 2005; Yu et al., 2007). Some studies have demonstrated that the changes in the exchange rates (K^{trans}, k_{ep}) or other heuristic parameters (such as percent maximum enhancement and AUC) showed significant differences between good responders and poor responders. However, in receiver operating characteristic (ROC) analysis, the AUC was not sufficiently high for them to serve as reliable response predictors. Furthermore, when compared to the early change in the tumor size, these kinetic parameters may not have a significantly higher predicting power.

The more attractive role of DCE-MRI is to evaluate the response of antiangiogenic or antivascular therapy (Bahri et al., 2009; Hahn et al., 2008; Mehta et al., 2011; Miller et al., 2005; Moreno-Aspitia et al., 2009; Padhani

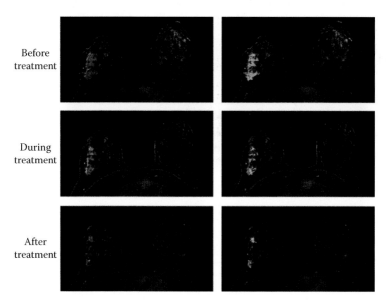

Before treatment

During treatment

After treatment

FIGURE 10.6
(See color insert.) An example of a non-mass-type breast cancer undergoing NAC treatment. The postcontrast fat sat images are shown. The tumor breaks up into scattered pieces. The pixel-by-pixel pharmacokinetic analysis is applied to obtain the parameters K^{trans} and k_{ep}. The right column shows color-coded K^{trans} values overlaid on the original images. It can be seen that although the tumor is responding well, some hot spots still exist in the tumor. The histogram of each parameter can be analyzed from the pixel-by-pixel fitting results and compared in these studies acquired before, during, and after chemotherapy.

and Leach, 2005; Wedam et al., 2006). For instance, the antiangiogenic agent, Bevacizumab (Avastin), is a monoclonal antibody that neutralizes the VEGF to inhibit angiogenesis, and it has been used for treating various types of cancers. DCE-MRI provides a means for assessing the treatment-induced vascular changes to investigate the early therapeutic response to this targeted drug. It may provide insightful information to evaluate the efficacy of drugs in early clinical trial phases and to guide the design for future studies. For this purpose, the changes in the entire tumor need to be evaluated, in contrast to the hotspot approach used for diagnosis. Since breast tumor is highly heterogeneous, there is a poor correlation between the parameters analyzed from the hot spot and the entire tumor ROI. The most useful analysis method is to perform pixel-by-pixel analysis of the enhancement kinetics from the entire tumor, and the obtained histograms can be compared between studies performed before and after therapy to evaluate the changes (Ah-See et al., 2008; Padhani et al., 2006). Since the parameters measured in different studies will be compared, other confounding factors that may affect the changes need to be considered. One of the most important factors is the difference in the AIF, as these therapies are very likely to affect the circulation of the patients. An international consensus panel for application of DCE-MRI in drug trials recommended that the AIF be measured on an individual basis and used as a reference to obtain quantitative parameters, such as K^{trans} and k_{ep}, or at least to provide a reference for normalization of lesion enhancements for evaluating changes (Leach et al., 2005). However, the subsequently published studies revealed a great difficulty in obtaining the AIF reliably and consistently (Ashton et al., 2008). If the measured AIF were problematic, using that as reference would lead to an even higher error in fitted parameters compared to using the AIF of the general population. Alternatively, if the variation of the measurements can be evaluated using a test–retest reproducibility study, any change that is greater than the measurement variation can be considered as true therapy-related changes (Ah-See et al., 2008).

10.6 Clinical Applications of DCE-MRI in Other Organs

DCE-MRI has been extensively applied in many other organs in addition to the breast. It is a suitable technique for studying solid tumors (or any mass lesions) in the brain, lung, liver, spinal cord, vertebrate, pancreas, kidney, colon and rectum, prostate, ovary, cervix, and musculoskeletal systems. The DCE kinetics from the lesion and the adjacent normal tissues can be measured and characterized following the methods described in Section 10.2.

In the brain, DCE-MRI may be used to distinguish between recurrent gliomas and radiation changes, which is critical for choosing the subsequent treatment strategies (Bisdas et al., 2011). K^{trans} and IAUC were found to be significantly higher in the recurrent glioma group than in the radiation necrosis group (Bisdas et al., 2011). In non-small-cell lung cancer patients receiving Sorafenib, a multikinase inhibitor targeting Raf and VEGF receptor (VEGFR), k_{ep} acquired from DCE-MRI was significantly associated with improved overall survival (OS) and progression-free survival (PFS) (Kelly et al., 2011).

In the liver, DCE-MRI has been applied for detection and characterization of liver lesions in patients with cirrhosis (Kim et al., 2004). There are dual vascular supplies in the liver, including the hepatic artery and the portal vein; therefore, two separate input functions have to be incorporated into the pharmacokinetic model (Koh et al., 2008). The capillary in the liver is the discontinuous type, which allows a lot of contrast agents to leak into the normal liver tissue very quickly. When using the low-molecular-weight Gd-contrast agents, the tumor is often detected based on the negative contrast with lower enhancements compared to the surrounding normal tissues. The enhancement at arterial and venous phases may further help characterize different lesions for diagnosis, and it is usually required to have a high temporal resolution compared to the application of DCE in other organs. If multiple sets of images can be acquired very quickly, the arterial phase image can be easily captured in one of the acquired sets without the need to time the acquisition based on contrast arrival time. The sensitivity of DCE-MRI at 1.5 T was less than 80% (Krinsky et al., 2001; Rode et al., 2001). At 3 T, by using a double-contrast-enhanced MRI with the liver-specific superparamagnetic iron oxide (SPIO) and the Gd-based contrast material, the detectability of small hepatocellular carcinoma (HCC, <1 cm) could be improved (Yoo et al., 2008).

DCE-MRI can be used for characterizing primary renal cell carcinoma (RCC) and monitoring antiangiogenic treatment (Notohamiprodjo et al., 2010). Papillary RCC could be differentiated from clear-cell variants by a distinct perfusion pattern (Notohamiprodjo et al., 2010). In RCC treated with anti-angiogenic therapy, the perfusion parameters were significantly decreased (Notohamiprodjo et al., 2010). Similar to the strong enhancements seen in normal liver tissue, the tissue parenchyma in the kidney also shows early and strong enhancements; therefore, another very important application of DCE-MRI is to provide multiple sets of postcontrast images for evaluation.

Currently, DCE-MRI is still not an established tool for the evaluation of human colorectal cancer (Ahn et al., 2011). A study was performed to correlate the pharmacokinetic parameters measured by DCE-MRI with angiogenesis in rectal carcinoma, and the results showed that DCE-MRI can predict angiogenesis measured by microvascular density (MVD) and VEGF expression and also help discriminate malignant from normal tissue (Zhang et al., 2008).

DCE-MRI of the prostate can be applied for detection, localization and staging of prostate cancer, detection of recurrence, assessment of biological aggressiveness, treatment monitoring, and prediction of prognosis (Engelbrecht et al., 2003; Franiel et al., 2011). The malignant cancer in the peripheral zone

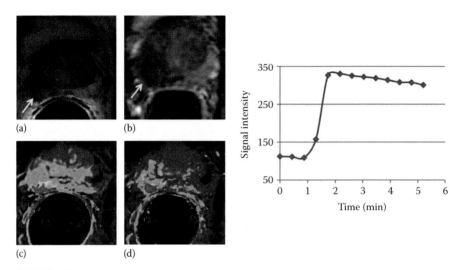

FIGURE 10.7
(See color insert.) An example of prostate cancer. A lesion is visible on T_2-weighted image (a) showing a hypointense area in the periphery zone compared to the contralateral side. It shows decreased ADC (b) due to increased cell density. The color-coded enhancement maps clearly show that the tumor has a higher IAUC (c) and also a higher wash-in slope (d) compared to the contralateral normal side. The DCE kinetics measured from the tumor shows the wash-out pattern.

shows earlier and stronger enhancement than the surrounding normal prostate tissue (Engelbrecht et al., 2003) as illustrated in Figure 10.7. For the detection of recurrence after prostatectomy, the addition of DCE-MRI led to an increase in sensitivity from 61% to 84% compared with using T_2-weighted imaging alone (Cirillo et al., 2009).

For the female pelvic diseases, adding DCE-MRI can improve the accuracy of anatomic MRI (mainly using T_2-weighted images) in staging of endometrial cancer (Sala et al., 2009). The depth of myometrial invasion was correctly diagnosed in 78% of cases on T_2-weighted images alone, which could be increased to 92% with the addition of DCE-MRI (Sala et al., 2009). DCE-MRI can be used for noninvasive monitoring of response to radiation therapy (RT) in cervical cancer (Mayr et al., 2010). Longitudinal monitoring has shown that tumor perfusion changes during RT and is associated with treatment outcome. Since the efficacy of RT is highly dependent on the oxygenation status of the tumor, a low perfusion in pre-RT, early RT, and mid-RT indicates a high risk of treatment failure, whereas in patients with initially high perfusion, a favorable outcome can be achieved (Mayr et al., 2010). DCE-MRI also helps differentiate tumor recurrence from radiation fibrosis in patients with cervical cancer (Sala et al., 2010). Quantitative DCE-MRI can be used as independent predictors of malignancy in a complex (solid, solid/cystic) ovarian mass (Dilks et al., 2010). There was a significant difference in the maximum enhancement, the percent maximum enhancement, and the wash-in rate between malignant and benign tumors (Dilks et al., 2010). DCE-MRI also improves characterization of cystic adnexal lesions and detection of small peritoneal implants in patients with ovarian cancer (Sala et al., 2010).

DCE-MRI has also been applied to evaluate tumors in the spinal cord. Myeloma, metastatic cancer, and lymphoma are commonly seen malignant cancers in the spine (examples shown in Figure 10.8). Their morphological appearance can be very similar on MRI, which makes it difficult to differentiate. The treatment and prognosis are different for these three types of tumor. A correct diagnosis based on imaging would help guide the biopsy and the subsequent treatment planning and prognosis. DCE-MRI has been applied to characterize the normal bone marrow and the tumor infiltration (Montazel et al., 2003; Rahmouni et al., 2003; Zha et al., 2010). Other areas include evaluating the treatment response (Lin et al., 2010) and predicting prognosis (Hillengass et al., 2007).

In addition to tumor imaging, DCE-MRI has also been proposed as a non-invasive tool to study the extent of plaque neovascularization. The parameters derived from DCE-MRI have been shown to correlate with plaque neovascularization in both patients and animal models (Kerwin et al., 2003,

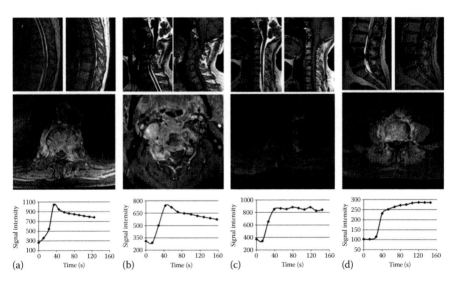

FIGURE 10.8
Four case examples of spinal cord tumors. The T_1-weighted, T_2-weighted, and contrast-enhanced images are shown. The diagnosis cannot be made based on their morphological appearance. (a) A myeloma: time to peak = 27 s, maximum SE = 300%, the steepest wash-in slope between two adjacent time points = 188%, the wash-out slope from maximum to the last time point = 62%, k_{ep} = 2.18 (1/min). (b) A metastatic thyroid cancer: time to peak = 30 s, maximum SE = 139%, steepest slope = 77%, wash-out slope = 54%, k_{ep} = 1.93 (1/min). (c) A metastatic breast cancer: time to peak = 28 s, maximum SE = 128%, steepest slope = 84%, wash-out slope = 6%, k_{ep} = 0.93 (1/min). (d) A lymphoma: time to peak = 30 s, maximum SE = 144%, steepest slope = 113%, wash-out slope = not available, k_{ep} = 0.72 (1/min).

2008). The presence of inflammatory infiltration and plaque neovascularization are both histological hallmarks of atherosclerotic plaque inflammation, which is considered a good marker of high-risk/vulnerable plaques (Calcagno et al., 2010).

References

Agrawal, G., Su, M.Y., Nalcioglu, O., Feig, S.A., Chen, J.H., 2009. Significance of breast lesion descriptors in the ACR BI-RADS MRI lexicon. *Cancer*. 115, 1363–1380.

Ahn, S.J., An, C.S., Koom, W.S., Song, H.T., Suh, J.S., 2011. Correlations of 3T DCE-MRI quantitative parameters with microvessel density in a human-colorectal-cancer xenograft mouse model. *Korean Journal of Radiology: Official Journal of the Korean Radiological Society*. 12, 722–730.

Ah-See, M.L.W., Makris, A., Taylor, N.J., Harrison, M., Richman, P.I., Burcombe, R.J., Stirling, J.J., et al., 2008. Early changes in functional dynamic magnetic resonance imaging predict for pathologic response to neoadjuvant chemotherapy in primary breast cancer. *Clinical Cancer Research*. 14, 6580–6589.

Ashton, E., Raunig, D., Ng, C., Kelcz, F., McShane, T., Evelhoch, J., 2008. Scan-rescan variability in perfusion assessment of tumors in MRI using both model and data-derived arterial input functions. *Journal of Magnetic Resonance Imaging*. 28, 791–796.

Bahri, S., Chen, J.H., Mehta, R.S., Carpenter, P.M., Nie, K., Kwon, S.Y., Yu, H.J., Nalcioglu, O., Su, M.Y., 2009. Residual breast cancer diagnosed by MRI in patients receiving neoadjuvant chemotherapy with and without bevacizumab. *Annals of Surgical Oncology*. 16, 1619–1628.

Balu-Maestro, C., Chapellier, C., Bleuse, A., Chanalet, I., Chauvel, C., Largillier, R., 2002. Imaging in evaluation of response to neoadjuvant breast cancer treatment benefits of MRI. *Breast Cancer Research and Treatment*. 72, 145–152.

Bassett, L.W., Dhaliwal, S.G., Eradat, J., Khan, O., Farria, D.F., Brenner, R.J., Sayre, J.W., 2008. National trends and practices in breast MRI. *American Journal of Roentgenology*. 191, 332–339.

Bisdas, S., Naegele, T., Ritz, R., Dimostheni, A., Pfannenberg, C., Reimold, M., Koh, T.S., Ernemann, U., 2011. Distinguishing recurrent high-grade gliomas from radiation injury: A pilot study using dynamic contrast-enhanced MR imaging. *Academic Radiology*. 18, 575–583.

Calcagno, C., Mani, V., Ramachandran, S., Fayad, Z.A., 2010. Dynamic contrast enhanced (DCE) magnetic resonance imaging (MRI) of atherosclerotic plaque angiogenesis. *Angiogenesis*. 13, 87–99.

Chen, J.H., Bahri, S., Mehta, R.S., Kuzucan, A., Yu, H.J., Carpenter, P.M., Feig, S.A., et al., 2011. Breast cancer: Evaluation of response to neoadjuvant chemotherapy with 3.0-T MR imaging. *Radiology*. 261, 735–743.

Chen, J.H., Feig, B., Agrawal, G., Yu, H., Carpenter, P.M., Mehta, R.S., Nalcioglu, O., Su, M.Y., 2008. MRI evaluation of pathological complete response and residual tumors in breast cancer after neoadjuvant chemotherapy. *Cancer*. 112, 17–26.

Chen, J.H., Feig, B.A., Hsiang, D.J.B., Butler, J.A., Mehta, R.S., Bahri, S., Nalcioglu, O., Su, M.Y., 2009. Impact of MRI-evaluated neoadjuvant chemotherapy response on change of surgical recommendation in breast cancer. *Annals of Surgery.* 249, 448–454.

Cheng, H.L., 2008. Investigation and optimization of parameter accuracy in dynamic contrast-enhanced MRI. *Journal of Magnetic Resonance Imaging.* 28, 736–743.

Choyke, P.L., Dwyer, A.J., Knopp, M.V., 2003. Functional tumor imaging with dynamic contrast-enhanced magnetic resonance imaging. *Journal of Magnetic Resonance Imaging.* 17, 509–520.

Cirillo, S., Petracchini, M., Scotti, L., Gallo, T., Macera, A., Bona, M., Ortega, C., Gabriele, P., Regge, D., 2009. Endorectal magnetic resonance imaging at 1.5 Tesla to assess local recurrence following radical prostatectomy using T2-weighted and contrast-enhanced imaging. *European Radiology.* 19, 761–769.

Daldrup-Link, H.E., Okuhata, Y., Wolfe, A., Srivastav, S., Oie, S., Ferrara, N., Cohen, R.L., Shames, D.M., Brasch, R.C., 2004. Decrease in tumor apparent permeability-surface area product to a MRI macromolecular contrast medium following angiogenesis inhibition with correlations to cytotoxic drug accumulation. *Microcirculation.* 11, 387–396.

De Los Santos, J., Bernreuter, W., Keene, K., Krontiras, H., Carpenter, J., Bland, K., Cantor, A., Forero, A., 2011. Accuracy of breast magnetic resonance imaging in predicting pathologic response in patients treated with neoadjuvant chemotherapy. *Clinical Breast Cancer.* 11, 312–319.

Dilks, P., Narayanan, P., Reznek, R., Sahdev, A., Rockall, A., 2010. Can quantitative dynamic contrast-enhanced MRI independently characterize an ovarian mass? *European Radiology.* 20, 2176–2183. doi: 10.1007/s00330-010-1795-6.

El Khouli, R.H., Macura, K.J., Jacobs, M.A., Khalil, T.H., Kamel, I.R., Dwyer, A., Bluemke, D.A., 2009. Dynamic contrast-enhanced MRI of the breast: Quantitative method for kinetic curve type assessment. *American Journal of Roentgenology.* 193, W295–W300.

Engelbrecht, M.R., Huisman, H.J., Laheij, R.J., Jager, G.J., van Leenders, G.J., Hulsbergen-Van De Kaa, C.A., de la Rosette, J.J., Blickman, J.G., Barentsz, J.O., 2003. Discrimination of prostate cancer from normal peripheral zone and central gland tissue by using dynamic contrast-enhanced MR imaging. *Radiology.* 229, 248–254.

Esserman, L., Kaplan, E., Partridge, S., Tripathy, D., Rugo, H., Park, J., Hwang, S., et al., 2001. MRI phenotype is associated with response to doxorubicin and cyclophosphamide neoadjuvant chemotherapy in stage III breast cancer. *Annals of Surgical Oncology.* 8, 549–559.

Franiel, T., Hamm, B., Hricak, H., 2011. Dynamic contrast-enhanced magnetic resonance imaging and pharmacokinetic models in prostate cancer. *European Radiology.* 21, 616–626.

He, D.F., Ma, D.Q., Jin, E.H., 2012. Dynamic breast magnetic resonance imaging: Pretreatment prediction of tumor response to neoadjuvant chemotherapy. *Clinical Breast Cancer.* 12, 94–101.

Hillengass, J., Wasser, K., Delorme, S., Kiessling, F., Zechmann, C., Benner, A., Kauczor, H.U., Ho, A.D., Goldschmidt, H., Moehler, T.M., 2007. Lumbar bone marrow microcirculation measurements from dynamic contrast-enhanced magnetic resonance imaging is a predictor of event-free survival in progressive multiple myeloma. *Clinical Cancer Research.* 13, 475–481.

Kelly, R.J., Rajan, A., Force, J., Lopez-Chavez, A., Keen, C., Cao, L., Yu, Y., et al., 2011. Evaluation of KRAS mutations, angiogenic biomarkers, and DCE-MRI in patients with advanced non-small-cell lung cancer receiving sorafenib. *Clinical Cancer Research: An Official Journal of the American Association for Cancer Research.* 17, 1190–1199.

Kerwin, W., Hooker, A., Spilker, M., Vicini, P., Ferguson, M., Hatsukami, T., Yuan, C., 2003. Quantitative magnetic resonance imaging analysis of neovasculature volume in carotid atherosclerotic plaque. *Circulation.* 107, 851–856.

Kerwin, W.S., Oikawa, M., Yuan, C., Jarvik, G.P., Hatsukami, T.S., 2008. MR imaging of adventitial vasa vasorum in carotid atherosclerosis. *Magnetic Resonance in Medicine.* 59, 507–514.

Kim, Y.K., Lee, J.M., Kim, C.S., 2004. Gadobenate dimeglumine-enhanced liver MR imaging: Value of dynamic and delayed imaging for the characterization and detection of focal liver lesions. *European Radiology.* 14, 5–13.

King, V., Brooks, J.D., Bernstein, J.L., Reiner, A.S., Pike, M.C., Morris, E.A., 2011. Background parenchymal enhancement at breast MR imaging and breast cancer risk. *Radiology.* 260, 50–60.

Koh, T.S., Thng, C.H., Lee, P.S., Hartono, S., Rumpel, H., Goh, B.C., Bisdas, S., 2008. Hepatic metastases: In vivo assessment of perfusion parameters at dynamic contrast-enhanced MR imaging with dual-input two-compartment tracer kinetics model. *Radiology.* 249, 307–320.

Krinsky, G.A., Lee, V.S., Theise, N.D., Weinreb, J.C., Rofsky, N.M., Diflo, T., Teperman, L.W., 2001. Hepatocellular carcinoma and dysplastic nodules in patients with cirrhosis: Prospective diagnosis with MR imaging and explantation correlation. *Radiology.* 219, 445–454.

Kuhl, C., 2007. The current status of breast MR imaging. Part I. Choice of technique, image interpretation, diagnostic accuracy, and transfer to clinical practice. *Radiology.* 244, 356–378.

Kuhl, C.K., Mielcareck, P., Klaschik, S., Leutner, C., Wardelmann, E., Gieseke, J., Schild, H.H., 1999. Dynamic breast MR imaging: Are signal intensity time course data useful for differential diagnosis of enhancing lesions? *Radiology.* 211, 101–110.

Kuhl, C.K., Schild, H.H., Morakkabati, N., 2005a. Dynamic bilateral contrast-enhanced MR imaging of the breast: Trade-off between spatial and temporal resolution. *Radiology.* 236, 789–800.

Kuhl, C.K., Schrading, S., Bieling, H.B., Wardelmann, E., Leutner, C.C., Koenig, R., Kuhn, W., Schild, H.H., 2007. MRI for diagnosis of pure ductal carcinoma in situ: A prospective observational study. *Lancet.* 370, 485–492.

Kuhl, C.K., Schrading, S., Leutner, C.C., Morakkabati-Spitz, N., Wardelmann, E., Fimmers, R., Kuhn, W., Schild, H.H., 2005b. Mammography, breast ultrasound, and magnetic resonance imaging for surveillance of women at high familial risk for breast cancer. *Journal of Clinical Oncology.* 23, 8469–8476.

Leach, M.O., Brindle, K.M., Evelhoch, J.L., Griffiths, J.R., Horsman, M.R., Jackson, A., Jayson, G.C., et al.; Pharmacodynamic/Pharmacokinetic technologies advisory committee, D.D.O.C.R.U.K., 2005. The assessment of antiangiogenic and antivascular therapies in early-stage clinical trials using magnetic resonance imaging: Issues and recommendations. *British Journal of Cancer.* 92, 1599–1610.

Lin, C., Luciani, A., Belhadj, K., Deux, J.F., Kuhnowski, F., Maatouk, M., Beaussart, P., Cuenod, C.A., Haioun, C., Rahmouni, A., 2010. Multiple myeloma treatment response assessment with whole-body dynamic contrast-enhanced MR imaging. *Radiology.* 254, 521–531.

Liney, G.P., Gibbs, P., Hayes, C., Leach, M.O., Turnbull, L.W., 1999. Dynamic contrast-enhanced MRI in the differentiation of breast tumors: User-defined versus semi-automated region-of-interest analysis. *Journal of Magnetic Resonance Imaging*. 10, 945–949.

Loo, C.E., Straver, M.E., Rodenhuis, S., Muller, S.H., Wesseling, J., Vrancken Peeters, M.J., Gilhuijs, K.G., 2011. Magnetic resonance imaging response monitoring of breast cancer during neoadjuvant chemotherapy: Relevance of breast cancer subtype. *Journal of Clinical Oncology*. 29, 660–666.

Loo, C.E., Teertstra, H.J., Rodenhuis, S., de Vijver, M.J.V., Hannemann, J., Muller, S.H., Peeters, M.J.V., Gilhuijs, K.G.A., 2008. Dynamic contrast-enhanced MRI for prediction of breast cancer response to neoadjuvant chemotherapy: Initial results. *American Journal of Roentgenology*. 191, 1331–1338.

Manton, D.J., Chaturvedi, A., Hubbard, A., Lind, M.J., Lowry, M., Maraveyas, A., Pickles, M.D., Tozer, D.J., Turnbull, L.W., 2006. Neoadjuvant chemotherapy in breast cancer: Early response prediction with quantitative MR imaging and spectroscopy. *British Journal of Cancer*. 94, 427–435.

Martincich, L., Montemurro, F., De Rosa, G., Marra, V., Ponzone, R., Cirillo, S., Gatti, M., et al., 2004. Monitoring response to primary chemotherapy in breast cancer using dynamic contrast-enhanced magnetic resonance imaging. *Breast Cancer Research and Treatment*. 83, 67–76.

Mayr, N.A., Wang, J.Z., Zhang, D., Grecula, J.C., Lo, S.S., Jaroura, D., Montebello, J., et al., 2010. Longitudinal changes in tumor perfusion pattern during the radiation therapy course and its clinical impact in cervical cancer. *International Journal of Radiation Oncology, Biology, Physics*. 77, 502–508.

Mehta, S., Hughes, N.P., Buffa, F.M., Li, S.P., Adams, R.F., Adwani, A., Taylor, N.J., et al., 2011. Assessing early therapeutic response to bevacizumab in primary breast cancer using magnetic resonance imaging and gene expression profiles. *Journal of the National Cancer Institute Monographs*. 2011, 71–74. doi: 10.1093/jncimonographs/lgr027.

Miller, J.C., Pien, H.H., Sahani, D., Sorensen, A.G., Thrall, J.H., 2005. Imaging angiogenesis: Applications and potential for drug development. *Journal of the National Cancer Institute*. 97, 172–187.

Montazel, J.L., Divine, M., Lepage, E., Kobeiter, H., Brell, S., Rahmouni, A., 2003. Normal spinal bone marrow in adults: Dynamic gadolinium-enhanced MR imaging. *Radiology*. 229, 703–709.

Moreno-Aspitia, A., Morton, R.F., Hillman, D.W., Lingle, W.L., Rowland, K.M. Jr., Wiesenfeld, M., Flynn, P.J., Fitch, T.R., Perez, E.A., 2009. Phase II trial of sorafenib in patients with metastatic breast cancer previously exposed to anthracyclines or taxanes: North Central Cancer Treatment Group and Mayo Clinic Trial N0336. *Journal of Clinical Oncology*. 27, 11–15. doi: 10.1200/JCO.2007.15.5242.

Newell, D., Nie, K., Chen, J.H., Hsu, C.C., Yu, H.J., Nalcioglu, O., Su, M.Y., 2010. Selection of diagnostic features on breast MRI to differentiate between malignant and benign lesions using computer-aided diagnosis: Differences in lesions presenting as mass and non-mass-like enhancement. *European Radiology*. 20, 771–781.

Nie, K., Chen, J.H., Yu, H.J., Chu, Y., Nalcioglu, O., Su, M.Y., 2008. Quantitative analysis of lesion morphology and texture features for diagnostic prediction in breast MRI. *Academic Radiology*. 15, 1513–1525.

Notohamiprodjo, M., Sourbron, S., Staehler, M., Michaely, H.J., Attenberger, U.I., Schmidt, G.P., Boehm, H., et al., 2010. Measuring perfusion and permeability in renal cell carcinoma with dynamic contrast-enhanced MRI: A pilot study. *Journal of Magnetic Resonance Imaging*. 31, 490–501.

Padhani, A.R., Hayes, C., Assersohn, L., Powles, T., Makris, A., Suckling, J., Leach, M.O., Husband, J.E., 2006. Prediction of clinicopathologic response of breast cancer to primary chemotherapy at contrast-enhanced MR imaging: Initial clinical results. *Radiology*. 239, 361–374.

Padhani, A.R., Leach, M.O., 2005. Antivascular cancer treatments: Functional assessments by dynamic contrast-enhanced magnetic resonance imaging. *Abdominal Imaging*. 30, 324–341.

Pickles, M.D., Lowry, M., Manton, D.J., Gibbs, P., Turnbull, L.W., 2005. Role of dynamic contrast enhanced MRI in monitoring early response of locally advanced breast cancer to neoadjuvant chemotherapy. *Breast Cancer Research and Treatment*. 91, 1–10.

Pinker, K., Grabner, G., Bogner, W., Gruber, S., Szomolanyi, P., Trattnig, S., Heinz-Peer, G., et al., 2009. A combined high temporal and high spatial resolution 3 Tesla MR imaging protocol for the assessment of breast lesions initial results. *Investigative Radiology*. 44, 553–558.

Raatschen, H.J., Simon, G.H., Fu, Y., Sennino, B., Shames, D.M., Wendland, M.F., McDonald, D.M., Brasch, R.C., 2008. Vascular permeability during antiangiogenesis treatment: MR imaging assay results as biomarker for subsequent tumor growth in rats. *Radiology*. 247, 391–399.

Rahmouni, A., Montazel, J.L., Divine, M., Lepage, E., Belhadj, K., Gaulard, P., Bouanane, M., Golli, M., Kobeiter, H., 2003. Bone marrow with diffuse tumor infiltration in patients with lymphoproliferative diseases: Dynamic gadolinium-enhanced MR imaging. *Radiology*. 229, 710–717.

Rieber, A., Brambs, H.J., Gabelmann, A., Heilmann, V., Kreienberg, R., Kuhn, T., 2002. Breast MRI for monitoring response of primary breast cancer to neo-adjuvant chemotherapy. *European Radiology*. 12, 1711–1719.

Rode, A., Bancel, B., Douek, P., Chevallier, M., Vilgrain, V., Picaud, G., Henry, L., et al., 2001. Small nodule detection in cirrhotic livers: Evaluation with US, spiral CT, and MRI and correlation with pathologic examination of explanted liver. *Journal of Computer Assisted Tomography*. 25, 327–336.

Sala, E., Crawford, R., Senior, E., Shaw, A., Simcock, B., Vrotsou, K., Palmer, C., Rajan, P., Joubert, I., Lomas, D., 2009. Added value of dynamic contrast-enhanced magnetic resonance imaging in predicting advanced stage disease in patients with endometrial carcinoma. *International Journal of Gynecological Cancer*. 19, 141–146. doi: 10.1111/IGC.0b013e3181995fd9.

Sala, E., Rockall, A., Rangarajan, D., Kubik-Huch, R.A., 2010. The role of dynamic contrast-enhanced and diffusion weighted magnetic resonance imaging in the female pelvis. *European Journal of Radiology*. 76, 367–385.

Saslow, D., Boetes, C., Burke, W., Harms, S., Leach, M.O., Lehman, C.D., Morris, E., et al., 2007. American cancer society guidelines for breast screening with MRI as an adjunct to mammography. *CA—A Cancer Journal for Clinicians*. 57, 75–89.

Schelfout, K., Van Goethem, M., Kersschot, E., Colpaert, C., Schelfhout, A.M., Leyman, P., Verslegers, I., et al., 2004. Contrast-enhanced MR imaging of breast lesions and effect on treatment. *European Journal of Surgical Oncology*. 30, 501–507.

Schorn, C., Fischer, U., Luftner-Nagel, S., Grabbe, E., 1999. Diagnostic potential of ultrafast contrast-enhanced MRI of the breast in hypervascularized lesions: Are there advantages in comparison with standard dynamic MRI? *Journal of Computer Assisted Tomography*. 23, 118–122.

Shimauchi, A., Giger, M.L., Bhooshan, N., Lan, L., Pesce, L.L., Lee, J.K., Abe, H., Newstead, G.M., 2011. Evaluation of clinical breast MR imaging performed with prototype computer-aided diagnosis breast MR imaging workstation: Reader study. *Radiology*. 258, 696–704.

Straver, M.E., Loo, C.E., Rutgers, E.J.T., Oldenburg, H.S.A., Wesseling, J., Peeters, M.J.T.F.D.V., Gilhuijs, K.G.A., 2010. MRI-model to guide the surgical treatment in breast cancer patients after neoadjuvant chemotherapy. *Annals of Surgery*. 251, 701–707.

Tillman, G.F., Orel, S.G., Schnall, M.D., Schultz, D.J., Tan, J.E., Solin, L.J., 2002. Effect of breast magnetic resonance imaging on the clinical management of women with early-stage breast carcinoma. *Journal of Clinical Oncology*. 20, 3413–3423.

Tofts, P.S., 1997. Modeling tracer kinetics in dynamic Gd-DTPA MR imaging. *Journal of Magnetic Resonance Imaging*. 7, 91–101.

Tofts, P.S., Kermode, A.G., 1991. Measurement of the blood-brain barrier permeability and leakage space using dynamic MR imaging. 1. Fundamental concepts. *Magnetic Resonance in Medicine*. 17, 357–367.

Tozaki, M., Fukuda, K., 2006. High-spatial-resolution MRI of non-masslike breast lesions: Interpretation model based on BI-RADS MRI descriptors. *American Journal of Roentgenology*. 187, 330–337.

Turnbull, L.W., 2009. Dynamic contrast-enhanced MRI in the diagnosis and management of breast cancer. *NMR in Biomedicine*. 22, 28–39.

Wedam, S.B., Low, J.A., Yang, S.X., Chow, C.K., Choyke, P., Danforth, D., Hewitt, S.M., et al., 2006. Antiangiogenic and antitumor effects of bevacizumab in patients with inflammatory and locally advanced breast cancer. *Journal of Clinical Oncology*. 24, 769–777.

Yoo, H.J., Lee, J.M., Lee, M.W., Kim, S.J., Lee, J.Y., Han, J.K., Choi, B.I., 2008. Hepatocellular carcinoma in cirrhotic liver: Double-contrast-enhanced, high-resolution 3.0T-MR imaging with pathologic correlation. *Investigative Radiology*. 43, 538–546.

Yu, H.J., Chen, J.H., Mehta, R.S., Nalcioglu, O., Su, M.Y., 2007. MRI measurements of tumor size and pharmacokinetic parameters as early predictors of response in breast cancer patients undergoing neoadjuvant anthracycline chemotherapy. *Journal of Magnetic Resonance Imaging*. 26, 615–623.

Zha, Y.F., Li, M.J., Yang, J.Y., 2010. Dynamic contrast enhanced magnetic resonance imaging of diffuse spinal bone marrow infiltration in patients with hematological malignancies. *Korean Journal of Radiology*. 11, 187–194.

Zhang, X.M., Yu, D., Zhang, H.L., Dai, Y., Bi, D., Liu, Z., Prince, M.R., Li, C., 2008. 3D dynamic contrast-enhanced MRI of rectal carcinoma at 3T: Correlation with microvascular density and vascular endothelial growth factor markers of tumor angiogenesis. *Journal of Magnetic Resonance Imaging*. 27, 1309–1316.

Zhang, Y., Fukatsu, H., Naganawa, S., Satake, H., Sato, Y., Ohiwa, M., Endo, T., Ichihara, S., Ishigaki, T., 2002. The role of contrast-enhanced MR mammography for determining candidates for breast conservation surgery. *Breast Cancer*. 9, 231–239.

11

Evaluation of Body MR Spectroscopy In Vivo

Rajakumar Nagarajan, Manoj K. Sarma, and M. Albert Thomas
University of California, Los Angeles

CONTENTS

11.1 Proton (¹H) Magnetic Resonance Spectroscopy

11.1.1 Introduction

Magnetic resonance imaging (MRI) provides anatomic images and morphometric characterization of pathological conditions, whereas magnetic resonance spectroscopy (MRS) provides metabolite/biochemical information about biological tissues *in vivo* and it is noninvasive. MRS has been used clinically for more than two decades. Other than brain, this advanced diagnostic tool is mainly used in the investigation of the diseases of prostate gland, calf muscle, spinal cord, breast, and liver. The aim of this chapter is to review the role of MRS in quantifying various metabolites in these organs and describe how this information could be used in the diagnosis of certain diseases.

11.1.2 Prostate Magnetic Resonance Spectroscopic Imaging

The incidence and mortality rate of prostate cancer (PCa) vary substantially worldwide, but it is the most common noncutaneous malignancy in the Western world. PCa is the most common type of cancer and is the second most common cause of cancer deaths among men in the United States (Jemal et al., 2010). Accurate diagnosis of this ailment at early stages is critical for these patients, which sometimes requires multimodality imaging techniques in addition to other clinical tests.

Thomas et al. demonstrated the first successful implementation of proton (^{1}H) MRS using topical localization enabled by a transrectal transmit/receive coil on a 2 T MRI scanner (1990). Currently, ^{1}H MRS is commercially available for the prostate using a three-dimensional (3D) magnetic resonance spectroscopic imaging (MRSI) technique (Brown et al., 1982). Prostate MRS has been performed predominantly using a multivoxel technique, with which intracellular metabolic information can be obtained from single or multiple sections. In ^{1}H MRS, the resonances for the prostate metabolites—choline (Cho), creatine (Cr), spermine (Spm), and citrate (Cit)—occur at distinct frequencies (~3.2, 3.0, 3.1, and 2.6 ppm, respectively). The areas under these peaks are related to the concentration of the respective metabolites, and changes in these concentrations can be used to identify cancer quantitatively (it should be noted that when MRS is performed at 1.5 T, the peaks for Cho, Cr, and Spm overlap and adversely affect such quantitation). In PCa, Cit and Spm levels decrease markedly (Kurhanewicz et al., 1996). PCa also leads to increased Cho because of higher cell membrane turnover, cell density, and phospholipid metabolism. As a result, the ratio of Cho to Cit is increased in cancer. Because the peak of Cr is close to that of Cho (separated by 0.15 ppm only), the ratio of Cho plus Cr to Cit (Cho + Cr/Cit) is typically measured in clinical MRS studies (Kurhanewicz et al., 1996). In the last decade, many publications have reported the advantages of the combined use of MRI and MRSI in order to overlay metabolic information directly on the corresponding anatomic display (Hoeks et al., 2011; Kim et al., 2009). A study conducted by Scheidler et al. showed a sensitivity of 95% and specificity of 91% for PCa detection, on a per-sextant basis, for combined MRSI and MRI (1999). However, the sensitivity and specificity was around 61–77% and 46–81% for MRI alone and 75% and 63% for MRSI alone, respectively. This study demonstrated that the addition of MRSI to conventional MRI significantly improved PCa diagnosis: for locations in the peripheral zone, positive predictive value (PPV) and negative predictive value (NPV) were 89–92% and 74–82%, respectively. Moreover, combined MRI–MRSI seems to be superior to sextant biopsy, with the largest increase in diagnostic accuracy at the apex of the prostate, which is difficult to accomplish by biopsy (Wefer et al., 2000). In two recent prospective studies conducted on patients with elevated prostate-specific antigen (PSA) levels, which used prostate biopsy as reference standard, it has been reported that combined MRI and

MRSI increased the accuracy in PCa detection and localization to 79% and 74.2%, respectively (Manenti et al., 2006; Reinsberg et al., 2007). However, Prando et al. found that prostate biopsy directed with endorectal MRSI might help increase the PCa detection rate in patients with an elevated PSA and a previous negative biopsy result (2005).

The most widely used grading system for PCa is called the Gleason system. This system uses a Gleason score (GS) of 2–10 to grade PCa, where higher grades represent more aggressive tumors (Gleason, 1966). The GS offers a good indication of the tumor's behavior as the prostate tumor with a low GS is likely to be slow-growing, while one with a high score is more likely to grow aggressively or to have already spread outside the prostate (metastasized). Nagarajan et al. showed that the metabolite ratio of (Cho + Cr/Cit) in MRSI measurement of PCa increased with an increase of two GS (GS 3 + 3 and GS 4 + 3) (2011). The MRSI of 67-year-old PCa patient with a GS of 3 + 3 is shown in Figure 11.1.

While MRS was demonstrated to be a valuable diagnostic tool, there are several limitations that need to be considered, such as long acquisition times, variability in results that stem from variations in postprocessing

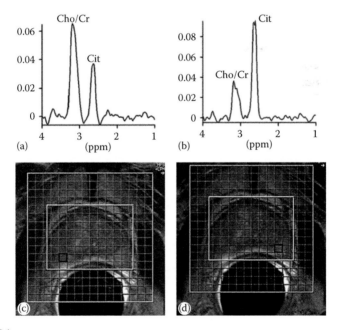

FIGURE 11.1
(Color version available from crcpress.com; see Preface.) MR spectroscopic imaging of a 67-year-old PCa patient with a GS of 3 + 3. (a) and (b) show the 1D spectra extracted from the 3D MRSI data recorded in the affected and contralateral area. (c) and (d) show the corresponding voxel locations. (From Nagarajan, R., et al., *Med Princ Pract.* 20, 444–448, 2011. With permission.)

techniques or B_0 shimming, and no direct visualization of the periprostatic anatomy. In addition, prostate biopsy done prior to MRS may lead to spectral degradation, which makes accurate assessment of the metabolite ratios unreliable.

11.1.3 Calf Muscle

MRS has been applied successfully to human muscle *in vivo* ever since magnets of an appropriate bore size became available. However, most of the research was aimed at short-term muscle energy metabolism (Avison et al., 1988; Blei et al., 1993; Chance, 1992; Schmitz et al., 2012) and used either phosphorous (^{31}P) MRS for the observation of high-energy phosphates or carbon (^{13}C) MRS to investigate muscular glycogen. Water-suppressed ^{1}H MR spectra of human muscle were presumed to feature few metabolite peaks dominated by a large hump of lipid resonances (Barany and Venkatasubramanian, 1989; Bruhn et al., 1991; Narayana et al., 1988; Pan et al., 1991; Williams et al., 1985). When Schick et al. compared the lipid resonances in calf muscle and in fat tissue, they observed two compartments of triglycerides with a resonance frequency shift of 0.2 ppm (1993). The authors assigned the resonance at 1.5 ppm to the CH protons of lipids in fat cells and supposed that the shifted resonance at 1.3 ppm could be attributed to lipids located inside muscle cells (i.e., in their cytoplasm), experiencing different bulk susceptibility. Lipids are either stored as subcutaneous or interstitial adipose tissue (extramyocellular lipids, EMCLs) or as intramyocellular lipids (IMCLs) in the form of liquid droplets in the cytoplasm of muscle cells, while free or protein-bound lipids in cytoplasm are of lower concentration. Fatty acids in the converted form of triacylglycerols are the most abundant long-term energy store of the human body. While EMCL (i.e., fat tissue) is metabolically relatively inert, there is evidence that IMCL can be mobilized and utilized with turnover times of several hours (Kayar et al., 1986). IMCL droplets are located in the cytoplasm in contact with mitochondria (Vock et al., 1996) and provide an important share of the energy supply during long-term endurance activities (Havel et al., 1964; Oberholzer et al., 1976). The MRS investigation of IMCL was done most commonly using single-voxel spectroscopy in humans (Boesch et al., 1997; Jacob et al., 1999; Krssak et al., 1999; Ramadan et al., 2010) as well as in animals (Dobbins et al., 2001; Jucker et al., 2003; Kuhlmann et al., 2003). MRI and MRS of soleus muscle from a 38-year-old healthy volunteer are shown in Figure 11.2. Apart from IMCL and EMCL, other metabolites were also detected in the human soleus muscle at 7 T [carnitine, taurine, phosphocholine (PCho), Cho, glycerophosphocholine (GPC), and carnosine]. WET water-suppression method was applied before the localization/acquisition sequence (Ogg et al., 1994). The 1D spectrum was Fourier transformed after zero filling to 4096 points and applying an exponential window function in MestReNova program (Mestralab Research S.L., Version: 5.0.2-2108, 2007).

There exist a few papers using MRSI methods of investigating IMCL (Hwang et al., 2001; Larson-Meyer et al., 2002; Shen et al., 2003). However, the relatively small number of studies performed using MRSI is probably due to the strong lipid signal from surrounding subcutaneous fat and bone marrow, which leads to contaminations of the spectra of interest due to signal bleeding at the low MRSI resolution. Volume-selective ^{1}H MRS/MRSI is currently the only methodology that allows noninvasive monitoring of IMCL levels *in vivo*. The main reason for the recently increased interest in the measurement of IMCL levels stems from the strong correlation observed between the IMCL concentration and insulin resistance. For example, Krssak et al. observed an inverse relationship between IMCL in the gastrocnemius muscle and insulin-stimulated glucose uptake in a clamp study of 23 normal subjects (1999). A similar relation was reported by Jacob et al. (1999) in insulin resistant, lean, nondiabetic offspring of patients with type-2 diabetes. This correlation between IMCL and insulin resistance in humans could also be reproduced in animals (Dobbins et al., 2001; Kuhlmann et al., 2003).

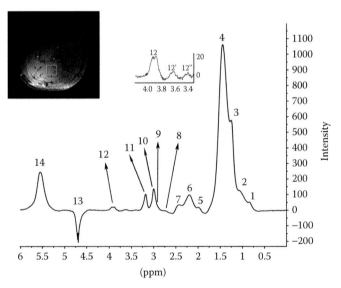

FIGURE 11.2
A localized 1D spectrum acquired from soleus muscle. Axial image of a human soleus muscle from a 38-year-old volunteer is shown inset and the voxel position from where the data were collected. About 3.3–4.3 ppm spectral region is expanded to show detail. See Table 11.1 for assignments. Acquisition parameters are spectral width, 4000 Hz; vector size, 2048 points; voxel size of 6 × 6 × 35 mm³; number of averages, 8; and repetition time, 2000 ms. The "WET" water-suppression method was applied before the acquisition sequence. The 1D spectrum was Fourier transformed after zero filling to 4096 points and applying an exponential window function in MestReNova program. (From Ramadan, S., et al., *J Magn Reson.* 204, 91–98, 2010. With permission.)

TABLE 11.1

Assignment of Resonance in the 1D MR Spectrum of Human Soleus Muscle Collected *In Vivo* at 7 T Shown in Figure 11.2

Molecules	Peak	Species	Chemical Shift (ppm)
Intramyocellular lipids	1	CH_3	0.84
Extramyocellular lipids	2	CH_3	1.05
Intramyocellular lipids	3	$(CH_2)_n$	1.25
Extramyocellular lipids	4	$(CH_2)_n$	1.44
Intramyocellular lipids	5	$-CH_2-CH=CH-$	2.00
Extramyocellular lipids	6	$-CH_2-CH=CH-$	2.20
Intramyocellular lipids	6	$-CH_2-(C=O)-OR$	2.20
Extramyocellular lipids	7	$-CH_2-(C=O)-OR$	2.43
Intramyocellular lipids	8	$-CH=CH-CH_2-CH=CH-$	2.76
Extramyocellular lipids	9	$-CH=CH-CH_2-CH=CH-$	2.93
Total creatine	10	CH_3	3.02
Carnitine/choline	11	$N(CH_3)_3$	3.18
Taurine	12"	$H_2N-CH_2-CH_2-SO_3$	3.41
Pcho, Cho, GPC	12'	$(CH_3)_3-N-CH_2-CH_2-OPO_3$	3.63
Total creatine	12	$CH_2(tCr)$	3.92
Residual water	13	H_2O	4.70
Total lipid	14	$-HC=CH-,-(C=O)-O-CHR_2$	5.55

Note: Creatine methyl peak at 3.02 ppm was used as a chemical shift reference.

11.1.4 Spinal Cord MRS

It is technically difficult to execute spectroscopy on the spinal cord. The spinal cord is small compared to the size of a voxel in a typical spectroscopy exam. Smaller voxels degrade the signal-to-noise ratio (SNR), increasing the errors in quantifying the metabolites. Moreover, its anatomical location is relatively deep within the surrounding tissue, which also reduces the available SNR due to reduced sensitivity of surface radio frequency (RF) coils. Despite these challenges, several groups have successfully conducted MRS on the spine.

For instance, MRS could provide valuable information in cervical spondylosis which occurs as a result of intervertebral disc degeneration (IVDD) (Baron and Young, 2007; Wilkinson, 1960). IVDD increases mechanical stresses at the endplates of the adjacent vertebral bodies (Kumaresan et al., 2001), eventually resulting in osteophytic bars that extend into the spinal canal (Parke, 1988). As a result, spinal cord compression and inflammation can occur resulting in myelopathy. Myelopathy from cervical spondylosis is the most common cause of nontraumatic spastic paresis (Moore and Blumhardt, 1997) and spinal dysfunction in the elderly (Young, 2000), and it is also the most common primary diagnosis for patients over 64 years who undergo cervical spine surgery (Wang et al., 2007). To prevent neurological deterioration, surgery is often performed shortly after diagnosis. Some studies have suggested that surgical intervention after neurological decline may result in worsened outcome (Sampath et al., 2000); therefore, surgeons are likely to operate on the basis of diagnostic imaging even in the absence of clinical symptoms (Irwin et al., 2005). The value of traditional imaging technologies in diagnosing and predicting early response to treatment, however, remains controversial.

Proton MRS has also been used in the investigation of other spinal disorders (Henning et al., 2008; Holly et al., 2009; Jayasundar et al., 1996; Layer et al., 1998). Due to technical challenges, quantitative spinal cord MRS has hardly ever been used and was mostly limited to the brain stem and the upper part of the cervical spine down to the C2 level (Blamire et al., 2007; Cooke et al., 2004; Edden et al., 2007; Gomez-Anson et al., 2000; Marliani et al., 2007). Preliminary studies indicate that metabolite concentrations are sufficiently high to be observed by MRS in the entire spinal cord, but spectral quality was insufficient for quantification in the middle and lower cervical regions and even more so in the thoracic and lumbar spinal cord (Dydak et al., 2005; Kim et al., 2004).

A recently published study reported quantitative single-voxel cervical spinal cord spectroscopy, presenting mean relative concentration ratios for N-acetylasparate (NAA), Cr, PCr, Cho, and mI in a group of 10 healthy volunteers by using a clinical 3 T MRI system (Marliani et al., 2007). The study applied the same acquisition and postprocessing protocol to quantify the main central nervous system (CNS) metabolites on the cervical spinal cord plaques of a group of 15 patients with relapsing-remitting multiple sclerosis (RRMS) and compared them with the metabolite content of a group of healthy volunteers, to investigate if there was a difference between the MR spectroscopic findings of the cervical spinal cord in patients with RRMS and volunteers. Among the few cervical spine spectroscopy studies (Ciccarelli et al., 2007; Cooke et al., 2004; Edden et al., 2007; Gomez-Anson et al., 2000; Henning et al., 2008; Kendi et al., 2004), only four provided data acquired from patients with MS and only one quantified the main metabolites on T_2-weighted hyperintense plaques. Kendi et al. measured NAA, Cr, Cho, mI, lipids, and lactate on the normal-appearing cervical cord of patients with MS, finding only NAA significantly reduced with respect to the corresponding values in healthy volunteers (2004). Blamire et al. recently quantified NAA, Cr, Cho, and mI in the cervical cord of patients with MS, but they did not verify the presence of potential T_2-weighted lesions on the measured cervical cord tract (2007). They also found that only NAA significantly reduced with respect to the corresponding values in healthy volunteers. The total NAA (tNAA) of the control was 6.7 mM and that of patient was 4.9 mM. An interesting study published by Henning et al. (2008) measured the main cervical spine metabolites of only one patient with MS, finding a decrease of NAA, and mI and an increase in Cho compared with a group of healthy volunteers.

11.1.5 Breast

Breast cancer is one of the most prevalent cancers in the female population, especially in the Western world. The screening of high-risk populations as well as early diagnoses and monitoring of early response to therapy are the key information in the prevention and effective treatment of breast cancer.

Breast spectra appear dominated by large lipid (1.3 ppm) and water (4.7 ppm) resonances, depending on tissue composition. The ^{1}H MR spectra of normal fatty breast tissues from healthy subjects typically exhibit large lipid resonances with little contribution from water, while the water resonance dominates the spectra of healthy glandular tissues and tumors with lower contributions from lipids (Mountford and Tattersall, 1987; Sharma et al., 2008; Sijens et al., 1988). For these reasons, several groups refer to using the fat/water ratio for a qualitative spectral analysis of breast tumors. For example, Kumar et al. adopted this method to monitor the therapeutic response of breast cancer patients to neoadjuvant chemotherapy by MRS using a 1.5 T system (2006). This study showed that the water/fat ratio was able to detect tumor response to therapy with higher sensitivity and specificity than tumor volume reduction. This simple characterization is not always considered acceptable, because the fibroglandular tissue in healthy controls can show large water and lipid resonances of varying amplitude (Haddadin et al., 2009). Moreover, tumor-containing voxels can include both fibroglandular and adipose tissues in breast examinations, as well as necrotic regions rich in lipids, leading to a wide variation of the relative amounts of water and lipid components.

Between water and lipid resonances, a total choline (tCho) signal is typically detected in breast tumors. Although this peak can also be visible in normal breast tissues at high magnetic field (Haddadin et al., 2009), and in breast of lactating mothers (Meisamy et al., 2004; Podo et al., 2007), most studies reported the detectability of tCho as a specific index of malignancy (Podo et al., 2007; Sardanelli et al., 2008; Sharma et al., 2008). Quantitative estimates of tCho, therefore, could be a very useful adjunct to longitudinal studies aimed at evaluating chemotherapy-induced effects on tumor size. This could also be used in comparative evaluation of examinations performed at different B_0 magnetic fields to assess the efficacy of the technique under different conditions of sensitivity and spatial resolution.

Roebuck et al. conducted an MRS study at 1.5 T to quantify tCho in breast tumors (1998). The use of water as internal standard for tCho quantification has been successfully applied to breast lesions (Baik et al., 2006; Bolan et al., 2003). This procedure has the advantage of being insensitive to the inclusion of adipose tissue in the voxel, because the water signal and the sideband resonance at 3.2 ppm from adipose tissue are small compared with those of most of lesions or fibroglandular tissue. The use of quantitative MRS as a diagnostic tool in breast cancer was reviewed by Haddadin et al. (2009). Figure 11.3 shows the fat-suppressed postcontrast image and MRS acquired at 4 T in a patient with an invasive ductal carcinoma. This study indicated that MRS alone could not be used for diagnosis because there are many tumors with low tCho and also few benign lesions with raised tCho. However, the combination of MRI and MRS might improve sensitivity, specificity, and accuracy of the diagnosis (Huang et al., 2004; Meisamy et al., 2005).

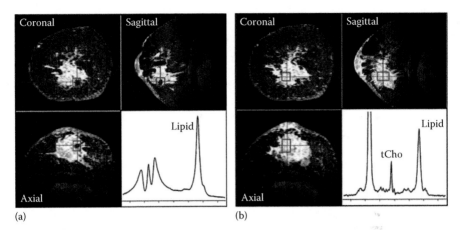

FIGURE 11.3
(Color version available from crcpress.com; see Preface.) Effect of radiographic markers on shim quality. Both (a) and (b) show a three-view multiplanar reconstruction of a 3D, fat-suppressed, postcontrast image acquired at 4 T in a patient with an invasive ductal carcinoma. The voxel ROI in (a) is placed directly over a metallic radiographic marker. Even after manual adjustment of the linear B_0 shims, the spectral quality is very poor. In (b), the voxel ROI was repositioned in the center of the lesion away from the marker, and a high-quality spectrum showing tCho and other metabolites could be obtained. (From Haddadin, I.S., et al., *NMR Biomed.* 22, 65–76, 2009. With permission.)

The increase in tCho signal measured in ^{1}H spectroscopy is believed to be the result of a switch in the phenotype of PCho that occurs as cells become malignant. Studies using human breast cell lines indicate that the GPC in normal cells is transformed to PCho as cells become malignant, with a stepwise increase in PCho as tissues progress through the stages of invasive breast cancer (Aboagye and Bhujwalla, 1999). As PCho increases, the tCho peak increases, as measured by ^{1}H spectroscopy. Quantitative criteria were used in a more recent study by Bartella et al. to assess the benefit of ^{1}H spectroscopy at 1.5 T for lesions larger than 1 cm in diameter (2006). A total of 56 patients with 57 suspicious lesions going to biopsy or biopsy-proven lesions were examined by single-voxel MRS.

More than 100 cases of benign breast lesions subjected to *in vivo* ^{1}H MRS have been reported in the literature (Jacobs et al., 2004; Kim et al., 2003; Tse et al., 2003; Yeung et al., 2001). No data quality problems were reported, and interpretable results were obtained in all cases. Included in this collection was about 40 normal breast tissues from normal age matched controls (Stanwell et al., 2005), as well as lactating breast (Stanwell et al., 2005; Tse et al., 2007). Genuinely benign breast lesions that had been studied were about 83. The reason for the high success rate probably relates to the large volume of interest, as the sizes of those lesions varied between 10 and 100 mm (mean 32.5 mm). In addition, when normal or lactating breast tissue was included in the test, there were no size constraints in the determination of the volume of interest. Fibroadenoma (FA) is a common benign fibroepithelial lesion, which is usually of low cellularity, and tends to occur in younger patients compared to the breast cancer demographic. FA can be easily distinguished from cancer usually by its circumscription on imaging and the lack of necrosis within the lesion. From the various series, a total of 46 FAs have been evaluated by *in vivo* ^{1}H MRS (Cecil et al., 2001; Jacobs et al., 2004; Kim et al., 2003; Kvistad et al., 1999; Roebuck et al., 1998; Tse et al., 2003; Yeung et al., 2001, 2002), with three false-positive cases constituting a false-positive rate of 6%. Fibrocystic change is another common breast lesion that occurs in younger women. It can be found in a range of sizes, and there are several different components to the disease, including epithelial hyperplasia, cystic change, apocrine metaplasia, fibrosis, inflammatory changes, and duct ectasia.

The most relevant MRS-detectable metabolite patterns reported to contribute to breast tumor diagnosis and monitoring are absolute or relative tCho content and fat/water spectral profile. These parameters may also be used to monitor tumor response to therapy, within the limitations imposed by large variability in water and lipid contents of mammary tissues (Sharma et al., 2011).

11.1.6 Liver

Liver ^{1}H MRS is currently used to detect and characterize fatty liver diseases, hepatic lesions, and evaluate changes due to chronic hepatitis. The evaluation of fatty infiltration of the liver using MR chemical shift imaging (CSI) was first demonstrated in 1984 by Dixon (1984) using a modified spin-echo technique that exploited the 3- to 4-ppm difference in resonant frequencies between fat and water. In this technique, two images are acquired, the first being a normal spin-echo image where the MR signals from water and fat were in phase and additive. The second image is collected at the same echo time (TE) but with the gradient echo shifted away from the Hahn spin echo such that the spin vectors of the water and the fat would be 180° out of phase. This results in an opposed-phase image where the MR signal was the difference between the fat and water signals. This technique was further refined by collecting a second echo that allowed the calculation of the percentage of fat in the liver corrected for T_2 relaxation (Heiken et al., 1985).

The lipid content of the liver can also be quantified using a volume-localized single-voxel technique (Longo et al., 1993, 1995; Thomsen et al., 1994). The lipid content measured by ^{1}H MRS has been shown to be better than that estimated by histopathologic analysis (Longo et al., 1995). However, liver lipid content measured using ^{1}H MRS without water suppression only measures two of the possible six lipid resonance peaks at best. The volume-localized ^{1}H MRS of the liver acquired with water suppression represents a powerful tool for the measurement of different triglyceride levels present in fatty liver disease. The advantage of the volume-localized spectroscopy techniques is that they provide a spectrum of the resonances within the volume element or voxel, allowing quantification of different lipid moieties.

11.1.6.1 Evaluation of Diffuse Liver Disease

Although diffuse liver fat, or steatosis (accumulation of fat within hepato-cytes), was previously considered to be a relatively benign and self-limiting entity, it is now recognized as a characteristic feature of nonalcoholic fatty liver disease (NAFLD), which often leads to necroinflammatory changes (known as nonalcoholic steatohepatitis) and even cirrhosis. In the United States, 40 million adults are thought to have NAFLD, which is strongly associated with obesity (Adams et al., 2005; Angulo, 2002; Festi et al., 2004; Ludwig et al., 1980). Furthermore, it is thought that progression of steatosis to nonalcoholic steatohepatitis and beyond is the primary cause of crypto-genic cirrhosis, which is the third most common indication for liver trans-plantation (Angulo, 2002). Liver steatosis is also a common feature of other types of diffuse liver disease, such as chronic viral hepatitis, especially that caused by hepatitis C infection, drug hepatotoxic effects (from antiretrovi-ral therapy or chemotherapeutic agents such as tamoxifen and methotrex-ate), and excessive alcohol consumption (Burke and Lucey, 2004; Limanond et al., 2004; Miller et al., 2001; Neuschwander-Tetri and Caldwell, 2003; Ristig et al., 2005).

^{1}H MRS has been shown to be effective for quantifying liver fat (Anderwald et al., 2002; Noworolski et al., 2009; Springer et al., 2010; Thomas et al., 2005). T_2-weighted MRI and MRS of a 22-year-old healthy control patient and a 51-year-old NAFLD patient are shown in Figure 11.4. In the liver, lipid peaks are identified at 0.9, 1.3, 2.0, 2.2, and 5.3 ppm. These peaks represent CH_3 (0.9 ppm), (CH_2) (1.3 ppm), CH_2 (2.0 and 2.2 ppm), and CH (5.3 ppm) lipids. The dominant lipid peaks are caused by the resonance of methyl ($-CH_3$) protons and methylene ($-CH_2$) in the triglyc-eride molecule, at 0.9–1.1 ppm and 1.3–1.6 ppm, respectively, along the frequency domain (Georgoff et al., 2012). Measurement of total hepatic triglycerides involves integrating all five peaks (at 0.9, 1.3, 2.0, 2.2, and 5.3 ppm). Measurement of total lipids commonly includes only the peaks attributed to CH_2 and CH_3, because the peak attributed to CH (5.3 ppm) is close to that of water. The ratio of the total lipid peak area to the water peak area is calculated for each patient by using the total lipids measure-ment from the suppressed water sequence and the unsuppressed water measurement from the unsuppressed water sequence. The total lipids are calculated as a percentage relative to water by summing the individual resonance peaks to obtain the total hepatic triglyceride peak area, then dividing this value by the sum of the total lipid and water peak areas. Recent studies have reported that altered liver metabolite profiles may be detected in patients with cirrhosis secondary to chronic liver disease of various causes with MRS at a field strength of 1.5 T (Cho et al., 2001; Lim et al., 2003). Although quantification of liver fat with ^{1}H MRS is becom-ing increasingly accepted, the metabolite changes indicative of inflamma-tion or fibrosis have not been clearly established. Cho et al. (2001) reported

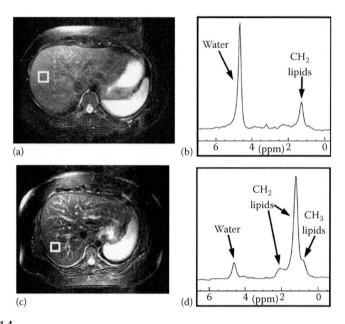

FIGURE 11.4

(a) and (b) are from a 22-year-old healthy female; (c) and (d) are from a 51-year-old female with NAFLD and Grade 2 steatosis. (a) and (c) are T_2-weighted axial images showing the locations of the MRS single voxels (white box); (b) and (d) are averaged, motion-corrected spectra showing water and lipid peaks. (From Bell, J.D., Bhakoo, K.K., *NMR Biomed.* 11, 354–359, 1998. With permission.)

findings of increased mean ratios of glutamine and glutamate complex to lipids and phosphomonoesters (PME) to lipids with ^{1}H MRS in patients with hepatitis B or hepatitis C viral infection ($n = 75$).

11.2 Phosphorous (^{31}P) MRS

11.2.1 Introduction

MRS is a nuclear magnetic resonance technique that has been frequently utilized for acquiring biochemical information *in vivo* from a defined volume of tissue. ^{31}P MRS also offers great potential in the diagnosis of various diseases. These include assessing metabolic disorders, both as a diagnostic aid and to monitor therapy, as well as the ability to conduct noninvasive research in human subjects with a variety of other diseases, such as Parkinson's, Alzheimer's, psychiatric illnesses, and diabetes (Agarwal et al., 2010; Gonzalez et al., 1996; Rango et al., 2007; Wu et al., 2012). The results of such research could lead not only to a better understanding of the disease but also to better and more accurate diagnosis and effective treatment.

11.2.2 Quantitation of Muscle Energy Metabolites by ^{31}P MRS

MRS has been widely used to study skeletal muscle metabolism. It provides a unique outlook on the *in situ* dynamics of tissue biochemistry by measuring the concentrations and/or turnover rates of metabolites (Ackerman et al., 1980; Chance et al., 1982; Hoult et al., 1974; Ross et al., 1981). This is primarily due to the noninvasive nature of the technique, which allows serial time course measurements while net changes in metabolite levels occur. ^{31}P spectroscopy is primarily a research tool, because there are several challenges that need to be addressed before widespread acceptance in clinical applications. ^{31}P MRS has inherently lower sensitivity than ^{1}H MRS because of the physical properties of the phosphorus nucleus. For studies in muscle, this is not a major concern because surface coils provide good SNR and adequate localization. In other organs such as brain, liver, and kidney, other localization techniques are required and currently come at the cost of lower SNRs.

^{31}P MRS has been extensively used to investigate noninvasively the energy metabolism of the human muscle since its first application using surface coils (Ackerman et al., 1980; Hoult et al., 1974). For over a decade, ^{31}P MRS was the workhorse for muscle physiologists who used this method particularly for studies of high-energy phosphates and intracellular pH (Binzoni et al., 1992; Chance et al., 1982; de Kerviler et al., 1991; Gruetter et al., 1990; Kemp et al., 1992; McCully et al., 1989; Meyer, 1988; Minotti et al., 1990; Quistorff et al., 1993; Ross et al., 1981; Roth and Weiner, 1991; Rothman et al., 1992; Yoshizaki et al., 1990).

The muscle examined is generally positioned in the center of the magnet, and a surface coil is placed next to the area of interest and used as both RF transmitter and receiver. Coil size and position as well as pulsing conditions determine the location and size of the volume of muscle investigated. Muscle is in many ways an ideal tissue for MRS studies because it can be stressed in the magnet with exercise while the biochemical response is monitored. ^{31}P MRS records signals from high-energy phosphate compounds which are central to energy metabolism *in vivo*. Phosphorus MR spectra from muscle contain seven resonance peaks: three arise from the phosphate groups of adenosine triphosphate (ATP); one from PCr and one from inorganic phosphates (Pi); and two additional smaller resonances can sometimes be observed from PME and phosphodiesters (PDE). The area under each resonance (i.e., the signal intensity) is proportional to the concentration of each metabolite, and concentrations of at least 1 mM provide sufficient SNR and give rise to peaks that are clearly visible. The concentration of the metabolically active adenosine diphosphate (ADP), which is important in the regulation of rates of mitochondrial ATP synthesis, does not produce a signal visible in the spectra, but can be calculated indirectly using the creatine kinase (CK) equilibrium equation. One of the most important features of ^{31}P MRS is its ability to monitor time-dependent changes of metabolites noninvasively. ^{31}P MRS is a sensitive and specific noninvasive method for the assessment of skeletal muscle mitochondrial ATP production. Observations at rest are not specific for mitochondrial disorders. During exercise, patients with mitochondrial myopathies display rapid PCr depletion. During recovery, ^{31}P MRS measurements are the most sensitive and the most specific indices used to assess skeletal muscle mitochondrial ATP production (Harkema and Meyer, 1997; Jeneson et al., 1996; Kemp, 2006; Smith et al., 2004). In normal exercising muscle, the hydrolysis of ATP releases the energy necessary for the sliding between myosin and actin proteins, the basis of muscle contraction. As a matter of consequence, one can observe in the MR spectra a PCr decrease stoechiometrically linked to Pi accumulation, while the ATP signal remains unchanged at least when exercise intensity is light or moderate.

Several methods have been used to quantify phosphorus-containing metabolites by ^{31}P MRS (Buchli and Boesiger, 1993; Doyle et al., 1997). One popular approach has been to estimate [PCr], say, from the ratio [PCr]/[ATP] (making appropriate correction for saturation effects), in effect using [ATP] as an internal standard (Taylor et al., 1983). The justification for this is the relatively small variation in [ATP] between individuals and the lack of difference among different fiber types.

^{31}P MRS has opened a window on bioenergetics during skeletal muscle exercise and recovery in a noninvasive manner and with a time resolution typically on the order of seconds. This has made a major contribution to the understanding of mammalian cell energy metabolism, its control, and the way in which it can be affected in disease. [31]P MR spectra of human muscle acquired serially at rest, during exercise and subsequent recovery. At the start of the energy challenge, hydrolyzed ATP is resynthesized from PCr breakdown and, depending on conditions, from anaerobic glycogenolysis. Thus, PCr levels decrease and Pi levels increase, while ATP levels remain constant. PCr hydrolysis consumes protons and generally results in a small rise in cellular pH (alkalinization) at the start of exercise. During aerobic exercise, most of the energy is subsequently provided by oxidative metabolism. Anaerobic glycogenolysis produces protons via lactate production and typically results in cellular acidification. After the energy challenge, the PCr buffer is restored and Pi levels normalize. The ATP used for resynthesis of PCr is mostly derived from oxidative phosphorylation. Several approaches to quantitative analysis and interpretation of ^{31}P MRS measurements of energy balance in muscle during and after several types of exercise have been proposed (Kemp and Radda, 1994). It is thus possible to estimate (a) the rates of glycogenolytic and aerobic ATP synthesis, (b) the oxidative capacity, (c) the proton efflux, and (d) the buffer capacity.

Although ^{31}P MRS is becoming more widely available from commercial vendors, it still remains largely as a research technique. Its utility lies in its ability to probe energy metabolism, as clearly demonstrated by Ko et al. (2008). Thus, its clinical utilization lies in examining other muscular diseases, metabolic diseases, neurodegenerative diseases, and, in some cases, cancer treatment (Garcia-Carrasco et al., 2007). Although ^{31}P MRS has shown promise as a diagnostic tool (Garcia-Carrasco et al., 2007), its efficacy in prognosis remains to be determined.

Ko et al. (2008) examined 14 healthy volunteers and 16 patients with myasthenia gravis (MG) by using ^{31}P MRS of the calf muscle. They used a surface RF coil and collected data before, during, and after a simple toe-flex exercise during a 10-min period. The following peak areas were measured and converted to concentration estimates: P_i; PCr; and γ, α, and β phosphate groups of ATP. The intracellular pH, PCr recovery, and maximum oxidative capacity were calculated from these concentrations. A follow-up ^{31}P MR spectroscopic examination was performed 6–9 months following thymectomy. There were no significant differences between control subjects and patients for resting muscle. There were no significant differences with exercise between control subjects and patients with mild forms of MG, nor did the treatment affect any measures in the mild-MG patients. However, for patients with more severe MG, several measures of metabolism were altered compared with those in control subjects and these measures included muscle pH, Pi, and oxidative capacity. These differences normalized following treatment in this group.

Various methods have been proposed for localizing the source of the ^{31}P MR signals to specific regions of the anatomy. The most promising MRS localization technique for acquiring spectroscopic data from small muscles of the foot is CSI. In CSI, the field of view in the imaging plane (slice) of the anatomy of interest is divided into a grid of voxels. In one study, ^{31}P CSI has been used to assess the diabetic foot (Suzuki et al., 2000). The study concluded that impaired ^{31}P energy metabolism was closely associated with motor nerve dysfunction and a high frequency of foot ulcers.

11.2.3 ^{31}P Liver MRS

Over the last two decades, the efficacy of ^{31}P MRS for the diagnosis of liver disease has been investigated as a noninvasive alternative to liver biopsy, which is still the gold standard (Lim et al., 2003). The *in vivo* spectra of the human liver reflect metabolic and biochemical alterations in disease. Several publications on liver function have addressed spectral changes related to the underlying liver disease. However, in order to be helpful as a diagnostic tool, MR spectra should provide both sensitive and specific information for different types of liver disease, such as hepatitis, steatosis, fibrosis, or cirrhosis.

The ^{31}P MRS spectrum provides information on phosphorylated compounds of hepatic metabolism: alpha-, beta-, and gamma peaks of nucleotide triphosphates (NTP), Pi, PME, and PDE. The NTP, PME, and PDE peaks are multicomponent, and individual resonances of their components cannot be distinguished by the majority of techniques that are currently used. PME contains information about components from glucose metabolism (gluconeogenesis and glycolysis) and cell membrane precursors such as phosphoethanolamine (PE) and PCho (Bell et al., 1993). The PDE peak contains information on cell membrane breakdown products, such as glycerophosphorylethanolamine (GPE) and glycerophosphorylcholine (GPC), and endoplasmic reticulum (ER) (Bailes et al., 1987). The diagnostic value of ^{31}P MRS for various types of diffuse liver disease has been investigated in several studies. Overall, diffuse liver disease was associated with increasing levels of PME and decreasing levels of PDE. These changes have been attributed to hepatocyte damage, increased phospholipid turnover in hepatocyte membranes, and/or altered glucose metabolism. In general, the magnitude of MRS changes increased significantly with increased disease severity and increased functional impairment. In one of the earliest reports, liver metabolite concentrations were studied in 24 patients with various types of diffuse liver disease as compared to healthy control subjects (Oberhaensli et al., 1990). The authors reported high PME and low PDE levels in patients with acute viral hepatitis, high PME levels in patients with alcoholic hepatitis, and decreased Pi and Pi/ATP ratios in primary biliary cirrhosis and in some patients with hepatitis (Oberhaensli et al., 1990). However, they noted that these changes were not present in all patients. Meyerhoff et al. used ^{31}P MRS to demonstrate that the livers of patients with alcoholic hepatitis and cirrhosis could be differentiated from normal liver and from each other on the basis of hepatic intracellular pH and absolute molar concentration of ATP (1989).

Cox et al. studied 49 patients with liver disease of varying etiology, including 25 patients with diffuse liver disease such as cirrhosis and nonhepatic malignancies (1992b). A nonspecific elevation in PME/PDE was observed in the ^{31}P MR spectra of 10 (40%) out of these 25 patients with mixed diffuse liver disease. Even though the spectral pattern did not distinguish between diseases of varying etiologies, there was a linear correlation between increasing PME/PDE and a reduction in plasma albumin concentrations ($P = 0.03$).

Taylor-Robinson et al. studied 14 patients with compensated cirrhosis (Pugh's score ≤ 7) and 17 with decompensated cirrhosis (Pugh's score ≥ 8) of various etiology and compared the findings with the data from healthy subjects (1997). Worsening liver function was associated with increased PME/NTP and decreased PDE/NTP ratios. In freeze-clamped tissue, elevated PE and PC and reduced GPE and GPC mirrored these *in vivo* changes, but no distinction was noted between compensated and decompensated cirrhosis. In contrast, electron microscopy showed that functional decompensation was associated with reduced ER in parenchymal liver disease, but elevated ER in biliary cirrhosis. In a more recent study in only 14 cirrhotic subjects, reduced ATP and elevated PME/PDE levels were detected in patients with decompensated cirrhosis only (Corbin et al., 2004).

In vivo ^{31}P MRS was performed on 28 healthy adult individuals and 32 patients with hepatic malignancies of varying histology, using CSI techniques (Bell and Bhakoo, 1998; Cox et al., 1992a). *In vivo* and *in vitro* spectra of the PME–PDE regions of normal and tumorous hepatic specimens are shown in Figures 11.5 and 11.6, respectively. Tumor tissue shows a significant change in the concentration of phospholipid metabolites (PCho, PE, GPC, and GPE) compared with normal tissue. Tumor tissue contains lower concentrations than the control liver of GPC (0.59 ± 0.15 vs. 2.46 ± 0.37; $p < .001$) and GPE (0.57 ± 0.17 vs. 2.25 ± 0.46; $p < .001$), and elevated levels of PCho

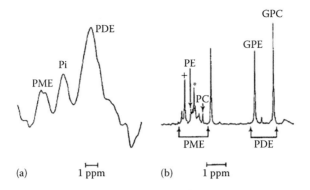

FIGURE 11.5
Hepatic ^{31}P spectra from a region of normal liver: (a) PME–PDE region of the *in vivo* spectrum (4D CSI); (b) PME–PDE region of the *in vitro* spectrum from an aqueous extract of histologically normal liver. (From Noworolski, S.M., et al., *Magn Reson Imaging*. 27, 570–576, 2009. With permission.)

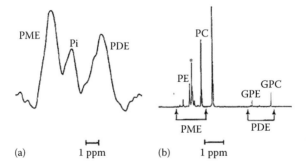

FIGURE 11.6
Hepatic ^{31}P spectra from a subject with hepatocellular carcinoma: (a) PME–PDE region of the *in vivo* spectrum with the tumor (4D CSI); (b) PME–PDE region of the *in vitro* spectrum from an aqueous extract of tumor obtained at the time of laparotomy. (From Noworolski, S.M., et al., *Magn Reson Imaging*. 27, 570–576, 2009. With permission.)

$(1.36 \pm 0.50$ vs. $0.17 \pm 0.11; p < .001)$ and PE $(2.47 \pm 0.84$ vs. $0.16 \pm 0.10; p < .001)$. No significant differences between primary and secondary hepatic tumors were observed by *in vitro* MRS.

11.2.4 Technical Challenges of ^{31}P MRS

A serious limitation of ^{31}P MRS is poor spatial and temporal resolution. Each spectrum of a CSI data set must be acquired separately in a serial fashion. Because the concentration of the ^{31}P metabolites is very small (at the millimolar level rather than at the molar level as in conventional MRI, which is based on the magnetization of the atomic nuclei of the highly abundant water molecules), each spectrum must be acquired many times and averaged to achieve an acceptable ratio of MR signal intensity to the random intensity fluctuations (noise). A method for acquiring ^{31}P metabolite information with improved spatial resolution that allows localization of metabolic activity to individual small foot muscles, with scan times that could be easily tolerated by patients, would make the ^{31}P MRS examination more feasible and acceptable by both clinicians and patients.

References

Aboagye, E.O., Bhujwalla, Z.M., 1999. Malignant transformation alters membrane choline phospholipid metabolism of human mammary epithelial cells. *Cancer Res*. 59, 80–84.

Ackerman, J.J., Grove, T.H., Wong, G.G., Gadian, D.G., Radda, G.K., 1980. Mapping of metabolites in whole animals by 31P NMR using surface coils. *Nature*. 283, 167–170.

Adams, L.A., Lymp, J.F., St Sauver, J., Sanderson, S.O., Lindor, K.D., Feldstein, A., Angulo, P., 2005. The natural history of nonalcoholic fatty liver disease: A population-based cohort study. *Gastroenterology*. 129, 113–121.

Agarwal, N., Port, J.D., Bazzocchi, M., Renshaw, P.F., 2010. Update on the use of MR for assessment and diagnosis of psychiatric diseases. *Radiology*. 255, 23–41.

Anderwald, C., Bernroider, E., Krssak, M., Stingl, H., Brehm, A., Bischof, M.G., Nowotny, P., Roden, M., Waldhausl, W., 2002. Effects of insulin treatment in type 2 diabetic patients on intracellular lipid content in liver and skeletal muscle. *Diabetes*. 51, 3025–3032.

Angulo, P., 2002. Nonalcoholic fatty liver disease. *N Engl J Med*. 346, 1221–1231.

Avison, M.J., Rothman, D.L., Nadel, E., Shulman, R.G., 1988. Detection of human muscle glycogen by natural abundance 13C NMR. *Proc Natl Acad Sci USA*. 85, 1634–1636.

Baik, H.-M., Su, M.-Y., Yu, H., Mehta, R., Nalcioglu, O., 2006. Quantification of choline-containing compounds in malignant breast tumors by 1H MR spectroscopy using water as an internal reference at 1.5 T. *MAGMA*. 19, 96–104.

Bailes, D.R., Bryant, D.J., Bydder, G.M., Case, H.A., Collins, A.G., Cox, I.J., Evans, P.R., et al., 1987. Localized phosphorus-31 NMR spectroscopy of normal and pathological human organs in vivo using phase-encoding techniques, *J Mag Reson*. 74, 158–170.

Barany, M., Venkatasubramanian, P.N., 1989. Volume-selective water-suppressed proton spectra of human brain and muscle in vivo. *NMR Biomed*. 2, 7–11.

Baron, E.M., Young, W.F., 2007. Cervical spondylotic myelopathy: A brief review of its pathophysiology, clinical course, and diagnosis. *Neurosurgery*. 60, S35–S41.

Bartella, L., Morris, E.A., Dershaw, D.D., Liberman, L., Thakur, S.B., Moskowitz, C., Guido, J., Huang, W., 2006. Proton MR spectroscopy with choline peak as malignancy marker improves positive predictive value for breast cancer diagnosis: Preliminary study. *Radiology*. 239, 686–692.

Bell, J.D., Bhakoo, K.K., 1998. Metabolic changes underlying 31P MR spectral alterations in human hepatic tumours. *NMR Biomed*. 11, 354–359.

Bell, J.D., Cox, I.J., Sargentoni, J., Peden, C.J., Menon, D.K., Foster, C.S., Watanapa, P., Iles, R.A., Urenjak, J., 1993. A 31P and 1H-NMR investigation in vitro of normal and abnormal human liver. *Biochim Biophys Acta*. 1225, 71–77.

Binzoni, T., Ferretti, G., Schenker, K., Cerretelli, P., 1992. Phosphocreatine hydrolysis by 31P-NMR at the onset of constant-load exercise in humans. *J Appl Physiol.* 73, 1644–1649.

Blamire, A.M., Cader, S., Lee, M., Palace, J., Matthews, P.M., 2007. Axonal damage in the spinal cord of multiple sclerosis patients detected by magnetic resonance spectroscopy. *Magn Reson Med.* 58, 880–885.

Blei, M.L., Conley, K.E., Kushmerick, M.J., 1993. Separate measures of ATP utilization and recovery in human skeletal muscle. *J Physiol.* 465, 203–222.

Boesch, C., Slotboom, J., Hoppeler, H., Kreis, R., 1997. In vivo determination of intra-myocellular lipids in human muscle by means of localized 1H-MR-spectroscopy. *Magn Reson Med.* 37, 484–493.

Bolan, P.J., Meisamy, S., Baker, E.H., Lin, J., Emory, T., Nelson, M., Everson, L.I., Yee, D., Garwood, M., 2003. In vivo quantification of choline compounds in the breast with 1H MR spectroscopy. *Magn Reson Med.* 50, 1134–1143.

Brown, T.R., Kincaid, B.M., Ugurbil, K., 1982. NMR chemical shift imaging in three dimensions. *Proc Natl Acad Sci USA.* 79, 3523–3526.

Bruhn, H., Frahm, J., Gyngell, M.L., Merboldt, K.D., Hanicke, W., Sauter, R., 1991. Localized proton NMR spectroscopy using stimulated echoes: Applications to human skeletal muscle in vivo. *Magn Reson Med.* 17, 82–94.

Buchli, R., Boesiger, P., 1993. Comparison of methods for the determination of absolute metabolite concentrations in human muscles by 31P MRS. *Magn Reson Med.* 30, 552–558.

Burke, A., Lucey, M.R., 2004. Non-alcoholic fatty liver disease, non-alcoholic steato-hepatitis and orthotopic liver transplantation. *Am J Transplant.* 4, 686–693.

Cecil, K.M., Schnall, M.D., Siegelman, E.S., Lenkinski, R.E., 2001. The evaluation of human breast lesions with magnetic resonance imaging and proton magnetic resonance spectroscopy. *Breast Cancer Res Treat.* 68, 45–54.

Chance, B., 1992. Noninvasive approaches to oxygen delivery and cell bioenergetics in functioning muscle. *Clin J Sport Med.* 2, 132–138.

Chance, B., Eleff, S., Bank, W., Leigh, J.S., Warnell, R., 1982. 31P NMR studies of control of mitochondrial function in phosphofructokinase-deficient human skeletal muscle. *Proc Natl Acad Sci USA.* 79, 7714–7718.

Cho, S.G., Kim, M.Y., Kim, H.J., Kim, Y.S., Choi, W., Shin, S.H., Hong, K.C., Kim, Y.B., Lee, J.H., Suh, C.H., 2001. Chronic hepatitis: In vivo proton MR spectroscopic evaluation of the liver and correlation with histopathologic findings. *Radiology.* 221, 740–746.

Ciccarelli, O., Wheeler-Kingshott, C.A., McLean, M.A., Cercignani, M., Wimpey, K., Miller, D.H., Thompson, A.J., 2007. Spinal cord spectroscopy and diffusion-based tractography to assess acute disability in multiple sclerosis. *Brain.* 130, 2220–2231.

Cooke, F.J., Blamire, A.M., Manners, D.N., Styles, P., Rajagopalan, B., 2004. Quantitative proton magnetic resonance spectroscopy of the cervical spinal cord. *Magn Reson Med.* 51, 1122–1128.

Corbin, I.R., Ryner, L.N., Singh, H., Minuk, G.Y., 2004. Quantitative hepatic phosphorus-31 magnetic resonance spectroscopy in compensated and decompensated cirrhosis. *Am J Phys Gastrointest Liver Physiol.* 287, G379–G384.

Cox, I.J., Bell, J.D., Peden, C.J., Iles, R.A., Foster, C.S., Watanapa, P., Williamson, R.C., 1992a. In vivo and in vitro 31P magnetic resonance spectroscopy of focal hepatic malignancies. *NMR Biomed.* 5, 114–120.

Cox, I.J., Menon, D.K., Sargentoni, J., Bryant, D.J., Collins, A.G., Coutts, G.A., Iles, R.A., et al., 1992b. Phosphorus-31 magnetic resonance spectroscopy of the human liver using chemical shift imaging techniques. *J Hepatol*. 14, 265–275.

de Kerviler, E., Leroy-Willig, A., Jehenson, P., Duboc, D., Eymard, B., Syrota, A., 1991. Exercise-induced muscle modifications: Study of healthy subjects and patients with metabolic myopathies with MR imaging and P-31 spectroscopy. *Radiology*. 181, 259–264.

Dixon, W.T., 1984. Simple proton spectroscopic imaging. *Radiology*. 153, 189–194.

Dobbins, R.L., Szczepaniak, L.S., Bentley, B., Esser, V., Myhill, J., McGarry, J.D., 2001. Prolonged inhibition of muscle carnitine palmitoyltransferase-1 promotes intramyocellular lipid accumulation and insulin resistance in rats. *Diabetes*. 50, 123–130.

Doyle, V.L., Payne, G.S., Collins, D.J., Verrill, M.W., Leach, M.O., 1997. Quantification of phosphorus metabolites in human calf muscle and soft-tissue tumours from localized MR spectra acquired using surface coils. *Phys Med Biol*. 42, 691–706.

Dydak, U., Kollias, S., Schar, M., Meier, D., Boesiger, P., 2005. MR spectroscopy in different regions of the spinal cord and in spinal cord tumors. *Proceedings of the 13th Annual Meeting of the International Society for Magnetic Resonance in Medicine*, May 7–13. Miami Beach, FL, USA, p. 813.

Edden, R.A.E., Bonekamp, D., Smith, M.A., Dubey, P., Barker, P.B., 2007. Proton MR spectroscopic imaging of the medulla and cervical spinal cord. *J Magn Reson Imaging*. 26, 1101–1105.

Festi, D., Colecchia, A., Sacco, T., Bondi, M., Roda, E., Marchesini, G., 2004. Hepatic steatosis in obese patients: Clinical aspects and prognostic significance. *Obes Rev*. 5, 27–42.

Garcia-Carrasco, M., Escarcega, R.O., Fuentes-Alexandro, S., Riebeling, C., Cervera, R., 2007. Therapeutic options in autoimmune myasthenia gravis. *Autoimmun Rev*. 6, 373–378.

Georgoff, P., Thomasson, D., Louie, A., Fleischman, E., Dutcher, L., Mani, H., Kottilil, S., et al., 2012. Hydrogen-1 MR spectroscopy for measurement and diagnosis of hepatic steatosis. *Am J Roentgenol*. 199, 2–7.

Gleason, D.F., 1966. Classification of prostatic carcinomas. *Cancer Chemother Rep*. 50, 125–128.

Gomez-Anson, B., MacManus, D.G., Parker, G.J., Davie, C.A., Barker, G.J., Moseley, I.F., McDonald, W.I., Miller, D.H., 2000. In vivo 1H-magnetic resonance spectroscopy of the spinal cord in humans. *Neuroradiology*. 42, 515–517.

Gonzalez, R.G., Guimaraes, A.R., Moore, G.J., Crawley, A., Cupples, L.A., Growdon, J.H., 1996. Quantitative in vivo 31P magnetic resonance spectroscopy of Alzheimer disease. *Alzheimer Dis Assoc Disord*. 10, 46–52.

Gruetter, R., Kaelin, P., Boesch, C., Martin, E., Werner, B., 1990. Non-invasive 31P magnetic resonance spectroscopy revealed McArdle disease in an asymptomatic child. *Eur J Pediatr*. 149, 483–486.

Haddadin, I.S., McIntosh, A., Meisamy, S., Corum, C., Styczynski Snyder, A.L., Powell, N.J., Nelson, M.T., Yee, D., Garwood, M., Bolan, P.J., 2009. Metabolite quantification and high-field MRS in breast cancer. *NMR Biomed*. 22, 65–76.

Harkema, S.J., Meyer, R.A., 1997. Effect of acidosis on control of respiration in skeletal muscle. *Am J Physiol*. 272, 491–500.

Havel, R.J., Carlson, L.A., Ekelund, L.G., Holmgren, A., 1964. Turnover rate and oxidation of different free fatty acids in man during exercise. *J Appl Physiol*. 19, 613–618.

Heiken, J.P., Lee, J.K., Dixon, W.T., 1985. Fatty infiltration of the liver: Evaluation by proton spectroscopic imaging. *Radiology*. 157, 707–710.

Henning, A., Schar, M., Kollias, S.S., Boesiger, P., Dydak, U., 2008. Quantitative magnetic resonance spectroscopy in the entire human cervical spinal cord and beyond at 3T. *Magn Reson Med*. 59, 1250–1258.

Hoeks, C.M.A., Barentsz, J.O., Hambrock, T., Yakar, D., Somford, D.M., Heijmink, S.W.T.P.J., Scheenen, T.W.J., et al., 2011. Prostate cancer: Multiparametric MR imaging for detection, localization, and staging. *Radiology*. 261, 46–66.

Holly, L.T., Freitas, B., McArthur, D.L., Salamon, N., 2009. Proton magnetic resonance spectroscopy to evaluate spinal cord axonal injury in cervical spondylotic myelopathy. *J Neurosurg Spine*. 10, 194–200.

Hoult, D.I., Busby, S.J., Gadian, D.G., Radda, G.K., Richards, R.E., Seeley, P.J., 1974. Observation of tissue metabolites using 31P nuclear magnetic resonance. *Nature*. 252, 285–287.

Huang, W., Fisher, P.R., Dulaimy, K., Tudorica, L.A., O'Hea, B., Button, T.M., 2004. Detection of breast malignancy: Diagnostic MR protocol for improved specificity. *Radiology*. 232, 585–591.

Hwang, J.H., Pan, J.W., Heydari, S., Hetherington, H.P., Stein, D.T., 2001. Regional differences in intramyocellular lipids in humans observed by in vivo 1H-MR spectroscopic imaging. *J Appl Physiol*. 90, 1267–1274.

Irwin, Z.N., Hilibrand, A., Gustavel, M., McLain, R., Shaffer, W., Myers, M., Glaser, J., Hart, R.A., 2005. Variation in surgical decision making for degenerative spinal disorders. Part II: Cervical spine. *Spine*. 30, 2214–2219.

Jacob, S., Machann, J., Rett, K., Brechtel, K., Volk, A., Renn, W., Maerker, E., et al., 1999. Association of increased intramyocellular lipid content with insulin resistance in lean nondiabetic offspring of type 2 diabetic subjects. *Diabetes*. 48, 1113–1119.

Jacobs, M.A., Barker, P.B., Bottomley, P.A., Bhujwalla, Z., Bluemke, D.A., 2004. Proton magnetic resonance spectroscopic imaging of human breast cancer: A preliminary study. *J Magn Reson Imaging*. 19, 68–75.

Jayasundar, R., Goyal, M., Sharma, R., Raghunathan, P., 1996. Proton MRS in Pott's spine—A case report. *Magn Reson Imaging*. 14, 691–695.

Jemal, A., Siegel, R., Xu, J., Ward, E., 2010. Cancer statistics, 2010. *CA: Cancer J Clin*. 60, 277–300.

Jeneson, J.A., Wiseman, R.W., Westerhoff, H.V., Kushmerick, M.J., 1996. The signal transduction function for oxidative phosphorylation is at least second order in ADP. *J Biol Chem*. 271, 27995–27998.

Jucker, B.M., Schaeffer, T.R., Haimbach, R.E., Mayer, M.E., Ohlstein, D.H., Smith, S.A., Cobitz, A.R., Sarkar, S.K., 2003. Reduction of intramyocellular lipid following short-term rosiglitazone treatment in Zucker fatty rats: An in vivo nuclear magnetic resonance study. *Metabolism*. 52, 218–225.

Kayar, S.R., Hoppeler, H., Howald, H., Claassen, H., Oberholzer, F., 1986. Acute effects of endurance exercise on mitochondrial distribution and skeletal muscle morphology. *Eur J Appl Physiol Occup Physiol*. 54, 578–84.

Kemp, G., 2006. Mitochondrial respiration in creatine-loaded muscle: Is there 31P-MRS evidence of direct effects of phosphocreatine and creatine in vivo? *J Appl Physiol*. 100, 1429–1430.

Kemp, G.J., Radda, G.K., 1994. Quantitative interpretation of bioenergetic data from 31P and 1H magnetic resonance spectroscopic studies of skeletal muscle: An analytical review. *Magn Reson Q*. 10, 43–63.

Kemp, G.J., Taylor, D.J., Radda, G.K., Rajagopalan, B., 1992. Bio-energetic changes in human gastrocnemius muscle 1-2 days after strenuous exercise. *Acta Physiol Scand*. 146, 11–14.

Kendi, A.T.K., Tan, F.U., Kendi, M., Yilmaz, S., Huvaj, S., Tellioglu, S., 2004. MR spectroscopy of cervical spinal cord in patients with multiple sclerosis. *Neuroradiology*. 46, 764–769.

Kim, J.K., Jang, Y.J., Cho, G., 2009. Multidisciplinary functional MR imaging for prostate cancer. *Korean J Radiol*. 10, 535–551.

Kim, J.-K., Park, S.-H., Lee, H.M., Lee, Y.-H., Sung, N.-K., Chung, D.-S., Kim, O.-D., 2003. In vivo 1H-MRS evaluation of malignant and benign breast diseases. *Breast*. 12, 179–182.

Kim, Y.-G., Choi, G.-H., Kim, D.-H., Kim, Y.-D., Kang, Y.-K., Kim, J.-K., 2004. In vivo proton magnetic resonance spectroscopy of human spinal mass lesions. *J Spinal Disord Tech*. 17, 405–411.

Ko, S.-F., Huang, C.-C., Hsieh, M.-J., Ng, S.-H., Lee, C.-C., Lee, C.-C., Lin, T.-K., Chen, M.-C., Lee, L., 2008. 31P MR spectroscopic assessment of muscle in patients with myasthenia gravis before and after thymectomy: Initial experience. *Radiology*. 247, 162–169.

Krssak, M., Falk Petersen, K., Dresner, A., DiPietro, L., Vogel, S.M., Rothman, D.L., Roden, M., Shulman, G.I., 1999. Intramyocellular lipid concentrations are correlated with insulin sensitivity in humans: A 1H NMR spectroscopy study. *Diabetologia*. 42, 113–116.

Kuhlmann, J., Neumann-Haefelin, C., Belz, U., Kalisch, J., Juretschke, H.-P., Stein, M., Kleinschmidt, E., Kramer, W., Herling, A.W., 2003. Intramyocellular lipid and insulin resistance: A longitudinal in vivo 1H-spectroscopic study in Zucker diabetic fatty rats. *Diabetes*. 52, 138–144.

Kumar, M., Jagannathan, N.R., Seenu, V., Dwivedi, S.N., Julka, P.K., Rath, G.K., 2006. Monitoring the therapeutic response of locally advanced breast cancer patients: Sequential in vivo proton MR spectroscopy study. *J Magn Reson Imaging*. 24, 325–332.

Kumaresan, S., Yoganandan, N., Pintar, F.A., Maiman, D.J., Goel, V.K., 2001. Contribution of disc degeneration to osteophyte formation in the cervical spine: A biomechanical investigation. *J Orthop Res*. 19, 977–984.

Kurhanewicz, J., Vigneron, D.B., Hricak, H., Narayan, P., Carroll, P., Nelson, S.J., 1996. Three-dimensional H-1 MR spectroscopic imaging of the in situ human prostate with high (0.24-0.7-cm3) spatial resolution. *Radiology*. 198, 795–805.

Kvistad, K.A., Bakken, I.J., Gribbestad, I.S., Ehrnholm, B., Lundgren, S., Fjosne, H.E., Haraldseth, O., 1999. Characterization of neoplastic and normal human breast tissues with in vivo (1)H MR spectroscopy. *J Magn Reson Imaging*. 10, 159–164.

Larson-Meyer, D.E., Newcomer, B.R., Hunter, G.R., 2002. Influence of endurance running and recovery diet on intramyocellular lipid content in women: A 1H NMR study. *Am J Physiol Endocrinol Metab*. 282, E95–E106.

Layer, G., Traber, F., Block, W., Braucker, G., Kretzer, S., Flacke, S., Schild, H., 1998. [1H MR spectroscopy of the lumbar spine in diffuse osteopenia due to plasmacytoma or osteoporosis]. *Rofo*. 169, 596–600.

Lim, A.K.P., Patel, N., Hamilton, G., Hajnal, J.V., Goldin, R.D., Taylor-Robinson, S.D., 2003. The relationship of in vivo 31P MR spectroscopy to histology in chronic hepatitis C. *Hepatology*. 37, 788–794.

Limanond, P., Raman, S.S., Lassman, C., Sayre, J., Ghobrial, R.M., Busuttil, R.W., Saab, S., Lu, D.S.K., 2004. Macrovesicular hepatic steatosis in living related liver donors: Correlation between CT and histologic findings. *Radiology*. 230, 276–280.

Longo, R., Pollesello, P., Ricci, C., Masutti, F., Kvam, B.J., Bercich, L., Croce, L.S., Grigolato, P., Paoletti, S., de Bernard, B., 1995. Proton MR spectroscopy in quantitative in vivo determination of fat content in human liver steatosis. *J Magn Reson Imaging*. 5, 281–285.

Longo, R., Ricci, C., Masutti, F., Vidimari, R., Croce, L.S., Bercich, L., Tiribelli, C., Dalla Palma, L., 1993. Fatty infiltration of the liver. Quantification by 1H localized magnetic resonance spectroscopy and comparison with computed tomography. *Invest Radiol*. 28, 297–302.

Ludwig, J., Viggiano, T.R., McGill, D.B., Oh, B.J., 1980. Nonalcoholic steatohepatitis: Mayo Clinic experiences with a hitherto unnamed disease. *Mayo Clin Proc*. 55, 434–438.

Manenti, G., Squillaci, E., Di Roma, M., Carlani, M., Mancino, S., Simonetti, G., 2006. In vivo measurement of the apparent diffusion coefficient in normal and malignant prostatic tissue using thin-slice echo-planar imaging. *Radiol Med*. 111, 1124–1133.

Marliani, A.F., Clementi, V., Albini-Riccioli, L., Agati, R., Leonardi, M., 2007. Quantitative proton magnetic resonance spectroscopy of the human cervical spinal cord at 3 Tesla. *Magn Reson Med*. 57, 160–163.

McCully, K.K., Boden, B.P., Tuchler, M., Fountain, M.R., Chance, B., 1989. Wrist flexor muscles of elite rowers measured with magnetic resonance spectroscopy. *J Appl Physiol*. 67, 926–932.

Meisamy, S., Bolan, P.J., Baker, E.H., Bliss, R.L., Gulbahce, E., Everson, L.I., Nelson, M.T., et al., 2004. Neoadjuvant chemotherapy of locally advanced breast cancer: Predicting response with in vivo (1)H MR spectroscopy—A pilot study at 4 T. *Radiology*. 233, 424–431.

Meisamy, S., Bolan, P.J., Baker, E.H., Pollema, M.G., Le, C.T., Kelcz, F., Lechner, M.C., et al., 2005. Adding in vivo quantitative 1H MR spectroscopy to improve diagnostic accuracy of breast MR imaging: Preliminary results of observer performance study at 4.0 T. *Radiology*. 236, 465–475.

Meyer, R.A., 1988. A linear model of muscle respiration explains monoexponential phosphocreatine changes. *Am J Physiol*. 254, 548–553.

Meyerhoff, D.J., Boska, M.D., Thomas, A.M., Weiner, M.W., 1989. Alcoholic liver disease: Quantitative image-guided P-31 MR spectroscopy. *Radiology*. 173, 393–400.

Miller, C.M., Gondolesi, G.E., Florman, S., Matsumoto, C., Munoz, L., Yoshizumi, T., Artis, T., et al., 2001. One hundred nine living donor liver transplants in adults and children: A single-center experience. *Ann Surg*. 234, 301–311.

Minotti, J.R., Johnson, E.C., Hudson, T.L., Zuroske, G., Fukushima, E., Murata, G., Wise, L.E., Chick, T.W., Icenogle, M.V., 1990. Training-induced skeletal muscle adaptations are independent of systemic adaptations. *J Appl Physiol*. 68, 289–294.

Moore, A.P., Blumhardt, L.D., 1997. A prospective survey of the causes of nontraumatic spastic paraparesis and tetraparesis in 585 patients. *Spinal Cord*. 35, 361–367.

Mountford, C.E., Tattersall, M.H., 1987. Proton magnetic resonance spectroscopy and tumour detection. *Cancer Surv*. 6, 285–314.

Nagarajan, R., Margolis, D., McClure, T., Raman, S., Thomas, M.A., 2011. Role of endorectal magnetic resonance spectroscopic imaging in two different Gleason scores in prostate cancer. *Med Princ Pract*. 20, 444–448.

Narayana, P.A., Hazle, J.D., Jackson, E.F., Fotedar, L.K., Kulkarni, M.V., 1988. In vivo 1H spectroscopic studies of human gastrocnemius muscle at 1.5 T. *Magn Reson Imag*. 6, 481–485.

Neuschwander-Tetri, B.A., Caldwell, S.H., 2003. Nonalcoholic steatohepatitis: Summary of an AASLD single topic conference. *Hepatology*. 37, 1202–1219.

Noworolski, S.M., Tien, P.C., Merriman, R., Vigneron, D.B., Qayyum, A., 2009. Respiratory motion-corrected proton magnetic resonance spectroscopy of the liver. *Magn Reson Imaging*. 27, 570–576.

Oberhaensli, R., Rajagopalan, B., Galloway, G.J., Taylor, D.J., Radda, G.K., 1990. Study of human liver disease with P-31 magnetic resonance spectroscopy. *Gut*. 31, 463–467.

Oberholzer, F., Claassen, H., Moesch, H., Howald, H., 1976. [Ultrastructural, biochemical and energy analysis of extreme duration performance (100km run)]. *Schweiz Z Sportmed*. 24, 71–98.

Ogg, R.J., Kingsley, R.B., Taylor, J.S., 1994. WET, a T1- and B1-insensitive water-suppression method for in vivo localized 1H NMR spectroscopy. *J Magn Reson B*. 104, 1–10.

Pan, J.W., Hamm, J.R., Hetherington, H.P., Rothman, D.L., Shulman, R.G., 1991. Correlation of lactate and pH in human skeletal muscle after exercise by 1H NMR. *Magn Reson Med*. 20, 57–65.

Parke, W.W., 1988. Correlative anatomy of cervical spondylotic myelopathy. *Spine*. 13, 831–837.

Podo, F., Sardanelli, F., Iorio, E., Canese, R., Carpinelli, G., Fausto, A., Canevari, S., 2007. Abnormal choline phospholipid metabolism in breast and ovary cancer: Molecular bases for noninvasive imaging approaches. *Curr Med Imaging Rev*. 3, 123–137.

Prando, A., Kurhanewicz, J., Borges, A.P., Oliveira, E.M., Figueiredo, E., 2005. Prostatic biopsy directed with endorectal MR spectroscopic imaging findings in patients with elevated prostate specific antigen levels and prior negative biopsy findings: Early experience. *Radiology*. 236, 903–910.

Quistorff, B., Johansen, L., Sahlin, K., 1993. Absence of phosphocreatine resynthesis in human calf muscle during ischaemic recovery. *Biochem J*. 291, 681–686.

Ramadan, S., Ratai, E.M., Wald, L.L., Mountford, C.E., 2010. In vivo 1D and 2D correlation MR spectroscopy of the soleus muscle at 7T. *J Magn Reson*. 204, 91–98.

Rango, M., Arighi, A., Biondetti, P., Barberis, B., Bonifati, C., Blandini, F., Pacchetti, C., Martignoni, E., Bresolin, N., Nappi, G., 2007. Magnetic resonance spectroscopy in Parkinson's disease and parkinsonian syndromes. *Funct Neurol*. 22, 75–79.

Reinsberg, S.A., Payne, G.S., Riches, S.F., Ashley, S., Brewster, J.M., Morgan, V.A., de Souza, N.M., 2007. Combined use of diffusion-weighted MRI and 1H MR spectroscopy to increase accuracy in prostate cancer detection. *Am J Roentgenol*. 188, 91–98.

Ristig, M., Drechsler, H., Powderly, W.G., 2005. Hepatic steatosis and HIV infection. *AIDS Patient Care STDS*. 19, 356–365.

Roebuck, J.R., Cecil, K.M., Schnall, M.D., Lenkinski, R.E., 1998. Human breast lesions: Characterization with proton MR spectroscopy. *Radiology*. 209, 269–275.

Ross, B.D., Radda, G.K., Gadian, D.G., Rocker, G., Esiri, M., Falconer-Smith, J., 1981. Examination of a case of suspected McArdle's syndrome by 31P nuclear magnetic resonance. *N Engl J Med*. 304, 1338–1342.

Roth, K., Weiner, M.W., 1991. Determination of cytosolic ADP and AMP concentrations and the free energy of ATP hydrolysis in human muscle and brain tissues with 31P NMR spectroscopy. *Magn Reson Med*. 22, 505–511.

Rothman, D.L., Shulman, R.G., Shulman, G.I., 1992. 31P nuclear magnetic resonance measurements of muscle glucose-6-phosphate. Evidence for reduced insulin-dependent muscle glucose transport or phosphorylation activity in non-insulin-dependent diabetes mellitus. *J Clin Invest*. 89, 1069–1075.

Sampath, P., Bendebba, M., Davis, J.D., Ducker, T.B., 2000. Outcome of patients treated for cervical myelopathy. A prospective, multicenter study with independent clinical review. *Spine*. 25, 670–676.

Sardanelli, F., Fausto, A., Podo, F., 2008. MR spectroscopy of the breast. *Radiol Med*. 113, 56–64.

Scheidler, J., Hricak, H., Vigneron, D.B., Yu, K.K., Sokolov, D.L., Huang, L.R., Zaloudek, C.J., Nelson, S.J., Carroll, P.R., Kurhanewicz, J., 1999. Prostate cancer: Localization with three-dimensional proton MR spectroscopic imaging—Clinicopathologic study. *Radiology*. 213, 473–480.

Schick, F., Eismann, B., Jung, W.I., Bongers, H., Bunse, M., Lutz, O., 1993. Comparison of localized proton NMR signals of skeletal muscle and fat tissue in vivo: Two lipid compartments in muscle tissue. *Magn Reson Med*. 29, 158–167.

Schmitz, J.P., Jeneson, J.A., van Oorschot, J.W., Prompers, J.J., Nicolay, K., Hilbers, P.A., van Riel, N.A., 2012. Prediction of muscle energy states at low metabolic rates requires feedback control of mitochondrial respiratory chain activity by inorganic phosphate. *PLoS One*. 7, e34118.

Sharma, U., Baek, H.M., Su, M.Y., Jagannathan, N.R., 2011. In vivo 1H MRS in the assessment of the therapeutic response of breast cancer patients. *NMR Biomed*. 24, 700–711.

Sharma, U., Sah, R.G., Jagannathan, N.R., 2008. Magnetic resonance imaging (MRI) and spectroscopy (MRS) in breast cancer. *Magn Reson Insights*. 2, 93.

Shen, W., Mao, X., Wang, Z., Punyanitya, M., Heymsfield, S.B., Shungu, D.C., 2003. Measurement of intramyocellular lipid levels with 2-D magnetic resonance spectroscopic imaging at 1.5 T. *Acta Diabetol*. 40 (Suppl. 1), 51–54.

Sijens, P.E., Wijrdeman, H.K., Moerland, M.A., Bakker, C.J., Vermeulen, J.W., Luyten, P.R., 1988. Human breast cancer in vivo: H-1 and P-31 MR spectroscopy at 1.5 T. *Radiology*. 169, 615–620.

Smith, S.A., Montain, S.J., Zientara, G.P., Fielding, R.A., 2004. Use of phosphocreatine kinetics to determine the influence of creatine on muscle mitochondrial respiration: An in vivo 31P-MRS study of oral creatine ingestion. *J Appl Physiol*. 96, 2288–2292.

Springer, F., Machann, J., Claussen, C.D., Schick, F., Schwenzer, N.F., 2010. Liver fat content determined by magnetic resonance imaging and spectroscopy. *World J Gastroenterol*. 16, 1560–1566.

Stanwell, P., Gluch, L., Clark, D., Tomanek, B., Baker, L., Giuffre, B., Lean, C., Malycha, P., Mountford, C., 2005. Specificity of choline metabolites for in vivo diagnosis of breast cancer using 1H MRS at 1.5 T. *Eur Radiol*. 15, 1037–1043.

Suzuki, E., Kashiwagi, A., Hidaka, H., Maegawa, H., Nishio, Y., Kojima, H., Haneda, M., et al., 2000. 1H- and 31P-magnetic resonance spectroscopy and imaging as a new diagnostic tool to evaluate neuropathic foot ulcers in type II diabetic patients. *Diabetologia*. 43, 165–172.

Taylor, D.J., Bore, P.J., Styles, P., Gadian, D.G., Radda, G.K., 1983. Bioenergetics of intact human muscle. A 31P nuclear magnetic resonance study. *Mol Biol Med*. 1, 77–94.

Taylor-Robinson, S.D., Sargentoni, J., Bell, J.D., Saeed, N., Changani, K.K., Davidson, B.R., Rolles, K., et al., 1997. In vivo and in vitro hepatic 31P magnetic resonance spectroscopy and electron microscopy of the cirrhotic liver. *Liver*. 17, 198–209.

Thomas, E.L., Hamilton, G., Patel, N., O'Dwyer, R., Dore, C.J., Goldin, R.D., Bell, J.D., Taylor-Robinson, S.D., 2005. Hepatic triglyceride content and its relation to body adiposity: A magnetic resonance imaging and proton magnetic resonance spectroscopy study. *Gut*. 54, 122–127.

Thomas, M.A., Narayan, P., Kurhanewicz, J., Jajodia, P., Weiner, M.W., 1990. 1H MR spectroscopy of normal and malignant human prostates in vivo. *J Magn Reson*. 87, 610–619.

Thomsen, C., Becker, U., Winkler, K., Christoffersen, P., Jensen, M., Henriksen, O., 1994. Quantification of liver fat using magnetic resonance spectroscopy. *Magn Reson Imaging*. 12, 487–495.

Tse, G.M., Yeung, D.K., King, A.D., Cheung, H.S., Yang, W.T., 2007. In vivo proton magnetic resonance spectroscopy of breast lesions: An update. *Breast Cancer Res Treat*. 104, 249–255.

Tse, G.M.K., Cheung, H.S., Pang, L.-M., Chu, W.C.W., Law, B.K.B., Kung, F.Y.L., Yeung, D.K.W., 2003. Characterization of lesions of the breast with proton MR spectroscopy: Comparison of carcinomas, benign lesions, and phyllodes tumors. *Am J Roentgenol*. 181, 1267–1272.

Vock, R., Hoppeler, H., Claassen, H., Wu, D.X., Billeter, R., Weber, J.M., Taylor, C.R., Weibel, E.R., 1996. Design of the oxygen and substrate pathways. VI. Structural basis of intracellular substrate supply to mitochondria in muscle cells. *J Exp Biol*. 199, 1689–1697.

Wang, M.C., Chan, L., Maiman, D.J., Kreuter, W., Deyo, R.A., 2007. Complications and mortality associated with cervical spine surgery for degenerative disease in the United States. *Spine*. 32, 342–347.

Wefer, A.E., Hricak, H., Vigneron, D.B., Coakley, F.V., Lu, Y., Wefer, J., Mueller-Lisse, U., Carroll, P.R., Kurhanewicz, J., 2000. Sextant localization of prostate cancer: Comparison of sextant biopsy, magnetic resonance imaging and magnetic resonance spectroscopic imaging with step section histology. *J Urol*. 164, 400–404.

Wilkinson, M., 1960. The morbid anatomy of cervical spondylosis and myelopathy. *Brain*. 83, 589–617.

Williams, S.R., Gadian, D.G., Proctor, E., Sprague, D.B., Talbot, D.F., Brown, F.F., Young, I.R., 1985. 1H nuclear-magnetic-resonance studies of muscle metabolism in vivo. *Biochem Soc Trans*. 13, 839–842.

Wu, F.-Y., Tu, H.-J., Qin, B., Chen, T., Xu, H.-F., Qi, J., Wang, D.-H., 2012. Value of dynamic 31P magnetic resonance spectroscopy technique in in vivo assessment of the skeletal muscle mitochondrial function in type 2 diabetes. *Chin Med J (Engl)*. 125, 281–286.

Yeung, D.K., Cheung, H.S., Tse, G.M., 2001. Human breast lesions: Characterization with contrast-enhanced in vivo proton MR spectroscopy—Initial results. *Radiology*. 220, 40–46.

Yeung, D.K.W., Yang, W.-T., Tse, G.M.K., 2002. Breast cancer: In vivo proton MR spectroscopy in the characterization of histopathologic subtypes and preliminary observations in axillary node metastases. *Radiology*. 225, 190–197.

Yoshizaki, K., Watari, H., Radda, G.K., 1990. Role of phosphocreatine in energy transport in skeletal muscle of bullfrog studied by 31P-NMR. *Biochim Biophys Acta*. 1051, 144–150.

Young, W.F., 2000. Cervical spondylotic myelopathy: A common cause of spinal cord dysfunction in older persons. *Am Fam Physician*. 62, 1064–1070, 1073.

12

Imaging Tissue Elasticity Using Magnetic Resonance Elastography

Yogesh K. Mariappan, Kiaran P. McGee, and Richard L. Ehman
Mayo Clinic

CONTENTS

12.1 Introduction

Magnetic resonance elastography (MRE) is a relatively novel MR imaging (MRI) technique that can quantitatively map the mechanical properties of soft tissues noninvasively (Muthupillai et al., 1995; Sack et al., 2001; Sinkus et al., 2012). It is clinically used for the assessment of liver fibrosis (Venkatesh et al., 2008b) and is rapidly evolving with a vast array of new applications being developed (Asbach et al., 2008; Chen et al., 2009a; Chopra et al., 2009; Dresner et al., 2001; Goss et al., 2006; Huwart et al., 2007; Kemper et al., 2004; Kruse et al., 2008, 2009; Litwiller et al., 2010a; Lopez et al., 2008; Mariappan et al., 2009a, 2011b; McGee et al., 2008; Ringleb et al., 2007; Shah et al., 2004; Sinkus et al., 2005a; Weaver et al., 2005; Xu et al., 2007). The principal motivation for the widespread interest in MRE is the well-documented, time-tested success of manual palpation to feel the difference in the mechanical properties of tissues and to differentiate between normal and abnormal tissues. For instance, the difference in hardness of malignant tumors in comparison to healthy breast tissue or benign nodules is still being used for the screening and diagnosis of breast cancer (Barton et al., 1999). The effectiveness of palpation can be understood when some liver tumors that have gone undetected in preoperative imaging with today's sophisticated medical imaging tools are still detected by simple touch at laparotomy (Elias et al., 2005). Unfortunately, manual palpation is subjective, less sensitive, and typically applicable only to superficial organs and for advanced stage diseases. None of the conventional medical imaging techniques, such as computed tomography (CT), MRI, or ultrasonography (US), can quantitatively measure the mechanical properties that are assessed by palpation.

Another important consideration for the success of MRE is the large dynamic range of the contrast parameter elastic modulus that is being assessed. Figure 12.1 shows the contrast mechanism used for MRE in comparison to the contrast parameters assessed by other medical imaging technologies.

It can be seen that the shear modulus of tissues at different physiological states and/or with different pathologies varies over five orders of magnitude,

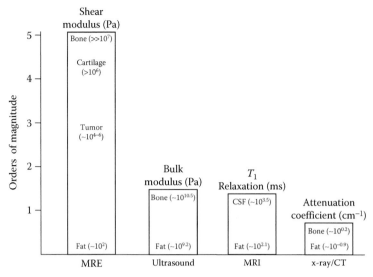

FIGURE 12.1
Imaging modality contrast mechanisms: the contrast mechanisms used for MRE and the three conventional imaging modalities. The shear modulus used to create contrast in MRE has the largest dynamic range with over five orders of magnitude variation among various physiological and pathological states of tissues. (Adapted from Mariappan, Y.K., et al., *Clinical Anatomy.* 23, 497–511, 2010.)

indicating the huge diagnostic potential of this technique. In contrast, the tissue properties assessed by other modalities vary over a much smaller scale.

In addition, MRE has received considerable attention due to the emergence of MRI as a widely available clinical tool, the flexibility of MR in acquiring arbitrary two-dimensional (2D) images and three-dimensional (3D) volumes, and its capability to measure motion in all three directions. The purpose of this chapter is to introduce MRE to the readers, to briefly discuss the essential components of this technology, to list some of the advantages and limitations, and to summarize a few clinical applications that are being investigated.

12.2 Elasticity Imaging

MRE belongs to a family of techniques collectively termed as "elasticity imaging" that are being actively investigated for imaging the mechanical properties of tissues. Due to its significant clinical potential, this field has been gaining significant traction, as can be visualized from Figure 12.2.

Figure 12.2a shows the number of citations (including journal articles, meeting abstracts, and patents) for the search term "elasticity imaging" in Google Scholar from 2001 to 2011, and the cumulative sum of citations from 2001

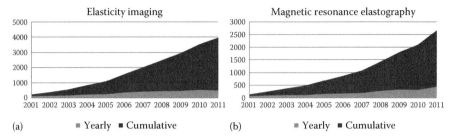

FIGURE 12.2
Growth of the field. Number of citations per year and yearly cumulative number of citations for the terms "elasticity imaging" (a) and "magnetic resonance elastography" (b) in Google Scholar from 2001 indicate the active research and development in the field.

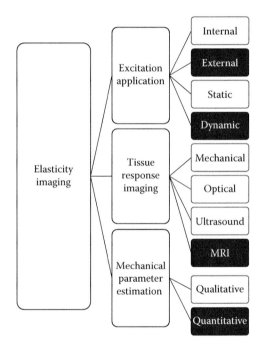

FIGURE 12.3
Approaches to elasticity imaging. The flowchart presents a bird's eye view classification of various elasticity imaging approaches. As highlighted, MRE typically uses external dynamic vibrations, uses MRI to image the shear wave propagation, and produces quantitative shear stiffness values.

onward. The significant growth in the number of citations in the recent years indicates the widespread interest in elasticity imaging techniques. This has resulted in multiple approaches to elasticity imaging; however, most of these techniques typically apply some kind of stress or mechanical excitation to the tissue, measure the tissue response to this stimulus, and calculate parameters that reflect the mechanical properties from this response. Figure 12.3 provides a bird's eye view classification of the different approaches based on these three essential steps in elasticity imaging.

Some of the primary methods of elasticity imaging are highlighted in the following discussion to provide the reader an appreciation for the breadth of the field and to build a foundation for the discussion of MRE.

12.2.1 Excitation Application

Application of a mechanical stimulus to the tissue of interest is typically the first step in elasticity imaging. The different approaches can be broadly classified based on either the source of motion as internal or external or the temporal characteristic as static or dynamic.

12.2.1.1 Source of Motion

12.2.1.1.1 Internal Motion

Endogenous physiological sources of motion such as the cardiac beating, respiration, or cerebrospinal fluid flow can be exploited for creating the stimulus for elasticity imaging. Varghese and Shi (2004) developed a method that uses the tissue response to respiratory diaphragmatic compressions to visualize the mechanical properties of radio frequency (RF)-ablated regions in the liver during and after thermal therapy. Mai and Insana (2002) imaged the strain distribution of the normal brachial artery due to internal pulsatile deformations caused by the cardiac output. Similarly, Bae et al. (2007) evaluated the use of carotid artery pulsation as a compression source for the differential diagnosis of thyroid nodules with ultrasound thyroid elastography. Kanai (2004) developed a noninvasive elasticity imaging technique that can measure the viscosity of the heart wall by imaging the minute vibrations within the heart wall produced by the closure of the aortic valve. Cardiac motion has also been investigated for the measurement of myocardial elasticity based on MR tagging methods (Lee et al., 2008; Zerhouni et al., 1988). One of the main advantages of using internal sources of motion is that they are readily available. However, this method is not suitable for all organs of interest since the amount of excitation the tissues receive could be very low. Also, it is not easy to control the excitation application with internal sources of motion.

12.2.1.1.2 External Motion

External drivers are easily controllable and can be designed to induce excitation in any organ of interest. For instance, Ophir et al. (1991) developed a technique (which was originally termed "elastography"), where static compression is applied with a conventional ultrasonic transducer and the tissue response is imaged with the same transducer. Krouskop et al. (1987) used an external tissue vibrator to induce low-frequency motion in soft. External drivers using bite blocks that shake the whole head to induce mechanical vibrations have been investigated by different groups for brain elasticity imaging (Kruse et al., 2008; Xu et al., 2007). There have also been investigations on using external sources to create motion remotely deep within the tissue; for instance, focused ultrasound beams have been investigated to create the necessary stimuli inside the tissue of interest (Chen et al., 2009b; Fatemi and Greenleaf, 1998; Nightingale et al., 2002; Wu et al., 2000). Similarly, focused ultrasound-based systems have been designed to create planar shear waves in media by moving the ultrasound beams at a supersonic speed (Bercoff et al., 2004).

12.2.1.2 Temporal Characteristics of Motion

12.2.1.2.1 Static

Manual palpation can be thought of as a static elasticity imaging technique. Static compression is a widely investigated and one of the earliest approaches for elasticity imaging. The techniques that use static forces include strain-encoding imaging (Osman, 2003), elastography (Ophir, et al., 1991, 1994), and stimulated echo elasticity imaging (Chenevert et al., 1998; Steele et al., 2000).

12.2.1.2.2 Dynamic

Dynamic excitation techniques include vibration sonoelastography (Lerner et al., 1990; Levinson et al., 1995) and MRE (Muthupillai et al., 1995; Sack et al., 2004). These techniques induce dynamic mechanical vibrations, usually in the range of 50–500 Hz and image the propagation of the waves produced by the excitation through the tissue of interest. Dynamic vibrations can be either transient (McCracken et al., 2005) or continuous harmonic (Yin et al., 2007a).

12.2.2 Tissue Response Measurement

The second and the most critical component of elasticity imaging is the measurement of the tissue response to the applied stimulus. Various approaches have been proposed and investigated for this purpose. Figure 12.3 lists the four main methods used, and a brief discussion of these methods follows.

12.2.2.1 Optical Methods

One of the initial works in elasticity imaging was performed by Von Gierke et al. (1952), who used visible light to image mechanical wave propagation on tissue surface and calculated the tissue elasticity and viscosity. Since then, sophisticated optical imaging-based elasticity imaging techniques such as optical coherence tomographic (OCT) elastography and tissue Doppler optical coherence elastography (tDOCE) have been developed. OCT uses low-coherence interferometry to produce images of optical scattering from internal tissue microstructures (Huang et al., 1991). OCT elastography measures the displacement and tissue strain with speckle modulation tracking (Rogowska et al., 2004; van Soest et al., 2007), and tDOCE measures the tissue displacement by using the Doppler effect (Ruikang et al., 2006).

12.2.2.2 Mechanical Methods

A recently developed technique called mechanical imaging (Egorov et al., 2006; Sarvazyan, 1998) uses pressure sensor arrays to compress the tissue, measures the stress patterns produced due to this compression, and reconstructs 3D tissue viscoelastic property maps with some help from anatomical atlases. Skin-mounted accelerometers have also been investigated for the calculation of the frequency and damping coefficients of free vibration of the tissues after impact (Wakeling and Nigg, 2001). Atomic force acoustic microscopy is a technique that uses a mechanical detection-based elasticity calculation and can be used for the elasticity imaging of dental enamels (Habelitz et al., 2001; Rabe et al., 2002); however, this technique is of extremely high resolution compared to the biomedical imaging techniques discussed later in this chapter.

12.2.2.3 Ultrasonic Methods

US has been widely investigated for elasticity imaging and can be considered to have rekindled the interest in elasticity imaging, with both the cross-correlation methods and the Doppler imaging methods being used to measure the tissue motion (Krouskop et al., 1987; Lerner et al., 1990; Nightingale et al., 2001; Yamakoshi et al., 1990). The term "elastography" as such was introduced to describe a technique that can provide a qualitative reflection of the stiffness of the tissue by measuring the tissue strain with ultrasound resulting from an external compression. Chen et al. have developed the technique called shear wave dispersion ultrasound vibrometry (SDUV) that can image the tissue elasticity and viscosity. A relatively novel method called transient elastography (TE) (de Ledinghen et al., 2007; Sandrin et al., 2003) induces a single transient shear wave into the tissue via a special transducer, images the propagation of this wave using ultrasound, and uses this information to calculate the Young's modulus of the tissue. In addition to ultrasound-based methods, a technique named ultrasound-stimulated vibro-acoustic spectrography records acoustic range emissions from objects in response to oscillatory radiation force, and the elastic maps obtained from these data have been shown to detect calcification in arteries (Fatemi and Greenleaf, 1998). Even though ultrasound-based techniques are fast, inexpensive, and widely available, they have limitations that include the need for a suitable acoustic window for ultrasound measurements and a limited imaging depth for the measurements due to the limited penetration of ultrasonic waves.

12.2.2.4 MRI Methods

One of the early implementations of MRI to measure tissue motion was for the assessment of cardiac function and pathologies using MR tagging techniques (Axel and Dougherty, 1989; Zerhouni et al., 1988). Recently, MR-based elasticity imaging techniques based on synchronous use of quasistatic compressions and motion-encoding gradients (MEGs) (Chenevert et al., 1998; Plewes et al., 1995) are being actively investigated. Muthupillai et al. (1995, 1996) developed MRE, the focus of this chapter, which involves inducing harmonic vibrations of acoustic range frequencies in tissue and imaging the propagation of these vibrations to calculate quantitative values for tissue mechanical parameters. Even though the expense of MRI is a significant limitation, due to the power and flexibility of MRI, these methods are being actively investigated, and hence, the applications are rapidly evolving.

12.2.3 Elasticity Parameter Calculation

The third step in elasticity imaging is to process the acquired data to estimate the mechanical properties of the tissue. We will briefly review the tissue mechanics underpinning the basis of this step. The mechanical response of a tissue is typically assumed to be linearly elastic, which defines the ratio of the applied stress to the resultant strain as indicated in the constitutive equation:

$$\sigma_{ij} = C_{ijkl} \gamma_{kl}, \tag{12.1}$$

where σ_{ij} is the applied stress tensor, γ_{kl} is the strain tensor, and C_{ijkl} is the elastic modulus tensor in the limit of an infinitesimally small deformation of the tissue as a result of an applied force (Auld, 1990). If the tissue can be considered isotropic, there are only two independent components of the elastic modulus tensor.

There are many parameters that describe the mechanical properties of materials. The bulk modulus (K) indicates a substance's resistance to uniform compression and is the component used to create contrast in conventional ultrasound imaging (in combination with the density of the material). The Young's modulus (E) and the shear modulus (μ) reflect the response of a substance to stress, respectively, which are often used for describing tissue mechanical properties and are used as contrast parameters in elasticity imaging. Poisson's ratio (ν), another widely used parameter, is the ratio of lateral strain to longitudinal strain. Among these parameters, as mentioned before, with the assumption of isotropy, only two parameters are independent, which implies that any parameter can be calculated from the knowledge of just two parameters. For instance, the shear modulus is related to Young's modulus by the equation:

$$\mu = \frac{E}{2(1+v)}.$$ (12.2)

Soft tissues are typically almost incompressible; hence, the Poisson's ratio can be assumed to be 0.5, implying that only one more mechanical parameter needs to be measured to derive any other parameter under the mentioned assumptions. For instance, Equation 12.2 can be written as $E = 3\mu$, which means that the calculation of Young's modulus or shear modulus provides the same information. Although bulk modulus and shear modulus are components of the elastic modulus tensor, since the changes in the bulk modulus among different tissues are not large (Figure 12.1), shear modulus is the preferred choice for tissue contrast.

Even though obtaining quantitative values of the elastic properties of a particular tissue would be the preferred option, obtaining a qualitative contrast between normal and abnormal tissues has been found to be helpful in some clinical applications. For instance, in the assessment of liver fibrosis, the staging of the disease may require an absolute value of an elasticity parameter, but the differentiation of a normal and a cirrhotic liver can be accomplished with a qualitative assessment. Most of the techniques that apply quasistatic excitation and measure the deformation occurring due to this stimulus calculate the displacement or strain information (Chenevert et al., 1998; Mai and Insana, 2002; O'Donnell et al., 1994; Osman, 2003; Plewes et al., 1995) as qualitative indicators of the underlying mechanical contrasts. To calculate quantitative values of shear modulus, accurate determination of the stress factors is necessary, which may be difficult to obtain due to complicated boundary conditions. On the other hand, with dynamic wave propagation techniques, quantitative shear modulus values can be calculated from the propagation of the shear waves using constitutive wave equations (Muthupillai et al., 1995; Sandrin et al., 2003).

Thus, there are various approaches and techniques that are in different stages of development and clinical implementation in the field of elasticity imaging. An exhaustive review of all these methods is beyond the scope of this chapter; interested readers are encouraged to refer to the detailed discussions found in the literature, such as in Greenleaf et al. (2003), Sarvazyan et al. (1995, 2011), and Wilson et al. (2000).

12.3 Magnetic Resonance Elastography

MRE is one of the rapidly developing elasticity imaging techniques, as is evident from Figure 12.2b, which indicates the number of yearly Google Scholar citations and the cumulative sum of citations from 2001 onward for the search term "magnetic resonance elastography." As can be seen in Figure 12.3, MRE typically induces dynamic harmonic vibrations using an external driver into the tissue of interest, images the propagation of shear waves with MRI, and calculates quantitative shear stiffness values of the tissue of interest. The following section highlights some of the key concepts of MRE.

12.3.1 Application of Shear Waves

Although the application of excitation to the tissue of interest can be either internal or external, and/or static or dynamic, conventional MRE typically uses dynamic vibrations of a single frequency (within the audio frequency range) induced by external drivers. The electrical signal for these drivers is created by a signal generator controlled by the MR pulse sequence. The signal is then amplified by an audio amplifier before being fed into the mechanical driver of choice. Over the years, several different mechanical driving mechanisms have been developed, each with their own advantages and limitations. Figure 12.4 schematically shows three of the popular driving systems.

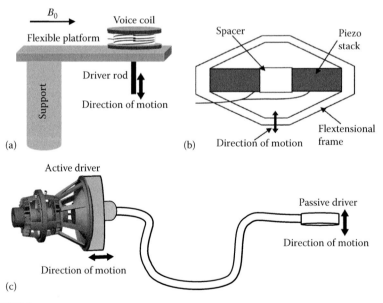

FIGURE 12.4
External driver systems. Schematic diagrams of (a) an electromechanical driver, (b) a piezo-electric stack, and (c) a pressure-activated driver system. The pressure-activated driver system is currently used for the clinical assessment of hepatic fibrosis in Mayo Clinic. (From Mariappan, Y.K., et al., *Clinical Anatomy*. 23, 497–511, 2010.)

12.3.1.1 Electromechanical Driver Systems

Electromechanical drivers are one of the initially developed driver systems. The main component of these systems is a voice coil, which induces a magnetic moment when an alternating current (AC) flows through it. When these driver systems are placed within the main magnet, the Lorentz force created between this magnetic moment created by the voice coils and the static magnetic field of the main magnet creates the motion (Braun et al., 2003; Muthupillai et al., 1995). Figure 12.4a shows a schematic diagram of one of the many different adaptations of an electromechanical driver system. With this orientation of the voice coil with respect to the static magnetic field B_0, when an AC passes through the voice coil, the local magnetic field created will be perpendicular to the main magnetic field of the magnet B_0, and the resultant magnetic moment tries to align with the main field, moving the drive shaft in one direction through the flexible platform. When the AC reverses, the local magnetic moment switches to the opposite direction, and the drive shaft deflects in the other direction. This back-and-forth motion of the drive shaft is coupled to the sample of interest using the driver rod, and the direction of motion is indicated in Figure 12.4a.

These drivers can be designed to create vibration forces in any direction; however, the voice coils have to be placed perpendicular to the static magnetic field to create significant motion. Also, the local magnetic field that these drivers create can interfere with the MRI process and hence can create image artifacts. In spite of these limitations, these drivers are widely used in a variety of applications (Chan et al., 2006; de Ledinghen et al., 2007; Green et al., 2008; Yin et al., 2007b) due to their high motion amplitudes, broad range of frequency applicability, adaptability, and inexpensiveness. Electromechanical drivers capable of producing high-frequency vibrations in the kilohertz range have been developed for special applications such as cartilage MRE (Lopez et al., 2008).

12.3.1.2 Piezoelectric Driver Systems

Another widely investigated mechanical driver system creates motion based on the reverse piezoelectric property, where a mechanical strain is generated from applied electricity. Depending on the application, either a single piezoelectric element (bender or extender) or multiple parallel element stacks can be used in MRE, and a schematic diagram of a piezoelectric stack-based driver system is shown in Figure 12.4b (Chen et al., 2006; Othman et al., 2005; Rossman et al., 2003; Uffmann et al., 2002). In this particular driver system, as an AC is applied, the piezo stacks expand and contract in the horizontal direction, which is converted into an amplified vertical motion due to the flextensional frame (made up of materials such as titanium, brass, aluminum, or plastic). This motion is coupled to the sample with a contact plate. Piezoelectric driver systems have also been used in combination with more than one driver component acting in cohesion to create the required motion (Weaver et al., 2005).

The main advantages of the use of piezoelectric driver technology are the ease of positioning of the driver in comparison to the electromechanical drivers and the avoidance of an additional magnetic field, leading to a reduction of artifacts. Also, these drivers can create more reliable waveforms with less ringing effects, which becomes especially useful in situations where a single transient wave is needed (McCracken et al., 2005).

12.3.1.3 Acoustic Speaker-Based Driver Systems

The most popular method of shear wave excitation used for MRE creates the required vibrations using the voice coil as used in acoustic speaker systems. The vibrations are again produced by the Lorentz force, but the static magnetic field is from a devoted permanent magnet present in an acoustic speaker such as housing, referred to as an active driver. With their own permanent magnets (that are much weaker than the static field B_0 of the MR system), the active drivers need to be placed away from the main magnet, necessitating an additional component to couple the vibrations to the tissue of interest. Two different approaches have been taken for this coupling:

1. Using a carbon rod 2–3 m long that directly transfers the motion of the speaker to the sample (Asbach et al., 2008). These drivers can be easily manipulated and have been shown to be capable of producing mechanical motions of multiple frequencies at the same time.

2. Using a connecting tube ending in a passive drum-like driver kept in contact with the tissue (Yin et al., 2008), where the vibrations are pneumatically transferred as longitudinal pressure waves through the air in the tube. This system is referred to as the pressure-activated driver system and is schematically shown in Figure 12.4c (Venkatesh et al., 2008b). Since the actual vibrations are produced by the active driver, the passive driver component can potentially be easily adapted to suit any organ of interest, such as the breast, brain, or prostate.

Even though these driver systems are gaining popularity and finding a wide variety of applications, these are optimally suited for a small range of frequencies. In addition, the pressure-activated driver is optimally suited for continuous vibration techniques due to the time lag needed for the vibrations produced at the active driver to pneumatically reach the tissue.

12.3.1.4 Focused Ultrasound-Based Drivers

Focused ultrasound-based radiation force is another technique that has been investigated to create the required mechanical motion in both static (Fatemi and Greenleaf, 1998; Nightingale et al., 2001) and dynamic (Bercoff et al., 2004) elasticity imaging strategies in a few ultrasound-based techniques. In a combination of ultrasound and MRI, focused ultrasound has been used to create shear waves deep within the tissue to be detected with MRI (Wu et al., 2000). This technique creates shear waves internally within the sample of interest with externally placed focused ultrasound transducers. One of the main advantages of focused ultrasound-based motion creation techniques is that the shear waves can be created very close to a particular region of interest, which could increase the confidence in the material property that is estimated. However, the sophistication required to operate the focused ultrasound in synchrony with the MRI and the fact that different amounts of energy need to be applied to create the same amount of displacement depending upon the stiffness of the tissue (i.e., a large amount of energy may be required to vibrate stiffer tissues) limit the usability of this approach.

12.3.2 Imaging the Shear Wave Propagation

In a landmark paper in *Science*, Muthupillai et al. (1995) described the dynamic MRE technique, where propagating shear waves in a sample are encoded into the phase of the MR images with the help of motion-sensitizing gradient cycles. A brief discussion on the basic mechanism of MRE motion encoding and the pulse sequence implementation is presented below.

12.3.2.1 Theory

The idea of motion sensitization that forms the basis of MRE has been in use for the quantitation of blood flow in MR angiography (Dumoulin and Hart, 1986) for many years. The phase induced in the transverse magnetization of spins (isochromats) moving within a magnetic field gradient is given by Moran (1982):

$$\phi(t_1) = \gamma \int_0^{t_1} \vec{G}_r(t) \bullet \vec{r}(t) dt, \tag{12.3}$$

where $\phi(t_1)$ is the time-dependent phase of the transverse magnetization, $\vec{G}_r(t)$ is the motion-sensitizing gradient, $\vec{r}(t)$ is the position vector of the moving spin, "$\bullet$" is the vector dot product, and γ is the gyromagnetic ratio $(\gamma/2\pi = 4257 \text{ Hz/G})$. When a continuous harmonic shear wave is induced in a sample, the position vector $\vec{r}(t)$ can be expressed as

$$\vec{r}(t) = \vec{r}_0 + \vec{\xi}_0 \exp[j(\vec{k} \bullet \vec{r} - \omega t + \theta)], \tag{12.4}$$

where $\vec{r}_0$ is the mean position of the isochromats, $\vec{\xi}_0$ is the peak amplitude, $\vec{k}$ is the wave number, ω is the angular frequency $(2\pi f)$ of the harmonic wave propagating through the tissue, and θ is an initial phase offset. If an MEG that is switched in polarity at the same frequency as the motion is applied during the motion, then the phase of the isochromat due to the motion and the applied magnetic field gradient can be written as

$$\varphi(\vec{r},\theta) = \frac{2\gamma NT(\vec{G}\bullet\vec{\xi}_0)}{\pi}\cos(\vec{k}\bullet\vec{r}+\theta),\qquad(12.5)$$

where N is the number of bipolar MEG cycles used to sensitize the motion and T is the period of a single MEG cycle (Muthupillai et al., 1995). This equation states that the phase of a harmonically vibrating isochromat is directly proportional to its displacement for a given MEG and phase relationship between the gradient and the motion. Equation 12.5 is the working equation of MRE used to image the propagating shear waves, through the use of MR pulse sequences that control the RF waveforms, the imaging gradients, and the MEGs.

12.3.2.2 Pulse Sequence

Since MRE uses MEGs inserted into conventional MRI pulse sequences to encode the motion, it can potentially be implemented with any MRI sequence. Thus, many different flavors of pulse sequences based on spin echo (SE), gradient-recalled echo (GRE), balanced steady-state free precision (bSSFP), and echo planar imaging (EPI) techniques exist (Bieri et al., 2006; Kruse et al., 2006; Maderwald et al., 2006; Rydberg et al., 2001), each with its own advantages, limitations, and applications.

Figure 12.5 shows a typical SE-based MRE pulse sequence with two RF pulses (90° and 180°), the slice-selection gradient, the phase-encoding gradient, and the frequency-encoding gradient.

The MEG (shown here only in the slice-selection direction) is placed in between the two pulses. Motion occurring in any direction can be encoded into the phase of the MR image by manipulating the axes in which the MEGs are placed. In this particular case, only the motion occurring in the slice-selection direction, which is orthogonal to the imaging plane, will be sensitized and encoded into the phase.

An MR image thus obtained containing the information of the propagating wave also includes other phase information including the background phase due to inhomogeneities in the magnetic field and the gradient fields (more severe in the case of gradient-echo acquisitions in comparison to SE acquisitions). Hence, generally two such images are collected with opposite polarities of MEG and a phase-difference image known as a wave image is calculated, similar to phase-contrast MRI. The solid and the dotted lines in Figure 12.5 indicate the MEGs with opposed polarities used sequentially to acquire the wave images. These wave images, in addition to being devoid of the non-motion-related phase in the image, have twice the motion sensitivity.

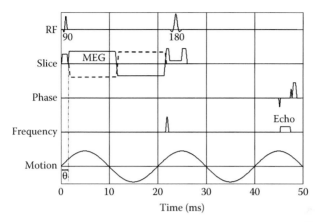

FIGURE 12.5
A SE MRE pulse sequence diagram. A typical bipolar MEG (solid line) as well as the negative MEG (dotted line) used for phase-contrast imaging are shown. The motion waveform and its temporal relationship (θ) with the MEG are also indicated.

The pulse sequence diagram (Figure 12.5) also schematically shows the induced continuous sinusoidal waveform and its temporal relationship with the MEG (θ). The motion is triggered by a standard transistor–transistor logic (TTL) pulse controlled by the pulse sequence through which the motion and the MEG are kept in synchrony. If the motion and the gradients are maintained and their temporal relation (θ) is altered, the phase of the MR images varies sinusoidally (refer to Equation 12.5). This behavior is utilized to acquire snapshots of the propagation of the waves (known as phase-offset images), typically at four to eight temporal samples spaced equally within a period of the motion to show the propagation of the wave in harmonic wave MRE experiments and to permit processing of the data through time. The temporal first harmonic calculated by a Fourier transformation of the acquired images is typically used in the subsequent image processing steps to remove residual static and higher harmonic phase information. Since multiple wave images are acquired within a single wave period, temporal interpolation of the wave data can be performed, usually used for display purposes. In some applications, instead of using harmonic shear waves, mechanical transient waves (McCracken et al., 2005) are being investigated to overcome issues such as wave reflection and interference. With these applications, a single period of a sinusoidal wave is delivered and, using the capability to change the temporal relationship between the motion and the MEG, a series of images with various delay values are acquired, showing the evolution of the transient wave over a certain time interval.

MRE pulse sequences can be designed either to have the MEG frequency matched to the frequency of motion [optimally sensitive for motion of that particular frequency (Muthupillai et al., 1995, 1996)], a particular multiple of the motion frequency (which has lower motion sensitivity) to have a lower echo time for applications with low-T_2 tissues [fractional encoding (Rump et al., 2007)], or to be sensitive to a broad range of frequencies (Asbach et al., 2008; Romano et al., 2003). The MEG waveform can also be gradient moment nulled (Grimm et al., 2007), which reduces the artifacts occurring due to constant velocity motion such as blood flow in the MR images.

12.3.3 Estimation of the Mechanical Parameters

The third component of MRE is the calculation of parameters that reflect the mechanical properties of the medium of interest typically using mathematical inversion algorithms (Manduca et al., 2001). Inversion algorithms based on the equations of motion, with simplifying assumptions such as isotropy, homogeneity, and incompressibility, enable us to calculate the mechanical properties of the tissue of interest that can be useful in clinical applications. This section provides a brief introduction to the underlying equations of motion, the popular approaches used for the calculation of mechanical parameters, and their advantages and limitations.

12.3.3.1 Theory

The frequency-domain constitutive equation of motion for a general, homogeneous, anisotropic, viscoelastic material relates an applied stress to the resultant strain and can be expressed as a rank 4 tensor (Auld, 1990). With the assumption of isotropy, the number of independent quantities reduces to the two Lamé constants λ and μ that control the longitudinal and shear strains, respectively. Among these, the second Lamé constant μ is the shear modulus that we are interested in calculating from the wave propagation images. Local homogeneity assumption makes the Lamé constants position independent and results in the following equation:

$$\mu\nabla^2 u + (\lambda + \mu)\nabla(\nabla\bullet u) = -\rho\omega^2 u, \tag{12.6}$$

where u is the displacement of the particle, ∇^2 is the Laplace operator, ∇ is the gradient operator, $\nabla\bullet$ is the divergence operator, ρ is the density (typically assumed to be 1000 kg/m^3 in MRE), and ω is the angular frequency. In soft tissues, the first Lamé parameter λ is usually much larger than the shear modulus μ, which makes solving the Equation 12.6 directly unstable in practice. There are two types of wave motion supported in the type of media described by Equation 12.6: shear waves (whose propagation is governed by μ) and longitudinal waves (whose propagation is governed predominantly by λ). The effect of λ can be simply neglected in some cases (if the excitation is primarily shear) or can be removed by filtering out longitudinal wave motion (since it is often of a very low spatial frequency). If incompressibility is assumed ($\nabla \bullet u = 0$), the equations of motion reduce to the simple Helmholtz equation:

$$\mu \nabla^2 u = -\rho \omega^2 u. \tag{12.7}$$

This also results in the components of motion in different orthogonal directions being effectively decoupled. This implies that motion in each direction satisfies the wave equation separately, and hence, μ of a sample can be calculated from MRE measurements in just one motion-sensitizing direction.

The shear modulus μ is a complex quantity, written as $\mu_r + i\mu_i$, where μ_r and μ_i are the real and imaginary parts of μ, respectively. μ_r indicates the storage modulus and μ_i reflects the loss modulus (attenuation) of a viscoelastic medium and are related to the viscoelastic parameters shear wave speed (V_s) and attenuation factor (α) at a given frequency of operation as (Auld, 1990)

$$V_s^2 = \frac{2(\mu_r^2 + \mu_i^2)}{\rho(\mu_r + \sqrt{\mu_r^2 + \mu_i^2})}, \tag{12.8}$$

$$\alpha^2 = \frac{\rho \omega^2 \sqrt{\mu_r^2 + \mu_i^2} - \mu_r}{2(\mu_r^2 + \mu_i^2)}. \tag{12.9}$$

In the initially developed inversion algorithms, the medium was considered attenuation-free, resulting in a simple equation:

$$\mu = \rho V_s^2, \tag{12.10}$$

since μ_i can be neglected in Equation 12.8. Advanced algorithms that were developed later were able to calculate both the real and imaginary parts of this shear modulus; however, some groups are still reporting shear modulus values as the product of density and squared wave speed, calculated from Equation 12.8, out of convention and convenience. It should be noted that since the shear modulus is frequency dependent and the attenuation and boundary effects are neglected, these values are often referred to as effective shear modulus or shear stiffness. The images of the mechanical properties of tissue calculated in MRE, that is, the shear stiffness maps, are often referred to as elastograms. Depending on the technique used, the resolution of the elastograms typically ranges from one-third to one-fifth of the native MRI resolution.

12.3.3.2 Inversion Techniques

12.3.3.2.1 Manual Estimation of Shear Wave Speed

As evident from Equation 12.10, the shear modulus can be calculated by measuring the shear wave speed, which can be written as $V_s = f_{mech}\lambda_{spat}$. Here, f_{mech} is the mechanical driving frequency and λ_{spat} is the spatial wavelength of the propagating shear wave. Hence, the shear stiffness of a tissue can be calculated manually from the wave propagation images by drawing a one-dimensional (1D) line profile in a direction normal to the propagating wave front, estimating the wave length and calculating the wave speed. This is one of the initially implemented techniques and is still used for certain applications. However, it is time consuming, subjective, and less reproducible due to the difference in selection of the region to extract the line profile. Also, this technique may not be applicable in situations of wave interference and heavy attenuation. However, this technique is widely used to get a first approximation for the stiffness to compare the performance of different automatic algorithms and in applications such as MRE of highly anisotropic skeletal muscles.

12.3.3.2.2 *Local Frequency Estimation*

The local frequency estimation (LFE) technique, one of the first automatic algorithms developed to calculate the shear modulus (Dresner et al., 2001), makes the assumption of an attenuation-free medium and measures the local spatial frequency $\left(f_{spat}\right)$ of the wave propagation images by combining local estimates of instantaneous frequency over several scales (Knutsson et al., 1994) derived by filters that are products of radial and directional components. The shear modulus can then be calculated by

$$\mu = \rho \frac{f_{mech}^{2}}{f_{spat}^{2}}. \tag{12.11}$$

The main limitation of the technique is its limited resolution, which makes accurate determination of sharp boundaries and small objects difficult. Also, since accurate quantitative estimates of the stiffness of a region are typically reached only half a wavelength into the region (Manduca et al., 2001), this algorithm often underestimates the stiffness of small stiff objects. Although μ_i cannot be directly calculated with this technique, viscosity information can be obtained with this technique by calculating the shear stiffness values using MRE data collected at several different excitation frequencies and fitting a curve using Equations 12.8 and 12.9 to calculate the real and imaginary parts of the shear modulus. Despite these limitations, this technique is still widely used to calculate the shear modulus of tissues due to its simplicity and robustness. This technique is less sensitive to image noise due to the multiscale data averaging in the calculation of the local frequency estimates and hence is a popular choice for applications where the SNR of the image is typically low (Mariappan et al., 2011a).

12.3.3.2.3 *Phase-Gradient Technique*

The phase-gradient technique quantifies the shear modulus by measuring the spatial gradient of the phase of the first harmonic component of motion, calculated by a temporal Fourier transform of the wave images. If the motion can be approximated by a simple plane wave, the phase gradient thus calculated is directly proportional to the local spatial frequency and hence can be used to calculate the shear modulus using Equation 12.11. This technique is applicable only when a plane-wave approximation is valid. Unfortunately, the plane-wave approximation is, in most cases, not applicable due to geometrical constraints (converging or diverging wave fields) or due to shear wave interference. With continuous-wave MRE, interference from reflected waves is often unavoidable and can create artifacts in the stiffness estimates of tissues derived using phase-gradient technique. This method can potentially be of higher spatial resolution than the LFE, but is more sensitive to noise. Since phase gradient can be more reliably calculated (compared to their spatial frequency content) in tissues with long shear waves, this technique is often being applied in some applications for high-stiffness materials such as cartilage MRE (Lopez et al., 2008).

12.3.3.2.4 Direct Inversion

One of the popular inversion techniques used for mechanical property estimation in MRE is called direct inversion (DI), where the shear modulus is directly calculated from Equation 12.7 as

$$\mu = -\rho\omega^2 \frac{u}{\nabla^2 u}. \tag{12.12}$$

A more general form of the solution can be obtained without making the incompressibility assumption and calculating both the Lamé constants (from Equation 12.6), but this necessitates motion information in all three orthogonal directions. This technique can have a very high spatial resolution and hence would be suited for regions with sharp stiffness boundaries. However, due to the calculation of derivatives, this technique is often sensitive to noise. Hence, regions with low wave amplitudes, due to attenuation, shadowing, or interference, will have high stiffness variance, and the technique usually requires data smoothing to reduce these artifacts. Despite these issues, DI is widely used because it can calculate both the real and imaginary parts of the shear modulus (i.e., it can directly calculate the viscosity of tissues).

12.3.3.2.5 Other Methods

Various other methods have also been developed to calculate the shear modulus by integrating the product of the noisy data with smooth test functions and to calculate potential functions that satisfy the equations of motion (Romano et al., 1998). Inversion algorithms based on finite element models (FEMs) have also been developed (Van Houten et al., 1999), where the tissue's mechanical properties are estimated through an iterative process by comparing the displacement profiles from the MRE data to FEM-based displacement profiles. FEM-based inversions are very powerful, flexible, and being widely investigated; however, the need for an accurate geometry and boundary conditions in addition to the higher computational expense limits the usability of this technique. A transversely isotropic model (Sinkus et al., 2005a) has been developed to calculate the anisotropic elastic properties of tissues for which the isotropy approximation is not valid, and it has been applied to breast tumors and highly anisotropic skeletal muscles. For the same problem of anisotropic wave propagation in muscle fibers, a sophisticated algorithm utilizing Helmholtz decomposition, hybrid radon transforms, and sliding window spatial Fourier transforms to calculate the wave dispersion along each fiber (portraying space-wave number profiles) (Romano et al., 2005) has been developed and tested to track waves in arbitrarily oriented fibers to be used for waveguide-constrained MRE. Thus, depending on the specific application, different inversion algorithms are being investigated and developed to provide the best possible shear stiffness estimations.

12.3.3.3 Preprocessing Steps

12.3.3.3.1 Phase Unwrapping

As evident from Equation 12.5, the phase of an MRE wave image is directly proportional to the displacement of the tissues. However, since phase information can only have unique values between $-\pi$ and π, high displacement amplitude can result in ambiguous phase wraps, which limits the applicable motion amplitude and the achievable motion sensitivity in order to avoid phase wraps. However, higher displacements and higher motion sensitivity are preferred to obtain more reliable stiffness estimates. Further, in some tissues with high attenuation values, to get detectable wave amplitude deep within the tissues (i.e., distant from the wave source), higher amplitude motion needs to be applied at the surface, resulting in phase wraps. Hence, ambiguous phase wraps are usually encountered in MRE wave data. However, within certain limits, most of these phase wraps can be unwrapped using standard phase unwrapping algorithms (Ghiglia and Pritt, 1998), and this has resulted in up to five times higher allowable motion amplitudes (Manduca et al., 2001).

12.3.3.3.2 Directional Filtering

Another preprocessing step that is being widely used is called directional filtering (Manduca et al., 2003a), which isolates waves propagating in different directions. It is used as a means to remove the artifacts present in the stiffness estimates of the tissues due to the interference of waves propagating in different directions. This filter uses both the temporal and spatial information in the MRE wave data and hence is referred to as spatiotemporal directional filter. The waves propagating in any direction satisfy the equations of motion. Hence, elastograms can be calculated from the waves propagating in each of these directions and a composite image is obtained by performing a weighted average of the elastograms based on the amplitude of motion present. The idea of using these filters to isolate waves propagating in different directions is also being used for the calculation of shear wave scattering at elastic interfaces used to measure the weldedness of interfaces (Papazoglou et al., 2007).

Other preprocessing steps used in some applications include the conventional low-pass filters to remove high-frequency noise and the high-pass filters to remove the effect of longitudinal waves. Curl filters have also been applied in many cases to remove longitudinal wave effects since displacement due to longitudinal motion is curl free (and the displacement due to shear motion is divergence free). As mentioned before, temporal first-harmonic filtering is applied in some applications to remove nonharmonic phase data, such as concomitant gradient field effects (Bernstein et al., 1998).

12.3.4 Phantom Experiment

We have so far summarized the essential components of MRE. To illustrate the concepts discussed above, an example data set obtained on a tissue-simulating phantom of known material parameters is provided in Figure 12.6.

The phantom demonstrated in this experiment had two gelatin layers of different stiffness values, made up of 7.5% and 15% Bovine gelatins, as indicated in the conventional magnitude image shown in Figure 12.6a as soft and hard layers, respectively. An electromechanical shear driver, shown schematically in Figure 12.6a, induced harmonic shear waves of frequency 100 Hz into the phantom in the direction indicated with the double-sided

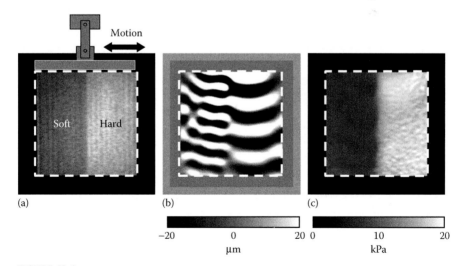

FIGURE 12.6
MRE on a simple two-layer phantom. (a) Conventional MR magnitude image with the soft and hard regions is indicated. A schematic diagram of a mechanical driver and the direction of motion are also indicated. (b) A single wave image from the MRE data and the longer shear wavelength within the harder region are visible. (c) The corresponding elastogram shows the stiff and soft regions clearly. (Data courtesy of Jennifer Kugel, Mayo Clinic, Scottsdale, AZ.)

arrow. The propagation of these waves in the phantom was imaged with a gradient-echo-based MRE pulse sequence with motion sensitivity in the direction of motion. One of the wave images is shown in Figure 12.6b; the wavelength difference between the soft and hard regions can be easily observed. The stiffness estimate of the phantom obtained using the LFE inversion algorithm with directional filtering is shown in Figure 12.6c; the higher stiffness of the hard gelatin compared to the soft gelatin is evident. Quantitatively, representative stiffness values of the two layers calculated by averaging the stiffness values present in regions of interest within the two layers were 2.5 kPa and 7.6 kPa, respectively. This experiment clearly illustrates the three steps in MRE and shows that MRE can quantitatively measure shear stiffness values.

12.4 Applications

The field of MRE has been rapidly evolving due to the flexibility, power, non-invasiveness, and potential clinical utility, with new applications emerging for various organs including liver, spleen, kidney, pancreas, brain, cartilage, prostate, heel fat pads, breast, heart, lungs, spinal cord, bone, eye, and muscle (Asbach et al., 2008; Chen et al., 2009a; Chopra et al., 2009; Dresner et al., 2001; Goss et al., 2006; Huwart et al., 2007; Kemper et al., 2004; Kolipaka et al., 2009; Kruse et al., 2008, 2009; Litwiller et al., 2010a; Lopez et al., 2008; Mariappan et al., 2009a, 2011b; McGee et al., 2008; Ringleb et al., 2007; Shah et al., 2004; Sinkus et al., 2005a; Weaver et al., 2005; Xu et al., 2007; Yin et al., 2007a). The following discussion summarizes some of the current applications of MRE that are being investigated.

12.4.1 Hepatic MRE

Application of MRE for the evaluation of the hepatic tissue is one of the earliest investigated applications and is currently gaining clinical acceptance for the assessment of hepatic fibrosis/cirrhosis (Huwart et al., 2007; Lomas et al., 1997; Venkatesh et al., 2008b). Hepatic cirrhosis is a significant health concern, is the second leading gastrointestinal cause of death and costs about 1.1 billion in direct costs within the United States alone (Shaheen et al., 2006). The major cause of cirrhosis is progressive fibrosis, which is a dynamic disease process and has been shown to be reversible if diagnosed and treated in an early stage. The main motivations behind the use of MRE for fibrosis assessment are the well-known inadequacies of biopsy (Bravo et al., 2001), the current gold standard of diagnosis of fibrosis and the fact that the mechanical properties change significantly with the progression of fibrosis (Yamanaka et al., 1985).

In its most accepted form, clinical hepatic MRE is performed at a frequency of 60 Hz using a pneumatic-based pressure-activated driver (Figure 12.4c). Wave data are acquired with four phase offsets and a modified direct-inversion algorithm with multi-scale capabilities is used for the estimation of the shear stiffness. Example data obtained from clinical MRE exams are shown in Figure 12.7.

Figure 12.7a–c shows a magnitude image, a single phase-offset wave image and the corresponding shear stiffness map obtained from a patient with a healthy liver. The dotted region of interest indicates the liver and the presence of shear waves within the ROI is easily visible. Figure 12.7d–f shows the corresponding data obtained from a patient with a stage 4 fibrosis. The presence of the disease cannot be conclusively detected from the magnitude image shown in Figure 12.7d; in contrast, if the wave image is observed the longer wavelength of the shear waves in Figure 12.7e compared to the wave image in Figure 12.7b is clearly visible. Figure 12.7f shows the stiffness map obtained from this data, and the high stiffness of the diseased liver in comparison to that of the healthy liver tissue is easily evident.

Figure 12.8 shows a graph adapted from (Yin et al., 2007a) indicating the stiffness values of normal and diseased livers at different stages of fibrosis. As the disease progresses, that is, worsens, the MRE derived shear stiffness of the liver tissues increases, indicating that hepatic tissue stiffness is directly related to the fibrosis stage. A Receiver operating characteristic (ROC) curve analysis provided a value of 2.93 kPa as an optimal threshold to distinguish healthy livers from fibrotic ones with a specificity of 98% and sensitivity of 99%. This high specificity and sensitivity in combination with the limitations of Biopsy has led to the technique gaining widespread and expedited clinical acceptance. For the particular data shown in Figure 12.7, the mean stiffness of the diseased liver was calculated to be 17.83 kPa, well above the critical threshold stiffness, in comparison to the stiffness of the healthy liver calculated to be 1.7 kPa, well below the 2.93 kPa.

In addition to the assessment of fibrosis, MRE is also being investigated for the diagnosis and characterization of hepatic tumors. It has been reported that malignant liver tumors are typically stiffer than their benign counterparts and healthy liver tissue (Venkatesh et al., 2008a). An example MRE data set obtained from a patient with a tumor is shown in Figure 12.7 above. This particular patient had a known carcinoid tumor in the liver whose boundaries are marked with the inner dotted region of interest in magnitude image, shown in Figure 12.7g. It can be seen from the wave image in Figure 12.7h that the tumor region shows an increased wavelength in comparison to other parts of the liver. In fact, the shear waves in the nontumor portions of the liver were also longer than the shear waves in the normal liver (Figure 12.7b) and an experienced observer can easily identify that the liver parenchyma itself is abnormal in this diseased liver. The corresponding stiffness estimate is shown in Figure 12.7i, the tumor can be seen to exhibit higher shear stiffness. It has been reported that a cutoff value of 5 kPa can

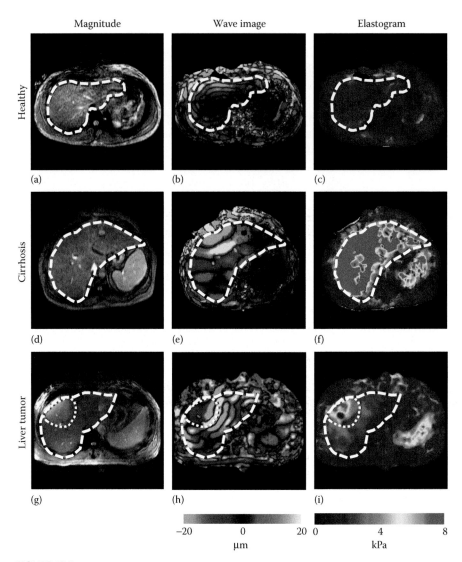

FIGURE 12.7

(See color insert.) Hepatic MRE. (a, b, c) The left, middle, and right columns show the magnitude image, a single wave image, and the shear stiffness map of a patient with a healthy liver, respectively. The boundaries of the liver are indicated with the dotted region and the stiffness was found to be 1.7 kPa. (d, e, f) The middle row shows the corresponding data from a patient with a cirrhotic liver. The longer shear wavelength and the higher shear stiffness (17.8 kPa) compared to the healthy liver are visible. (g, h, i) The bottom row shows data obtained from a patient with a malignant tumor in the liver, whose borders are indicated with the inner dotted region of interest. The high stiffness of the tumor is well evident from the elastogram (i). The shear stiffness of the tumor was found to be 9.23 kPa, well above the malignancy cutoff value of 5 kPa. (From Mariappan, Y.K., et al., *Clinical Anatomy*. 23, 497–511, 2010.)

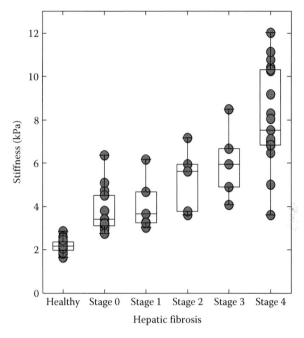

FIGURE 12.8

Hepatic MRE results. MRE data were obtained at 60 Hz from one of the initial studies, where hepatic MRE was performed on 35 healthy volunteers and 50 patients with varying levels of fibrosis. The stiffness of the healthy livers was very similar and the stiffness value and the intragroup variability increase linearly with the progression of the disease. A cutoff of 2.93 kPa is currently clinically used to differentiate between healthy and diseased livers. (Data courtesy of Meng Yin, Mayo Clinic, Scottsdale, AZ.)

differentiate malignant tumors from benign tumors and normal liver paren-chyma, and in this particular case, the stiffness of the carcinoid tumor was found to be 9.23 kPa, suggesting its malignancy.

12.4.2 MRE of Other Abdominal Organs

In addition to the application of MRE to the liver, other abdominal organs including spleen, kidney, and pancreas are also being investigated. The elas-tograms shown in Figure 12.7 also show the stiffness maps of the spleens, which are more evident in Figure 12.7f and i. It has been reported that there is a significant correlation between the MRE derived shear stiffness of the liver and the spleen (Talwalkar et al., 2009). Further, preliminary data from a canine model have indicated a significant correlation between splenic stiff-ness and the portal vein pressure gradient (Nedredal et al., 2011), thus pro-viding a potential noninvasive imaging tool to screen for and diagnose portal hypertension. Renal MRE is being investigated for the diagnosis of kid-ney diseases that include glomerulonephropathy, renal fibrosis, interstitial nephritis, and so on where invasive biopsy remains the most applicable clinical diagnostic tool. Preliminary data are promising where MRE derived shear stiffness values correlated well with the duration of exposure to ethylene glycol-tainted drinking water in a rat model of renal fibrosis (Shah et al., 2004). A point to note is the relationship between the kidney stiffness and other hemodynamic factors, which indicate that kidney turgor needs to be considered in interpretation of elasticity measurements (Warner et al., 2011). Recent human trials have proved the feasibility of human renal MRE and have shown that it is highly reproducible (Rouviere et al., 2011). Another abdominal organ that is being investigated to be assessed with MRE is the pancreas, where, again, noninvasive clinical tools for the diagnosis of chronic pancreatitis or pancreatic tumors are limited. Recent preliminary data have indicated that the pancreas affected with chronic pancreatitis exhibits higher stiffness in comparison to the healthy pancreas (Mariappan et al., 2011b).

12.4.3 Brain MRE

MRE of the brain is another area of significant research and clinical interest (Green et al., 2008; Kruse et al., 2008; Sack et al., 2009a). While there is no clinical precedent for "brain palpation," emerging data suggest that there is a huge diagnostic potential for the stiffness information of the brain since it could be related to multiple disease processes such as Alzheimer's disease (AD) and multiple sclerosis (Murphy et al., 2011; Wuerfel et al., 2010). Also, this information could be useful in biomechanical studies of brain trauma and in the development of neurosurgery techniques. While it would be difficult to use ultrasound-based approaches to assess brain mechanical properties, MRE is well suited for such an application (Green et al., 2008; Kruse et al., 2008; Murphy et al., 2011; Sack et al., 2009a; Wuerfel et al., 2010; Xu et al., 2007) and hence is being actively investigated. An example brain MRE data collected from a healthy volunteer using a multislice single-shot SE-based MRE sequence is shown in Figure 12.9.

An axial anatomical image and a single MRE wave image at a frequency of 60 Hz are shown in Figure 12.9a and b, respectively. Here, the vibrations were introduced into the brain using a pressure-activated driver system (Murphy et al., 2011), where the passive driver is made up of a flexible pillow-like component that included a soft, inelastic fabric cover over a porous springy mesh. The elastogram obtained from the wave data with a 3D LFE technique is shown in Figure 12.9c, indicating the stiffness map of brain, where a very good correlation of the magnitude image and the stiffness information is evident.

One of the many applications that MRE of the brain is being investigated is for the diagnosis and characterization of AD, and a portion of data obtained from a pilot study (Murphy et al., 2011) in comparing the brain stiffness of AD patients at a frequency of 60 Hz to age- and gender-matched healthy volunteers is shown in Figure 12.10.

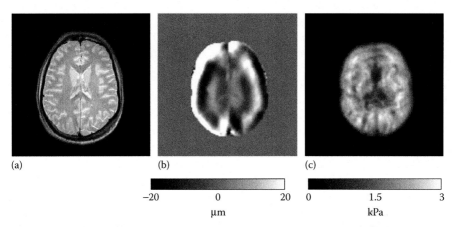

(a)   (b)   (c)

FIGURE 12.9
Brain MRE. (a) An axial MR magnitude image of the healthy brain; (b) a single wave image from a multislice MRE performed with 60 Hz shear vibrations, showing good shear wave propagation throughout the brain; (c) the shear stiffness map calculated from these data with 3D LFE, good correlation between the magnitude, and the elastogram can be observed. (Adapted from Mariappan, Y.K., et al., *Clinical Anatomy.* 23, 497–511, 2010.)

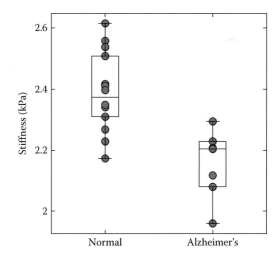

FIGURE 12.10
Brain MRE results. Comparison of shear stiffness values obtained with MRE in subjects with no diagnosed brain disease and age- and gender-matched patients indicates that the shear stiffness reduces with AD. (Data courtesy of Dr. Matt Murphy, Mayo Clinic, Scottsdale, AZ.)

It is easily evident from this figure that the stiffness of the brain in patients with AD (2.2 kPa) is statistically significantly lower than that of the healthy brain tissue (2.4 kPa). Many research groups are involved in investigating the technique for different brain applications, and emerging data indicate that there are significant changes in the brain shear stiffness in multiple sclerosis, cancer, and normal pressure hydrocephalus (Streitberger et al., 2010; Wuerfel et al., 2010; Xu et al., 2007). In addition to the clinical pathophysiological studies, it has also been reported in the literature that the brain stiffness changes with age and gender (Sack et al., 2009a).

12.4.4 Breast MRE

Breast cancer is one of the leading causes of death in women, and the current diagnostic methods do not offer the desired sensitivity and specificity. The most commonly used MRI technique is contrast-enhanced MRI (CE-MRI), which introduces MR contrast agents into the blood and images the blood perfusion in the selected region of interest (Heywang et al., 1986) as a function of time. The application of this technique to the breast has proven to have a very high sensitivity for the detection of tumor nodules, but the specificity can be quite low (Heywang-Kobrunner et al., 1997), leading to false positives and thus to unnecessary biopsies. Breast tumors are appreciated to be quite stiffer than normal breast tissue and, hence, manual palpation is a recommended part of routine screening and helps in the detection of tumors (Barton et al., 1999). Thus, it is natural that MRE is being investigated (McKnight et al., 2002; Sinkus et al., 2005a) with great interest in breast imaging. An example data obtained from a 45-year-old female with a 5 cm malignant adenocarcinoma is shown in Figure 12.11.

An anatomical MR magnitude image of the breast is shown in Figure 12.11a, with the boundaries of the breast and the tumor delineated. In this particular case, shear waves of frequency 100 Hz were introduced into the tissue using electromechanical drivers and wave images were obtained with a

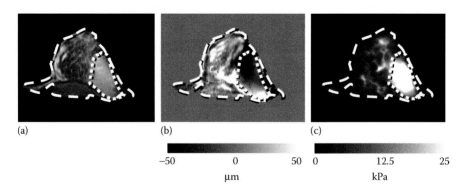

(a)　　　　　(b)　　　　　(c)

−50　　　　0　　　50　0　　　12.5　　25

μm　　　　　kPa

FIGURE 12.11
Breast MRE. (a) Magnitude image of the breast of a patient volunteer with a large, palpable adenocarcinoma; the boundary of the tumor is indicated with the inner dotted curve. (b) A single wave image from MRE performed at 100 Hz, longer shear wavelengths can be observed within the tumor. (c) The corresponding shear stiffness map is shown. As expected, the higher stiffness of the tumor is easily visible. (Adapted from Mariappan, Y.K., et al., *Clinical Anatomy.* 23, 497–511, 2010.)

gradient-echo-based MRE sequence. The wave image shown in Figure 12.11b indicates that the wavelengths are longer within the tumor compared to the normal glandular tissues. Shear stiffness maps were obtained using LFE algorithm in combination with directional filtering, and the calculated elastogram is shown in Figure 12.11c. As expected from observation of the wave data, the tumor is significantly stiffer than the healthy breast tissue.

Figure 12.12 shows data obtained from a pilot breast MRE study performed on a group of 12 subjects, 6 healthy controls, and 6 patients with biopsy-proven, palpable breast malignancies.

The mean shear stiffness measurements of the noncancerous breast tissue and the tumors were found to be 7.5 kPa and 25.9 kPa, respectively. This huge difference in the shear stiffness values and the limitations of the current diagnostic techniques suggest that there is huge clinical potential for the use of MRE in breast imaging. MRE is being proposed as complementary technique to interrogate the suspicious regions as indicated by CE-MRI, and preliminary evidence suggests that the combined technique may increase the diagnostic specificity (Huwart et al., 2007).

In addition to the MRE studies performed with conventional low-frequency waves (60–200 Hz) that can quantitatively image the shear stiffness of the tumors, a novel approach termed as high-frequency mode conversion (HFMC) MRE, is also being investigated wherein higher frequency vibrations are induced within the breast and the presence of the tumors can be qualitatively detected due to differential attenuation of shear waves (Mariappan et al., 2009c).

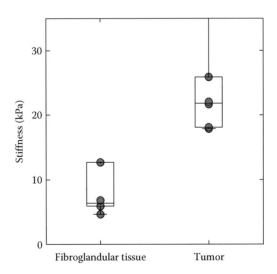

FIGURE 12.12
Breast MRE results. Comparison of MRE-derived stiffness values obtained from healthy fibroglandular breast tissue and tumors. The shear stiffness value not visible in this figure was 93 kPa. As expected, the stiffness value of the tumors was significantly higher than that of healthy breast tissue. (Data courtesy of Jennifer Kugel, Mayo Clinic, Scottsdale, AZ.)

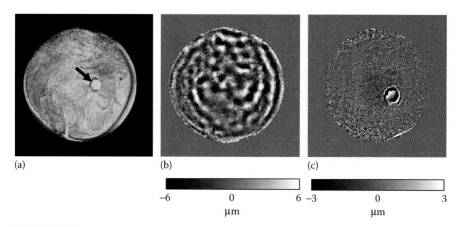

FIGURE 12.13
HFMC MRE. (a) Magnitude image of an *ex vivo* breast specimen with a focused ultrasound-induced stiff lesion, indicated with the arrow. A single wave image obtained from MRE performed at low frequency (b, 200 Hz) and at high frequency (c, 1000 Hz). The presence of the stiff region can be directly detected from the wave image at the higher frequency. (Adapted from Mariappan, Y.K., et al., *Magnetic Resonance in Medicine*. 62, 1457–1465, 2009c.)

Preliminary investigations have provided promising data and an example data obtained from an *ex vivo* breast tissue specimen with a focused ultrasound-induced stiff region is shown in Figure 12.13.

Figure 12.13a shows the magnitude image and the stiff region is indicated with the arrow. Figure 12.13b shows the shear wave displacement field at the conventional frequency of 200 Hz. It can be seen that at this frequency, shear waves can be visualized throughout the specimen. However, the presence of the tumor cannot be detected based on the displacement data alone; shear stiffness maps need to be calculated to identify the tumor. In the shear wave displacement field at the higher frequency of 1000 Hz shown in Figure 12.13c, the tumor alone exhibits shear waves and hence can be easily identified. With the combined knowledge of the contrast-enhancing suspicious regions obtained from CE-MRI and the stiff regions obtained from HFMC, the number of unnecessary biopsies can potentially be reduced. Clinical testing of this technique in combination with CE-MRI is being performed to assess the advantages of combined technique.

12.4.5 Skeletal Muscle MRE

It is well appreciated that the mechanical properties of skeletal muscles change significantly depending on their physiological and/or pathological state. For instance, it has been reported that the dynamic Young's modulus of muscle is almost directly proportional to the tension and dependent upon the presence or absence of neuromuscular disease (Basford et al., 2002;

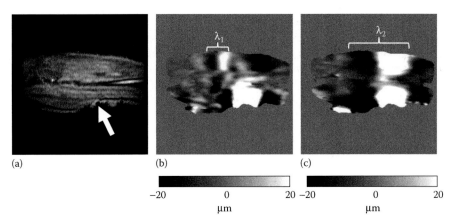

(a) (b) (c)

-20   0   20   -20   0   20
μm   μm

FIGURE 12.14
Skeletal muscle MRE. (a) A sagittal magnitude image of the calf muscle and the location of the driver indicated by the arrow. A single 100 Hz MRE wave image of the muscle while exerting (b) 0 N/m and (c) 10 N/m force, respectively. λ_1 and λ_2 indicate the shear wavelengths at these two states, the shear wavelength is higher when the muscle is active and exerting a force. (Adapted from Mariappan, Y.K., et al., *Clinical Anatomy.* 23, 497–511, 2010.)

Dresner et al., 2001; Mason, 1977). Hence, MRE has been extensively investigated to study the stiffness of skeletal muscles and a representative data set from a calf muscle is shown in Figure 12.14.

The magnitude image of the slice of interest is shown in Figure 12.14a. An electromechanical driver was used to induce shear waves at a frequency of 100 Hz into the muscle and the position of this driver is indicated by the arrow. MRE was performed with the muscle at its relaxed state and when the muscle was exerting a 10 N/m force in a custom-built leg press. Figure 12.14b shows a single wave image of the muscle at its relaxed state and the wavelength of the shear waves is indicated as λ_1. The corresponding wave image with the working muscle is shown in Figure 12.14c and the shear wavelength here is indicated as λ_2. It can be seen that the wavelength is longer as the muscle is exerting a force ($\lambda_2 > \lambda_1$). Due to the anisotropy of the muscles, the stiffness of muscle is typically calculated by manual measurement of 1D wave profiles in the direction of wave propagation, and the values for the data shown in the image are found to be 16 kPa and 96 kPa, respectively.

MRE of the skeletal muscles has been successfully applied to assess the stiffness of various healthy muscles including biceps brachii, gastrocnemius, soleus, tibialis anterior, flexor digitorum profundus (FDP), thigh muscles, vastus lateralis, vastus medialis, and sartorius (Ringleb et al., 2007). Various applications investigating the change in stiffness of the muscle due to pathological conditions are also being studied (Basford et al., 2002; Brauck et al., 2007; Ringleb et al., 2007). It has been reported that the stiffness of gastrocnemius, soleus, and tibialis anterior muscles was significantly higher in patients with poliomyelitis, flaccid paraplegia, and spastic paraplegia.

Similarly, it was found that patients with Grave's disease (hyperthyroidism) had a faster increase in muscle stiffness after treatment in comparison to healthy volunteers. It has also been reported that the stiffness of the soleus muscle of patients with hypogonadism can be monitored with MRE and it changes with therapy. Thus, the assessment of the stiffness of muscles can have huge impact on the diagnosis and prognosis of many skeletal muscle pathologies.

12.4.6 Cardiac MRE

Cardiovascular disease is a major factor affecting health and is the largest killer in the United States. MRE is being investigated for various cardiovascular diseases including heart failure, myocardial infarction, and hypertension (Kolipaka et al., 2009, 2011; Sack et al., 2009b). Recent data have shown that the MRE-derived shear stiffness of abdominal aorta in patients with hypertension is higher than in normotensive subjects (Kolipaka et al., 2011).

There are many challenges for the successful implementation of MRE of the heart primarily because it is a complex moving organ. Recent work suggests that it is possible to image propagating shear waves using electrocardiogram (ECG)-gated MRE pulse sequences. There are mainly two schools of thought being investigated for MRE of the heart: one way is to evaluate the image pattern to calculate the shear stiffness map from the obtained wave images (Kolipaka et al., 2009), similar to conventional MRE. The other idea is to measure the amplitude of the shear wave and calculate the amplitude ratio with reference to the wave amplitude at the chest surface (WAV-MRE), which is directly proportional to the shear modulus (Sack et al., 2009b). Both of these methods have shown promise and have provided data, suggesting their utility for the diagnosis of heart diseases.

An example data obtained from a recent, conventional human MRE study performed on a healthy volunteer is shown in Figure 12.15.

Figure 12.15a shows the anatomical image of the left ventricle (LV) of the heart and the boundaries of the myocardium are indicated. Shear waves of 80 Hz were induced into the volunteer using the pressure-activated driver system, with the passive driver positioned on the anterior chest wall. Waves propagating in the three orthogonal directions of motion were sensitized, and Figure 12.15b shows data with motion sensitized in the through-plane direction and the presence of shear waves is clearly evident. An important issue for the cardiac MRE is the inverse problem of estimation of the shear stiffness from the wave images due to the complicated wave propagation pattern and the complex anatomy. Significant effort is being dedicated for the development of an optimal wave inversion approach; however, conventional inversion algorithms can provide "effective" shear stiffness maps. Such an elastogram calculated from the shown wave data is indicated in Figure 12.15c, showing mostly homogenous tissue stiffness for this healthy volunteer.

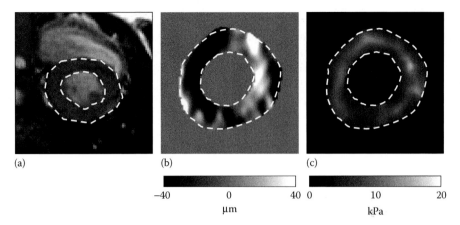

(a)   (b)   (c)

FIGURE 12.15
Cardiac MRE. (a) A short-axis view magnitude image of a human heart in end-systole *in vivo*; the boundaries of the ventricles are indicated. (b) A single 80 Hz MRE wave image showing shear wave propagation within the myocardium. (c) The corresponding shear stiffness map of the ventricular wall, indicating mostly homogeneous stiffness values with a mean value of 7.7 kPa. (Data courtesy of Dr. Kolipaka, Ohio State University Medical Center, Columbus, OH.)

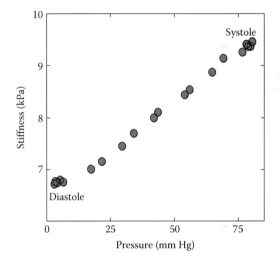

FIGURE 12.16
Cardiac MRE: stiffness vs. pressure. A plot of stiffness vs. pressure showing a good correlation between the two variables. The stiffness of the myocardium at the systole was higher than at the diastole. (Data courtesy of Dr. Kolipaka, Ohio State University Medical Center, Columbus, OH.)

Figure 12.16 shows data obtained from a pig study (Kolipaka et al., 2010), where both pressure data and MRE data were obtained on the LV. Pressure values were obtained from a pressure catheter placed within the LV cavity and MRE data were obtained at 20 cardiac phases using a cine gradient-echo retrospectively gated MRE sequence at a frequency of 80 Hz. The stiffness of the myocardium increased gradually with the LV pressure, and it can be seen that the myocardial stiffness is higher in the systole than in the diastole. Cardiac MRE is still in its early stages, but has provided promising data to indicate that this could have a huge potential in the clinical management of many cardiovascular diseases.

12.4.7 Pulmonary MRE

It is appreciated that diseases of the lung are typically associated with significant topographically heterogeneous changes in the mechanical properties. For instance, restrictive lung diseases such as interstitial lung disease increase the shear modulus significantly, and obstructive lung diseases result in significantly decreased stiffness. This indicates that the knowledge of mechanical properties of lung parenchyma may have a significant impact on clinical diagnosis and management of many lung diseases. Toward this end, MRE has been under intense investigation to be applied for the lungs (Maître et al., 2009; Mariappan et al., 2011a; McGee et al., 2008). The main challenges for successful implementation of MRE include the low MR signal within the lungs with conventional proton MRI due to the low tissue density, ultrashort T_2^* of lung parenchyma as well as cardiac- and respiratory-induced motion artifacts. There have been efforts in using contrast agents like hyperpolarized helium to improve the MR signal and they have provided very promising results (McGee et al., 2008). However, due to the expense, the scarcity of the contrast agent (Cho, 2009), and the sophistication needed for these techniques, conventional proton-based MRE is being investigated and has provided promising results.

To address the low signal within the lungs, specialized pulse sequences that provide high signal and have sufficient motion sensitivity have been developed (Mariappan et al., 2011a), and Figure 12.17 shows example preliminary data acquired from human lungs. Figure 12.17a shows the sagittal magnitude image obtained from one of the experiments; the dashed region of interest indicates the lungs that can be seen to have very low signal compared to the signal of the liver and muscle tissues. However, even with this low signal, the presence of shear wave propagation can be seen in the wave image obtained at 50 Hz shown in Figure 12.17b. Figure 12.17c shows the corresponding shear stiffness map calculated with the help of LFE algorithm. Validation of the developed technique was performed by measuring known changes in lung stiffness that occurs at varying levels of inflation/respiration. Figure 12.17a–c shows the data obtained at the residual volume (RV) and Figure 12.17d–f shows the corresponding data obtained at the total lung capacity (TLC). The shear wavelengths were longer and the shear stiffness values were higher at the TLC in comparison to the RV, in agreement with the literature (Hajji et al., 1979). A validation study was performed on a series of 10 healthy volunteers and preliminary results are shown in Figure 12.18.

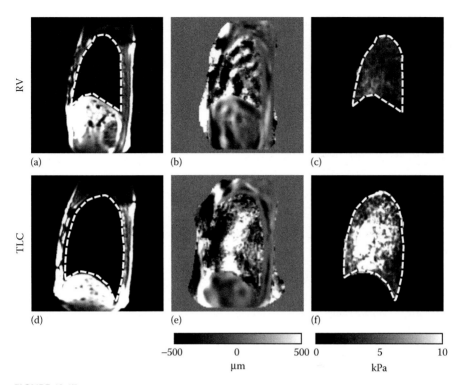

FIGURE 12.17
Lung MRE. (a) Magnitude image of the human lungs, with the volunteer holding the breath at the residual capacity (RV). (b) A single wave image at 50 Hz indicating the presence of shear waves within the lungs and (c) the corresponding shear stiffness map. (d–f) Corresponding data obtained, with the subject holding the breath at the TLC. The longer wave lengths and the higher shear stiffness values at the TLC in comparison to those at RV are easily observable.

MRE data were again obtained at both the RV and the TLC, and as expected, the RV stiffness values were lower than their counterparts at the TLC. The density values are expected to change between different states of respiration and will affect the stiffness estimation with MRE. Hence, relative density values measured from independent volumetric MR acquisition were incorporated into the measured stiffness values to calculate the density-corrected shear moduli values shown in the figure. It can be seen that the density-corrected shear stiffness values showed the expected increase in shear stiffness from RV to TLC.

12.4.8 Shear Stiffness Values of Select Tissues

As can be inferred from the above discussion, applications of MRE are rapidly expanding and exhaustive description of all these applications is

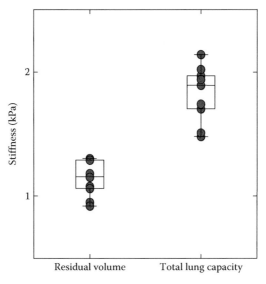

FIGURE 12.18

Lung MRE results. Results from a pilot study of 10 healthy volunteers comparing the density-corrected stiffness values at the RV and the TLC. The stiffness values at the TLC were statistically significantly higher than at the RV, in agreement with the literature. (Adapted from Mariappan, Y.K., et al., *Journal of Magnetic Resonance Imaging*. 33, 1351–1361, 2011a.)

beyond the scope of this chapter. However, the following tabulation provides example MRE-derived shear stiffness values (and the frequency at which they were obtained) for a select set of organs to indicate the development and the breadth of the field:

Tissue	Shear stiffness (kPa)	Frequency of operation (Hz)	References
Liver			
Healthy	2.2	60	Yin et al. (2007a)
Cirrhotic	8.9		
Spleen			
Healthy	3.6	60	Talwalkar et al. (2009)
Compensated cirrhosis and esophageal varices	12.6		
Pancreas			
Healthy	0.5	40	Mariappan et al. (2011b)
Chronic pancreatitis	0.7		
Prostate			
Central	2.2	65	Kemper et al. (2004)
Peripheral	3.3		

(*continued*)

Tissue	Shear stiffness (kPa)	Frequency of operation (Hz)	References
Brain			
Normal	2.4	60	Murphy et al. (2011)
Alzheimer's	2.2		
Spinal cord			
Superior portion	12.1	80	Kruse et al. (2009)
Cauda equina	2.7		
Lung			
Residual volume	1.1	40	Mariappan et al. (2011a)
Total lung capacity	1.8		
Heart			
Diastole	6.7	80	Kolipaka et al. (2010)
Systole	9.6		
Abdominal aorta			
Normotensive	3.7	60	Kolipaka et al. (2011)
Hypertensive	9.3		
Breast			
Adipose tissue	3.3		
Fibroglandular tissue	7.5	100	McKnight et al. (2002)
Tumor	25		
Muscle			
Healthy	16.6	150	Basford et al. (2002)
Neuromuscular disease	38.4		
Eye			
Vitreous humor	0.01	10	
Cornea	13.3	300	Litwiller et al. (2010a, 2010b)
Sclera	3300	300	
Kidney	2.2	45	Rouviere et al. (2011)
Heel fat pad	8.7	100	Weaver et al. (2005)
Cartilage	2000	5000	Lopez et al. (2008)
Bone	0.8×10^6	1500	Chen et al. (2009a)

12.4.9 Variational Applications with MRE

MRE is being primarily developed as an imaging tool that can quantitate the mechanical properties of soft tissues; however, the core component of displacement measurement with MRI can be used to address other clinical questions as well. Such applications include vibration imaging, HFMC MRE, shear line imaging, and compliance-weighted imaging (Mariappan et al., 2009a, 2009b, 2009c; Papazoglou et al., 2007, 2009). HFMC MRE was introduced previously where high-frequency longitudinal vibrations are induced in an object to detect the presence of stiff regions (Figure 12.13) (Mariappan et al., 2009c). Shear line imaging aims to assess the functional connectivity of slip interfaces and can be potentially used for the diagnosis of abdominal adhesions and staging of prostate tumors (Mariappan et al., 2009a).

As an example of these variational applications, vibration imaging is discussed here. Vibration imaging is a technique that can locate the functionally finger-specific, yet anatomically indistinct, subcompartments of the forearm flexor muscles (Mariappan et al., 2009c). There are two extrinsic forearm flexor muscles [flexor digitorum superficialis (FDS) and FDP] that can independently control the flexion of the index, middle, ring, and little fingers. These muscles control specific fingers through selective activation of functional subcompartments specific for each finger (Bhadra et al., 1999; Fleckenstein et al., 1992; Jeneson et al., 1990). The knowledge of the location of these subcompartments can be obtained by vibrating a specific finger and measuring the amount of motion within the forearm musculature with the displacement measurement as used in MRE.

Figure 12.19 shows an example data obtained from one of the experiments. Figure 12.19a indicates the magnitude image of an axial slice of the forearm and Figure 12.19b indicates the displacement field when the four fingers are vibrated simultaneously with known phase relation. From these displacement data, the functional subcompartments of the two forearm muscles could be identified and are indicated in Figure 12.19c for the four fingers. This technique can potentially have a role in studying the hand and wrist biomechanics.

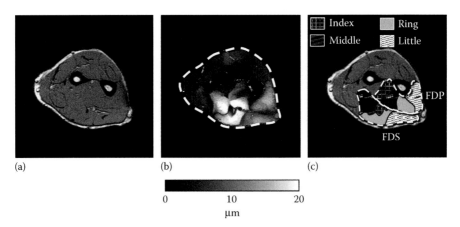

(a)   (b)   (c)

FIGURE 12.19
Vibration imaging. (a) An axial magnitude image of the forearm, indicating no discernible anatomical boundaries between the functional subcompartments of the flexor muscles. (b) Motion amplitude map obtained, with all four fingers being vibrated simultaneously with a known phase delay. (c) Overlay image of the segmented functional subcompartments of both flexor muscles for the four fingers. (Adapted from Mariappan, Y.K., et al., *Journal of Magnetic Resonance Imaging.* 31, 1395–1401, 2009a.)

12.5 Vibration Safety

MRE has significant potential to be utilized in various clinical applications, and the number of human imaging trials is expanding significantly. For clinical translation of MRE, an important factor that needs to be addressed is the safety of the proposed technique, specially the mechanical vibration. To investigate and establish the safety of MRE, the amplitude of the mechanical vibrations applied in a few applications was compared to the European Union (EU) whole-body and extremity vibration limits (Ehman et al., 2008). An excerpt of these data is shown in Figure 12.20.

Experimentally determined mean maximum amplitudes (expressed in microns) obtained from MRE exams of the liver, kidney, breast, muscle, and brain at different frequencies are shown. Frequency-weighted maximum permissible displacement values according to the EU whole-body and extremity vibration limits are shown by the solid line and the dotted lines, respectively. It can be seen that the vibrations applied for MRE are well within the safety limits. The logarithmic scale on the displacement axis, the typical locally applied vibration, and the typical <5 min vibration exposure times all underline the safety of MRE studies.

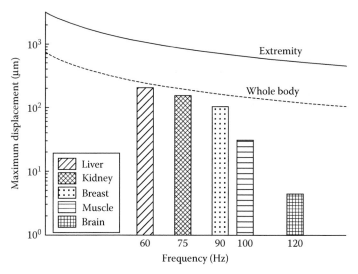

FIGURE 12.20

Vibration safety. Experimental maximum amplitudes obtained from MRE exams of liver, kidney, breast, muscle, and brain at different frequencies. The extremity and whole-body vibration limits according to the EU are indicated by the solid and the dotted lines, respectively. This figure indicates that the typical localized vibrations applied with MRE are well within the safety limits. (Adapted from Ehman, E.C., et al., *Physics in Medicine and Biology.* 53, 925–935, 2008.)

12.6 Future Directions

MRE has undergone a lot of developments and achieved clinical acceptance for hepatic fibrosis assessment since its introduction in 1995. However, there are still many challenges and hence opportunities for further technical development. Various improvements are being proposed and different application dependent improvements are being investigated in all the three steps of MRE.

From a physics perspective, the spatial resolution of the stiffness maps calculated with MRE is dependent upon the frequency of the probing shear waves; the spatial resolution increases as the frequency of the applied waves is increased. However, this also increases the attenuation of the shear waves resulting in lesser penetration depth and less reliable stiffness estimation in regions exhibiting less wave energy. Thus, there is always a trade-off between spatial resolution, the shear wave penetration depth, and the accuracy of the stiffness maps. To address this issue, improved driver technology for MRE, such as the use of arrays of multiple vibration sources, is under investigation (Mariappan et al., 2009d). In addition to the frequency-spatial resolution trade-off, the operational frequency of shear vibrations is dependent upon the stiffness of the tissue of interest; for instance, stiffer tissues such as bone, tendon, and cartilage require much higher vibration frequencies (in the kilohertz range) (Chen et al., 2009a; Lopez et al., 2008). The ability to create high-amplitude shear waves at these higher frequencies is a challenge that needs to be and are being addressed. Further, investigations are also on to improve the flexibility of the driver systems for easier translation into clinical practice (Chen et al., 2010).

The mechanical wave propagation within the human body is typically complicated by the presence of both longitudinal waves and shear waves travelling in multiple directions. While simple 2D wave imaging as performed in the clinically accepted protocol of hepatic MRE has been shown to be sufficient (Yin et al., 2009) for the assessment of the diffuse fibrotic diseases of the large, relatively homogeneous liver, many other applications require acquisition of wave data propagating in all the three orthogonal directions of motion from an entire 3D volume. This type of acquisition may be prohibitively long using conventional sequences, but research is being carried out to use faster pulse sequences such as EPI and/or parallel imaging techniques and hence are becoming more practical (Glaser et al., 2006; Kruse et al., 2006). Further, the current MR scanner gradient hardware configuration limits the ability to reliably encode high-frequency shear waves, which may be addressed in the future with specialized hardware solutions (Lopez et al., 2008).

The mathematical inversion algorithms used to calculate the stiffness maps from the displacement information as obtained from MRE have improved significantly in recent years. But with the development of newer applications and the push toward wave imaging in the 3D domain, there are many opportunities for improvement. For instance, with MRE of the cardiac muscle and the eyeball, shear waves propagate in a thin shell and the inversion algorithms need to be adapted to accommodate these complex wave propagation patterns. Another example would include MRE of the lungs, where the lung parenchyma is made up of millions of alveoli filled with air exhibiting a lot of air–tissue interfaces. To address this problem, a poroelastographic model is being developed (Perrinez et al., 2010). Further, from the information-rich MRE wave propagation data, additional tissue characterization parameters such as estimates of mechanical anisotropy, nonlinearity, and viscoelastic behavior can be determined which can potentially have clinical utility. Advanced inversion algorithms that can assess these parameters are currently under development (Asbach et al., 2008; Manduca et al., 2003b; Romano et al., 2005; Sack et al., 2002, 2004; Sinkus et al., 2005a, 2005b).

12.7 Conclusion

Dynamic MRE is a relatively novel, noninvasive, phase-contrast MR-based quantitative imaging technique, capable of quantifying the mechanical properties of soft tissues *in vivo*. It is rapidly evolving and becoming available as an optional product on conventional MRI scanners. It is currently used for the clinical assessment of hepatic fibrosis and has had significant impact on the management of fibrosis, especially in reducing the number of unnecessary biopsies. Due to the flexibility of MRI and the potential clinical value of quantitative mechanical properties, a number of other clinical applications are being developed and will have more and more impact on managing numerous diseases in the future.

References

Asbach, P., Klatt, D., Hamhaber, U., Braun, J., Somasundaram, R., Hamm, B., Sack, I., 2008. Assessment of liver viscoelasticity using multifrequency MR elastography. *Magnetic Resonance in Medicine*. 60, 373–379.

Auld, B.A., 1990. *Acoustic Fields and Waves in Solids*. Malabar, FL: R.E. Krieger.

Axel, L., Dougherty, L., 1989. MR imaging of motion with spatial modulation of magnetization. *Radiology*. 171, 841–845.

Bae, U., Dighe, M., Dubinsky, T., Minoshima, S., Shamdasani, V., Kim, Y., 2007. Ultrasound thyroid elastography using carotid artery pulsation: Preliminary study. *Journal of Ultrasound in Medicine*. 26, 797–805.

Barton, M.B., Harris, R., Fletcher, S.W., 1999. Does this patient have breast cancer?: The screening clinical breast examination: Should it be done? How? *JAMA: The Journal of the American Medical Association.* 282, 1270–1280.

Basford, J.R., Jenkyn, T.R., An, K.N., Ehman, R.L., Heers, G., Kaufman, K.R., 2002. Evaluation of healthy and diseased muscle with magnetic resonance elastography. *Archives of Physical Medicine and Rehabilitation.* 83, 1530–1536.

Bercoff, J., Tanter, M., Fink, M., 2004. Supersonic shear imaging: A new technique for soft tissue elasticity mapping. *IEEE Transactions on Ultrasonics, Ferroelectrics and Frequency Control.* 51, 396–409.

Bernstein, M.A., Zhou, X.J., Polzin, J.A., King, K.F., Ganin, A., Pelc, N.J., Glover, G.H., 1998. Concomitant gradient terms in phase contrast MR: Analysis and correction. *Magnetic Resonance in Medicine.* 39, 300–308.

Bhadra, N., Keith, M.W., Peckham, P.H., 1999. Variations in innervation of the flexor digitorum profundus muscle. *The Journal of Hand Surgery.* 24, 700–703.

Bieri, O., Maderwald, S., Ladd, M.E., Scheffler, K., 2006. Balanced alternating steady-state elastography. *Magnetic Resonance in Medicine.* 55, 233–241.

Brauck, K., Galban, C.J., Maderwald, S., Herrmann, B.L., Ladd, M.E., 2007. Changes in calf muscle elasticity in hypogonadal males before and after testosterone substitution as monitored by magnetic resonance elastography. *European Journal of Endocrinology/European Federation of Endocrine Societies.* 156, 673–678.

Braun, J., Braun, K., Sack, I., 2003. Electromagnetic actuator for generating variably oriented shear waves in MR elastography. *Magnetic Resonance in Medicine.* 50, 220–222.

Bravo, A.A., Sheth, S.G., Chopra, S., 2001. Liver biopsy. *The New England Journal of Medicine.* 344, 495–500.

Chan, Q.C.C., Li, G., Ehman, R.L., Grimm, R.C., Li, R., Yang, E.S., 2006. Needle shear wave driver for magnetic resonance elastography. *Magnetic Resonance in Medicine.* 55, 1175–1179.

Chen, J., McGregor, H., Glaser, K., Mariappan, Y., Kolipaka, A., Ehman, R., 2009a. Magnetic resonance elastography in trabecular bone: Preliminary results. In: *Proceedings of the 17th Scientific Meeting of the International Society for Magnetic Resonance in Medicine.* 18–24 April, Honolulu, HI, p. 847.

Chen, J., Ni, C., Zhuang, T., 2006. Imaging mechanical shear waves induced by piezoelectric ceramics in magnetic resonance elastography. *Chinese Science Bulletin.* 51, 755–760.

Chen, J., Stanley, D., Glaser, K.J., Yin, M., Manduca, A., Ehman, R.L., 2010. Ergonomic flexible drivers for MR Elastography. In: *Proceedings of the 18th Scientific Meeting of the International Society for Magnetic Resonance in Medicine.* 1–7 May, Stockholm, Sweden, p. 1052.

Chen, S., Urban, M.W., Pislaru, C., Kinnick, R., Zheng, Y., Yao, A., Greenleaf, J.F., 2009b. Shearwave dispersion ultrasound vibrometry (SDUV) for measuring tissue elasticity and viscosity. *IEEE Transactions on Ultrasonics, Ferroelectrics, and Frequency Control.* 56, 55–62.

Chenevert, T.L., Skovoroda, A.R., O'Donnell, M., Emelianov, S.Y., 1998. Elasticity reconstructive imaging by means of stimulated echo MRI. *Magnetic Resonance in Medicine.* 39, 482–490.

Cho, A., 2009. Physics. Helium-3 shortage could put freeze on low-temperature research. *Science.* 326, 778–779.

Chopra, R., Arani, A., Huang, Y., Musquera, M., Wachsmuth, J., Bronskill, M., Plewes, D., 2009. In vivo MR elastography of the prostate gland using a trans-urethral actuator. *Magnetic Resonance in Medicine*. 62, 665–671.

de Ledinghen, V., Le Bail, B., Rebouissoux, L., Fournier, C., Foucher, J., Miette, V., Castera, L., et al., 2007. Liver stiffness measurement in children using FibroScan: Feasibility study and comparison with fibrotest, aspartate transaminase to platelets ratio index, and liver biopsy. *Journal of Pediatric Gastroenterology and Nutrition*. 45, 443–450.

Dresner, M.A., Rose, G.H., Rossman, P.J., Muthupillai, R., Manduca, A., Ehman, R.L., 2001. Magnetic resonance elastography of skeletal muscle. *Journal of Magnetic Resonance Imaging*. 13, 269–276.

Dumoulin, C.L., Hart, H.R., Jr., 1986. Magnetic resonance angiography in the head and neck. *Acta Radiologica Supplement*. 369, 17–20.

Egorov, V., Ayrapetyan, S., Sarvazyan, A.P., 2006. Prostate mechanical imaging: 3-D image composition and feature calculations. *IEEE Transactions on Medical Imaging*. 25, 1329–1340.

Ehman, E.C., Rossman, P.J., Kruse, S.A., Sahakian, A.V., Glaser, K.J., 2008. Vibration safety limits for magnetic resonance elastography. *Physics in Medicine and Biology*. 53, 925–935.

Elias, D., Sideris, L., Pocard, M., de Baere, T., Dromain, C., Lassau, N., Lasser, P., 2005. Incidence of unsuspected and treatable metastatic disease associated with operable colorectal liver metastases discovered only at laparotomy (and not treated when performing percutaneous radiofrequency ablation). *Annals of Surgical Oncology*. 12, 298–302.

Fatemi, M., Greenleaf, J.F., 1998. Ultrasound-stimulated vibro-acoustic spectrography. *Science*. 280, 82–85.

Fleckenstein, J.L., Watumull, D., Bertocci, L.A., Parkey, R.W., Peshock, R.M., 1992. Finger-specific flexor recruitment in humans: Depiction by exercise-enhanced MRI. *Journal of Applied Physiology*. 72, 1974–1977.

Ghiglia, D.C., Pritt, M.D., 1998. *Two-Dimensional Phase Unwrapping: Theory, Algorithms and Software*. New York: Wiley.

Glaser, K.J., Felmlee, J.P., Ehman, R.L., 2006. Rapid MR elastography using selective excitations. *Magnetic Resonance in Medicine*. 55, 1381–1389.

Goss, B.C., McGee, K.P., Ehman, E.C., Manduca, A., Ehman, R.L., 2006. Magnetic resonance elastography of the lung: Technical feasibility. *Magnetic Resonance in Medicine*. 56, 1060–1066.

Green, M.A., Bilston, L.E., Sinkus, R., 2008. In vivo brain viscoelastic properties measured by magnetic resonance elastography. *NMR in Biomedicine*. 21, 755–764.

Greenleaf, J.F., Fatemi, M., Insana, M., 2003. Selected methods for imaging elastic properties of biological tissues. *Annual Review of Biomedical Engineering*. 5, 57–78.

Grimm, R.C., Yin, M., Ehman, R.L., 2007. Gradient moment nulling in MR elastography of the liver. In: *Proceedings of the 15th Scientific Meeting of the International Society for Magnetic Resonance in Medicine*. 19–25 May, Berlin, Germany, p. 961.

Habelitz, S., Marshall, S.J., Marshall, G.W., Jr., Balooch, M., 2001. Mechanical properties of human dental enamel on the nanometre scale. *Archives of Oral Biology*. 46, 173–183.

Hajji, M.A., Wilson, T.A., Lai-Fook, S.J., 1979. Improved measurements of shear modulus and pleural membrane tension of the lung. *Journal of Applied Physiology: Respiratory, Environmental and Exercise Physiology*. 47, 175–181.

Heywang, S.H., Hahn, D., Schmidt, H., Krischke, I., Eiermann, W., Bassermann, R., Lissner, J., 1986. MR imaging of the breast using gadolinium-DTPA. *Journal of Computer Assisted Tomography*. 10, 199–204.

Heywang-Kobrunner, S.H., Viehweg, P., Heinig, A., Kuchler, C., 1997. Contrast-enhanced MRI of the breast: Accuracy, value, controversies, solutions. *European Journal of Radiology*. 24, 94–108.

Huang, D., Swanson, E.A., Lin, C.P., Schuman, J.S., Stinson, W.G., Chang, W., Hee, M.R., et al., 1991. Optical coherence tomography. *Science*. 254, 1178–1181.

Huwart, L., Sempoux, C., Salameh, N., Jamart, J., Annet, L., Sinkus, R., Peeters, F., ter Beek, L.C., Horsmans, Y., Van Beers, B.E., 2007. Liver fibrosis: Noninvasive assessment with MR elastography versus aspartate aminotransferase-to-platelet ratio index. *Radiology*. 245, 458–466.

Jeneson, J.A., Taylor, J.S., Vigneron, D.B., Willard, T.S., Carvajal, L., Nelson, S.J., Murphy-Boesch, J., Brown, T.R., 1990. 1H MR imaging of anatomical compartments within the finger flexor muscles of the human forearm. *Magnetic Resonance in Medicine*. 15, 491–496.

Kanai, H., 2004. Viscoelasticity measurement of heart wall in in vivo. In: *Ultrasonics Symposium, 2004 IEEE*. 24–27 August, Montreal, QC, Canada, Vol. 1, pp. 482–485.

Kemper, J., Sinkus, R., Lorenzen, J., Nolte-Ernsting, C., Stork, A., Adam, G., 2004. MR elastography of the prostate: Initial in vivo application. *RoFo; Fortschritte auf dem Gebiete der Rontgenstrahlen und der Nuklearmedizin*. 176, 1094–1099.

Knutsson, H., Westin, C.F., Granlund, G., 1994. Local multiscale frequency and bandwidth estimation. In: *Proceedings of the IEEE International Conference on Image Processing*. 13–16 November, Austin, TX, Vol. 1, pp. 36–40.

Kolipaka, A., Araoz, P.A., McGee, K.P., Manduca, A., Ehman, R.L., 2010. Magnetic resonance elastography as a method for the assessment of effective myocardial stiffness throughout the cardiac cycle. *Magnetic Resonance in Medicine*. 64, 862–870.

Kolipaka, A., McGee, K.P., Araoz, P.A., Glaser, K.J., Manduca, A., Romano, A.J., Ehman, R.L., 2009. MR elastography as a method for the assessment of myocardial stiffness: Comparison with an established pressure-volume model in a left ventricular model of the heart. *Magnetic Resonance in Medicine*. 62, 135–140.

Kolipaka, A., Woodrum, D., Araoz, P.A., Ehman, R.L., 2011. MR elastography of the in vivo abdominal aorta: A feasibility study for comparing aortic stiffness between hypertensives and normotensives. *Journal of Magnetic Resonance Imaging*. 35, 582–586.

Krouskop, T.A., Dougherty, D.R., Vinson, F.S., 1987. A pulsed Doppler ultrasonic system for making noninvasive measurements of the mechanical properties of soft tissue. *Journal of Rehabilitation Research and Development*. 24, 1–8.

Kruse, S., Kolipaka, A., Manduca, A., Ehman, R., 2009. Feasibility of evaluating the spinal cord with MR elastography. In: *Proceedings of the 17th Scientific Meeting of the International Society for Magnetic Resonance in Medicine*. 18–24 April, Honolulu, HI, p. 629.

Kruse, S.A., Grimm, R.C., Lake, D.S., Manduca, A., Ehman, R.L., 2006. Fast EPI based 3D MR elastography of the brain. In: *Proceedings of the 14th Scientific Meeting of the International Society for Magnetic Resonance in Medicine*. 6–12 May, Seattle, WA, p. 3385.

Kruse, S.A., Rose, G.H., Glaser, K.J., Manduca, A., Felmlee, J.P., Jack, C.R., Ehman, R.L., 2008. Magnetic resonance elastography of the brain. *NeuroImage*. 39, 231–237.

Lee, W.N., Qian, Z., Tosti, C.L., Brown, T.R., Metaxas, D.N., Konofagou, E.E., 2008. Preliminary validation of angle-independent myocardial elastography using MR tagging in a clinical setting. *Ultrasound in Medicine and Biology*. 34, 1980–1997.

Lerner, R.M., Huang, S.R., Parker, K.J., 1990. "Sonoelasticity" images derived from ultrasound signals in mechanically vibrated tissues. *Ultrasound in Medicine and Biology*. 16, 231–239.

Levinson, S.F., Shinagawa, M., Sato, T., 1995. Sonoelastic determination of human skeletal-muscle elasticity. *Journal of Biomechanics*. 28, 1145–1154.

Litwiller, D.V., Lee, S.J., Kolipaka, A., Mariappan, Y.K., Glaser, K.J., Pulido, J.S., Ehman, R.L., 2010a. MR elastography of the ex vivo bovine globe. *Journal of Magnetic Resonance Imaging*. 32, 44–51.

Litwiller, D.V., Mariappan, Y., Ehman, R.L., 2010b. MR elastography of the ocular vitreous body. In: *Proceedings of the 18th Scientific Meeting of the International Society for Magnetic Resonance in Medicine*. 1–7 May, Stockholm, Sweden, p. 2414.

Lomas, D.J., Graves, N.J., Muthupillai, R., Ehman, R.L., 1997. MR elastography of human liver specimens: Initial results. *British Journal of Radiology*. 70, 124.

Lopez, O., Amrami, K.K., Manduca, A., Ehman, R.L., 2008. Characterization of the dynamic shear properties of hyaline cartilage using high-frequency dynamic MR elastography. *Magnetic Resonance in Medicine*. 59, 356–364.

Maderwald, S., Uffmann, K., Galban, C.J., de Greiff, A., Ladd, M.E., 2006. Accelerating MR elastography: A multiecho phase-contrast gradient-echo sequence. *Journal of Magnetic Resonance Imaging*. 23, 774–780.

Mai, J.J., Insana, M.F., 2002. Strain imaging of internal deformation. *Ultrasound in Medicine and Biology*. 28, 1475–1484.

Maître, X., Sinkus, R., Santarelli, R., Sarracanie, M., Dubuisson, R., Boriasse, E., Durand, E., Darrasse, L., Bittoun, J., 2009. In vivo lung elastography with hyperpolarized helium-3 MRI. In: *Proceedings of the 17th Scientific Meeting of the International Society for Magnetic Resonance in Medicine*. 18–24 April, Honolulu, HI, p. 6.

Manduca, A., Lake, D.S., Kruse, S.A., Ehman, R.L., 2003a. Spatio-temporal directional filtering for improved inversion of MR elastography images. *Medical Image Analysis*. 7, 465–473.

Manduca, A., Lake, D.S., Kugel, J.L., Rossman, P.J., Ehman, R.L., 2003b. Dispersion measurements from simultaneous multi-frequency MR elastography. In: *Proceedings of the 11th Scientific Meeting of the International Society for Magnetic Resonance in Medicine*. 10–16 July, Toronto, ON, Canada, p. 550.

Manduca, A., Oliphant, T.E., Dresner, M.A., Mahowald, J.L., Kruse, S.A., Amromin, E., Felmlee, J.P., Greenleaf, J.F., Ehman, R.L., 2001. Magnetic resonance elastography: Non-invasive mapping of tissue elasticity. *Medical Image Analysis*. 5, 237–254.

Mariappan, Y.K., Glaser, K.J., Ehman, R.L., 2010. Magnetic resonance elastography: A review. *Clinical Anatomy*. 23, 497–511.

Mariappan, Y.K., Glaser, K.J., Hubmayr, R.D., Manduca, A., Ehman, R.L., McGee, K.P., 2011a. MR elastography of human lung parenchyma: Technical development, theoretical modeling and in vivo validation. *Journal of Magnetic Resonance Imaging*. 33, 1351–1361.

Mariappan, Y.K., Glaser, K.J., Manduca, A., Ehman, R.L., 2009a. Cyclic motion encoding for enhanced MR visualization of slip interfaces. *Journal of Magnetic Resonance Imaging*. 30, 855–863.

Mariappan, Y.K., Glaser, K.J., Manduca, A., Ehman, R.L., 2009b. Vibration imaging for functional analysis of flexor muscle compartments. *Journal of Magnetic Resonance Imaging*. 31, 1395–1401.

Mariappan, Y.K., Glaser, K.J., Manduca, A., Romano, A.J., Venkatesh, S.K., Yin, M., Ehman, R.L., 2009c. High-frequency mode conversion technique for stiff lesion detection with magnetic resonance elastography (MRE). *Magnetic Resonance in Medicine*. 62, 1457–1465.

Mariappan, Y.K., Glaser, K.J., Takahashi, N., Young, P., Ehman, R.L., 2011b. Assessment of chronic pancreatitis with MR elastography. In: *Proceedings of the 19th Scientific Meeting of the International Society for Magnetic Resonance in Medicine*. 7–13 May, Montreal, QC, Canada, p. 812.

Mariappan, Y.K., Rossman, P.J., Glaser, K.J., Manduca, A., Ehman, R.L., 2009d. Magnetic resonance elastography with a phased-array acoustic driver system. *Magnetic Resonance in Medicine*. 61, 678–685.

Mason, P., 1977. Dynamic stiffness and crossbridge action in muscle. *Biophysics Structure Mechanics*. 4, 15–25.

McCracken, P.J., Manduca, A., Felmlee, J.P., Ehman, R.L., 2005. Mechanical transient-based magnetic resonance elastography. *Magnetic Resonance in Medicine*. 53, 628–639.

McGee, K.P., Hubmayr, R.D., Ehman, R.L., 2008. MR elastography of the lung with hyperpolarized ^{3}He. *Magnetic Resonance in Medicine*. 59, 14–18.

McKnight, A.L., Kugel, J.L., Rossman, P.J., Manduca, A., Hartmann, L.C., Ehman, R.L., 2002. MR elastography of breast cancer: Preliminary results. *American Journal of Roentgenology*. 178, 1411–1417.

Moran, P.R., 1982. A flow velocity zeugmatographic interlace for NMR imaging in humans. *Magnetic Resonance Imaging*. 1, 197–203.

Murphy, M.C., Huston, J., III, Jack, C.R., Jr., Glaser, K.J., Manduca, A., Felmlee, J.P., Ehman, R.L., 2011. Decreased brain stiffness in Alzheimer's disease determined by magnetic resonance elastography. *Journal of Magnetic Resonance Imaging*. 34, 494–498.

Muthupillai, R., Lomas, D.J., Rossman, P.J., Greenleaf, J.F., Manduca, A., Ehman, R.L., 1995. Magnetic resonance elastography by direct visualization of propagating acoustic strain waves. *Science*. 269, 1854–1857.

Muthupillai, R., Rossman, P.J., Lomas, D.J., Greenleaf, J.F., Riederer, S.J., Ehman, R.L., 1996. Magnetic resonance imaging of transverse acoustic strain waves. *Magnetic Resonance in Medicine*. 36, 266–274.

Nedredal, G.I., Yin, M., McKenzie, T., Lillegard, J., Luebke-Wheeler, J., Talwalkar, J., Ehman, R., Nyberg, S.L., 2011. Portal hypertension correlates with splenic stiffness as measured with MR elastography. *Journal of Magnetic Resonance Imaging*. 34, 79–87.

Nightingale, K., Soo, M.S., Nightingale, R., Trahey, G., 2002. Acoustic radiation force impulse imaging: In vivo demonstration of clinical feasibility. *Ultrasound in Medicine and Biology*. 28, 227–235.

Nightingale, K.R., Palmeri, M.L., Nightingale, R.W., Trahey, G.E., 2001. On the feasibility of remote palpation using acoustic radiation force. *The Journal of the Acoustical Society of America*. 110, 625–634.

O'Donnell, M., Skovoroda, A.R., Shapo, B.M., Emelianov, S.Y., 1994. Internal displacement and strain imaging using ultrasonic speckle tracking. *IEEE Transactions on Ultrasonics, Ferroelectrics and Frequency Control*. 41, 314–325.

Ophir, J., Cespedes, I., Ponnekanti, H., Yazdi, Y., Li, X., 1991. Elastography: A quantitative method for imaging the elasticity of biological tissues. *Ultrasonic Imaging.* 13, 111–134.

Ophir, J., Miller, R.K., Ponnekanti, H., Cespedes, I., Whittaker, A.D., 1994. Elastography of beef muscle. *Meat Science.* 36, 239–250.

Osman, N.F., 2003. Detecting stiff masses using strain-encoded (SENC) imaging. *Magnetic Resonance in Medicine.* 49, 605–608.

Othman, S.F., Xu, H., Royston, T.J., Magin, R.L., 2005. Microscopic magnetic resonance elastography (mMRE). *Magnetic Resonance in Medicine.* 54, 605–615.

Papazoglou, S., Hamhaber, U., Braun, J., Sack, I., 2007. Horizontal shear wave scattering from a nonwelded interface observed by magnetic resonance elastography. *Physics in Medicine and Biology.* 52, 675–684.

Papazoglou, S., Xu, C., Hamhaber, U., Siebert, E., Bohner, G., Klingebiel, R., Braun, J., Sack, I., 2009. Scatter-based magnetic resonance elastography. *Physics in Medicine and Biology.* 54, 2229–2241.

Perrinez, P.R., Kennedy, F.E., Van Houten, E.E., Weaver, J.B., Paulsen, K.D., 2010. Magnetic resonance poroelastography: An algorithm for estimating the mechanical properties of fluid-saturated soft tissues. *IEEE Transactions on Medical Imaging.* 29, 746–755.

Plewes, D.B., Betty, I., Urchuk, S.N., Soutar, I., 1995. Visualizing tissue compliance with MR imaging. *Journal of Magnetic Resonance Imaging.* 5, 733–738.

Rabe, U., Amelio, S., Kopycinska, M., Hirsekorn, S., Kempf, M., Göken, M., Arnold, W., 2002. Imaging and measurement of local mechanical material properties by atomic force acoustic microscopy. *Surface and Interface Analysis.* 33, 65–70.

Ringleb, S.I., Bensamoun, S.F., Chen, Q., Manduca, A., An, K.N., Ehman, R.L., 2007. Applications of magnetic resonance elastography to healthy and pathologic skeletal muscle. *Journal of Magnetic Resonance Imaging.* 25, 301–309.

Rogowska, J., Patel, N.A., Fujimoto, J.G., Brezinski, M.E., 2004. Optical coherence tomographic elastography technique for measuring deformation and strain of atherosclerotic tissues. *Heart.* 90, 556–562.

Romano, A.J., Abraham, P.B., Rossman, P.J., Bucaro, J.A., Ehman, R.L., 2005. Determination and analysis of guided wave propagation using magnetic resonance elastography. *Magnetic Resonance in Medicine.* 54, 893–900.

Romano, A.J., Rossman, P.J., Grimm, R.C., Bucaro, J.A., Ehman, R.L., 2003. Broad-spectrum beam magnetic resonance elastography. In: *Proceedings of the 16th Scientific Meeting of the International Society for Magnetic Resonance in Medicine.* 3–9 May, Toronto, ON, Canada, p. 1083.

Romano, A.J., Shirron, J.J., Bucaro, J.A., 1998. On the noninvasive determination of material parameters from a knowledge of elastic displacements: Theory and numerical simulation. *IEEE Transactions on Ultrasonics Ferroelectrics and Frequency Control.* 45, 751–759.

Rossman, P., Glaser, K., Felmlee, J., Ehman, R., 2003. Piezoelectric bending elements for use as motion actuators in MR elastography. In: *Proceedings of the 16th Scientific Meeting of the International Society for Magnetic Resonance in Medicine.* 3–9 May, Toronto, ON, Canada, p. 1075.

Rouviere, O., Souchon, R., Pagnoux, G., Menager, J.M., Chapelon, J.Y., 2011. Magnetic resonance elastography of the kidneys: Feasibility and reproducibility in young healthy adults. *Journal of Magnetic Resonance Imaging.* 34, 880–886.

Ruikang, K.W., Zhenhe, M., Sean, J.K., 2006. Tissue Doppler optical coherence elastography for real time strain rate and strain mapping of soft tissue. *Applied Physics Letters*. 89, 144103.

Rump, J., Klatt, D., Braun, J., Warmuth, C., Sack, I., 2007. Fractional encoding of harmonic motions in MR elastography. *Magnetic Resonance in Medicine*. 57, 388–395.

Rydberg, J., Grimm, R., Kruse, S., Felmlee, J., McCracken, P., Ehman, R., 2001. Fast spin-echo magnetic resonance elastography of the brain. In: *Proceedings of the 9th Scientific Meeting of the International Society for Magnetic Resonance in Medicine*. 21–27 April, Glasgow, Scotland, p. 1647.

Sack, I., Beierbach, B., Wuerfel, J., Klatt, D., Hamhaber, U., Papazoglou, S., Martus, P., Braun, J., 2009a. The impact of aging and gender on brain viscoelasticity. *NeuroImage*. 46, 652–657.

Sack, I., Bernarding, J., Braun, J., 2002. Analysis of wave patterns in MR elastography of skeletal muscle using coupled harmonic oscillator simulations. *Magnetic Resonance Imaging*. 20, 95–104.

Sack, I., Buntkowsky, G., Bernarding, J., Braun, J., 2001. Magnetic resonance elastography: A method for the noninvasive and spatially resolved observation of phase transitions in gels. *Journal of the American Chemical Society*. 123, 11087–11088.

Sack, I., Mcgowan, C.K., Samani, A., Luginbuhl, C., Oakden, W., Plewes, D.B., 2004. Observation of nonlinear shear wave propagation using magnetic resonance elastography. *Magnetic Resonance in Medicine*. 52, 842–850.

Sack, I., Rump, J., Elgeti, T., Samani, A., Braun, J., 2009b. MR elastography of the human heart: Noninvasive assessment of myocardial elasticity changes by shear wave amplitude variations. *Magnetic Resonance in Medicine*. 61, 668–677.

Sandrin, L., Fourquet, B., Hasquenoph, J.M., Yon, S., Fournier, C., Mal, F., Christidis, C., et al., 2003. Transient elastography: A new noninvasive method for assessment of hepatic fibrosis. *Ultrasound in Medicine and Biology*. 29, 1705–1713.

Sarvazyan, A., 1998. Mechanical imaging—A new technology for medical diagnostics. *International Journal of Medical Informatics*. 49, 195–216.

Sarvazyan, A., Hall, T.J., Urban, M.W., Fatemi, M., Aglyamov, S.R., Garra, B.S., 2011. An overview of elastography—An emerging branch of medical imaging. *Current Medical Imaging Reviews*. 7, 255–282.

Sarvazyan, A.P., Skovoroda, A.R., Emelianov, S.Y., Fowlkes, J.B., Pipe, J.G., Adler, R.S., Buxton, R.B., Carson, P.L., 1995. Biophysical bases of elasticity imaging. In: *Acoustical Imaging*. Vol. 21, J.P. Jones, ed. New York: Plenum Press, pp. 223–240.

Shah, N.S., Kruse, S.A., Lager, D.J., Farell-Baril, G., Lieske, J.C., King, B.F., Ehman, R.L., 2004. Evaluation of renal parenchymal disease in a rat model with magnetic resonance elastography. *Magnetic Resonance in Medicine*. 52, 56–64.

Shaheen, N.J., Hansen, R.A., Morgan, D.R., Gangarosa, L.M., Ringel, Y., Thiny, M.T., Russo, M.W., Sandler, R.S., 2006. The burden of gastrointestinal and liver diseases, 2006. *The American Journal of Gastroenterology*. 101, 2128–2138.

Sinkus, R., Daire, J.-L., Vilgrain, V., Van Beers, B.E., 2012. Elasticity imaging via MRI: Basics, overcoming the waveguide limit, and clinical liver results. *Current Medical Imaging Reviews*. 8, 56–63.

Sinkus, R., Tanter, M., Catheline, S., Lorenzen, J., Kuhl, C., Sondermann, E., Fink, M., 2005a. Imaging anisotropic and viscous properties of breast tissue by magnetic resonance elastography. *Magnetic Resonance in Medicine*. 53, 372–387.

Sinkus, R., Tanter, M., Xydeas, T., Catheline, S., Bercoff, J., Fink, M., 2005b. Viscoelastic shear properties of in vivo breast lesions measured by MR elastography. *Magnetic Resonance Imaging.* 23, 159–165.

Steele, D.D., Chenevert, T.L., Skovoroda, A.R., Emelianov, S.Y., 2000. Three-dimensional static displacement, stimulated echo NMR elasticity imaging. *Physics in Medicine and Biology.* 45, 1633–1648.

Streitberger, K.J., Wiener, E., Hoffmann, J., Freimann, F.B., Klatt, D., Braun, J., Lin, K., et al., 2010. In vivo viscoelastic properties of the brain in normal pressure hydro-cephalus. *NMR in Biomedicine.* 24, 385–392.

Talwalkar, J.A., Yin, M., Venkatesh, S., Rossman, P.J., Grimm, R.C., Manduca, A., Romano, A., Kamath, P.S., Ehman, R.L., 2009. Feasibility of in vivo MR elasto-graphic splenic stiffness measurements in the assessment of portal hypertension. *AJR. American Journal of Roentgenology.* 193, 122–127.

Uffmann, K., Abicht, C., Grote, W., Quick, H.H., Ladd, M.E., 2002. Design of an MR-compatible piezoelectric actuator for MR elastography. *Concepts in Magnetic Resonance.* 15, 239–254.

Van Houten, E.E.W., Paulsen, K.D., Miga, M.I., Kennedy, F.E., Weaver, J.B., 1999. An overlapping subzone technique for MR-based elastic property reconstruction. *Magnetic Resonance in Medicine.* 42, 779–786.

van Soest, G., Mastik, F., de Jong, N., van der Steen, A.F., 2007. Robust intravascular optical coherence elastography by line correlations. *Physics in Medicine Biology.* 52, 2445–2458.

Varghese, T., Shi, H., 2004. Elastographic imaging of thermal lesions in liver in-vivo using diaphragmatic stimuli. *Ultrasonic Imaging.* 26, 18–28.

Venkatesh, S.K., Yin, M., Glockner, J.F., Takahashi, N., Araoz, P.A., Talwalkar, J.A., Ehman, R.L., 2008a. MR elastography of liver tumors: Preliminary results. *AJR. American Journal of Roentgenology.* 190, 1534–1540.

Venkatesh, S.K., Yin, M., Talwalkar, J.A., Ehman, R.L., 2008b. Application of liver MR elastography in clinical practice. In: *Proceedings of the 16th Scientific Meeting of the International Society for Magnetic Resonance in Medicine.* 3–9 May, Toronto, ON, Canada, p. 2611.

Von Gierke, H.E., Oestreicher, H.L., Franke, E.K., Parrack, H.O., Wittern, W.W., 1952. Physics of vibrations in living tissues. *Journal of Applied Physiology.* 4, 886–900.

Wakeling, J.M., Nigg, B.M., 2001. Soft-tissue vibrations in the quadriceps measured with skin mounted transducers. *Journal of Biomechanics.* 34, 539–543.

Warner, L., Yin, M., Glaser, K.J., Woollard, J.A., Carrascal, C.A., Korsmo, M.J., Crane, J.A., Ehman, R.L., Lerman, L.O., 2011. Noninvasive in vivo assessment of renal tissue elasticity during graded renal ischemia using MR elastography. *Investigative Radiology.* 46, 509–514.

Weaver, J.B., Doyley, M., Cheung, Y., Kennedy, F., Madsen, E.L., Van Houten, E.E.W., Paulsen, K., 2005. Imaging the shear modulus of the heel fat pads. *Clinical Biomechanics.* 20, 312–319.

Wilson, L.S., Robinson, D.E., Dadd, M.J., 2000. Elastography—The movement begins. *Physics in Medicine Biology.* 45, 1409–1421.

Wu, T., Felmlee, J.P., Greenleaf, J.F., Riederer, S.J., Ehman, R.L., 2000. MR imaging of shear waves generated by focused ultrasound. *Magnetic Resonance in Medicine.* 43, 111–115.

Wuerfel, J., Paul, F., Beierbach, B., Hamhaber, U., Klatt, D., Papazoglou, S., Zipp, F., Martus, P., Braun, J., Sack, I., 2010. MR-elastography reveals degradation of tissue integrity in multiple sclerosis. *NeuroImage*. 49, 2520–2525.

Xu, L., Lin, Y., Xi, Z.N., Shen, H., Gao, P.Y., 2007. Magnetic resonance elastography of the human brain: A preliminary study. *Acta Radiologica*. 48, 112–115.

Yamakoshi, Y., Sato, J., Sato, T., 1990. Ultrasonic imaging of internal vibration of soft tissue under forced vibration. *IEEE Transactions on Ultrasonics, Ferroelectrics, and Frequency Control*. 37, 45–53.

Yamanaka, N., Okamoto, E., Toyosaka, A., Ohashi, S., Tanaka, N., 1985. Consistency of human liver. *The Journal of Surgical Research*. 39, 192–198.

Yin, M., Glaser, K.J., Talwalkar, J., Manduca, A., Ehman, R.L., 2009. Validity of a 2D wavefield model in MR elastography of the liver. In: *Proceedings of the 17th Scientific Meeting of the International Society for Magnetic Resonance in Medicine*. 18–24 April, Honolulu, HI, p. 709.

Yin, M., Rouviere, O., Glaser, K.J., Ehman, R.L., 2008. Diffraction-biased shear wave fields generated with longitudinal magnetic resonance elastography drivers. *Magnetic Resonance Imaging*. 26, 770–780.

Yin, M., Talwalkar, J.A., Glaser, K.J., Manduca, A., Grimm, R.C., Rossman, P.J., Fidler, J.L., Ehman, R.L., 2007a. Assessment of hepatic fibrosis with magnetic resonance elastography. *Clinical Gastroenterology and Hepatology*. 5, 1207–1213.

Yin, M., Woollard, J., Wang, X.F., Torres, V.E., Harris, P.C., Ward, C.J., Glaser, K.J., Manduca, A., Ehman, R.L., 2007b. Quantitative assessment of hepatic fibrosis in an animal model with magnetic resonance elastography. *Magnetic Resonance in Medicine*. 58, 346–353.

Zerhouni, E.A., Parish, D.M., Rogers, W.J., Yang, A., Shapiro, E.P., 1988. Human heart: Tagging with MR imaging—A method for noninvasive assessment of myocardial motion. *Radiology*. 169, 59–63.

13

Imaging Conductivity and Permittivity of Tissues Using Electric Properties Tomography

Ulrich Katscher and Tobias Voigt

Philips Technologie GmbH

CONTENTS

13.1 Introduction

Electric properties, that is, electric conductivity and permittivity, exhibit a large variety across the different types of human tissues (see, e.g., Gabriel et al., 1996). Mapping these electric properties yields images rich in contrast and information. Since this contrast is found in both healthy and pathologic tissues, the knowledge of tissue electric properties has great potential to help answer a broad range of diagnostic questions. Prominent examples of possible diagnostic applications are staging of tumors (Haemmerich et al., 2003; Joines et al., 1994; Schepps and Foster, 1980; Surowiec et al., 1988), changes in brain tissue after stroke (Liu et al., 2006), and heart muscle viability after myocardial infarction (Schaefer et al., 2002). Numerous other potential applications can be considered, such as control of radio frequency (RF) ablation or RF hyperthermia (McRae and Esrick, 1993) and conductivity of brain and skull for electrocardiographic (EEG) and magnetoencephalographic (MEG) mapping (van Oosterom, 1991).

Harvesting diagnostic information provided by the electric properties is hampered by the inability to accurately measure these parameters noninvasively *in vivo*. The majority of published medical electric properties data have been derived from *ex vivo* tissue samples or invasive animal studies. This chapter discusses a novel approach to measure electric properties *in vivo* and noninvasively using MRI. The quantitative mapping of electric conductivity and permittivity using MRI was first suggested in 1991 (Haacke et al., 1991). However, only recently systematic research began to exploit this idea via electric properties tomography (EPT) (Katscher et al., 2006, 2009; Wen, 2003). EPT allows quantitative determination of the conductivity and permittivity by postprocessing the RF transmit field map acquired on a standard MR scanner.

MR electrical impedance tomography (MREIT) is an alternative MR-based method for imaging tissue conductivity (Muftuler et al., 2004; Seo et al., 2005). MREIT takes advantage of the spatial encoding capabilities of MRI; however, this technique applies external currents to the patient via mounted electrodes. Another MR-based approach, called RF current density imaging (RF-CDI), eliminates the need for mounting electrodes by applying a separate RF pulse for current induction (Beravs et al., 1999; Scott et al., 1995; Wang et al., 2009). These methods differ from the EPT approach presented in this chapter. EPT uses a standard MR system, requires neither electrode mounting nor the application of additional RF energy, but postprocesses the 3D map of the RF transmit pulse of the imaging sequence (so-called B_1-map). The transmit field is characteristically distorted according to the electric properties of the patient's tissues. By measuring the distorted transmit field, this allows reconstruction of the underlying electric properties responsible for the distortions observed. To illustrate this feature, Figure 13.1 shows the correlation between the field of a quadrature RF coil, distorted by a spherical object, and the undistorted field of the empty coil. This correlation is close to 100% for lower values of the electric properties and goes down in a characteristic manner for increasing values of the electric properties. For the electric properties of human tissues, the correlation is typically around 70%. Thus, a measurable effect can be expected when investigating patients with EPT.

Conductivity and permittivity depend on the frequency of the applied fields. For all biological tissue types, conductivity increases with frequency and permittivity decreases with frequency. For MRI, fields have

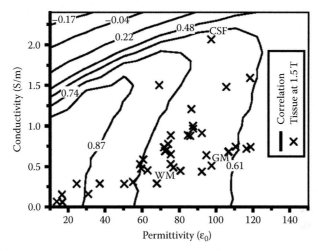

FIGURE 13.1
Mean flip angle alteration caused by a homogeneous sphere at 1.5 T. This provides the opportunity that measuring the distorted RF transmit field allows one to reconstruct the tissue's electric properties, which is the basic idea behind EPT. (From Voigt, T., Quantitative MR imaging of the electric properties and local SAR based on improved RF transmit field mapping (band 11). PhD dissertation. KIT Scientific Publishing, Karlsruhe, Germany, 2011.)

to be applied at Larmor frequency according to the main field strength of the MR system. Electric properties determined by EPT correspond to this Larmor frequency, which is significantly higher than the frequency used in "classical" bioimpedance studies. However, besides their potential diagnostic value, electric properties are of crucial importance in the area of RF safety. Potential tissue heating is a major problem at high field MR, particularly in the framework of parallel RF transmission (Graesslin et al., 2006; Seifert et al., 2007; Zelinski et al., 2008). Maximum allowed specific absorption rate (SAR), directly related to tissue heating, often limits the parameter space for specific MR sequences. Thus, an accurate patient–specific SAR determination is highly desirable, and EPT might provide a step toward such patient–specific SAR determination with sufficient accuracy and short computation time (Katscher et al., 2009; Voigt et al., 2009). For SAR calculation, it is essential to determine the electric properties at the MR system's Larmor frequency, and this can be accomplished using EPT.

EPT is based on the measurement of the magnetic RF field with MRI. This is a nontrivial task, and comes with a couple of technical and physical problems. Different workarounds for these problems are possible, leading to different versions of EPT. We begin this chapter with the discussion of the mathematical background for different versions of EPT. Different model assumptions of EPT are presented and possibilities to overcome these assumptions are shown. Lastly, EPT results of research groups around the world are summarized, including phantom simulations and experiments as well as studies of volunteers and patients.

13.2 Theory

13.2.1 Relation between Magnetic Field and Electric Properties

EPT helps quantify the electrical properties (conductivity and permittivity) of the tissues, which are derived from MRI. The transmitted RF pulse is affected by these electrical properties of the tissues being imaged. The corresponding relation is obtained by suitably rearranging Maxwell's equations, which can be done in many different ways. In the following, the method is presented as given in Voigt et al. (2011a).

The magnetic field strength vector $\mathbf{H}$ and the electric field vector $\mathbf{E}$, corresponding to the RF fields of the MR system, are assumed to be time-harmonic $\mathbf{H}, \mathbf{E} \sim \exp(i\omega t)$. With this assumption, Ampere's law is integrated along a closed path ∂A, which is given by the border of the area A:

$$i\omega \oint_{\partial A} \nabla \times \mathbf{H}(\mathbf{r}) \cdot d\mathbf{l} = \oint_{\partial A} \kappa(\mathbf{r}) \mathbf{E}(\mathbf{r}) \cdot d\mathbf{l}. \tag{13.1}$$

Here, $\kappa = \varepsilon - i\sigma/\omega$ denotes the complex permittivity, with ε being the scalar permittivity, σ the electric conductivity, and ω the Larmor frequency, and $d\mathbf{l}$ denotes a line element. The spatial position is given by $\mathbf{r}$. For the time being, an isotropic κ is assumed. Then, Faraday's law is

$$-i\omega\mu \int_A \mathbf{H}(\mathbf{r}) \cdot d\mathbf{s} = \oint_{\partial A} \mathbf{E}(\mathbf{r}) \cdot d\mathbf{l}, \tag{13.2}$$

with μ being the permeability (assumed to be constant) and $d\mathbf{s}$ a surface element. Dividing Equation 13.1 by Equation 13.2 relates κ to $\mathbf{H}$:

$$\frac{\oint_{\partial A} \nabla \times \mathbf{H}(\mathbf{r}) \cdot d\mathbf{l}}{\omega^2\mu \int_A \mathbf{H}(\mathbf{r}) \cdot d\mathbf{s}} = \frac{\oint_{\partial A} \kappa(\mathbf{r}) \mathbf{E}(\mathbf{r}) \cdot d\mathbf{l}}{\oint_{\partial A} \mathbf{E}(\mathbf{r}) \cdot d\mathbf{l}} \approx \kappa(\mathbf{r}). \tag{13.3}$$

In the last step of Equation 13.3, it is assumed that the spatial variation of κ along ∂A is significantly smaller than the variation of $\mathbf{E}$; for example, if ∂A lies inside compartments with constant κ. Equation 13.3 represents the EPT approach as presented in Katscher et al. (2009). Then, using Stokes theorem and the vector identity $\nabla \times \nabla \times \mathbf{H} = -\Delta\mathbf{H}$, Equation 13.3 becomes

$$\frac{\int_A \nabla \times \nabla \times \mathbf{H}(\mathbf{r}) \cdot d\mathbf{s}}{\omega^2\mu \int_A \mathbf{H}(\mathbf{r}) \cdot d\mathbf{s}} = \frac{-\int_A \Delta\mathbf{H}(\mathbf{r}) \cdot d\mathbf{s}}{\omega^2\mu \int_A \mathbf{H}(\mathbf{r}) \cdot d\mathbf{s}} \approx \kappa(\mathbf{r}). \tag{13.4}$$

In the following, positive and negative circularly polarized coordinates are employed, indexed by p (for plus) and m (for minus), respectively, and a z-component. The new basis vectors $\mathbf{e}_p$, $\mathbf{e}_m$, $\mathbf{e}_z$ are related to the Cartesian basis vectors $\mathbf{e}_x$, $\mathbf{e}_y$, $\mathbf{e}_{z,\text{cart}}$ via $\mathbf{e}_p = \frac{1}{\sqrt{2}}(\mathbf{e}_x + i\mathbf{e}_y)$, $\mathbf{e}_m = \frac{1}{\sqrt{2}}(\mathbf{e}_x - i\mathbf{e}_y)$, and $\mathbf{e}_z = \mathbf{e}_{z,\text{cart}}$, and are thus mutually orthogonal. The magnetic field vector becomes $\mathbf{H} = (H_p, H_m, H_z)^T$, with $H_p = \frac{1}{\sqrt{2}}(H_x + iH_y)$, $H_m = \frac{1}{\sqrt{2}}(H_x - iH_y)$, and $H_z = H_{z,\text{cart}}$. All three components of this vector enter Equation 13.4; however, only H_p is directly accessible via MR due to its coupling to the spins of the protons. In Equation 13.4, A is an arbitrarily oriented surface inside the imaged object. This additional degree of freedom can be used to eliminate the inaccessible H_m and H_z from Equation 13.4 as described below.

The surface element $d\mathbf{a} = \mathbf{e}_p dm\,dz$ with its normal parallel to $\mathbf{e}_p$ replaces $\mathbf{H}$ by H_p in numerator and denominator of Equation 13.4. This unitary coordinate transformation does not alter the scalar operator Δ. An additional integration using line element dp along the direction of the positively circulating component $\mathbf{e}_p$ leads to the volume integral:

$$\frac{-\int_{\varphi^{-1}(l)}\left[\int_{\varphi^{-1}(A)}\Delta H_p(\varphi(\mathbf{r}))\,idm\,dz\right]dp}{\omega^2\mu\int_{\varphi^{-1}(l)}\left[\int_{\varphi^{-1}(A)}H_p(\varphi(\mathbf{r}))\,idm\,dz\right]dp} = \frac{-\int_{\varphi^{-1}(V)}\Delta H_p(\varphi(\mathbf{r}))\,dV}{\omega^2\mu\int_{\varphi^{-1}(V)}H_p(\varphi(\mathbf{r}))\,dV} \quad (13.5)$$

$$\approx \kappa(\varphi(\mathbf{r})),$$

with φ being the described change of basis and $dV = dp\,dm\,dz$ a volume element. The volume integrals of Equation 13.5 can be transformed back to Cartesian space. Finally, Δ is replaced by applying Gauss' theorem:

$$-\frac{\int_V \Delta H_p(\mathbf{r})\,dV}{\omega^2\mu\int_V H_p(\mathbf{r})\,dV} = -\frac{\oint_{\partial V}\nabla H_p(\mathbf{r})\cdot d\mathbf{a}}{\omega^2\mu\int_V H_p(\mathbf{r})\,dV} \approx \kappa(\mathbf{r}), \quad (13.6)$$

where V denotes the integration volume with a surface ∂V and a surface element $d\mathbf{a}$. The integration in Equation 13.6 can be performed in Cartesian space. In contrast to the approach presented in Katscher et al. (2009), Equation 13.6 is based only on the measurable, positively rotating component H_p of the transmit field, similar to Bulumulla et al. (2009), Cloos and Bonmassar (2009), Haacke et al. (1991), and Wen (2003). However, in contrast to the references cited above, Equation 13.6 avoids the explicit calculation of the second spatial derivative, which is important for the stability of the numerical reconstruction.

Equations 13.3 through 13.6 provide κ in absolute units. The division of the magnetic field's magnitude in Equation 13.6 cancels out any scaling factors.

Instead of the RF signal emitted by the RF coil, the depicted approach can alternatively be applied to the RF signal emitted by the relaxing spins, that is, to the spatial sensitivity profile H_m of the receive coil. However, no methods are yet published determining receive sensitivities exactly, and thus, EPT is typically based on RF transmit sensitivities as presented above.

13.2.2 Phase-Based Conductivity Imaging and Amplitude-Based Permittivity Imaging

This section discusses an approximate version of the EPT, based on the "full" EPT as given in the previous section. This approximation considerably simplifies data acquisition and reconstruction while introducing an acceptable error. Again, the section exemplarily follows the approach of Voigt et al. (2011a).

Explicitly taking into account that $H_p = \mu^{-1}B_p\exp(i\varphi_p)$ in the differentiation of Equation 13.6 allows us to obtain separate equations for σ and ε. Here, $B_p(\varphi_p)$ is the magnitude (phase) of the RF transmit field's positively rotating component, which is the "active" component responsible for the spin excitation. Thus, Equation 13.6 can be rewritten as

$$
\kappa(\mathbf{r}) \approx \frac{-\oint_{\partial V} \nabla H_p(\mathbf{r}) \cdot d\mathbf{a}}{\omega^2 \mu \int_V H_p(\mathbf{r}) dV}
$$

$$
= \frac{-\oint_{\partial V} e^{i\varphi_p(\mathbf{r})} \nabla B_p(\mathbf{r}) \cdot d\mathbf{a}}{\omega^2 \mu \int_V B_p(\mathbf{r}) e^{i\varphi_p(\mathbf{r})} dV} - i \frac{\oint_{\partial V} B_p(\mathbf{r}) e^{i\varphi_p(\mathbf{r})} \nabla \varphi_p(\mathbf{r}) \cdot d\mathbf{a}}{\omega^2 \mu \int_V B_p(\mathbf{r}) e^{i\varphi_p(\mathbf{r})} dV}
$$

(13.7)

or

$$
\kappa(\mathbf{r}) \omega^2 \mu \int_V B_p(\mathbf{r}) e^{i\varphi_p(\mathbf{r})} dV
$$

(13.8)

$$
\approx -\oint_{\partial V} e^{i\varphi_p(\mathbf{r})} \nabla B_p(\mathbf{r}) \cdot d\mathbf{a} - i \oint_{\partial V} B_p(\mathbf{r}) e^{i\varphi_p(\mathbf{r})} \nabla \varphi_p(\mathbf{r}) \cdot d\mathbf{a}.
$$

Now, Green's identity is applied to both parts on the right-hand side of Equation 13.8:

$$
\oint_{\partial V} e^{i\varphi_p(\mathbf{r})} \nabla B_p(\mathbf{r}) \cdot d\mathbf{a} = \int_V e^{i\varphi_p(\mathbf{r})} \Delta B_p(\mathbf{r}) dV + i \int_V e^{i\varphi_p(\mathbf{r})} \nabla \varphi_p(\mathbf{r}) \cdot \nabla B_p(\mathbf{r}) dV \quad (13.9)
$$

and

$$i \oint_{\partial V} B_P(\mathbf{r}) e^{i\varphi_P(\mathbf{r})} \nabla \varphi_P(\mathbf{r}) \cdot d\mathbf{a} = i \int_V B_P(\mathbf{r}) e^{i\varphi_P(\mathbf{r})} \Delta \varphi_P(\mathbf{r}) dV + i \int_V \nabla \left(B_P(\mathbf{r}) e^{i\varphi_P(\mathbf{r})} \right) \cdot \nabla \varphi_P(\mathbf{r}) dV$$

$$= i \int_V B_P(\mathbf{r}) e^{i\varphi_P(\mathbf{r})} \Delta \varphi_P(\mathbf{r}) dV + i \int_V e^{i\varphi_P(\mathbf{r})} \nabla B_P(\mathbf{r}) \cdot \nabla \varphi_P(\mathbf{r}) dV \tag{13.10}$$

$$- i \int_V e^{i\varphi_P(\mathbf{r})} B_P(\mathbf{r}) \left(\nabla \varphi_P(\mathbf{r}) \right)^2 dV.$$

Since the volume integrals in all parts of Equations 13.8 through 13.10 are identical, the factor $e^{i\varphi_P(\mathbf{r})}$ can be dropped. Separating real and imaginary parts gives

$$\varepsilon(\mathbf{r}) = \left(\omega^2 \mu \int_V B_P(\mathbf{r}) dV \right)^{-1} \left[\int_V B_P(\mathbf{r}) \left(\nabla \varphi_P(\mathbf{r}) \right)^2 dV - \int_V \Delta B_P(\mathbf{r}) dV \right], \tag{13.11}$$

$$\sigma(\mathbf{r}) = \left(\omega \mu V \right)^{-1} \int_V \Delta \varphi_P(\mathbf{r}) dV + 2 \int_V \frac{\nabla B_P(\mathbf{r})}{B_P(\mathbf{r})} \cdot \nabla \varphi_P(\mathbf{r}) dV. \tag{13.12}$$

Gauss theorem is applied to the Δ-terms yielding

$$\sigma(\mathbf{r}) = \left(\omega \mu V \right)^{-1} \left(\oint_{\partial V} \nabla \varphi_P(\mathbf{r}) \cdot d\mathbf{a} + 2 \int_V \left[\nabla \ln \left(B_P(\mathbf{r}) \right) \cdot \nabla \varphi_P(\mathbf{r}) \right] dV \right), \tag{13.13}$$

$$\varepsilon(\mathbf{r}) = \left(\omega^2 \mu \int_V B_P(\mathbf{r}) dV \right)^{-1} \left[\int_V \left(\nabla \varphi_P(\mathbf{r}) \right)^2 B_P(\mathbf{r}) dV - \oint_{\partial V} \nabla B_P(\mathbf{r}) \cdot d\mathbf{a} \right]. \tag{13.14}$$

As explained above, these equations are valid for compartments with constant κ. Next, approximate equations for σ and ε can be derived from Equations 13.13 through 13.14 by approximating the exponential functions by first-order Taylor series and assuming that the variations of $\nabla \varphi_P$ along the boundary ∂V are larger than the variations of B_P. In this way, σ is related only to the phase:

$$\sigma(\mathbf{r}) \approx \frac{1}{\mu_0 \omega V} \int_{\partial V} \nabla \varphi_P(\mathbf{r}) \cdot d\mathbf{a} \tag{13.15}$$

and the permittivity is related only to the magnitude:

$$\varepsilon(\mathbf{r}) \approx \frac{-\oint_{\partial V} \nabla B_{\mathrm{p}}(\mathbf{r}) \cdot \mathrm{d}a}{\mu_0 \omega^2 \int_V B_{\mathrm{p}}(\mathbf{r}) \mathrm{d}V}.$$

(13.16)

Under certain circumstances, Equations 13.15 and 13.16 can be obtained by cancelling identical terms in the fractions of Equation 13.7. Thus, assuming that $\exp(i\varphi_{\mathrm{p}})$ cancels out yields a quantity approximating ε. Similarly, assuming that $B_{\mathrm{p}} \exp(i\varphi_{\mathrm{p}})$ cancels out yields a quantity approximating σ.

Plane waves in infinite, homogeneous media as a simple model for the RF penetration in human body provide further motivation to utilize Equations 13.15 and 13.16. Here, the plane wave $H_{\mathrm{p}} = \mu^{-1} B_{\mathrm{p}} e^{ikz}$ propagating in z-direction is a solution to Equation 13.6 with the wave number k. Comparing the coefficients gives $ik = \pm \sqrt{\omega^2 \mu \varepsilon - i\omega\mu\sigma}$. The root can be written as a sum of real and imaginary parts (using only the positive term for the sake of simplicity):

$$ik = \sqrt{\frac{\omega\mu}{2}} \left[-\sqrt{\omega|\kappa| + \omega\varepsilon} + i\sqrt{\omega|\kappa| - \omega\varepsilon} \right].$$

(13.17)

The transmit phase φ_{p} is described by the imaginary part and the magnitude B_{p} by the real part of Equation 13.17. In the following discussion, two applicability regimes A and B of Equations 13.15 and 13.16 are investigated.

(A) The approximations Equations 13.15 and 13.16 are feasible for the regime $\sigma \approx \omega\varepsilon$, as found for most human tissue types. The impact of σ and ε on the amplitude B_{p} is estimated by differentiating the real part of Equation 13.17:

$$\frac{\partial(\Re(ik))}{\omega\partial\varepsilon} = -\sqrt{\frac{\omega\mu}{8}} \frac{1}{\sqrt{\omega|\kappa| + \omega\varepsilon}} \left(\frac{\omega\varepsilon}{\omega|\kappa|} + 1 \right),$$

(13.18)

$$\frac{\partial(\Re(ik))}{\partial\sigma} = -\sqrt{\frac{\omega\mu}{8}} \frac{\sigma}{\omega|\kappa|\sqrt{\omega|\kappa| + \omega\varepsilon}}.$$

(13.19)

With $\sigma = \omega\varepsilon$, the ratio of the norms of Equations 13.18 and 13.19 becomes

$$\left| \frac{\partial(\Re(ik))}{\omega\partial\varepsilon} \right|_{\sigma=\omega\varepsilon} \Bigg/ \left| \frac{\partial(\Re(ik))}{\partial\sigma} \right|_{\sigma=\omega\varepsilon} = 1 + \sqrt{2} > 1.$$

(13.20)

Thus, changes in the real part (and hence in B_{p}) are mainly caused by ε. Similarly, differentiating the imaginary part of Equation 13.17 gives

$$\frac{\partial\left(\Im(ik)\right)}{\partial\sigma} = \sqrt{\frac{\omega\mu}{8}}\frac{\sigma}{\omega|\kappa|\sqrt{\omega|\kappa|-\omega\varepsilon}}, \tag{13.21}$$

$$\frac{\partial\left(\Im(ik)\right)}{\omega\partial\varepsilon} = \sqrt{\frac{\omega\mu}{8}}\frac{1}{\sqrt{\omega|\kappa|-\omega\varepsilon}}\left[\frac{\varepsilon\omega}{\omega|\kappa|}-1\right], \tag{13.22}$$

and, according to Equation 13.20, the ratio of Equations 13.21 and 13.22 for $\sigma = \omega\varepsilon$ is

$$\left.\left|\frac{\partial\left(\Im(ik)\right)}{\partial\sigma}\right|\right|_{\sigma=\omega\varepsilon}\bigg/\left.\left|\frac{\partial\left(\Im(ik)\right)}{\omega\partial\varepsilon}\right|\right|_{\sigma=\omega\varepsilon} = \frac{1}{\left|1-\sqrt{2}\right|} > 1. \tag{13.23}$$

Thus, changes in the imaginary part (and hence in φ_p) are mainly caused by σ.

(B) The model of plane waves can also motivate the increased accuracy of the approximations Equations 13.15 and 13.16 for the regimes $\sigma \gg \omega\varepsilon$ and $\omega\varepsilon \gg \sigma$. For $\omega\varepsilon \gg \sigma$, the approximation $\omega|\kappa| = \sqrt{\omega^2\varepsilon^2+\sigma^2} \approx \omega\varepsilon$ can be used in Equation 13.17, yielding

$$\Re(ik) \approx -\omega\sqrt{\mu\varepsilon}. \tag{13.24}$$

This equation provides high accuracy for amplitude-based permittivity imaging for $\omega\varepsilon \gg \sigma$, since its real part (and thus the amplitude of H_p) is only influenced by ε. On the other hand, for $\sigma \gg \omega\varepsilon$, the approximation $\omega|\kappa|-\omega\varepsilon \approx \sigma$ yields

$$\Im(ik) \approx \sqrt{\frac{\omega\mu\sigma}{2}}. \tag{13.25}$$

This equation demonstrates the high accuracy of phase-based conductivity imaging for $\sigma \gg \omega\varepsilon$, since its imaginary part (and thus the phase of H_p) is only influenced by σ.

Phase-based conductivity imaging typically results in a bias toward higher values of σ. This bias can be explained by the convex magnitude and the concave phase distribution frequently observed for RF transmission fields, yielding antiparallel gradients of B_p and φ_p. The resulting negative dot product in Equation 13.13 leads to a negative contribution of the neglected component in Equation 13.15. On the other hand, magnitude-based permittivity reconstructions result in lower values than the exact approach, since the transition from Equation 13.14 to Equation 13.16 omits a quadratic term, and thus always results in a positive component.

In Figures 13.2 and 13.3, the validity of the approximation Equations 13.15 and 13.16 is demonstrated using a spherical phantom (diameter = 20 cm) with uniform conductivity $0\,\text{S/m} < \sigma < 2.5\,\text{S/m}$ and permittivity $0\,\text{S/m} < \omega\varepsilon < 1.2\,\text{S/m}$,

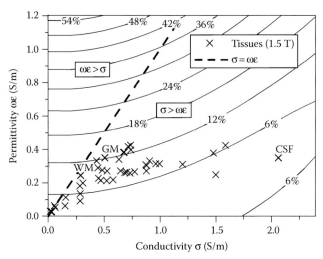

FIGURE 13.2
Conductivity NRMSE introduced by phase-based approximation for a spherical, homogeneous phantom. (Adapted from Voigt, T., et al., *Magnetic Resonance in Medicine*. 66, 456–66, 2011a.) The dashed line separates the two regions $\omega\varepsilon > \sigma$ and $\sigma > \omega\varepsilon$. Highest errors in reconstructed conductivity are found for $\omega\varepsilon \gg \sigma$. Dielectric properties of human tissue are indicated by crosses. (Adapted from Gabriel, S., et al., *Physics in Medicine and Biology*. 41, 2271–93, 1996.) For most tissues, NRMSE is on the order of 10%.

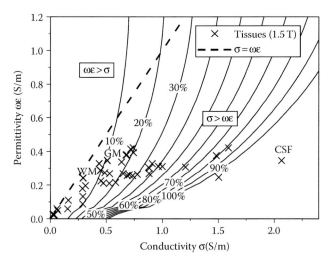

FIGURE 13.3
Permittivity NRMSE introduced by magnitude-based approximation for a spherical, homogeneous phantom. (Adapted from Voigt, T., et al., *Magnetic Resonance in Medicine*. 66, 456–66, 2011a.) The dashed line separates the two regions $\omega\varepsilon > \sigma$ and $\sigma > \omega\varepsilon$. Highest errors in reconstructed permittivity are found for $\sigma \gg \omega\varepsilon$. Dielectric properties of human tissue are indicated by crosses. (Adapted from Gabriel, S., et al., *Physics in Medicine and Biology*. 41, 2271–93, 1996.) For most tissues, except for highly conductive body fluids such as blood and CSF, NRMSE is on the order of 20%.

which covers the whole range of electric properties of human tissue at 64 MHz. The figures are based on the comparison between true and reconstructed σ and ε, benchmarked by the normalized root mean square error (NRMSE):

$$\text{NRMSE} = \frac{1}{\sigma_{\text{true}}} \sqrt{N^{-1} \sum_{n \leq N} \left(\sigma_{\text{recon}}(\mathbf{r}_n) - \sigma_{\text{true}} \right)^2},$$

with N being the number of voxels in the investigated volume of interest (analogously for ε). In accordance with the discussion above, Figures 13.2 and 13.3 show that phase-based conductivity imaging is more valid in the regime $\sigma \gg \omega\varepsilon$ and magnitude-based permittivity imaging is more valid in the regime $\omega\varepsilon \gg \sigma$. Roughly, $2\omega\varepsilon < \sigma < 10\omega\varepsilon$ can be found for biological tissue values at 1.5 T. Thus, the error of phase-based conductivity imaging is only of the order of 5–15%. The error of magnitude-based permittivity imaging is only of the order of 10–20% for most tissues. However, the error increases significantly for highly conductive tissues such as cerebrospinal fluid (CSF) or blood.

Equations 13.15 and 13.16 are particularly advantageous, since imaging conductivity and permittivity can be separated into two different measurements. For permittivity imaging, only the magnitude has to be determined, skipping an additional phase measurement. For conductivity imaging, only the transmit phase has to be determined, skipping an additional magnitude measurement. The phase measurement can be significantly faster than the magnitude measurement. For example, a phase measurement based on a steady-state free precession sequence conductivity study was reported which allowed measuring the conductivity of solving salt in water in real time (Stehning et al., 2011).

13.2.3 Spatially Varying Electric Properties

The full (inhomogeneous) Helmholtz equation is given by

$$\Delta \mathbf{H}(\mathbf{r}) = \mu_0 \omega^2 \kappa(\mathbf{r}) \mathbf{H}(\mathbf{r}) + i\omega \left(\nabla \kappa(\mathbf{r}) \right) \times \mathbf{E}. \tag{13.26}$$

The first term on the right-hand side of Equation 13.26 is the leading term for high frequencies. It is equivalent to Equation 13.6, the "full" EPT reconstruction. Equations 13.6 and 13.26 differ by the explicit second spatial derivative, which deteriorates the numerical reconstruction stability. Nevertheless, several EPT studies are based on this "truncated" Helmholtz equation (Bulumulla et al., 2009; Cloos and Bonmassar, 2009; Haacke et al., 1991; Van Lier et al., 2010; Wen, 2003):

$$\frac{\Delta \mathbf{H}(\mathbf{r})}{\mu_0 \omega^2 \mathbf{H}(\mathbf{r})} = \frac{\Delta H_{\text{p}}(\mathbf{r})}{\mu_0 \omega^2 H_{\text{p}}(\mathbf{r})} = \kappa(\mathbf{r}). \tag{13.27}$$

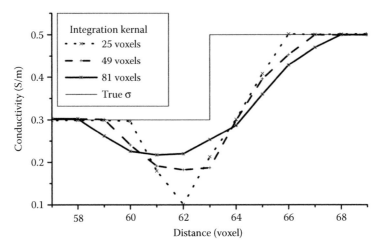

FIGURE 13.4
Simulations using a phantom with two adjacent hemispheres of different conductivities. The variation of κ inside the integration area leads to an error along the hemisphere interface, which depends on the applied numerical kernel size. Profiles of the reconstructed conductivity along the central axis are plotted. (Adapted from Katscher, U., et al., *IEEE Transactions on Medical Imaging*. 28, 1365–74, 2009.)

The second term of Equation 13.26 describes a spatially varying κ. Thus, neglecting this term typically results in artifacts along boundaries between compartments with different κ, usually an unphysical over- or undershooting of the reconstructed κ. Figure 13.4 illustrates such artifacts in the framework of a numerical phantom study. The peaks of the artifacts can be lowered by increasing the kernel size of the numerical calculus operations, however, simultaneously increasing the affected area (Katscher et al., 2009).

These boundary artifacts can be avoided by performing separate reconstructions on different body compartments, segmented prior to reconstruction. Alternatively, the reconstruction can be based on the full Helmholtz equation (Equation 13.26). In this case, the occurring partial derivatives of κ are equivalent to additional unknowns. In Zhang et al. (2010), it was suggested that these additional unknowns could be determined by a "dual excitation algorithm" via parallel transmission systems. The use of parallel transmission systems enables the determination of the unknown partial derivatives of κ by a comparison of different excitations of the same subject. It was reported that this approach reduces the artifacts along compartment boundaries significantly. However, it was also reported that the adverse noise figure of the "dual excitation algorithm" requires a large signal-to-noise ratio (SNR) of the input data to obtain reasonable results (Zhang et al., 2010).

In Nachman et al. (2007), an alternative solution to cope with varying κ was proposed:

$$\frac{\Delta \mathbf{H}(\mathbf{r}) \cdot (\nabla \times \mathbf{H}(\mathbf{r}))}{\mu_0 \omega^2 \mathbf{H}(\mathbf{r}) \cdot (\nabla \times \mathbf{H}(\mathbf{r}))} = \kappa(\mathbf{r}). \tag{13.28}$$

However, all three spatial components of the magnetic field are required to solve this equation. Moreover, Equation 13.28 suffers from singularities at

$$\mathbf{H}(\mathbf{r}) = 0 \tag{13.29}$$

and

$$\nabla \times \mathbf{H}(\mathbf{r}) = 0. \tag{13.30}$$

The singularity indicated by Equation 13.29 also appears in Equation 13.26 and (mitigated by integration) in Equation 13.6. However, the occurrence of this singularity is equivalent to a signal void in the underlying MR images [e.g., air/bone or outside the field of view (FOV) of the RF coil applied] and thus seems to be acceptable. The second singularity Equation 13.30 is equivalent to $\mathbf{E} = 0$. This is usually the case around the center of a quadrature volume coil (QVC) (i.e., the center of all standard MR images), which seems to be less acceptable than the first-mentioned singularity Equation 13.29.

13.2.4 Anisotropy

In its general form, κ is a rank-2 tensor, describing the potential anisotropy of the conductivity and permittivity. Anisotropic cases can be found *in vivo* in tissue showing a preferred cell direction, for instance, in nerves and muscles (Epstein and Foster, 1983; Gabriel et al., 1996). Thus, anisotropic tissue electric properties characterize the underlying cell structure and could be useful for diagnosis. In the theory of EPT as presented above, κ was assumed to be isotropic. A possible way to estimate the anisotropy of κ via EPT was outlined in Katscher et al. (2010b). In this approach, Equation 13.3 is applied to an acquired B_1 map multiple times, however, with different orientations of the integration area A. Let us assume for demonstration purposes that κ has only a single spatial component κ_{long}. In this case, the reconstructed κ depends on the orientation of A relative to κ_{long}:

$$\frac{\oint_{\partial A} \nabla \times \mathbf{H}(\mathbf{r}) d\mathbf{r}}{\mu_0 \omega^2 \int_A \mathbf{H}(\mathbf{r}) da} = \frac{\oint_{\partial A} \kappa(\mathbf{r}) \mathbf{E}(\mathbf{r}) d\mathbf{r}}{\oint_{\partial A} \mathbf{E}(\mathbf{r}) d\mathbf{r}} \approx \kappa_{\text{long}} \cos\left(\angle \kappa_{\text{long}}, A\right) f(\mathbf{E}). \tag{13.31}$$

For instance, the orientation $A = A_{yz}$ perpendicular to κ_{xx} yields zero conductivity if $\kappa_{long} = \kappa_{xx}$:

$$\frac{\int_{\partial Ayz} \{\kappa_{xx} E_x, 0, 0\} d\{0, y, z\}}{\int_{\partial Ayz} E d\vec{r}} = 0. \tag{13.32}$$

For $A = A_{xy}$ or $A = A_{xz}$ parallel to κ_{xx}, nonzero results are expected:

$$\frac{\int_{\partial Axz} \{\kappa_{xx} E_x, 0, 0\} d\{x, 0, z\}}{\int_{\partial Axz} E d\vec{r}} \approx \kappa_{xx} f(\mathbf{E}). \tag{13.33}$$

In Equation 13.33, the weighting function $f(\mathbf{E})$ approaches $f(\mathbf{E}) = 1$ for isotropic (and "sufficiently constant") κ. Unfortunately, Equation 13.31 requires the unknown magnetic field components H_m and H_z. For a QVC, a suitable modeling would be $H_m = H_z = 0$, which of course introduces additional errors in the reconstruction. Nevertheless, Katscher et al. (2010b) reported the measurement of anisotropic effects in a straw phantom based on Equation 13.31.

13.3 Local SAR

Particularly at high main fields, tissue heating is potentially hazardous, specifically in the framework of parallel RF transmission. This heating is related directly to the RF energy absorbed during an MR examination, given by the SAR, thus requiring reliable estimation of SAR prior to scanning. Current, model-based SAR estimation methods neither are patient-specific nor take into account patient position or posture. This current practice of SAR simulations using standardized subjects introduces several disadvantages. First, SAR simulations based on standardized subjects show limited accuracy (Buchenau et al., 2009), since patient size (Liu et al., 2005) and position (Chen et al., 2010) affect the actual local SAR pattern significantly. Additional uncertainties are introduced by the nonspecific assignment of dielectric properties in the patient model. According to Hurt et al. (2000), the exact value of electric properties is crucial for local SAR determination. Thus, it seems to be strongly recommended to base the determination of local SAR on patient-specific data. Another potential source of errors for SAR simulations is given by idealizing the RF transmit chain including the RF transmit coil. Malfunctions or imperfections along the transmit chain can lead to different SAR patterns in real study settings. The remaining model errors necessitate large safety margins and therefore might lead to wasted imaging capabilities with respect to SNR and/or imaging speed. To fully utilize the system's imaging capabilities, an exact patient-specific determination of local SAR would be advantageous.

As an alternative to model-based SAR estimations, B_1-based SAR estimations using EPT have been performed in the framework of simulated data (Cloos and Bonmassar, 2009) and *in vivo* experiments (Voigt et al., 2009) and validated via finite-difference time-domain (FDTD) simulations with a realistic coil model as well as a realistic patient model based on individual patient data. For B_1-based SAR estimation, Ampere's law is used to remove the electric field from the SAR definition:

$$\text{SAR}(\mathbf{r}) \sim \sigma(\mathbf{r})\mathbf{E}^2(\mathbf{r}) = \sigma(\mathbf{r})\left(\frac{\nabla \times \mathbf{H}(\mathbf{r})}{\omega\kappa(\mathbf{r})}\right)^2. \tag{13.34}$$

To solve this equation, κ can be determined with EPT as described above. In the second step, a model has to be chosen for the missing field components H_m and H_z. The approximation $H_m = H_z = 0$ can be applied for QVCs. To get the absolute scaling of the local SAR, the absolute scaling of $\mathbf{H}$ is required (in contrast to the EPT reconstruction of σ and ε). Standard MR scanners usually provide this scaling, which is typically in the range of several microtesla.

Reconstructed local SAR for phantom (Katscher et al., 2009) and *in vivo* studies (Voigt et al., 2010a) are shown in Figure 13.5. An FDTD simulation

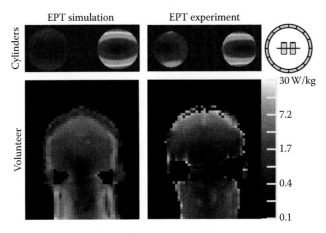

FIGURE 13.5
(**Color version available from crcpress.com; see Preface.**) Local SAR distribution for phantom (top), including sketch of experimental setup (Adapted from Katscher, U., et al., *IEEE Transactions on Medical Imaging*. 28, 1365–74, 2009.), and for *in vivo* study (bottom). (Adapted from Voigt, T., et al., *Proceedings of the 18th Annual Meeting of ISMRM*, 1–7 May, Stockholm, Sweden, 2010a, p. 3876.) The results exhibit a high correlation between measured and simulated SAR, enabling the identification and localization of local SAR hot spots. However, the absolute scaling of the local SAR appears to be reduced as a consequence of neglecting unknown magnetic field components.

based on the segmented 3D MR image of the individual volunteer is shown for comparison. In Figure 13.5, SAR corresponds to an RF duty cycle of 100% and $B_1 = 9$ μT. The results exhibit a high correlation between simulated and measured SAR, thus allowing the identification and localization of local SAR hot spots. The local SAR's absolute scaling appears to be slightly reduced, which is a consequence of the assumption $H_m = H_z = 0$. B_1-based SAR estimation can also be performed using the phase-based conductivity reconstruction, assuming a constant permittivity. It was reported that this approach still maintains the overall spatial shape of the local SAR, however, it introduces further reduction of its scale (Voigt et al., 2010a).

13.4 Measurement Methods

In EPT, the equations discussed above are applied to the RF transmit field generated by the RF coil used in a standard MR system. This section summarizes methods to measure this (complex) RF transmit field.

The MR signal is affected by the positive and negative rotating magnetic field components, $H_p = |H_p| \exp(i\varphi_p)$ and $H_m = |H_m| \exp(i\varphi_m)$. Assuming a repetition time TR and echo time TE of $TR \gg T_1 \gg TE$, the image signal S depends on H_p and H_m via (Hoult, 2000)

$$S(\mathbf{r}) = V_1 M_0(\mathbf{r}) H_m(\mathbf{r}) \exp\left(i\varphi_p(\mathbf{r})\right) \sin\left(V_2 \alpha |H_p(\mathbf{r})|\right)$$

$$= V_1 M_0(\mathbf{r}) |H_m(\mathbf{r})| \exp\left(i\left(\varphi_p(\mathbf{r}) + \varphi_m(\mathbf{r})\right)\right) \sin\left(V_2 \alpha |H_p(\mathbf{r})|\right) \quad (13.35)$$

$$\equiv V_1 M_0(\mathbf{r}) |H_m(\mathbf{r})| \exp\left(i\tilde{\varphi}(\mathbf{r})\right) \sin\left(V_2 \alpha |H_p(\mathbf{r})|\right).$$

Here, α is the nominal flip angle as chosen in the scanner user interface; V_1 and V_2 are system-dependent constants; and $\tilde{\varphi} = \varphi_p + \varphi_m$ is the transceive phase. Furthermore, M_0 contains the (noncomplex) image contrast with respect to relaxation effects and spin density. The image signal is proportional to H_m and the phase factor $\exp(i\varphi_p)$; the modulus $|H_p|$ enters the argument of $\sin(\cdot)$. It is assumed that the main magnetic field points in negative z-direction, leading to the effective transmit sensitivity H_p and the effective receive sensitivity H_m.

13.4.1 Measurement of RF Transmit Magnitude

The nonlinear effect of the transmit magnitude $|H_p|$ on S allows straightforward measurement of it, and numerous techniques have been published that describe various such approaches (so-called B_1 mapping; see, e.g., Akoka et al., 1993; Sacolick et al., 2010; Stollberger and Wach, 1996; Yarnykh, 2007).

In principle, EPT can be based on any one of these B_1-mapping methods. The accuracy of EPT depends on the accuracy of this mapping, that is, the most accurate B_1-mapping method results in the most accurate EPT reconstruction. The topic of optimum B_1 mapping has been discussed by many studies independent of EPT (see, e.g., Morrell and Schabel, 2010).

13.4.2 Measurement of RF Transmit Phase

EPT also requires the knowledge of the absolute transmit phase φ_p. No standard method has been published yet to determine φ_p exactly. The measured transceive phase $\tilde{\varphi}$ contains both the transmit φ_p and the receive phase φ'_m (see Equation 13.35). The prime indicates that the coil used for reception is not necessarily identical (or identically driven) with the transmit coil.

To determine φ_p, a QVC is advantageous, for example, a quadrature body or head coil. A QVC usually switches the polarization between transmission and reception, because the QVC's structure imposes that, without switching H_p is substantially larger than H_m. As explained in the following (Hoult, 2000; Katscher et al., 2009), switching leads to a receive sensitivity H'_m close to H_p.

The fields $\mathbf{H}_1$ and $\mathbf{H}_2$ of the two ports of a QVC are given by

$$\mathbf{H}_1 = H_{x1}\mathbf{x} + H_{y1}\mathbf{y},$$
$$\mathbf{H}_2 = H_{x2}\mathbf{x} + H_{y2}\mathbf{y}, \tag{13.36}$$

using Cartesian basis vectors $\mathbf{x}$ and $\mathbf{y}$. For RF transmission, these two ports are driven with a phase shift of 90°, which is switched for reception. The resulting fields $\mathbf{H}$ and $\mathbf{H}'$ are

$$\mathbf{H} = \mathbf{H}_1 + i\mathbf{H}_2,$$
$$\mathbf{H}' = i\mathbf{H}_1 + \mathbf{H}_2. \tag{13.37}$$

The positive circularly polarized component of $\mathbf{H}_T$ is the active component during transmission and the negative circularly polarized component of $\mathbf{H}_R$ is the active component during reception:

$$H_p = (\mathbf{H}_1 + i\mathbf{H}_2)_x + i\left(\mathbf{H}_1 + i\mathbf{H}_2\right)_y = H_{x1} - H_{y2} + iH_{x2} + iH_{y1},$$
$$H'_m = (i\mathbf{H}_1 + \mathbf{H}_2)_x - i\left(i\mathbf{H}_1 + \mathbf{H}_2\right)_y = i(H_{x1} - H_{y2} - iH_{x2} - iH_{y1}). \tag{13.38}$$

The two ports are designed to transmit two different Cartesian components in an empty coil, for example, $H_{y1} = H_{x2} = 0$. Hence, the two fields differ only by a constant phase factor of 90°:

$$H_p = iH'_m. \tag{13.39}$$

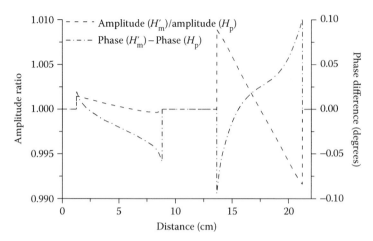

FIGURE 13.6
Comparison of H_p and H'_m for a bicylindrical phantom (see Figure 13.5). The amplitude ratio and phase difference are plotted along the central axis. The marginal differences between the two fields justify the assumption underlying Equation 13.40. (Adapted from Katscher, U., et al., *IEEE Transactions on Medical Imaging.* 28, 1365–74, 2009.)

Loading the coil with material of finite κ (e.g., a patient) leads to finite orthogonal components H_{y1} and H_{x2}. In this case, Equation 13.39 is satisfied only approximately or in case of certain symmetries in the measured subject/object (Lee et al., 2010). In general, the magnitude of these components increases with the main field applied and the electric properties of the material. Figure 13.6 shows a comparison of H_p and H'_m for a bicylindrical phantom, exhibiting only marginal differences at 64 MHz. Up to 300 MHz, this error increases only slightly, as documented in Van Lier et al. (2011a). Now, assuming $H_p \approx H'_m$ for a switched QVC, it follows that

$$\tilde{\varphi} = \varphi_p + \varphi'_m \approx 2\varphi_p \Rightarrow \varphi_p \approx \frac{\tilde{\varphi}}{2}. \tag{13.40}$$

Before the transceive phase is divided by half, it has to be unwrapped. This unwrapping in the three spatial dimensions can be facilitated by performing it separately for each differentiation in the three spatial directions.

The assumption of Equation 13.40 for φ_p is applied in most of the published EPT studies. However, in the case of phase-based conductivity imaging, any arbitrary superposition of two physical phases just leads to twice the reconstructed conductivity due to the linearity of Equation 13.15. This justifies the assumption of Equation 13.40 for arbitrary combinations of transmit and receive RF coils, for example, surface coils.

Using a parallel RF transmission system, it seems to be possible to distinguish φ_p and φ_m by comparing reconstruction results from different transmit channels (Katscher et al., 2011; Sodickson, 2010). Different transmit channels must yield identical electric properties, which gives a means to separate φ_p and φ_m of the different channels. To achieve this, the EPT equation for the conductivity (Equation 13.13) is applied for Tx array element i:

$$\sigma_i(\mathbf{r}) = (\mu_0 \omega V)^{-1} \left(\oint_{\partial V} \nabla \varphi_p^i(\mathbf{r}) \, d\mathbf{a} + 2 \int_V \left[\nabla \left(\ln B_p^i(\mathbf{r}) \right) \cdot \nabla \varphi_p^i(\mathbf{r}) \right] dv \right). \quad (13.41)$$

Next, φ_p^i is replaced by half the transceive phase:

$$\tilde{\varphi}^i = \frac{\varphi_p^i + \varphi_m}{2}. \quad (13.42)$$

This choice represents the RF transmission with a single channel i and an arbitrary (e.g., quadrature) reception with phase φ_m used in all individual Tx channel measurements. Using $\tilde{\varphi}^i$ instead φ_p^i, a contaminated conductivity $\tilde{\sigma}_i$ is obtained:

$$\tilde{\sigma}_i(\mathbf{r}) = (\mu_0 \omega V)^{-1} \left(\int_{\partial V} \nabla \tilde{\varphi}^i(\mathbf{r}) \, d\mathbf{a} + 2 \int_V \left[\nabla \left(\ln B_p^i(\mathbf{r}) \right) \cdot \nabla \tilde{\varphi}^i(\mathbf{r}) \right] dv \right). \quad (13.43)$$

Since this contamination depends on the individual Tx channel, the difference between two channels i and j is not zero as for the correct conductivity $\sigma_i - \sigma_j = 0$ but

$$
\begin{aligned}
\tilde{\sigma}_i - \tilde{\sigma}_j &= \frac{1}{\mu_0 \omega V} \left\{
\begin{array}{l}
\oint_{\partial V} \nabla \left(\tilde{\varphi}^i - \tilde{\varphi}^j \right) d\mathbf{a} + 2 \int_V \left[\nabla \left(\ln B_p^i \right) \cdot \nabla \left(\tilde{\varphi}^i \right) \right] dv \\
\qquad\qquad - 2 \int_V \left[\nabla \left(\ln B_p^j \right) \cdot \nabla \left(\tilde{\varphi}^j \right) \right] dv
\end{array}
\right\} \\[2ex]
&= \frac{1}{\mu_0 \omega V} \left\{
\begin{array}{l}
\dfrac{1}{2} \sigma_i - \dfrac{1}{2} \sigma_j + \int_V \left[\nabla \left(\ln B_p^i \right) \cdot \nabla \left(\varphi_m \right) \right] dv \\
\qquad\qquad - \int_V \left[\nabla \left(\ln B_p^j \right) \cdot \nabla \left(\varphi_m \right) \right] dv
\end{array}
\right\} \\[2ex]
&= \frac{1}{\mu_0 \omega V} \int_V \left[\nabla \left(\ln \frac{B_p^i}{B_p^j} \right) \cdot \nabla \left(\varphi_m \right) \right] dv.
\end{aligned}
\quad (13.44)
$$

The obtained equation allows calculation of the absolute Rx phase φ_m from the measured B_1 maps and the $\tilde{\varphi}^i$-based conductivities. Afterward, the correct Tx phase φ_p^i can easily be derived from φ_m and $\tilde{\varphi}^i$ using Equation 13.42. Similarly, parallel RF transmission can be utilized to determine the two phase contributions via Gauss' law of magnetism (Zhang et al., 2011).

Before using the transceive phase according to Equation 13.40 or Equation 13.42, the following issues have to be taken into account.

(A) The transceive phase must not contain any contributions from B_0, that is, any off-resonance effects. This is in contrast to quantitative susceptibility mapping, which uses only the off-resonance phase and removes all effects from B_1. Spin-echo-based sequences exclude off-resonance effects automatically due to its refocusing pulses. The transceive phase of field-echo-based sequences includes off-resonance effects, which can be removed by any kind of B_0 mapping. The easiest approach would be to measure the transceive phase at two different TEs and to extrapolate it back to TE = 0. Of course, more sophisticated B_0 maps, for example, obtained in the framework of Dixon techniques (Glover and Schneider, 1991), can also be applied.

(B) The transceive phase must contain only contributions from eddy currents arising from the MR RF pulse. Any other spurious eddy currents, particularly those induced in the tissue by gradient switching, have to be removed. This is achieved by averaging two separate measurements with inverted gradient polarization (Bernstein et al., 2004) or using sequences with balanced gradients (Stehning et al., 2011).

13.5 Reconstruction Methods

The numerical treatment of calculus operations is required to solve the EPT equations presented in Section 13.2. Summing up voxels in a square region around the target voxel and suitable scaling is equivalent to the numerical computation of the integrals. For numerical differentiation, a standard method is given by Savitzky–Golay filtering (Savitzky and Golay, 1964), that is, the application of a set of convolution coefficients obtained by a local polynomial regression. In through-plane direction, the applied number of convolution coefficients is typically lower than in in-plane directions. To reduce CPU time, the resulting 3D kernel can be applied in k-space. However, a fixed kernel size potentially leads to undefined voxels along region-of-interest (ROI) boundaries. To minimize these undefined voxels, the kernel size could be dynamically adjusted in the vicinity of the ROI boundaries.

A small kernel size causes a low SNR and a high spatial resolution of the reconstructed κ. On the other hand, a large kernel size causes a high SNR and a low spatial resolution of the reconstructed κ, advantageous for a low SNR of the input B_1 map. In typical reconstructions, two to six voxels in all spatial directions around the target voxel are involved. The described trade-off is shown in Figure 13.7 using a phantom study (Katscher et al., 2009).

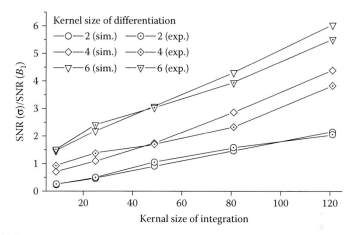

FIGURE 13.7
Impact of numerical calculus operation on resulting SNR. The SNR of the reconstructed conductivity (normalized to the SNR of the underlying B_1 map) increases with the applied numerical kernel size of both the differentiation and the integration. This holds for simulated data (open symbols) as well as for experimental data (dotted symbols). (Adapted from Katscher, U., et al., *IEEE Transactions on Medical Imaging.* 28, 1365–74, 2009.)

The revealed noise figure causes the "effective" spatial resolution of the conductivity/permittivity image to be clearly lower than the B_1 map's spatial resolution.

At low main field strengths and disregarding kernel size effects, the SNR of the reconstructed permittivity is significantly lower than that of the reconstructed conductivity. The following two reasons are responsible for this feature:

1. The conductivity's impact on the phase φ_p is much stronger than the permittivity's impact on the amplitude B_p.
2. The SNR of B_p maps is typically much lower than that of phase images.

To mitigate this problem, SNR could be increased by lowering the spatial resolution. However, a low spatial resolution precludes image segmentation prior to reconstruction, which could have been used to test the assumption of negligible spatial variation of κ. Thus, a preferred solution of the SNR problem could be to increase the main field B_0. Several effects have to be taken into account for performing EPT at higher B_0. First, with higher B_0, conductivities increase and permittivities decrease according to the 4-Cole–Cole model (Gabriel et al., 1996). Second, increasing B_0 and thus ω_0 leads to stronger influence of dielectric properties on RF fields. The effect is linear in conductivity and quadratic in permittivity (c.f. Equations 13.13 and 13.14), which overcompensates aforementioned permittivity decrease. This improves the sensitivity of conductivity and permittivity measurements at high B_0.

For gray matter (GM) and white matter (WM), $\sigma \approx \omega\varepsilon$ is observed at $B_0 \approx 3$ T, which implies a sufficient accuracy of magnitude-based permittivity imaging (Equation 13.16). For higher B_0, the impact of ε on B_p further increases, eventually leading to $\omega\varepsilon > \sigma$ for most tissue types. This supports magnitude-based permittivity imaging but simultaneously hampers phase-based conductivity imaging.

A systematic investigation of the impact of B_0 on the EPT reconstruction and its model assumptions is given in Van Lier et al. (2011a).

13.6 Example Results—Simulations

Example results of an FDTD study using a human head model are shown in Figures 13.8 and 13.9 (Voigt et al., 2011a). The human head was placed at the isocenter of a quadrature body coil at 1.5 T with dielectric properties according to Gabriel et al. (1996). The magnetic fields were simulated with an isotropic

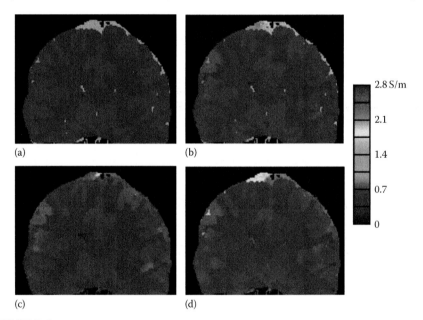

(a)   (b)

(c)   (d)

FIGURE 13.8
(See color insert.) Conductivity reconstruction of FDTD simulation of a realistic head model. (a) Conductivity distribution used in FDTD simulations. (b) Exact reconstruction using Equation 13.13, that is based on complex H_p. (c) Approximated reconstruction using Equation 13.15, that is based only on the transmit phase φ_p. (d) Approximated reconstruction based on the transceive phase assumption (Equation 13.40). A corresponding, quantitative analysis can be found in Table 13.1. (From Voigt, T., et al., Quantitative conductivity and permittivity imaging of the human brain using electric properties tomography. 2011a, 456–66, Copyright Wiley-VCH Verlag GmbH & Co. KGaA. Reproduced with permission.)

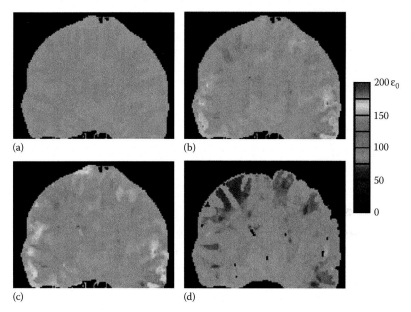

FIGURE 13.9
(See color insert.) Permittivity reconstruction of FDTD simulation data of a realistic head model. (a) Permittivity distribution used in FDTD simulations. (b) Exact reconstruction using Equation 13.14, that is, based on complex H_p. (c) Same as (b), however, based on the transceive phase assumption Equation 13.40. (d) Approximated reconstruction using Equation 13.16, that is, based only on B_1 magnitude. CSF is not shown in (d) since magnitude-based reconstruction of CSF permittivity violates $\omega\varepsilon \gg \sigma$, and thus does not yield a meaningful value (cf. Figure 13.3). A corresponding, quantitative analysis can be found in Table 13.1. (From Voigt, T., et al., Quantitative conductivity and permittivity imaging of the human brain using electric properties tomography. 2011a, 456–66, Copyright Wiley-VCH Verlag GmbH & Co. KGaA. Reproduced with permission.)

resolution of $1 \times 1 \times 1$ mm³. The resulting RF transmit fields, B_p and φ_p, were used for exact and approximate reconstruction according to Equations 13.13 and 13.14 and Equations 13.15 and 13.16, respectively. An additional reconstruction was performed with the phase assumption (Equation 13.40), which mimics the switched coil polarization performed for experimental signal acquisition.

Small deviation is visible when full and approximated reconstructions are compared qualitatively (Figures 13.8 and 13.9). Residual errors in the exact reconstruction compared to the "true" solution are due to numerical effects. Fitting a second-order polynomial is particularly suitable for constant electric properties, since in this case solving the inverse of Equations 13.15 and 13.16 yields a quadratic phase φ_p and magnitude B_p. This explains the higher accuracy achieved in simulations of homogeneous than heterogeneous objects such as the head model. In the latter case, results could be improved by using higher order polynomials, which suffer from higher noise sensitivity. The approximations introduce an NRMSE of roughly 10% for conductivity and 15% for permittivity reconstructions. Quantitative values of conductivity and permittivity for full and approximate reconstructions are shown in Table 13.1.

TABLE 13.1

Analysis of FDTD Results

ROI	Conductivity (S/m)		Permittivity [ε_0]		Input value	
	Exact	Approximate	Exact	Approximate	σ	ε
WM	0.30 ± 0.03	0.33 ± 0.04	68 ± 4	62 ± 6	0.29	67.8
GM	0.52 ± 0.06	0.57 ± 0.07	100 ± 9	89 ± 11	0.51	97.4
CSF	2.09 ± 0.12	2.19 ± 0.08	104 ± 12	n/a	2.07	97.3

Source: Voigt, T., et al., *Magnetic Resonance in Medicine.* 66, 456–66, 2011a.
Notes: CSF, cerebrospinal fluid; GM, gray matter; ROI, region of interest; WM, white matter. Quantitative comparison of average values of conductivity and permittivity inside different ROIs. Quantities were reconstructed from FDTD data. Corresponding images are shown in Figures 13.9 and 13.10.

These results confirm that approximate EPT seems to be feasible at the expense of an acceptable error of roughly 10% as expected from theory. These expectations are also confirmed by experimental results as will be described in the next section.

13.7 Experimental Results

13.7.1 Phantom Measurements

An example phantom study on a standard clinical MR system demonstrates the feasibility and accuracy of EPT (Figure 13.10) (Katscher et al., 2009, 2010a). An isocentric, bicylindrical phantom with different electric conductivities/permittivities was placed in a quadrature body coil at $B_0 = 1.5$ T. The electric conductivity was set to $0 < \sigma < 2$ S/m via the saline concentration and determined with an independent probe. Permittivity was set to $18 < \varepsilon_r < 80$ via 2-propanol and determined via the mixing ratios applied. These electric properties roughly cover the physiological range (Gabriel et al., 1996). B_1 maps were acquired using modified versions of actual flip angle imaging (AFI) (Nehrke, 2009; Voigt et al., 2010b; Yarnykh, 2007). AFI derives B_1 maps from analyzing two fast field-echo (FFE) images acquired in an interleaved fashion with the same flip angle, but at two different repetition times. This makes AFI independent of imperfect flip angle scaling arising from nonlinearities in the RF transmit chain. The phase φ^+ was obtained from the transceive phase of the long TR FFE image acquired for AFI (see Equation 13.40). This phase was corrected for susceptibility artifacts and main field inhomogeneities via an additionally acquired B_0 map using a dual echo sequence (ΔTE $= 10$ ms), subtracting the corresponding B_0-induced phase from the transceive phase. Conductivity was reconstructed according to Equations 13.13 and 13.15 using a kernel of five voxels for in-plane derivatives and four voxels for through-plane derivatives. Permittivity was reconstructed according to Equations 13.14 and 13.16 using eight voxels in-plane and six

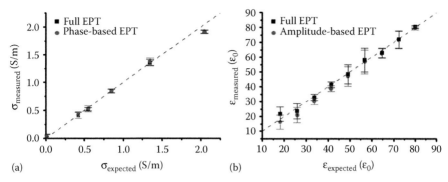

FIGURE 13.10
(Color version available from crcpress.com; see Preface.) EPT phantom experiments. For sketch of experimental setup, see Figure 13.5. A high correlation between *a priori* determined and EPT-measured values is visible for (a) conductivity and (b) permittivity for both the full (black data points) and approximate reconstructions (light gray data points).

voxels through-plane. Reconstructed electric properties were averaged inside a circular ROI and compared with the expected values. For both conductivity and permittivity, the quantitative results exhibit a correlation of more than 99% between reconstructed and expected values as well as for both full and approximate EPT reconstructions.

13.7.2 *In Vivo* Measurements

The following sections outline example studies applying EPT *in vivo* in the human head investigating the electric properties of the brain. First, experiments are described with volunteers at $B_0 = 1.5$ T.

In one of the volunteers, five different ROIs were defined based on the anatomic image: GM, WM, CSF, cerebellum, and corpus callosum (Voigt et al., 2009). Inside these compartments, average values and standard deviation were calculated from the reconstructed electric conductivity according to the exact formula (Equation 13.13). A good agreement between reconstructed, quantitative conductivity values and literature values (Gabriel et al., 1996; Sekino et al., 2004) was found (see Table 13.2). Of course, one has to keep in mind that values reported in the literature are obtained from animal and/ or *ex vivo* data, hampering a direct comparison with the human *in vivo* data.

In another study, three different ROIs were defined based on the anatomic image: GM, WM, and CSF (Voigt et al., 2011a). Inside these compartments, average values and standard deviation were calculated from the reconstructed conductivity and permittivity (Figures 13.11 and 13.12). Reconstructions were performed according to the exact formulae (Equations 13.13 and 13.14) and the approximate formulae (Equations 13.15 and 13.16). Again, the obtained quantitative conductivity and permittivity values (see Table 13.3) were in good agreement with the values reported in the literature (Gabriel et al., 1996). The considerable permittivity NRMSE arises from

TABLE 13.2

Volunteer's Brain Conductivity

Region of interest	EPT (S/m)	Literature (S/m)
Cerebrospinal fluid	1.85 ± 0.87	2.07
Cerebellum	0.57 ± 0.15	0.72
Corpus callosum	0.24 ± 0.09	0.21
White matter	0.32 ± 0.12	0.29
Gray matter	0.48 ± 0.07	0.51

Source: Voigt, T., et al., *Proceedings of the 17th Annual Meeting of ISMRM*, 18–24 April, Honolulu, HI, 2009, p. 4513.

Notes: Mean values in five different compartments roughly match with literature values (Gabriel et al., 1996; Sekino et al., 2004).

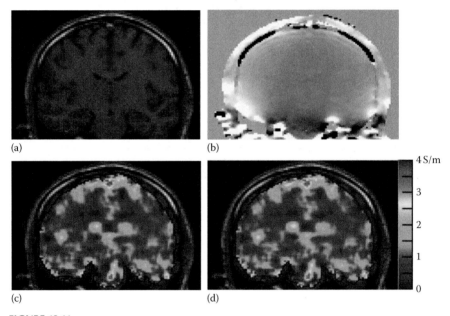

(a)   (b)   (c)   (d)   4 S/m   3   2   1   0

FIGURE 13.11

(Color version available from crcpress.com; see Preface.) Brain conductivity study. (a) Magnitude and (b) phase of spin-echo image acquired to determine φ_p. (c) Exact conductivity reconstruction using Equation 13.13, that is, based on complex Hp. (d) Approximated conductivity reconstruction using Equation 13.15, that is, based only on the transmit phase φ_p. (From Voigt, T., et al., Quantitative conductivity and permittivity imaging of the human brain using electric properties tomography. 2011a, 456–66, Copyright Wiley-VCH Verlag GmbH & Co. KGaA. Reproduced with permission.)

the large-scale calculus operations applied over different compartments. The phase-based reconstruction increases the conductivity results by roughly 5–10% and the magnitude-based reconstruction decreases the permittivity values by roughly 10–15%, which is in accordance with phantom and brain simulations (Section 13.6).

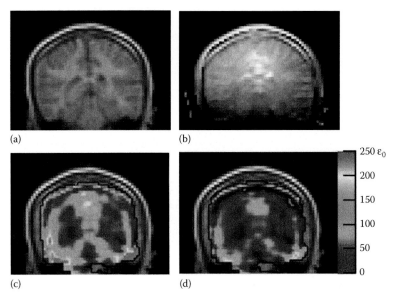

FIGURE 13.12
(Color version available from crcpress.com; see Preface.) Brain permittivity study. (a) FFE image acquired to determine B_p. (b) Corresponding reconstruction of B_p. (c) Exact permittivity reconstruction using Equation 13.14, that is, based on complex H_p. (d) Approximated permittivity reconstruction using Equation 13.16, that is, based only on B_p magnitude. (From Voigt, T., et al., Quantitative conductivity and permittivity imaging of the human brain using electric properties tomography. 2011a, 456–66, Copyright Wiley-VCH Verlag GmbH & Co. KGaA. Reproduced with permission.)

TABLE 13.3

Analysis of *In Vivo* Study

	Conductivity (S/m)		Permittivity [ε_0]	
Region of interest	Exact formula	Phase-based approximation	Exact formula	Magnitude-based approximation
White matter	0.39 ± 0.15	0.43 ± 0.15	72 ± 64	63 ± 66
Gray matter	0.69 ± 0.14	0.72 ± 0.15	103 ± 69	91 ± 70
Cerebrospinal fluid	1.75 ± 0.34	1.82 ± 0.37	104 ± 21	98 ± 20

Source: Voigt, T., et al., *Magnetic Resonance in Medicine.* 66, 456–66, 2011a.
Notes: Quantitative comparison of average values of conductivity and permittivity inside different regions of interest. Values were reconstructed from measured data of a healthy volunteer. Corresponding images are shown in Figures 13.12 and 13.13.

The intersubject conductivity variability was investigated in another study, based on the ROIs of GM, WM, and CSF as before (Voigt et al., 2011a). Once again, average values and standard deviation were calculated inside these compartments from the reconstructed conductivity maps. Reconstructions were performed according to the approximate formula (Equation 13.15).

458 Quantifying Morphology and Physiology of the Human Body Using MRI

TABLE 13.4

Intersubject Conductivity Variability

Region of interest	White matter (S/m)	Gray matter (S/m)	Cerebrospinal fluid (S/m)
Volunteer 1	0.43 ± 0.15	0.72 ± 0.15	1.82 ± 0.37
Volunteer 2	0.33 ± 0.11	0.63 ± 0.25	1.00 ± 0.45
Volunteer 3	0.37 ± 0.15	0.74 ± 0.33	1.26 ± 0.28
Volunteer 4	0.37 ± 0.23	0.85 ± 0.3	1.85 ± 0.72
Volunteer 5	0.30 ± 0.18	0.70 ± 0.26	1.54 ± 0.59
Mean (%)	0.36 ± 12.2	0.73 ± 9.8	1.49 ± 21.8

Source: Voigt, T., et al., *Magnetic Resonance in Medicine.* 66, 456–66, 2011a.
Notes: The higher variability in CSF seems to be caused by the significantly lower number of voxels available for averaging conductivity values. These intersubject variabilities contain two components, measurement errors and physiological differences between subjects.

TABLE 13.5

Field Strength Dependence of Permittivity

Region of interest	1.5 T		3 T	
	EPT	Literature	EPT	Literature
White matter	72 ± 64	67.8	55 ± 27	52.5
Gray matter	103 ± 69	97.4	79 ± 25	73.5
Cerebrospinal fluid	104 ± 21	97.3	102 ± 26	84.0

Source: Katscher, U., et al., *Proceedings of the 17th Annual Meeting of ISMRM,* 1–7 May, 2010b, p. 2866.
Notes: Values of a volunteer's brain reconstructed with EPT match literature expectations (Gabriel et al., 1996) of a permittivity slightly higher at 1.5 T than at 3 T.

The observed intersubject variability was roughly 10% for GM and WM and roughly 20% for CSF (see Table 13.4). The higher CSF variability seems to be caused by the much lower number of voxels available for averaging conductivity values. Of course, the intersubject variability observed contains measurement errors as well as physiological differences. Future studies should focus on separating these two components.

The permittivity's field strength dependence has been investigated by scanning a volunteer at 1.5 T and 3 T (Katscher et al., 2010a). GM, WM, and CSF were defined based on the anatomic image, and average values and standard deviation inside these compartments were calculated from the permittivity reconstructed according to the exact formula (Equation 13.14). The observed decrease of permittivity with increasing frequency is in accordance with literature (Table 13.5). However, large uncertainties remain and these results have to be confirmed by further studies.

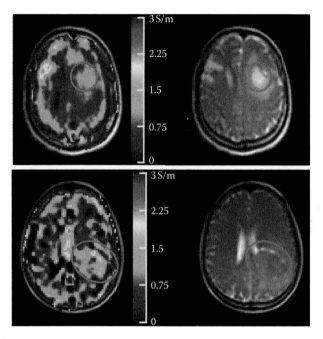

FIGURE 13.13
(See color insert.) Tumor pilot study. Tumor conductivity appears to be pathologically increased by a factor of 2–3 compared with the surrounding WM. Quantitative values are given in Table 13.6. (From Voigt, T., et al., *Proceedings of the 19th Annual Meeting of ISMRM*, 7–13 May, Montreal, QC, Canada, 2011b, p. 127.)

TABLE 13.6

Tumor Pilot Study according to Figure 13.13

Patient	White matter (S/m)	Glioma (S/m)
1	0.36 ± 0.12	0.97 ± 0.18
2	0.38 ± 0.22	1.02 ± 0.37

Source: Voigt, T., et al., *Proceedings of the 19th Annual Meeting of ISMRM*, 7–13 May, Montreal, QC, Canada, 2011b, p. 127.
Notes: Tumor conductivity appears to be pathologically increased by a factor of 2–3 compared with the surrounding WM.

Finally, initial EPT studies have been conducted on two patients with glioma (Van Lier et al., 2011b; Voigt et al., 2011b). Results from the two studies conducted at 1.5 T are shown in Figure 13.13 and Table 13.6 (Voigt et al., 2011b). Based on the anatomic image, tumor and WM were delineated in each patient. Average values and standard deviation inside these compartments were calculated from the electric conductivity, which was reconstructed according to the approximate formula (Equation 13.15). Compared with the surrounding WM, the tumor conductivity seems to be increased pathologically by a factor of 2–3. Similar results have been reported at higher field strengths (Van Lier et al., 2011b).

The factor of 2–3 is in the range of conductivity measurements obtained from surgically excised glioma samples (Lu et al., 1992).

Initial EPT results have also been reported for breast tumors (Borsic, 2010). It is expected that more clinical studies will investigate the electric properties in tumors and other lesions in the near future.

13.8 Discussion

MR-EPT, which has been recently developed to determine electric conductivity and permittivity of tissues *in vivo* using a standard MR system and sequences, was proven to be feasible. Numerous simulations and experiments confirmed that EPT is able to map electric properties with high accuracy and spatial resolution. The EPT reconstruction algorithm is relatively trivial, that is, it contains no hidden iteration, patient modeling, solving ill-posed inverse problems, or other issues. However, some technical aspects of EPT and its assumptions are still under investigation:

1. For accurate reconstruction of κ, EPT requires the knowledge of φ_p. However, standard MR methods can only map the transceive phase. The extraction of φ_p from this transceive phase is a central issue for EPT. Using a volume coil, φ_p can be estimated by dividing the transceive phase in half. Approximate EPT depends linearly on the phase, and thus, it is sufficient to halve the reconstructed conductivity independently of the applied RF coils. An analytical determination of φ_p seems possible by using a parallel RF transmission system. However, applicability and stability still have to be proven for this method.

2. EPT reconstructions based on the truncated Helmholtz equation are not applicable in regions with strongly varying κ. Erroneous oscillations of the reconstructed κ are observed along boundaries between compartments of different κ. These oscillations can be avoided by separate reconstructions in the different compartments. Alternatively, EPT can be based on the full Helmholtz equation, again utilizing parallel RF transmission to obtain the missing partial derivatives of κ.

3. The knowledge of the nonmeasureable field components H_m and H_z is not required to determine (constant) κ. However, it can be used to determine a nonconstant κ as an alternative to the determination of the missing partial derivatives of κ as discussed above. Furthermore, the accurate determination of local SAR requires the knowledge of H_m and H_z.

Future research will focus on the treatment of these technical issues. Moreover, researchers should expand the functionality of EPT, searching for additional information available with this technique. Here, it also seems worthwhile to continue the work on the outlined method of measuring anisotropic κ. Also methods to map electrical properties at frequencies other than the Larmor frequency are needed.

On the other hand, several clinical studies are being conducted to evaluate the diagnostic value of EPT. In order to become a clinically accepted method, EPT must be able to answer diagnostic questions better than existing methods, for example, by providing better sensitivity and specificity in diagnosing various diseases. Currently, there is a lack of knowledge about *in vivo* electric properties of normal tissues and pathologies in humans because no clinical modality exists to measure κ *in vivo*. Clinical EPT studies will enable researchers to explore the biochemical relation between the investigated pathology and the observed electric properties. This makes EPT a particularly interesting and exciting tool for future studies.

Acknowledgments

The authors thank for many fruitful discussions with Christian Findeklee, Christian Stehning, and Jochen Keupp.

References

Akoka, S., Franconi, F., Seguin, F., Le Pape, A., 1993. Radiofrequency map of an NMR coil by imaging. *Magnetic Resonance Imaging.* 11, 437–41.

Beravs, K., Frangez, R., Gerkis, A.N., Demsar, F., 1999. Radiofrequency current density imaging of kainate-evoked depolarization. *Magnetic Resonance in Medicine.* 42, 136–40.

Bernstein, M.A., King, K.F., Zhou, X.J., 2004. *Handbook of MRI Pulse Sequences.* Elsevier Academic Press, Amsterdam, the Netherlands.

Borsic, A., 2010. Imaging electrical properties of the breast—Current experiences at Dartmouth. *International Workshop on MR-Based Impedance Imaging,* 8–10 December, Seoul, Korea.

Buchenau, S., Haas, M., Hennig, J., Zaitsev, M., 2009. A comparison of local SAR using individual patient data and a patient template. In: *Proceedings of the 17th Annual Meeting of ISMRM,* 18–24 April, Honolulu, HI, p. 4798.

Bulumulla, S., Yeo, T., Zhu, Y., 2009. Direct calculation of tissue electrical parameters from B1 maps. In: *Proceedings of the 17th Annual Meeting of ISMRM,* 18–24 April, Honolulu, HI, p. 3043.

Chen, X., Hamamura, Y., Steckner, M., 2010. Numerical simulation of SAR for 3T whole body coil: Effect of patient loading positions on local SAR hotspot. In: *Proceedings of the 18th Annual Meeting of ISMRM,* 1–7 May, Stockholm, Sweden.

Cloos, M., Bonmassar, G., 2009. Towards direct B1 based local SAR estimation. In: *Proceedings of the 17th Annual Meeting of ISMRM,* 18–24 April, Honolulu, HI, p. 3037.

Epstein, B.R., Foster, K.R., 1983. Anisotropy in the dielectric properties of skeletal muscle. *Medical & Biological Engineering & Computing.* 21, 51–5.

Gabriel, S., Lau, R.W., Gabriel, C., 1996. The dielectric properties of biological tissues: III. Parametric models for the dielectric spectrum of tissues. *Physics in Medicine and Biology.* 41, 2271–93.

Glover, G.H., Schneider, E., 1991. Three-point Dixon technique for true water/ fat decomposition with B0 inhomogeneity correction. *Magnetic Resonance in Medicine.* 18, 371–83.

Graesslin, I., Falaggis, K., Vernickel, P., Röschmann, P., Leussler, C., Zhai, Z., Morich, M., Katscher, U., 2006. Safety considerations concerning SAR during RF amplifier malfunctions in parallel transmission. In: *Proceedings of the 14th Annual Meeting of ISMRM,* 6–12 May, Seattle, WA.

Haacke, E., Petropoulos, L., Nilges, E., Wu, D., 1991. Extraction of conductivity and per-mittivity using magnetic resonance imaging. *Physics in Medicine and Biology.* 36, 723.

Haemmerich, D., Staelin, S.T., Tsai, J.Z., Tungjitkusolmun, S., Mahvi, D.M., Webster, J.G., 2003. In vivo electrical conductivity of hepatic tumours. *Physiological Measurement.* 24, 251–60.

Hoult, D.I., 2000. The principle of reciprocity in signal strength calculations—A math-ematical guide. *Concepts in Magnetic Resonance.* 12, 173–87.

Hurt, W.D., Ziriax, J.M., Mason, P.A., 2000. Variability in EMF permittivity values: Implications for SAR calculations. *IEEE Transactions on Biomedical Engineering.* 47, 396–401.

Joines, W.T., Zhang, Y., Li, C., Jirtle, R.L., 1994. The measured electrical properties of normal and malignant human tissues from 50 to 900 MHz. *Medical Physics.* 21, 547–50.

Katscher, U., Findeklee, C., Voigt, T., 2011. Single element SAR measurements in a multi-transmit system. In: *Proceedings of the 19th Annual Meeting of ISMRM,* 7–13 May, p. 494.

Katscher, U., Hanft, M., Vernickel, P., Findeklee, C., 2006. Electric properties tomogra-phy (EPT) via MRI. In: *Proceedings of the 14th Annual Meeting of ISMRM,* 6–12 May, Seattle, WA, pp. 3035–7.

Katscher, U., Karkowski, P., Findeklee, C., 2010a. Permittivity determination via phantom and in vivo B1 mapping. In: *Proceedings of the 17th Annual Meeting of ISMRM,* 1–7 May, p. 239.

Katscher, U., Voigt, T., Findeklee, C., 2010b. Estimation of the anisotropy of elec-tric conductivity via B1 mapping. In: *Proceedings of the 17th Annual Meeting of ISMRM,* 1–7 May, p. 2866.

Katscher, U., Voigt, T., Findeklee, C., Vernickel, P., Nehrke, K., Dossel, O., 2009. Determination of electric conductivity and local SAR via B1 mapping. *IEEE Transactions on Medical Imaging,* 8–10 December, 28, 1365–74.

Lee, S., Bulumulla, S., Dixon, W., Yeo, D., 2010. B1+ phase mapping for MR-based elec-trical property measurement of a symmetric phantom. *International Workshop on MR-Based Impedance Imaging,* 8–10 December, Seoul, Korea.

Liu, L., Dong, W., Ji, X., Chen, L., He, W., Jia, J., 2006. A new method of noninvasive brain-edema monitoring in stroke: Cerebral electrical impedance measurement. *Neurological Research.* 28, 31–7.

Liu, W., Collins, C., Smith, M., 2005. Calculations of B1 distribution, specific energy absorption rate, and intrinsic signal-to-noise ratio for a body-size birdcage coil loaded with different human subjects at 64 and 128 MHz. *Applied Magnetic Resonance*. 29, 5–18.

Lu, Y., Li, B., Xu, J., Yu, J., 1992. Dielectric properties of human glioma and surrounding tissue. *International Journal of Hyperthermia*. 8, 755–60.

McRae, D.A., Esrick, M.A., 1993. Changes in electrical impedance of skeletal muscle measured during hyperthermia. *International Journal of Hyperthermia*. 9, 247–61.

Morrell, G.R., Schabel, M.C., 2010. An analysis of the accuracy of magnetic resonance flip angle measurement methods. *Physics in Medicine and Biology*. 55, 6157–74.

Muftuler, L.T., Hamamura, M., Birgul, O., Nalcioglu, O., 2004. Resolution and contrast in magnetic resonance electrical impedance tomography (MREIT) and its application to cancer imaging. *Technology in Cancer Research & Treatment*. 3, 599–609.

Nachman, A., Wang, D., Ma, W., Joy, M., 2007. A local formula for inhomogeneous complex conductivity as a function of the RF magnetic field. Unsolved problems and unmet needs session. In: *Proceedings of the 15th Annual Meeting of ISMRM*, 19–25 May, Berlin, Germany.

Nehrke, K., 2009. On the steady-state properties of actual flip angle imaging (AFI). *Magnetic Resonance in Medicine*. 61, 84–92.

Sacolick, L.I., Wiesinger, F., Hancu, I., Vogel, M.W., 2010. B1 mapping by Bloch-Siegert shift. *Magnetic Resonance in Medicine*. 63, 1315–22.

Savitzky, A., Golay, M.J.E., 1964. Smoothing and differentiation of data by simplified least squares procedures. *Analytical Chemistry*. 36, 1627–39.

Schaefer, M., Gross, W., Ackemann, J., Gebhard, M.M., 2002. The complex dielectric spectrum of heart tissue during ischemia. *Bioelectrochemistry*. 58, 171–80.

Schepps, J.L., Foster, K.R., 1980. The UHF and microwave dielectric properties of normal and tumour tissues: Variation in dielectric properties with tissue water content. *Physics in Medicine and Biology*. 25, 1149–59.

Scott, G.C., Joy, M.L., Armstrong, R.L., Henkelman, R.M., 1995. Rotating frame RF current density imaging. *Magnetic Resonance in Medicine*. 33, 355–69.

Seifert, F., Wubbeler, G., Junge, S., Ittermann, B., Rinneberg, H., 2007. Patient safety concept for multichannel transmit coils. *Journal of Magnetic Resonance Imaging*. 26, 1315–21.

Sekino, M., Inoue, Y., Ueno, S., 2004. Magnetic resonance imaging of mean values and anisotropy of electrical conductivity in the human brain. *Neurology & Clinical Neurophysiology*. 2004, 55.

Seo, J., Kwon, O., Woo, E., 2005. Magnetic resonance electrical impedance tomography (MREIT): Conductivity and current density imaging. *Journal of Physics*. 12, 140.

Sodickson, D., 2010. Addressing the challenges of electrical property mapping with magnetic resonance. *International Workshop on MR-Based Impedance Imaging*, 8–10 December, Seoul, Korea.

Stehning, C., Voigt, T., Katscher, U., 2011. Real-time conductivity mapping using balanced SSFP and phase-based reconstruction. In: *Proceedings of the 19th Annual Meeting of ISMRM*, 7–13 May, p. 128.

Stollberger, R., Wach, P., 1996. Imaging of the active B1 field in vivo. *Magnetic Resonance in Medicine*. 35, 246–51.

Surowiec, A.J., Stuchly, S.S., Barr, J.B., Swarup, A., 1988. Dielectric properties of breast carcinoma and the surrounding tissues. *IEEE Transactions on Bio-Medical Engineering*. 35, 257–63.

Van Lier, A., Hoogduin, J.M., Polders, D.L., Boer, V.O., Hendrikse, J., Robe, P.A., Woerdeman, P.A., Lagendijk, J.J., Luijten, P.R., van den, C.A., 2011b. Electrical conductivity imaging of brain tumours. In: *Proceedings of the 19th Annual Meeting of ISMRM*, 7–13 May, Montreal, QC, Canada, p. 4464.

Van Lier, A., Raaijmakers, A., Brunner, D., Klomp, D., Pruessmann, K., Lagendijk, J., van den Berg, C., 2010. Propagating RF phase: A new contrast to detect local changes in conductivity. In: *Proceedings of the 18th Annual Meeting of ISMRM*, 1–7 May, Stockholm, Sweden, p. 2864.

Van Lier, A., Voigt, T., Katscher, U., van den Berg, C., 2011a. Comparing electric properties tomography at 1.5, 3 and 7 T, B0. In: *Proceedings of the 19th Annual Meeting of ISMRM*, 7–13 May, Montreal, QC, Canada.

van Oosterom, A., 1991. History and evolution of methods for solving the inverse problem. *Journal of Clinical Neurophysiology*. 8, 371–80.

Voigt, T., 2011. Quantitative MR imaging of the electric properties and local SAR based on improved RF transmit field mapping (band 11). PhD dissertation. KIT Scientific Publishing, Karlsruhe, Germany. Available from: http://digbib.ubka.uni-karlsruhe.de/volltexte/1000020953.

Voigt, T., Doessel, O., Katscher, U., 2009. Imaging conductivity and local SAR of the human brain. In: *Proceedings of the 17th Annual Meeting of ISMRM*, 18–24 April, Honolulu, HI, p. 4513.

Voigt, T., Homann, H., Katscher, U., Doessel, O., 2010a. Patient-specific in vivo local SAR estimation and validation. In: *Proceedings of the 18th Annual Meeting of ISMRM*, 1–7 May, Stockholm, Sweden, p. 3876.

Voigt, T., Katscher, U., Doessel, O., 2011a. Quantitative conductivity and permittivity imaging of the human brain using electric properties tomography. *Magnetic Resonance in Medicine*. 66, 456–66.

Voigt, T., Nehrke, K., Doessel, O., Katscher, U., 2010b. T-1 corrected B-1 mapping using multi-TR gradient echo sequences. *Magnetic Resonance in Medicine*. 64, 725–33.

Voigt, T., Väterlein, O., Stehning, C., Katscher, U., Fiehler, J., 2011b. In vivo glioma characterization using MR conductivity imaging. In: *Proceedings of the 19th Annual Meeting of ISMRM*, 7–13 May, Montreal, QC, Canada, p. 127.

Wang, D., De Monte, T.P., Ma, W., Joy, M.L., Nachman, A.I., 2009. Multislice radio-frequency current density imaging. *IEEE Transactions on Medical Imaging*. 28, 1083–92.

Wen, H., 2003. Noninvasive quantitative mapping of conductivity and dielectric distributions using RF wave propagation effects in high-field MRI. In: *Proceedings of SPIE*, Vol. 5030, 15 February, San Diego, CA, pp. 471–7.

Yarnykh, V.L., 2007. Actual flip-angle imaging in the pulsed steady state: A method for rapid three-dimensional mapping of the transmitted radiofrequency field. *Magnetic Resonance in Medicine*. 57, 192–200.

Zelinski, A.C., Angelone, L.M., Goyal, V.K., Bonmassar, G., Adalsteinsson, E., Wald, L.L., 2008. Specific absorption rate studies of the parallel transmission of inner-volume excitations at 7T. *Journal of Magnetic Resonance Imaging*. 28, 1005–18.

Zhang, X., van de Moortele, P.-F., Schmitter, S., He, B., 2011. Imaging electrical properties of the human brain using a 16-channel transceiver array coil at 7 T. In: *Proceedings of the 19th Annual Meeting of ISMRM*, 7–13 May, Montreal, QC, Canada, p. 126.

Zhang, X.T., Zhu, S.A., He, B., 2010. Imaging electric properties of biological tissues by RF field mapping in MRI. *IEEE Transactions on Medical Imaging*. 29, 474–81.

14

Quantitative Susceptibility Mapping to Image Tissue Magnetic Properties

Yi Wang and Tian Liu

Cornell University

CONTENTS

14.1 Introduction

It is well known that spatial variations in tissue magnetic susceptibility cause magnetic field inhomogeneity, leading to signal loss due to spin dephasing or hypointensity that is particularly prominent in gradient echo MRI (GRE MRI). In certain imaging protocols, the susceptibility source is a contrast agent and there is great desire to know how much contrast agent has arrived in particular regions. One option is to look at the percentage of hypointensity relative to a control region of known susceptibility. Percentage of hypointensity is a quantity, but how is this quantity related to the amount of susceptibility?

Another option is to look at the signal phase, which has been regarded as the local inhomogeneity measurement; but how is this phase related to the underlying susceptibility quantity?

To answer these questions about the relation between MRI signal hypointensity (or phase) and tissue magnetic susceptibility, we need to examine the physics of magnetic susceptibility, generation of field inhomogeneity, and MRI signal detection. Then the determination of susceptibility from MRI signal can be formulated quantitatively, and algorithms can be developed to solve for the susceptibility.

14.2 Tissue Magnetism as a Biomarker for Diseases

Tissue magnetic susceptibility χ reflects the electromagnetic response of biomaterials to the main magnetic field B_0 to generate magnetization M:

$$M = \frac{\chi B_0}{\mu_0}. \tag{14.1a}$$

When tissue becomes magnetized in an MR scanner, it generates its own magnetic field that affects MRI signal. The comprehensive quantitative treatment of tissue magnetism requires advanced quantum theory; but for a conceptual understanding, we will review the simpler classical description of the tissue magnetization process. Tissue magnetic susceptibility can be decomposed into paramagnetic and diamagnetic components.

The paramagnetic part comes from contributions of unpaired electrons in biomaterial molecules that have intrinsic magnetic moment, μ, which is thousand times larger than nuclear magnetic moments. These magnetic moments tend to align with the B_0 field to remain at the lowest energy state. Thermal motion at temperature T disrupts this alignment, reaching an equilibrium paramagnetic magnetization M described by Boltzmann's law for n_s spins per unit volume:

$$M = \frac{n_s \mu^2 B_0}{k_B T}, \tag{14.1b}$$

where μ is the magnetic moment and k_B is the Boltzmann's constant. The diamagnetic component comes from the response of the orbiting electrons, even though their spins are paired evenly up and down producing no net spin. The simple electron ($\#n_e$, charge q_e, mass m_e) orbit (radius r) model provides a quantum mechanically correct result for explaining tissue diamagnetism:

$$M = -\frac{n_e q_e^2}{6 m_e} <r^2> B_0. \tag{14.1c}$$

In general, for complex electron clouds without spherical symmetry, the orbit interpretation becomes difficult, as an electron cloud state consists of different orbits mixed by the magnetic field—electron orbit coupling. These virtual orbit effects need to be taken into consideration for a proper quantum mechanical account of diamagnetism.

Most biomaterials do not have unpaired electrons and are weakly diamagnetic. When unpaired electrons are present, they contribute a paramagnetic magnetization that may dominate over the diamagnetic part, resulting in a net paramagnetic magnetization. Hemoglobin is of great biomedical interest because it changes susceptibility from diamagnetic to paramagnetic as it becomes deoxygenated. The oxygen binds cooperatively with iron(III) in hemoglobin and pairs all electrons, making oxyhemoglobin diamagnetic. Dissociation of oxygen from hemoglobin leads to unpaired electrons and therefore to paramagnetism. Since oxygen consumption is related to organ functions, such as neural activation, magnetism effects provide an important means for investigating organ functions and form the foundation of brain functional MRI (fMRI).

Deoxyhemoglobin, hemosiderin, iron deposit, myelin and calcification, and contrast agents all posess prominent magnetic properties that may reflect underlying pathophysiology.

14.3 MRI Signal Phase and Susceptibility Mapping

MRI signal phase allows estimation of field that is a nonlocal convolution of tissue magnetic properties with the dipole kernel, but MRI signal magnitude has a much more complex relation with tissue magnetization.

The major effect of tissue magnetization is that it generates a local magnetic field. Consequently, tissue magnetization directly affects MR signal phase, and indirectly affects MR signal magnitude through spin-phase dispersion within a voxel.

To quantify these effects, tissue magnetization can be considered as a collection of magnetic moments of dipole origin. A dipole moment m located at the origin produces a magnetic field b at observation location r according to the following equation:

$$b(r) = \frac{\mu_0}{4\pi} \frac{3m \cdot \hat{r} - m}{r^3}. \tag{14.2a}$$

Only the z-component (parallel with B_0) affects the spin precession rate, so it is useful to have the following dipole z-field:

$$d(r) = \frac{1}{4\pi} \frac{3\cos^2\theta - 1}{r^3}. \tag{14.2b}$$

The magnetization $M(r)$ is a collection of magnetic moments and will produce a magnetic field b in space that is the sum of contributions from all magnetic moments, with the following z-component:

$$b_z(r) = \left(\frac{\mu_0}{4\pi}\right)\int d^3 r'\left(3\cos^2\theta_{rr'} - 1\right)\frac{M(r')}{|r-r'|^3} = d(r) \otimes M(r). \qquad (14.2c)$$

For a biomaterial of given susceptibility $\chi \ll 1$, it gains a magnetization M in an external field B_0, which is given as $M = \chi B_0/\mu_0$. Scaling the tissue-induced inhomogeneity field $b_z(r)$ to B_0 as δ_b, we have

$$\delta_b(r) = d(r) \otimes \chi(r). \qquad (14.3)$$

This convolution illustrates that the induced field is a nonlocal effect. The inhomogeneity field at a location is not directly the result of tissue susceptibility at that location. If we want to know the susceptibility of tissues at particular locations, we need to perform deconvolution over the measured magnetic field inhomogeneity.

Gradient echo imaging is very sensitive to susceptibility effects, which manifest as T_2^* hypointensity in magnitude images and local variation in phase images; T_2^* hypointensity represents intravoxel dephasing effects of field variations within a voxel:

$$\int_0^{\Delta r} d^3 r\rho(r)e^{-i\omega_0\delta_b(r)TE} \cong e^{-TE/T_2^*}. \qquad (14.4)$$

This hypointensity strongly depends on imaging parameters including echo time TE, voxel size and orientation Δr, spin distribution within the voxel $\rho(r)$, and field strength B_0 (Nandigam et al., 2009; Vernooij et al., 2008). Even when the imaging parameters are fixed, hypointensity still varies with the geometric relation between susceptibility source and voxel location. Furthermore, the hypointensities caused by iron (paramagnetic $\chi > 0$), calcium (diamagnetic $\chi < 0$), and air (no spin $\rho(r) = 0$) look similar, making hypointensity contrast nonspecific.

Phase divided by echo time and the gyromagnetic ratio uniquely determines the magnetic field of a susceptibility source without the need for any calibration. This relationship between MRI signal phase and local magnetic field has been commonly used in MRI literature, but it requires careful examination in quantitative MRI. To be rigorous, this relationship is true for a single spin; for spins in a voxel, this commonly used approximation is good to the third order under many imaging conditions:

$$e^{-i\omega_0\delta_b(\mathbf{r})t} = 1 - i\omega_0\delta_b(\mathbf{r})t - \frac{\left(\omega_0\delta_b(\mathbf{r})t\right)^2}{2} + o\left(\left(\omega_0\delta_b(\mathbf{r})t\right)^3\right)$$

(14.5)

$$\text{phase}\left\{\int_0^{\Delta r} d^3r\rho(\mathbf{r})e^{-i\omega_0\delta_b(\mathbf{r})t}\right\} = \omega_0 < \delta_b(\mathbf{r}) > t + o\left(\left(\omega_0\delta_b(\mathbf{r})t\right)^3\right).$$

This simple relationship between phase and the average field in a voxel $< \delta_b(\mathbf{r}) >$ represents an important and powerful quantification approach. Unfortunately, phase cannot be used for direct tissue characterization, because it is a convolution of a unit dipole field (d) with the surrounding magnetic susceptibility distribution. So, phase cannot be used to quantify tissue susceptibility such as local iron content, as clearly demonstrated in experiments (Yao et al., 2009). A deconvolution of the dipole kernel or the inversion from field to source is required.

14.4 Quantitative Susceptibility Mapping

Tissue magnetic susceptibility can be determined by solving the field-to-source inverse problem termed quantitative susceptibility mapping (QSM). The forward problem of source-to-field convolution in Equation 14.3 can be expressed in the Fourier domain as a point-wise multiplication:

$$\Delta_b(k) = D(k)X(k).$$

(14.6)

The dipole kernel in Fourier domain $D \equiv (1/3 - k_z^2/k^2)$ has a pair of zero cone surfaces at the magic angle $\pm 55°$, which causes susceptibility X to be undersampled on the cone surface. Consequently, the inverse problem from measured magnetic field inhomogeneity to tissue susceptibility source is ill-posed and has an infinite number of possible solutions.

Various solutions have been proposed to solve this ill-posed inverse problem. Noticing that the undersampling pattern of the dipole kernel has a conical structure pointing at the B_0 direction, resampling the same object at a different orientation with respect to the B_0 field effectively rotates the cone and recovers the missing data, leading to a unique and exact solution (Liu et al., 2009). Although this *calculation of susceptibility through multiple orientation sampling* (COSMOS) approach maintains the full fidelity to the measured data, and may be regarded as the reference standard for QSM, it requires multiple scans and rotating the subject in a magnet that is often impractical.

There are also generic regularization methods for solving ill-posed inverse problems. Different regularization strategies have been tried by imposing on the solution a selective mathematical property such as smoothness, sparsity, and piecewise constancy, or by using zero padding in Fourier space division. Nevertheless, these approaches suffer from systematic bias (errors and artifacts) because these mathematical properties do not, in general, agree with physical reality (de Rochefort et al., 2008, 2010; Kressler et al., 2010; Li and Leigh, 2004; Li et al., 2011; Shmueli et al., 2009; Wharton et al., 2010; Wu et al., 2012).

14.4.1 Morphology-Enabled Dipole Inversion Algorithm

Morphology-enabled dipole inversion (MEDI) algorithm is a practical and accurate QSM solution that uses the magnitude images as the constraint. If we regard the dipole kernel zeroes as factors that cause slight undersampling, additional information other than phase is needed for the full reconstruction of tissue susceptibility. In MRI, the magnitude image conveniently provides this information. According to Maxwell's equation, the inhomogeneous field disturbance in an otherwise uniform magnetic field is caused by the interface between different susceptibility sources (Jackson, 1999). This often coincides with the tissue boundaries seen in the magnitude images, because both of them arise from the interfaces between different tissues. Therefore, for a physically meaningful solution, interfaces in susceptibility that are not accompanied by interfaces in the magnitude images should be rare.

This observation inspired the design of a MEDI method. In MEDI, a weighted L1-norm minimization is employed to sparsify the differences between the interfaces of a candidate susceptibility map and the interfaces determined from the magnitude images, while the candidate susceptibility map also has to be able to induce the field inhomogeneity determined from the phase images in GRE MRI. Mathematically, this constrained convex optimization problem is identical to an unconstrained Lagrangian problem, for which many efficient solvers have been developed in applied mathematics (de Rochefort et al., 2010; Liu et al., 2012):

$$\chi^*(r) = \mathrm{argmin}_{\chi(r)} \sum_r \left\| w(r)\left[\delta_b(r) - d(r) \otimes \chi(r) \right] \right\|_2^2 + \lambda \sum_r \left| m(r)G\left[\chi(r) \right] \right|_1^{\sim}, \quad (14.7)$$

where a spatial weighting $w(r)$ is used to whiten noise in the phase measurement (Conturo and Smith, 1990; Kressler et al., 2008); the value of λ may be chosen such that the residue (first term in Equation 14.7) matches the expected noise level (Lustig et al., 2007); $m(r)$ encodes spatial prior information, and G is the gradient operator. We have proved that the error propagation in Equation 14.7 is linearly bounded by error in the MRI measurement and the accuracy of the prior information about the edges, suggesting effective means to improve the precision of the susceptibility solution.

Simulations (Figure 14.1) suggest that susceptibility is mostly determined by phase data and the magnitude images help eliminate artifacts. This

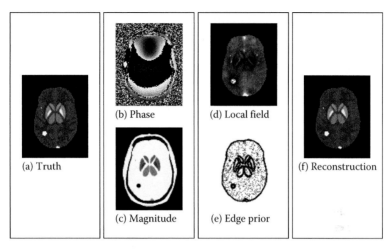

FIGURE 14.1

Illustration of the MEDI approach to the field-to-source inverse problem. MR imaging is simulated on a numerical brain (a) and both phase (b) and magnitude (c) images were obtained. (d) Local field is estimated from phase after removing background field, and prior information about the edges (e) is generated from the magnitude image by inverting its gradient. (f) Susceptibility map is obtained by fitting susceptibility with local field data via Maxwell's equation and constraining susceptibility interfaces with tissue interfaces via their least discordance.

means that the constrained minimization does not force susceptibility to be constant within uniform regions of magnitude images. For example, the gray–white matter interface is not reflected in the magnitude image, but it is recovered in QSM due to the changes in phase.

Although a spatial weighting term $w(r)$ is employed in the data-fidelity term to account for spatially varying noise, it is still assumed that phase noise is normally distributed. This assumption breaks down in signal void regions such as air or hemorrhage. In this case, the data fitting may be formulated using complex exponential functions in the complex plane, where the noise in real and imaginary channels might still be regarded as normally distributed (Liu et al., 2013).

14.4.2 Removing Background Fields without Affecting Local Magnetic Fields

Accurate mapping of the local magnetic field induced by tissue susceptibility is necessary to minimize errors in the MRI measurement, so the background field induced by air–tissue interface has to be removed. However, previous methods including geometrical modeling (Neelavalli et al., 2009) and the commonly used high-pass filtering (Haacke et al., 2009; Langham et al., 2009) might lead to large errors due to imperfect geometrical information or spectral overlap between local and background fields. We have observed that the dipole field from a source in a region of interest (ROI) is approximately orthogonal to any dipole field from a source outside the ROI, except near the ROI boundaries, so the background field and the local field are approximately orthogonal. Accordingly, we developed a *projection onto dipole fields* (PDF) method for the optimal estimation of the background field (f_b) by projecting the measured total field (f) onto the subspace spanned by the external dipole source (χ_e) as in the Hilbert space projection theorem:

$$\chi_e^*(r) = \text{argmin}_{\chi_e(r)} \sum_{r \in \text{ROI}} \left\| w(r) \left[\delta_b(r) - d(r) \otimes \chi_e(r) \right] \right\|_2^2. \tag{14.8}$$

The background field f_b is simply the field generated by background susceptibility distribution $d(r) \otimes \chi_e(r)$. Our preliminary data on phantom of known background field and *in vivo* imaging indicate this effective dipole fitting provides accurate background removal without affecting local field, far superior to the geometric modeling and high-pass filtering method (Liu et al., 2011).

Another method for background separation is the *sophisticated harmonic artifact reduction on phase data* (SHARP) approach (Schweser et al., 2010). It is noticed that the background field component, which is a harmonic function inside the predefined ROI, can be eliminated by the spherical mean value operation. This method has an elegant analytical formulation if the boundary conditions of the local field were known. In practice, digitization error is introduced in the method. However, high quality *in vivo* results have been shown with the SHARP method (Schweser et al., 2011).

14.4.3 Preliminary Experimental Validation of MEDI-Based QSM

Phantom validation. Gadolinium (Gd) solutions at different concentrations were imaged using a 1.5T scanner. Due to the saturation effect, all the vials had similar intensity (Figure 14.2a). The local field map revealed dipole patterns with different strengths (Figure 14.2b), and it highlighted the nonlocal effect of the dipole fields. The signal intensity on the reconstructed QSM linearly correlated with the [Gd], and there was an excellent agreement between estimated and prepared [Gd] (regression slope close to unity, $R^2 = 1$).

In vivo validation by comparing with COSMOS and values from published literature. QSM images of healthy subjects ($n = 9$) generated from MEDI and COSMOS showed similar qualitative results (Figure 14.3a). The linear regression (Figure 14.3b) between MEDI and COSMOS reconstructed QSM demonstrated an excellent quantitative agreement.

The MEDI method was also validated in quantifying endogenous iron and deoxyhemoglobin in human brain imaging ($n = 9$) by comparing with literature values (Riederer et al., 1989). MEDI showed that the susceptibility of the venous blood was 0.267 ± 0.02 parts per million (ppm) higher than

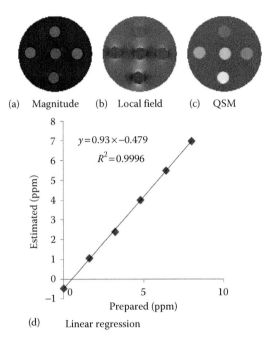

(a) Magnitude (b) Local field (c) QSM

(d) Linear regression

FIGURE 14.2
(a) Magnitude, (b) local field, and (c) QSM of a water phantom consisting of Gd vials at various concentrations. (d) A linear agreement between QSM-estimated Gd and prepared Gd was observed.

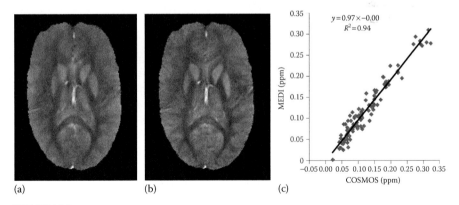

(a) (b) (c)

FIGURE 14.3
MEDI (a) and COSMOS (b) generate similar QSM, as confirmed by linear regression (c). Voxel values from caudate nucleus, putamen, globus pallidus, substantia nigra, red nucleus, and vein of galen were extracted for the linear regression illustrated in (c).

that of the arterial blood, corresponding to a net reduction in oxygenation saturation of $29.5 \pm 2\%$, which was in good agreement with $31.7 \pm 3.9\%$ measured by invasive catheterization in an age-matched (22.9 ± 2.4 years) group of volunteers (Gibbs et al., 1942). Furthermore, susceptibility allows estimation of $[Fe^{3+}]$, assuming it is the dominant contributor to the susceptibility

TABLE 14.1

Comparison between MEDI-Estimated [Fe^{3+}] and Literature Values

Structure	MEDI (mg/g)	Literature (mg/g)
Red nucleus	21.2 ± 14.2	
Putamen	22.8 ± 7.0	31 ± 5.6
Caudate nucleus	25.0 ± 6.0	
Substantia nigra	33.2 ± 9.5	16 ± 4.2
Globus pallidus	56.0 ± 5.7	53 ± 12

Notes: Values are reported in mean $\pm$ SD.
Unit is $\mu g/g$.

(Table 14.1). The MEDI estimated [Fe^{3+}] in the substantia nigra, putamen, and globus pallidus were in the same range as those in postmortem measurements (Riederer et al., 1989).

14.5 Clinical Applications of QSM

14.5.1 QSM of Cerebral Microbleeds

The susceptibility values in QSM reconstructed images from all echoes for a given repetition time (TR) was fairly independent of TE (Figure 14.4), in sharp contrast with the sensitive TE dependence of hypointensity in the magnitude and susceptibility weighted imaging (SWI) images, and moderate dependence in R_2^* maps (Figure 14.4a). While both iron and calcium deposits appeared hypointense in magnitude SWI images and positive in R_2^* map, the foci of positive susceptibility on QSM (red arrows in Figure 14.4a) indicated iron deposits in cerebral microbleeds, CMB (Greenberg et al., 2009), and the foci of negative susceptibility in the atria of the lateral ventricles (green arrows in Figure 14.4a) were interpreted as deposits of highly concentrated calcification, an experimentally confirmed endogenous biomarker with negative susceptibility (Liu et al., 2009). This patient study suggests that QSM is a reliable biomarker that is independent of imaging parameters, and it can serve as a standardized measure of CMB for definitive cross-study comparisons.

14.5.2 QSM for Measuring Hematoma Volume

Hematoma volume is a powerful predictor of 30-day mortality risk after spontaneous intracerebral hemorrhage (ICH). However, the volume measured on gradient echo MRI is inaccurate, generally larger than the volume measured on computed tomography (CT). It is also observed that the volume depends

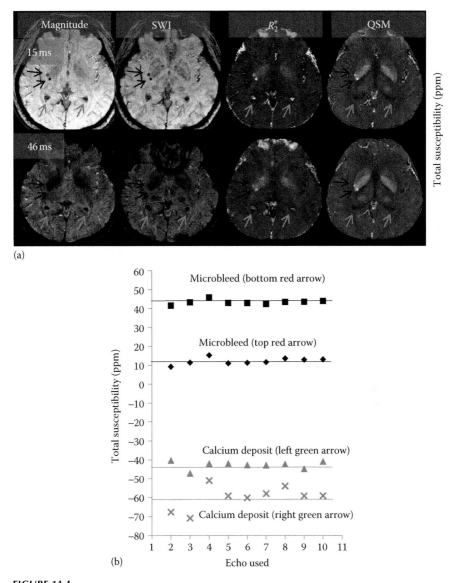

FIGURE 14.4
(See color insert.) (a) CMB appearances (red arrows) change with TE drastically in magnitude, SWI, and moderately R_2^* map but little in QSM. (b) QSMs reconstructed using various echoes confirm its invariance with TE.

on TE (Figure 14.5a). However, this TE dependence is eliminated in QSM. This is because there is increased dephasing in the later echoes, reducing the signal intensity in magnitude images. But the larger dephasing also presents opportunities to better estimate the magnetic field, which leads to more accurate QSM.

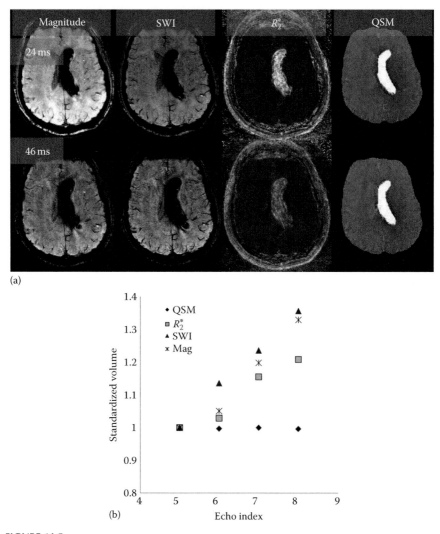

(a)

(b)

FIGURE 14.5
(Color version available from crcpress.com; see Preface.) (a) Hemorrhage volume increase with TE in magnitude, SWI, and R_2^* map but not in QSM. (b) QSMs reconstructed using various echoes confirm its invariance with TE.

14.5.3 QSM Calcification Quantification Validated by Computed Tomography

Calcifications can be unambiguously identified on QSM as they possess diamagnetism. *In vivo* validations including QSM and CT images obtained within a week of each other showed high agreement between QSM and CT (Figure 14.6a) for calcification detection. Calcium deposits measured by the total susceptibility on QSM and the total Hounsfield unit on CT showed excellent agreement (Figure 14.6b).

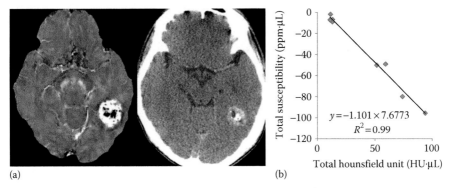

(a) (b)

FIGURE 14.6

Mixture of calcification and hemorrhage in QSM (left) and CT (right) (a). The hypointense core inside a bright rim on QSM indicates calcification surrounded by hemorrhage, which is supported by the CT scan. (b) The total susceptibility measured in QSM is in good agreement with Hounsfield unit measured in CT.

14.6 Susceptibility Tensor Imaging

For susceptibility tensor imaging (STI), Equation 14.2 can be generalized using the tensor form to describe the magnetic field generated by orientation-dependent magnetization of biomaterials. Using Maxwell's equations we get:

$$\Delta B = \frac{\overline{X} \cdot \vec{B}_0}{3} - \vec{k} \cdot \vec{B}_0 \frac{\vec{k} \cdot \left(\overline{X} \cdot \vec{B}_0 \right)}{k^2}, \tag{14.9}$$

where $\overline{X}$ is the second rank tensor of the magnetic susceptibility in the Fourier domain, ΔB is the relative difference field, $\vec{B}_0$ is the applied field direction, and $\vec{k}$ is the Fourier space vector. In STI, this problem is solved by combining multiple acquisitions of the subject with different $\vec{B}_0$ directions (Liu, 2010):

$$\overline{X}^*(k) = \text{argmin}_{\overline{X}(k)} \sum_{p,r} \left\| w_p(r) \left[\begin{array}{c} FT^{-1}\left(\Delta B_p(k) \right) \\ -FT^{-1}\left(\dfrac{\overline{X} \cdot \vec{B}_{0p}}{3} - \vec{k} \cdot \vec{B}_{0p} \dfrac{\vec{k} \cdot \left(\overline{X} \cdot \vec{B}_{0p} \right)}{k^2} \right) \end{array} \right] \right\|_2^2, \tag{14.10}$$

where w_p is a weighting based on the noise in the magnitude image and p denotes the index of the acquired orientation. The solution of this equation gives the distribution of tensor that satisfies the data through the convolution with the dipole kernel in Fourier space. Unfortunately, this STI technique is not practical for most current MRI scanners because many orientations ($n = 12$) are needed to obtain sufficient data to reconstruct the susceptibility tensor. Reasonable constraints such as cylindrical symmetry in this model might help condition the problem and reduce the number of model parameters needed for the reconstruction making it more stable when less data are available ($n = 3$) (Li et al., 2012). The anisotropy of the susceptibility tensor can be displayed by the degree of anisotropy defined by Jelinek (1981) as the ratio of the maximum to minimum eigenvalue, λ_1/λ_3.

14.7 Conclusion

Tissue magnetism can be studied using gradient echo MRI. The local field generated by tissue magnetic susceptibility can be measured from the phase image. The inverse problem of deconvolving magnetic field to derive the susceptibility map can be solved uniquely using prior information about tissue structure, such as edge location information estimated from the magnitude image. QSM is a novel technique with great potential for a broad range of clinical applications including measuring iron deposition in neurodegenerative diseases, evaluating the burden of CMB and hematoma volume, enabling calcification assessment using MRI, assessing oxygen transpiration defects in stroke patients, and defining anatomic target in deep brain for surgical intervention.

References

Conturo, T.E., Smith, G.D., 1990. Signal-to-noise in phase angle reconstruction: Dynamic range extension using phase reference offsets. *Magn Reson Med.* 15, 420–37.

de Rochefort, L., Brown, R., Prince, M.R., Wang, Y., 2008. Quantitative MR susceptibility mapping using piece-wise constant regularized inversion of the magnetic field. *Magn Reson Med.* 60, 1003–9.

de Rochefort, L., Liu, T., Kressler, B., Liu, J., Spincemaille, P., Lebon, V., Wu, J., Wang, Y., 2010. Quantitative susceptibility map reconstruction from MR phase data using Bayesian regularization: Validation and application to brain imaging. *Magn Reson Med.* 63, 194–206.

Gibbs, E.L., Lennox, W.G., Nims, L.F., Gibbs, F.A., 1942. Arterial and cerebral venous blood—Arterial-venous differences in man. *J Biol Chem.* 144, 325–32.

Greenberg, S.M., Vernooij, M.W., Cordonnier, C., Viswanathan, A., Al-Shahi Salman, R., Warach, S., Launer, L.J., Van Buchem, M.A., Breteler, M.M., 2009. Cerebral microbleeds: A guide to detection and interpretation. *Lancet Neurol.* 8, 165–74.

Haacke, E.M., Mittal, S., Wu, Z., Neelavalli, J., Cheng, Y.C., 2009. Susceptibility-weighted imaging: Technical aspects and clinical applications, part 1. *AJNR Am J Neuroradiol.* 30, 19–30.

Jackson, J.D., 1999. *Classical Electrodynamics*, 3rd ed, John Wiley & Sons, New York.

Jelinek, V., 1981. Characterization of the magnetic fabric of rocks. *Tectonophysics*. 79, T63–T67.

Kressler, B., de Rochefort, L., Liu, T., Spincemaille, P., Jiang, Q., Wang, Y., 2010. Nonlinear regularization for per voxel estimation of magnetic susceptibility distributions from MRI field maps. *IEEE Trans Med Imaging*. 29, 273–81.

Kressler, B., de Rochefort, L., Spincemaille, P., Liu, T., Wang, Y., 2008. Estimation of sparse magnetic susceptibility distributions from MRI using non-linear regularization. *Proceedings of the 16th Scientific Meeting of the International Society of Magnetic resonance in Medicine*. 3–9 May, Toronto, ON, Canada, p. 1514.

Langham, M.C., Magland, J.F., Floyd, T.F., Wehrli, F.W., 2009. Retrospective correction for induced magnetic field inhomogeneity in measurements of large-vessel hemoglobin oxygen saturation by MR susceptometry. *Magn Reson Med*. 61, 626–33.

Li, L., Leigh, J.S., 2004. Quantifying arbitrary magnetic susceptibility distributions with MR. *Magn Reson Med*. 51, 1077–82.

Li, W., Wu, B., Liu, C., 2011. Quantitative susceptibility mapping of human brain reflects spatial variation in tissue composition. *Neuroimage*. 55, 1645–56.

Li, X., Vikram, D.S., Lim, I.A., Jones, C.K., Farrell, J.A., van Zijl, P.C., 2012. Mapping magnetic susceptibility anisotropies of white matter in vivo in the human brain at 7T. *Neuroimage*. 62, 314–30.

Liu, C., 2010. Susceptibility tensor imaging. *Magn Reson Med*. 63, 1471–7.

Liu, J., Liu, T., de Rochefort, L., Ledoux, J., Khalidov, I., Chen, W., Tsiouris, A.J., et al., 2012. Morphology-enabled dipole inversion for quantitative susceptibility mapping using structural consistency between the magnitude image and the susceptibility map. *Neuroimage*. 59, 2560–8.

Liu, T., Khalidov, I., de Rochefort, L., Spincemaille, P., Liu, J., Tsiouris, A.J., Wang, Y., 2011. A novel background field removal method for MRI using projection onto dipole fields (PDF). *NMR Biomed*. 24, 1129–36.

Liu, T., Spincemaille, P., de Rochefort, L., Kressler, B., Wang, Y., 2009. Calculation of susceptibility through multiple orientation sampling (COSMOS): A method for conditioning the inverse problem from measured magnetic field map to susceptibility source image in MRI. *Magn Reson Med*. 61, 196–204.

Liu, T., Wisnieff, C., Lou, M., Chen, W., Spincemaille, P., Wang, Y., 2013. Nonlinear formulation of the magnetic field to source relationship for robust quantitative susceptibility mapping. *Magn Reson Med*. 69, 467–76.

Lustig, M., Donoho, D., Pauly, J.M., 2007. Sparse MRI: The application of compressed sensing for rapid MR imaging. *Magn Reson Med*. 58, 1182–95.

Nandigam, R.N., Viswanathan, A., Delgado, P., Skehan, M.E., Smith, E.E., Rosand, J., Greenberg, S.M., Dickerson, B.C., 2009. MR imaging detection of cerebral microbleeds: Effect of susceptibility-weighted imaging, section thickness, and field strength. *AJNR Am J Neuroradiol*. 30, 338–43.

Neelavalli, J., Cheng, Y.C., Jiang, J., Haacke, E.M., 2009. Removing background phase variations in susceptibility-weighted imaging using a fast, forward-field calculation. *J Magn Reson Imaging*. 29, 937–48.

Riederer, P., Sofic, E., Rausch, W.D., Schmidt, B., Reynolds, G.P., Jellinger, K., Youdim, M.B., 1989. Transition metals, ferritin, glutathione, and ascorbic acid in parkinsonian brains. *J Neurochem*. 52, 515–20.

Schweser, F., Deistung, A., Lehr, B.W., Reichenbach, J.R., 2011. Quantitative imaging of intrinsic magnetic tissue properties using MRI signal phase: An approach to in vivo brain iron metabolism? *Neuroimage*. 54, 2789–807.

Schweser, F., Lehr, B.W., Deistung, A., Reichenbach, J.R., 2010. A novel approach for separation of background phase in SWI phase data utilizing the harmonic function mean value property. *Proceedings of the 18th Scientific Meeting of the International Society of Magnetic resonance in Medicine*. 1–7 May, Stockholm, Sweden. p. 142.

Shmueli, K., de Zwart, J.A., van Gelderen, P., Li, T.Q., Dodd, S.J., Duyn, J.H., 2009. Magnetic susceptibility mapping of brain tissue in vivo using MRI phase data. *Magn Reson Med*. 62, 1510–22.

Vernooij, M.W., Ikram, M.A., Wielopolski, P.A., Krestin, G.P., Breteler, M.M., van der Lugt, A., 2008. Cerebral microbleeds: Accelerated 3D T2*-weighted GRE MR imaging versus conventional 2D T2*-weighted GRE MR imaging for detection. *Radiology*. 248, 272–7.

Wharton, S., Schafer, A., Bowtell, R., 2010. Susceptibility mapping in the human brain using threshold-based k-space division. *Magn Reson Med*. 63, 1292–304.

Wu, B., Li, W., Guidon, A., Liu, C., 2012. Whole brain susceptibility mapping using compressed sensing. *Magn Reson Med*. 67, 137–47.

Yao, B., Li, T.Q., Gelderen, P., Shmueli, K., de Zwart, J.A., Duyn, J.H., 2009. Susceptibility contrast in high field MRI of human brain as a function of tissue iron content. *Neuroimage*. 44, 1259–66.

Index